Library of
Davidson College

Davidson's
The Biochemistry
of the Nucleic Acids

Davidson's
The Biochemistry of the Nucleic Acids

Revised by
R. L. P. Adams
R. H. Burdon
A. M. Campbell
R. M. S. Smellie
*Department of Biochemistry
University of Glasgo*w

EIGHTH EDITION

1976
ACADEMIC PRESS
NEW YORK, SAN FRANCISCO
A subsidiary of Harcourt Brace Jovanovich, publishers

First published 1950
in Methuen's Monographs on Biochemical Subjects
Second edition, 1953
Third edition, 1957
Fourth edition, 1960
Fifth edition, 1965
Sixth edition, 1969
Seventh edition, 1972
First published as a Science Paperback, 1972
Eighth edition, 1976

© 1976, Chapman and Hall

Printed in Great Britain by
Richard Clay (*The Chaucer Press*), Ltd.,
Bungay, Suffolk

ISBN 0 412 21380 X (*cased edition*)
ISBN 0 412 21390 7 (*Science Paperback*)

This title is available in both hardbound and paperback editions. The paperback edition is sold subject to the condition that it shall not, by way of trade or otherwise, be lent, re-sold, hired out, or otherwise circulated without the publisher's prior consent in any form of binding or cover other than that in which it is published and without a similar condition including this condition being imposed on the subsequent purchaser.

All rights reserved. No part of this book may be reprinted, or reproduced or utilized in any form or by any electronic, mechanical or other means, now known or hereafter invented, including photocopying and recording, or in any information storage and retrieval system, without permission in writing from the Publisher.

Published in the United States by
ACADEMIC PRESS, INC.,
111 *Fifth Avenue, New York, N.Y.* 10003
Library of Congress Catalog Card Number 76–25737
For Academic Press, Inc.: ISBN 0–12–205360–5

Contents

	Preface	*page* vii
	Abbreviations and Nomenclature	ix
1	Introduction	1
2	The Occurrence of Nucleic Acids	7
3	Chemical Constituents of Nucleic Acids	37
4	Isolation and Characterization of Nucleic Acids	50
5	The Structure of DNA	83
6	The Structure of RNA	114
7	Nucleic Acids in Viruses and Plasmids	137
8	Nucleases and Related Enzymes	163
9	The Metabolism of Nucleotides	201
10	The Genetic Function of DNA	223
11	Replication of DNA	243
12	The Biosynthesis of RNA: Transcription	318
13	The Biological Function of RNA: Protein Synthesis	359
14	Nucleic Acids and the Regulation of Protein Synthesis	396
	Index	415

Plates

Between pages 84 *and* 85

    I Thread of DNA emerging from an osmotically ruptured coliphage particle

   II Electron micrograph illustrating the fine structure of a mammalian cell

  III Replicating molecules of SV40 DNA

  IV Two forms of PM2 virus DNA

   V Model of Yeast tRNA^{Phe}

  VI Electron micrograph of transcription from DNA

 VII Electron Micrograph of 45S Ribosomal Precursor RNA

VIII Chromatin Fibres from chicken erythrocytes

Preface

When the first edition of this book was published in 1950, it set out to present an elementary outline of the state of knowledge of nucleic acid biochemistry at that time and it was the first monograph on the subject to appear after Levene's book on Nucleic Acids in 1931. The fact that an eighth edition is required after only twenty-five years and that this contains little of the content of the original book provides some measure of the speed with which knowledge has advanced in this field.

'The Child's Guide to the Nucleic Acids' as it is known within the Department in Glasgow is still intended primarily as an introduction to the subject for advanced undergraduates in biochemistry, for chemists seeking to find some understanding of the more biological aspects of the subject and for biologists who require some knowledge of the chemical and molecular aspects.

The first seven editions emerged from the pen of the late J. N. Davidson who died in September 1972 shortly after completing the Seventh Edition. The new edition has been revised extensively by four of his colleagues who recognize the need for a book giving reasonably comprehensive coverage of the field at an up to date, but elementary level. In undertaking this we have endeavoured to retain something of the character and structure of earlier editions while at the same time introducing new ideas and concepts and eliminating some of the more out-dated material. Some chapters, particularly those dealing with structure, function, biosynthesis, and control have been very extensively rewritten: the order of presentation of the chapters has been changed, some previous chapters have been combined and some new chapters have been included. One particularly valuable feature of earlier editions, was the very extensive reference list and this again we have endeavoured to retain but with the elimination of older material and its replacement by reference to more recent reviews and original papers.

Our late colleague Professor Davidson took the greatest of pains

PREFACE

in the preparation of earlier editions and we hope that in revising this volume we have succeeded in our objectives as outlined above and that the eighth edition will live up to the very high standards set by its predecessors.

It is a pleasure to express our thanks for those who have allowed us to reproduce figures, diagrams, and plates. In particular we should like to thank Dr R. J. Britten and Dr D. E. Kohne for Fig. 5.11, Dr Leslie Coggins for Plate IV, Dr R. S. Gilmour for advice about Fig. 14.5, Dr J. P. Goddard for Plate V, Professor D. Lang for Plate I, Dr O. L. Miller for Plate VI, Dr D. E. Olins and Dr A. L. Olins for Plate VIII, Dr N. P. Salzman for Plate III, Dr Tuneko Okazaki for making available data from her late husband's laboratory, Dr P. G. Toner for Plate II and Dr P. K. Wellauer for Plate VII.

We should also like to thank the editors of *Nature* for permission to reproduce Fig. 10.1, the American Society for Microbiology, the editors of *Science* for permission to reproduce Plate VI and the editors of *Proceedings of the National Academy of Sciences*, USA, for permission to reproduce Plate VII.

We are particularly grateful to Mr Robin Callander who has been responsible for the preparation of diagrams and figures and Miss Nancy Golder, Mrs Golda Morrison, and Mrs Mary Wilson for valuable secretarial assistance.

RLPA
RHB
AMC
RMSS

January 1976

Abbreviations and Nomenclature

The abbreviations employed in this book are those approved by the Commission on Biochemical Nomenclature (CBN) of the International Union of Pure and Applied Chemistry (IUPAC) and the International Union of Biochemistry (IUB).

Nucleosides

A	adenosine
G	guanosine
C	cytidine
U	uridine
ψ	pseudouridine
I	inosine
X	xanthosine
T	ribosylthymine (ribothymidine)
N	unspecified nucleoside
R	unspecified purine nucleoside
Y	unspecified pyrimidine nucleoside
dA	2′-deoxyribosyladenosine
dG etc.	2′-deoxyribosylguanosine etc.
dT	thymidine

Minor nucleosides (when in sequences)

m^1I	1-methylinosine
m^1G	1-methylguanosine
m^2G	N^2-methylguanosine
m^2_2G	N^2-dimethylguanosine
Gm	2′-O-methylguanosine
mA	methyladenosine
m^6_2A	N^6-dimethyladenosine
iA	N^6-isopentenyladenosine
hU	5,6-dihydrouridine
m^5C	5-methylcytidine
ac^4C	N^4-acetylcytidine

THE BIOCHEMISTRY OF THE NUCLEIC ACIDS

Um	$2'$-O-methyluridine
Cm	$2'$-O-methylcytidine

Nucleotides

AMP	adenosine $5'$-monophosphate
GMP etc.	guanosine $5'$-monophosphate etc.
dAMP	$2'$-deoxyribosyladenine $5'$-monophosphate
dGMP etc.	$2'$-deoxyribosylguanine $5'$-monophosphate etc.
dTMP	thymidine $5'$-monophosphate
$2'$-AMP, $3'$-AMP, ($5'$-AMP) etc.	$2'$-, $3'$-, (and $5'$-, where needed for contrast) phosphate of adenosine etc.
ADP, etc.	$5'$-(pyro) diphosphate of adenosine, etc.
ATP, etc.	$5'$-(pyro) triphosphate of adenosine, etc.

Polynucleotides

RNA	ribonucleic acid or ribonucleate
DNA	deoxyribonucleic acid or deoxyribonucleate
mRNA; rRNA	messenger RNA; ribosomal RNA
nRNA	nuclear RNA
tRNA	transfer RNA (RNA that accepts and transfers amino acids; amino acid-accepting RNA)
Alanine tRNA or tRNA^{Ala}, etc.	the transfer RNA molecule that normally accepts alanine, etc.
Alanyl-tRNA^{Ala} or Ala-tRNA^{Ala} or Ala-tTNA	the same, with alanyl residue covalently linked
poly (N), or (N)$_n$, or (rN)$_n$	polymer of ribonucleotide N
poly (dN), or (dN)$_n$	polymer of deoxyribonucleotide N
poly (N-N'), or r(N-N')$_n$ or (rN-rN')$_n$	copolymer of N-N'-N-N'- in regular, alternating, *known* sequence

poly d(N-N'), or d(N-N')$_n$ or (dN-dN')$_n$	copolymer of dN-dN'-dN-dN'- in regular, alternating, *known* sequence
poly (N,N') or (N,N')$_n$	copolymer of N and N' in *random* sequence
poly (A)·poly (B) or (A)$_n$·(B)$_n$	two chains, generally or completely associated
poly (A), poly (B) or (A)$_n$, (B)$_n$	two chains, association unspecified or unknown
poly (A) + poly (B) or (A)$_n$ + (B)$_n$	two chains, generally or completely unassociated

Miscellaneous

RNase, DNase	ribonuclease, deoxyribonuclease
P_i, PP_i	inorganic orthophosphate and pyrophosphate

Amino acids

Ala	Alanine
Arg	Arginine
Asp	Aspartic acid
Asn	Asparagine
Cys	Cysteine
Glu	Glutamic acid
Gln	Glutamine
Gly	Glycine
His	Histidine
Ile	Isoleucine
Leu	Leucine
Lys	Lysine
Met	Methionine
fMet	formylMethionine
Phe	Phenylalanine
Pro	Proline
Ser	Serine
Thr	Threonine
Trp	Tryptophan
Tyr	Tyrosine
Val	Valine

Further details of the recommendations of the Commission on Biochemical Nomenclature are printed in the *J. Biol. Chem.*, **246,**

4894 (1971), *Biochim. Biophys. Acta*, **247,** 1 (1971). *Biochemistry*, **5,** 1445 (1966), *Arch. Biochem. Biophys.*, **115,** 1 (1966), *J. mol. Biol.*, **55,** 299 (1971), *Biochem. J.*, **126,** 1 (1972), *European J. Biochem.*, **15,** 203 (1970), and *Progress in Nucleic Acid Research and Molecular Biology*, **15** (1975).

In naming enzymes the recommendations of the Commission on Enzymes of the International Union of Biochemistry (1965) are followed as far as possible. The numbers recommended by the Commission are inserted in the text after the name of each enzyme.

CHAPTER 1

Introduction

The fundamental investigations which led to the discovery of the nucleic acids were made by Friedrich Miescher [1] (1844–95), who may be regarded as the founder of our knowledge of the chemistry of the cell nucleus. In early work carried out in 1868, in the laboratory of Hoppe-Seyler in Tübingen, he isolated the nuclei from pus cells obtained from discarded surgical bandages and showed that the nuclear material contained an unusual phosphorus compound named by him 'nuclein', which we now know to have been nucleoprotein. Miescher's investigations were continued in Basel, where most of his working life was spent. There he became interested in salmon sperm as a source of nuclear material, and in 1872 he showed that isolated sperm heads contained an acidic compound now recognized as nucleic acid, and a base to which the name 'protamine' was given. It was subsequently shown that nucleic acids were normal constituents of all cells and tissues which were examined, and Miescher's investigations of the nucleic acids were continued by Altmann, who in 1889 described a method for the preparation of protein-free nucleic acids from animal tissues and from yeast. The work was continued later by Kossel in Heidelberg, Jones in Baltimore, Levene in New York, Hammarsten in Stockholm, Gulland in Nottingham, and many others [2–7].

One of the best animal sources of nucleic acid was found to be the thymus gland, and it is not surprising therefore that most work was concentrated on nucleic acid from this source. On hydrolysis it was found to yield the purine bases adenine and guanine, the pyrimidine bases cytosine and thymine, a sugar which was eventually shown to be D(−)-2-deoxyribose, and phosphoric acid. It has come to be known as deoxyribonucleic acid or DNA. On the other hand, the nucleic acid from yeast on hydrolysis yielded adenine, guanine, cytosine, uracil, a pentose sugar which was eventually shown to be D(−)-ribose, and phosphoric acid. It therefore differed from thymus nucleic acid in containing uracil in place of thymine and ribose in place of deoxyribose and has come to be known as ribonucleic acid or RNA. Since most nucleic acids from animal sources appeared to resemble thymus nucleic acid, and since the only other nucleic acid which had at that time (1920) been prepared in reasonable quantities from a plant source appeared to be very similar to yeast nucleic acid, the impression grew up that deoxypentosenucleic acid of the thymus type was characteristic of animal tissues, and pentosenucleic acid of the yeast type was characteristic of plant tissues [5]. Thus Jones, in 1920, stated categorically: 'we come to understand quite clearly that there are only two nucleic acids in nature, one obtainable from the nuclei of animal cells and the other from the nuclei of plant cells' [6].

It was not long before the validity of this conception was questioned. It had been known since early times that pentose derivatives were present in animal tissues. For example, the so-called β-nucleoprotein, which was originally prepared from mammalian pancreas by O. Hammarsten [7] in 1894, was known to contain a pentose sugar, and Jorpes [8] eventually prepared from this material a nucleic acid of the pentose type which he showed to resemble yeast nucleic acid and to be abundant in pancreatic tissue. The presence of pentosenucleic acids in the mammary gland was also suggested by the work of Odenius [9] and of Mandel and Levene [10]. Pentosenucleotide derivatives were also demonstrated in chick embryo pulp by Calvery [11], in spleen and the liver by Jones and Perkins [12] and by Thomas and Berariu [13], and in sea urchin eggs by Blanchard [14]. It thus appeared probable that pentosenucleic acids were normal constituents of animal tissues as well as of plant cells, and Jones and Perkins [12] expressed the view: 'the distinction between plant and animal nucleic acids will in future not be so definitely drawn'.

It was not until the early 1940's that unequivocal evidence was obtained by the ultraviolet spectrophotometric results of Caspersson [15], by the histochemical experiments of Brachet [16], and by the chemical analytical results of Davidson [17, 18] that RNA is a general constituent of animal, plant, and bacterial cells.

These advances established the biology of the nucleic acids on a new foundation. The use of new techniques in cytochemistry and cell fractionation showed that DNA and RNA are normal constituents of all cells, plant and animal, DNA being confined to the nucleus while RNA is found also in the cytoplasm [17–22].

INTRODUCTION

The development of techniques of subcellular fractionation and for the isolation of nuclei [23–28] made possible chemical measurements of the distribution of DNA and RNA amongst the subcellular fractions of various cell types, and led ultimately to the recognition of RNA in the nuclear, ribosomal, and soluble fractions of cells (Chapter 2) and to the demonstration of the constancy in the average amount of DNA per nucleus in the somatic cells of any given species [29] (see Chapter 2).

These studies, combined with measurements of the incorporation of labelled precursors into DNA and the RNA's of different subcellular fractions [30, 31], demonstrated that there were considerable differences in the metabolic activity of the various classes of RNA (Chapter 12).

Many of the earliest contributions to our understanding of the structure of nucleic acids arose out of the work of Levene and Jacobs [3, 32, 33] who established the presence of D-ribose, hypoxanthine, and phosphorus in inosinic acid from muscle, and later the presence of the ribonucleotides of adenine, guanine, cytosine, and uracil in yeast nucleic acid. They also recognized the occurrence of thymine in place of uracil in thymus nucleic acid. The presence of these nucleotides in approximately equimolar proportions led to the development of the tetranucleotide hypothesis for both DNA and RNA, in which both nucleic acids were considered to be polymeric structures containing equivalent amounts of mononucleotides derived from each of the four purine and pyrimidine bases linked together in repeating units. It was not until much later that the sugar in thymus nucleic acid was positively identified as D(−)-2-deoxyribose.

This concept of nucleic acid structure survived until the late 1940's despite the fact that evidence for it was not strong [34]. It was only when methods had been developed for the quantitative analysis of nucleic acids [35, 36] that the tetranucleotide hypothesis was finally abandoned as a consequence of the demonstration that the various nucleotides did not occur in equimolar proportions [37]. However, Chargaff [38] drew attention to certain regularities in the composition of DNA, namely, that the sum of the purines was equal to the sum of the pyrimidines, that the sum of the amino bases (adenine and cytosine) was equal to the sum of the keto bases (guanine and thymine), and that adenine and thymine, and guanine and cytosine, were present in equivalent amounts (Chapter 5). These observations were to be of crucial importance in the subsequent interpretation of X-ray crystallographic analyses.

The elucidation of the detailed structure of nucleosides and nucleotides can largely be attributed to Todd and his collaborators (for review see [39]), who established the nature of the β-glycosidic linkage between the sugar residues and the purine or pyrimidine bases and the nature of the phosphate ester bonds. Their work, taken together with the studies of Cohn and his colleagues [40], provided final confirmation of the nature of the internucleotide linkage in both DNA and RNA and made it possible for clear concepts of the primary structure of the two types of nucleic acid to be put forward (Chapters 5 and 6).

At the time that these new developments were taking place in the understanding of the chemical structure of nucleotides and polynucleotides, progress was also being made in X-ray crystallographic studies of DNA. Arising from the work of Astbury [41], Pauling and Corey [42], Wilkins and his colleagues [43], and Franklin and Gosling [44], Watson and Crick [45] proposed their now famous double-helical structure made up of specifically hydrogen-bonded base pairs (Chapter 5) and which suggested 'a possible copying mechanism for the genetic material'.

The elucidation of the main features of the primary structure o DNA and RNA and of the secondary structure of DNA provided the foundation for the subsequent rapid increase in our understanding of the mechanisms of synthesis of DNA and RNA and of the roles of these substances in living systems [46]. The following chapters of this book are concerned with more detailed consideration of the various aspects of structure, function, and biosynthesis of nucleic acids which have been touched upon above [47–63].

REFERENCES

[1] Miescher, F. (1879) *Die Histochemischen und Physiologischen Arbeiten.* Leipzig
[2] Fruton, J. S. (1972) *Molecules and Life.* New York: Wiley-Interscience
[3] Levene, P. A. and Bass, L. W. (1931). *Nucleic Acids.* New York: Chemical Catalog Co.
[4] Altmann, R. (1889) *Nucleinsäuren, Arch. Anat. Physiol.*, 524
[5] Levene, P. A. (1921) *J. Biol. Chem.*, **48**, 119
[6] Jones, W. (1920) *The Nucleic Acids.* London: Longmans
[7] Hammarsten, O. (1894) *Hoppe-Seyler's Ztschr.*, **19**, 19
[8] Jorpes, E. (1924) *Biochem. Ztschr.*, **151**, 227
   (1928) *Acta Med. Scand.*, **68**, 253, 503
   (1934) *Biochem. J.*, **28**, 2102

INTRODUCTION

[9] Odenius, R. (1900) *Jahresber. Fortschr. Thierchem.*, **30**, 39
[10] Mandel, J. A. and Levene, P. A. (1905). *Hoppe-Seyler's Ztschr.*, **46**, 155
[11] Calvery, H. O. (1928) *J. Biol. Chem.*, **77**, 489, 497
[12] Jones, W. and Perkins, M. E. (1924–5) *J. Biol. Chem.*, **62**, 290
[13] Thomas, P. and Berariu, C. (1924) *Compt. Rend. Soc. Biol.*, **91**, 1470
[14] Blanchard, K. C. (1935) *J. Biol. Chem.*, **108**, 251
[15] Caspersson, T. (1950) *Cell Growth and Cell Function.* New York: Norton
[16] Brachet, J. (1950) *Chemical Embryology.* New York: Interscience
[17] Davidson, J. N. and Waymouth, C. (1944) *Biochem. J.*, **38**, 39
[18] Davidson, J. N. and Waymouth, C. (1944–5) *Nutrition Abs. Rev.*, **14**, 1
[19] Feulgen, R. and Rossenbeck, H. (1924) *Hoppe-Seyler's Ztschr.*, **135**, 203
[20] Kiesel, A. and Belozerski, A. N. (1934) *Hoppe-Seyler's Ztschr.*, **229**, 160
[21] Belozerski, A. N. (1936) *Biochimia*, **1**, 253
      (1939) *Compt. Rend. Acad. Sci. U.R.S.S.*, **25**, 751
[22] Behrens, M. (1938) *Hoppe-Seyler's Ztschr.*, **253**, 185
[23] Claude, A. (1946) *J. Exp. Med.*, **84**, 51
[24] Hogeboom, C., Schneider, W. C. and Palade, G. E. (1948) *J. Biol. Chem.*, **172**, 619
[25] Dounce, A. L. (1943) *J. Biol. Chem.*, **147**, 685
      (1943) *ibid.*, **151**, 221, 235
[26] Mirsky, A. E. and Pollister, A. W. (1943) *J. Gen. Physiol.*, **30**, 117
[27] Behrens, M. (1938) *Aberhalden's Hanbuch der Biologische Arbeitsmethoden*, Sect. 5, Part 10, p. 1363
[28] Dounce, A. L. (1955) *The Nucleic Acids*, Vol. 2, p. 93 (E. Chargaff and J. N. Davidson, Eds.). New York: Academic Press
[29] Vendrely, R. (1955) *The Nucleic Acids*, Vol. 2, p. 155 (E. Chargaff and J. N. Davidson, Eds.). New York: Academic Press
[30] Marshak, A. and Calvet, F. (1949) *J. Cell Comp. Physiol.*, **34**, 451.
[31] Smellie, R. M. S. (1955) *The Nucleic Acids*, Vol. 2, p. 393 (E. Chargaff and J. N. Davidson, Eds.). New York: Academic Press
[32] Levene, P. A. and Jacobs, W. A. (1908) *Ber. Deutsch. Chem. Ges.*, **41**, 2703
      (1909) *ibid.*, **42**, 335
[33] Levene, P. A. and Jacobs, W. A. (1912) *J. Biol. Chem.*, **12**, 411
[34] Gulland, J. H. (1947) *Symp. Soc. Exp. Biol.*, **1**, 1
[35] Vischer, E. and Chargaff, E. (1947) *J. Biol. Chem.*, **168**, 781
[36] Wyatt, G. R. (1955) *The Nucleic Acids*, Vol. 1, p. 243 (E. Chargaff and J. N. Davidson, Eds.). New York: Academic Press
[37] Chargaff, E. (1950) *Experientia*, **6**, 201
[38] Chargaff, E. (1955) *The Nucleic Acids*, Vol. 1, p. 307 (E. Chargaff and J. N. Davidson, Eds.). New York: Academic Press
[39] Brown, D. M. and Todd, A. R. (1955) *The Nucleic Acids*, Vol. 1, p. 409 (E. Chargaff and J. N. Davidson, Eds.). New York: Academic Press
[40] Cohn, W. E. (1956) *Currents in Biochemical Research*, p. 460 (D. E. Green, Ed.). New York: Interscience
[41] Astbury, W. T. (1947) *Symp. Soc. Exp. Biol.*, **1**, 66
[42] Pauling, L. and Corey, R. B. (1953) *Proc. Nat. Acad. Sci.*, **39**, 84.
[43] Wilkins, M. F. H., Stokes, A. R. and Wilson, H. R. (1953) *Nature*, **171**, 738
[44] Franklin, R. E. and Gosling, R. G. (1953) *Nature*, **171**, 740; **172**, 156
[45] Watson, J. D. and Crick, F. H. C. (1953) *Nature*, **171**, 737
[46] Kornberg, A. (1974) *DNA Synthesis.* San Francisco: Freeman

[47] Jordan, D. O. (1960) *Chemistry of the Nucleic Acids.* London: Butterworths
[48] Potter, V. R. (1960) *Nucleic Acid Outlines.* Minneapolis: Burgess Publishing Company
[49] Steiner, R. F. and Beers, R. F. (1961) *Polynucleotides.* Amsterdam: Elsevier
[50] Allen, F. W. (1962) *Ribonucleoproteins and Ribonucleic Acids.* Amsterdam: Elsevier
[51] Perutz, M. F. (1962) *Proteins and Nucleic Acids.* Amsterdam: Elsevier
[52] Chargaff, E. (1963) *Essays on Nucleic Acids.* Amsterdam: Elsevier
[53] Davidson, J. N. and Cohn, W. E. (Eds.) (1963–72) *Progress in Nucleic Acid Research and Molecular Biology,* Vols. 1–15. New York: Academic Press
[54] Michelson, A. M. (1963) *The Chemistry of Nucleosides and Nucleotides.* New York: Academic Press.
[55] Synthesis and Structure of Macromolecules (1963). *Cold Spring Harb. Symp. Quant. Biol.* Vol. 28
[56] Taylor, J. H. (Ed.) (1963) *Molecular Genetics,* Part I; (1967) Part II. New York: Academic Press
[57] Vogel, H. J., Bryson, V. and Lampen, J. O. (Eds.) (1963) *Informational Macromolecules.* New York: Academic Press
[58] Steiner, R. F. (1965) *The Chemical Foundations of Molecular Biology.* Princeton: Van Nostrand
[59] Jukes, T. H. (1966) *Molecules and Evolution.* New York: Columbia University Press
[60] Kendrew, J. (1966) *The Thread of Life.* London: Bell
[61] Watson, J. D. (1968) *The Double Helix.* New York: Atheneum
[62] Cantoni, G. L. and Davies, D. R. (Eds.) (1966) *Procedures in Nucleic Acid Research.* New York: Harper and Row
[63] Grossman, L. and Moldave, K. (Eds.) (1967) Part A; (1968) Part B; (1971) Parts C and D, *Nucleic Acids,* being Vols. 12, 20, and 21 of *Methods in Enzymology* (S. P. Colowick and N. O. Kaplan, Eds.). New York: Academic Press
[64] Parish, J. H. (1972) *Principles and Practice of Experiments with Nucleic Acids.* London: Longmans
[65] Kornberg, A. (1974) *DNA Synthesis.* San Francisco: Freeman
[66] Lewin, B. (1974) *Gene Expression,* Vols. I and II. London: Wiley

CHAPTER 2

The Occurrence of Nucleic Acids

2.1 Introduction

It has been pointed out in Chapter 1 that the object of the research which led to the discovery of the nucleic acids was the chemical investigation of the cell nucleus. The assumption that the nucleic acids were essentially nuclear constituents was explicitly or tacitly accepted until about 1930. It was first seriously challenged in 1938 when Behrens by elaborate and laborious methods separated plant tissues into a nuclear and a cytoplasmic fraction and showed that the latter contained substantial quantities of RNA. It was finally overthrown by the histochemical methods described in Section 2.9. In 1938 Caspersson and Schultz [1] demonstrated that the cytoplasm of certain rapidly proliferating cells was rich in material which absorbed ultraviolet light intensely, giving an absorption curve characteristic of the nucleic acids, but which on account of its Feulgen negative properties appeared to be RNA. The development of the ribonuclease histochemical test brought confirmation of the presence of RNA in the cell cytoplasm, and final proof has been given by the isolation of RNA from cytoplasmic material free from nuclear contamination. Both DNA and RNA have been found in all eukaryotic and prokaryotic cells but many viruses contain either DNA or RNA but not both.

2.2 The animal cell

Developments in the biochemistry of RNA have been profoundly influenced by simultaneous advances in cytology. Consequently consideration of the place of RNA in the life of the cell must be prefaced by a very brief outline of some current ideas on the fine structure of the cytoplasm.

A schematic diagram of a typical animal cell is shown in Fig. 2.1. Inside the cell membrane is the *cytoplasm* in which are suspended numerous inclusions, the largest of which is the more or less centrally placed *nucleus* bounded by a double membrane pierced by a number of pores. If the cytoplasm of a living cell is carefully examined, preferably in the phase-contrast microscope, a number of rod-like bodies, the *mitochondria*, will be clearly seen. Provided that a suitable fixative has been chosen the mitochrondria can also be demonstrated in fixed preparations by the use of special staining methods such as that of Altmann. While the mitochondria are easily demonstrated by appropriate methods in the living cell, they do not show up after

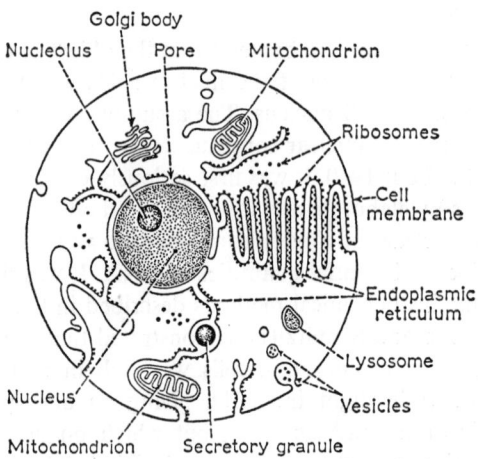

Fig. 2.1 *Schematic representation of a typical animal cell*

the tissue has been treated with any of the commonly employed histological fixatives such as formalin.

The cytoplasm of certain cells such as those from the pancreas also contains a number of more or less spherical granules, the *secretory* granules, which vary in amount with the secretory state of the cell. A number of *vacuoles* may also be seen and some globules of *fat*. Liver cells contain granules of *glycogen*.

In tissue sections fixed and stained by the usual methods inclusions of basophilic material can often be seen in the cytoplasm [2]. It is believed that these represent endoplasmic reticulum and ribosomes (see below) coagulated by the fixative.

A much more detailed picture of the cell has been obtained by the techniques of electron microscopy [3–8]. In electron micrographs (Plate II) the mitochrondria (dimensions 0.5–5 μm $\times$ 0.3–0.7 μm) appear as oval profiles. Each mitochondrion is bounded by an outer and an inner membrane about 5 nm in thickness, the inner membrane being connected with a series of incomplete partitions, the *cristae mitochondriales*, which project into the interior of the organelle dividing it into a series of interconnecting compartments.

One of the most interesting cytoplasmic components as revealed by electron microscopy is a complex mixture of strands and vesicles which Porter [3] termed the *endoplasmic reticulum* (Plate II). It is limited by a membrane about 5 nm thick separating the content of the tubules and vesicles from the general matrix of the cytoplasm and giving the whole component the character of a finely divided vacuolar system, which has been identified with the basophilic or chromophilic component (the *ergastoplasm*) of the cytoplasm. It is believed that the tubules of the endoplasmic reticulum form a series of canals leading, according to some authorities, from the exterior of the cell to the nucleus (Fig. 2.1) [9, 10]. A dense area of the endoplasmic reticulum is associated with the *Golgi* body [11] and the *centrosome* and *centrioles*.

The structure of the endoplasmic reticulum in the liver cell has been intensively studied in very thin sections in the electron microscope. Palade and Siekevitz [12] found that it could be represented by numerous profiles of circular, oval, or elongated shape with smaller diameter measuring 40–150 nm. They are bounded by a very fine membrane and have an apparently homogeneous content. Two types of profile can be distinguished, smooth and rough surfaced. The smooth-surfaced profiles (40–100 nm in diameter) are circular, oval, or irregular in shape and correspond to vesicles and contorted tubules linked together in a tightly meshed reticulum. The rough-surfaced profiles are more numerous and are of length 50 nm to 5 μm with a fairly constant diameter of about 50 nm. They frequently occur in more or less parallel arrays separated from each other at fairly regular intervals (Plate II). The rough surface is due to the presence of small round electron-dense particles (10–20 nm diameter) attached to the outside surface of the limiting membrane (Fig. 2.1). These particles are known as *ribosomes* (p. 21). They are also found free in the cytoplasm especially in rapidly proliferating cells and they may occur in the nucleus and even in the mitochondria [13]. They are present in bacterial and plant cells as well as in mammalian cells.

The fine structure of the cell is described in numerous reviews [6, 9, 10, 14–22].

THE BIOCHEMISTRY OF THE NUCLEIC ACIDS

2.3 The bacterial cell

Microorganisms are of such diverse complexity and varied morphology that it is not possible to give more than the briefest outline of their structure, but the following description of the common rod-shaped bacterial cell may serve to indicate the principal features (Fig. 2.2). The bacterial cell consists of a *protoplast* including

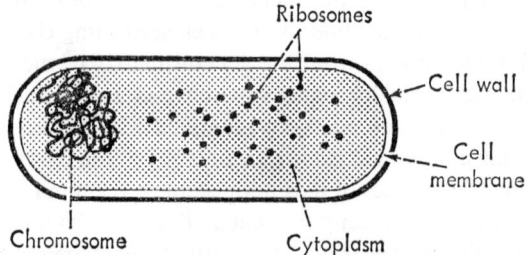

Fig. 2.2 *Diagrammatic representation of the structure of the bacterial cell*

nucleus and cytoplasm bounded by a delicate cytoplasmic membrane. The membrane itself is in close contact with the rigid *cell wall* of characteristic shape. In many bacteria the cell wall is surrounded by a much wider capsule which usually consists of complex polysaccharides. The cytoplasm contains granules of various types, some of which are reserve food materials, but the most important and abundant cytoplasmic particles are the bacterial ribosomes. The bacterial nucleus, which has not the easily definable structure of the animal cell nucleus, contains DNA as basic genetic material.

When the cell wall is eliminated by digestion with the enzyme lysozyme, the membrane and its contents are released as the osmotically sensitive protoplast. Gram-positive organisms yield protoplasts which are free of wall constituents, but gram-negative organisms yield osmotically sensitive spheres known as *spheroplasts* which retain fragments of the wall.

The *mycoplasmas* are very simple microorganisms which possess no cell walls. They are the smallest known type of free living organism.

Full descriptions of bacterial morphology are to be found in standard textbooks of microbiology [23, 24].

2.4 The separation of cytoplasmic components

While the composition of cellular constituents may be studied *in situ* with the aid of some of the elegant spectrophotometric techniques mentioned later, from the chemical point of view most information has been obtained by the study of cell components separated by the process of *differential centrifugation* from cells disrupted in bulk in a suitable medium.

The original classical experiments on the separation of subcellular fractions, carried out by Bensley [25] and by Claude [26–28], resulted in the separation of liver tissue disrupted in saline solution into a nuclear fraction, a fraction of large granules including mitochondria, a fraction of small granules which Claude termed *microsomes*, and a supernatant fraction devoid of sedimentable material.

The isolation of *cell nuclei* was originally carried out from nucleated avian red blood cells by laking with water [29], by freezing and thawing [30], with lysolecithin [31], with tyrothricin [32], or, most conveniently, with saponin [33].

From tissues other than blood, nuclei are sometimes prepared by procedures involving treatment of the finely divided tissue with a weak acid, such as citric acid, followed by differential centrifugation and washing with very dilute acid. It will be recalled that the original method employed by Miescher in isolating nuclei was to treat pus cells with dilute acetic acid. The citric acid method has been developed and improved by Dounce [34, 35] and by Mirsky and Pollister [36], in whose papers full details are given. Dounce [34] has isolated nuclei by the citric acid method at different pH values and has pointed out that nuclei prepared at pH values much below 3·0 undoubtedly lose much of their histone content and so give high analytical values for nucleic acid and lipid when analysed in bulk. On the other hand, nuclei isolated at pH 6·0–6·2 appear to lose some RNA and probably also some protein. Since most methods for the preparation of clean nuclei free from contamination with cytoplasmic residues involve repeated washing either with dilute sodium chloride solution or with dilute citric acid, it is not surprising that, as Mirsky and his colleagues [37] point out, the protein content of such isolated nuclei, as determined by gross chemical analysis, is considerably below the values found for similar nuclei isolated from non-aqueous media. Such nuclei were originally prepared by Behrens [38] by a method in which powdered freeze-dried tissue was allowed to sediment out in columns of organic solvents of graded density. This method has subsequently been modified and improved [36, 39, 40] and the nuclei so obtained have the advantage of retaining all their acid-soluble constituents as well as all the nuclear proteins.

More recently glycerol has replaced the petroleum ether originally used as the homogenization medium [41]. Useful methods for isolating nuclei in sucrose solutions have also been described [42, 43]. A commonly employed procedure is that of Schneider and Peterman [44] in which 0·25M-sucrose containing 0·0018M-CaCl$_2$ is used. Philpot and Stanier [45] have employed for the isolation of rat liver nuclei a medium containing 0·3M-sucrose, 0·065M-potassium glycerophosphate, 0·001M-MgCl$_2$, and 40 per cent glycerol. Widnell and Tata [46] have prepared metabolically active nuclei by homogenizing the tissue in 0·32M-sucrose–3mM-MgCl$_2$ with subsequent purification in 2·2M-sucrose–1mM-MgCl$_2$.

The criteria for assessing the quality of preparations of nuclei are: (i) their morphological appearance in the ordinary light microscope, in the phase-contrast microscope, and in the electron microscope; (ii) the presence of enzymes such as NAD pyrophosphorylase (E.C.2.7.7.1) which are known to be exclusively of nuclear origin; (iii) the absence of cytoplasmic enzymes such as cytochrome oxidase (E.C.1.9.3.1) or glucose-6-phosphatase (E.C.3.1.3.9).

Such criteria are met in preparations of nuclei made by quick and simple methods involving the use of detergents such as Triton X-100 (a member of the octylphenoxyethanol series) [47], Tween 80 (polyoxethylene sorbitan monoleate) [48], or sodium deoxycholate [49].

Metabolically active nuclei can also be prepared by gradient centrifugation in sucrose and dextran [50] or Ficoll (a high polymer of sucrose) [51].

Methods of isolation have been discussed in several reviews [35, 42, 43, 52–57]. Even the cleanest preparations, as judged by the standards of conventional microscopy, may show some adhering cytoplasmic debris when examined in the electron microscope [58]. The major problem, as with the isolation of all subcellular fractions, is that in the attempt to obtain microscopically clean nuclei damage may occur with consequent loss of normal intranuclear components. It must be emphasized that the cell nucleus is not a homogeneous structure. Therefore, while gross chemical analysis on bulk material can give in general terms the nature and relative amounts of the various constituents, it tells us nothing about their distribution in the various regions of the nucleus which are studied by the cytologist [59, 60].

Much effort has gone into the subfractionation of nuclei, and methods are available for the isolation of purified nucleoli [61–67], nuclear membrane fragments [68, 69, 275], centrioles [70], mitotic spindles [71], and individual metaphase chromosomes [72].

Nuclei are sedimented from a cell homogenate by low-speed centrifugation, and this leaves the *mitochondria* as the largest particles remaining in the supernatant fraction. In order to prevent aggregation of the smaller particles it is essential to include sucrose in the homogenization medium. In 0·88M-sucrose the mitochondria retain their rod-like morphological characteristics and their ability to stain supravitally with the dye Janus Green B. At this concentration of sucrose, however, on account of the high viscosity and density of the medium, very high centrifugal speeds have to be employed to sediment the subcellular fractions, and a compromise is to use 0·25M-sucrose as a medium in which aggregation of granules does not occur and in which mitochondria can readily be prepared with the same biochemical properties as those obtained in 0·88M-sucrose although they no longer stain with Janus Green B and are spherical rather than elongated in shape. For the separation by differential centrifugation of the subcellular fractions from a homogenate in 0·25M-sucrose prepared in a Potter–Elvehjem homogenizer, a force of 700 g is employed to remove nuclei and general cell debris, including unbroken cells. After removal of the nuclear fraction the extract is centrifuged at 8500 g for 10 minutes to bring down the mitochondria and at 100 000 g for 60 minutes to bring down the microsomes. The supernatant fraction is said to be derived from the *cell sap* and contains no easily sedimentable material. Many variations of this scheme of differential centrifugation are, of course, possible and the different schemes available have been extensively reviewed [18, 42, 43, 73–86]. General methods of cell disruption for the isolation of nucleic acids are discussed further in Chapter 4.

The *microsomes* (diameter 16–150 nm) isolated by such procedures [84, 87] are too small to be resolved by the light microscope and were at first 'cytochemical concepts without any known morphological counterpart in the intact cell'. In an electron microscope study of sections of the microsome pellet from liver tissue Palade and Siekevitz [12] demonstrated that the predominant structural element is represented by membrane-bound profiles recalling those found in the endoplasmic reticulum in sections in the intact liver cell. These profiles appear to correspond in three dimensions to tubules or cisternae and may be smooth surfaced, though the majority carry on their surface small dense particles similar to those observed in electron micrographs of whole cells (p. 9). The microsomes therefore are not artefacts introduced by homogenization of the tissue but, as fragments of the endoplasmic reticulum, they represent cytoplasmic structures known to pre-exist in the intact cell.

When suspensions of microsomes are treated with sodium deoxycholate they are disrupted into an unsedimentable portion derived from the membranous component and containing most of the protein and nearly all of the phospholipid, pigment, and enzymes, and a particulate portion sedimentable at 100 000 g containing almost 20 per cent of the protein and nearly all of the RNA of the microsomes which is itself the bulk of the RNA of the cytoplasm. These small particles, which must be clearly distinguished from the microsomes themselves, contain approximately equal amounts of RNA and protein [88–92] and are in fact isolated *ribosomes* [93, 94].

While this description refers essentially to animal cells, it should be kept in mind that bacterial cells also contain ribosomes which can be isolated by differential centrifugation [95–97]. Contaminating nucleases may be removed by chromatography of the ribosomes on columns of Sephadex G200.

The mixture of subcellular constituents produced on cell disruption can conveniently be separated into fractions differing in specific gravity by the method of *density gradient separation* which involves centrifuging them through a suitable density gradient until each species finds its own density level [76, 98–101] (see Chapter 4).

2.5 The cell nucleus

For a full account of the cytology of the cell nucleus the reader should consult one of the standard texts [20, 22, 102, 274] or the three-volume work edited by Busch [103]. The structure, biochemistry, and functions of the nuclear envelope have been recently reviewed [275].

In most animal cells the nucleus is a circular body lying in the cytoplasm and bounded by a double nuclear membrane (Fig. 2.1) in which pores are visible in electron micrographs [60, 104]. In liver cells the nucleus accounts for only 10–18 per cent of the mass of the cell, whereas in tissues such as thymus it may represent 60 per cent of the total cell mass. Within the resting nucleus there may be seen one or more spherical *nucleoli* and a meshwork of strongly basophilic

THE OCCURRENCE OF NUCLEIC ACIDS

chromatin [279]. The rest of the nucleus is filled with weakly staining material forming the nuclear sap.

The cell nucleus contains the following chief constituents:

(a) Deoxyribonucleic acid (DNA).
(b) Basic proteins (protamines or histones).
(c) Acidic proteins [105–108].
(d) Enzymes [56, 109].
(e) Ribonucleic acid (RNA).
(f) Lipids, other phosphorus compounds [110], and inorganic elements, including potassium, calcium, and magnesium [111].

The earliest investigations of cell nuclei were made by Miescher, using pus cell nuclei and the heads of salmon spermatozoa. Stedman [112] pointed out that much of Miescher's work, which was published after his death by Schmiedeberg, has given rise to the impression that the head of the salmon sperm consists of 96 per cent protamine nucleate, the remainder being accounted for partly as inorganic material. Although Miescher's own earlier writings indicated that in his opinion the salmon sperm heads contained material other than nucleic acid and protamine, the impression grew up that cell nuclei were essentially nucleoprotamines or nucleohistones. In 1942, however, Mayer and Gulick [113] showed that cell nuclei contained proteins other than protamines or histones, and Caspersson and his colleagues, as a result of their photometric measurements in the ultraviolet, concluded that the nucleus contained, in addition to DNA and basic proteins of the histone type, a higher protein of the globulin type and a small amount of RNA [114]. Such non-histone proteins are of considerable importance to the cell [108] (see also p. 407).

2.5.1 *Protamines and histones*

It has been known for a long time that cell nuclei contain simple basic proteins of the histone and protamine type with molecular weights as low as 2000. The protamines are simple basic proteins rich in arginine but lacking in tyrosine and tryptophan. Histones, on the other hand, are rather more complex in structure; they contain tyrosine but little or no tryptophan. The two basic proteins can therefore be distinguished by means of Millon's reaction, which is positive with histones and negative with protamine.

Protamines [115] are found in the sperm heads of salmon and

herring, but not of all fish, since cod sperm contains histone. Histones, in general, are found in the nuclei of somatic cells in amounts ranging from 3 per cent in certain tumours to 24 per cent in fowl erythrocytes [116].

There are five main classes of histone differing in their relative content of lysine and arginine (Table 2.1) [117]. The complete amino acid sequence has been worked out for several of them from different sources [118-122] and it is found to have been remarkably conserved throughout evolution. This implies that these proteins serve a common function in species from classes as far apart as mammals, fish, and plants. The basic amino acids are clustered at the N-terminal end of the molecule where hydrophobic regions are absent, and it has been suggested that this part of the molecule binds to DNA where it serves a structural role (see below). Despite the

TABLE 2.1

Composition of histones

Histone	Amino acid composition	Molecular weight	Number of residues total	basic
Very lysine rich H1, I, f1	27% lysine 24% alanine 2% arginine	~21 500	~210	~64
Lysine rich H2b, IIb$_2$, f2b	16% lysine 6% arginine	13 774	125	30
Arginine/lysine rich H2a, IIb$_1$, f2a$_2$	11% lysine 9% arginine 11% glycine	14 004	129	30
Arginine rich H3, III, f3	10% lysine 15% arginine	15 324	135	32
Glycine/arginine rich H4, IV, F2a$_1$	10% lysine 14% arginine 15% glycine	11 282	102	28

highly conserved primary structure there is considerable modification of these basic proteins by phosphorylation, acetylation, and methylation reactions, and this may dramatically affect their affinity for DNA [123, 274].

There is a close temporal relationship between the time of synthesis of DNA and the histones [124]. The five different histones are all made on a group of small polysomes from which histone messenger RNA has been isolated [125, 126]. Synthesis occurs only during S-phase and is inhibited by inhibitors of DNA synthesis [127].

THE OCCURRENCE OF NUCLEIC ACIDS

In cell nuclei and in sperm heads the bulk of the DNA, which is of course negatively charged, is held in structural association by electrostatic forces with histones and protamines which are positively charged [128]. From the results of X-ray diffraction studies on nucleoprotamines it has been suggested that the protamine peptide chain lies along the shallow groove of the DNA double helix (Fig. 5.6) with the positively charged basic ends of the arginine side-chains held by electrostatic forces to the negatively charged phosphate groups of the DNA [128-130].

The nucleoproteins of the cell nucleus, while insoluble in 0·14M-sodium chloride, are soluble in more concentrated solutions [36]. Isolated nuclei are first washed thoroughly in 0·14M-sodium chloride to remove cytoplasmic material. When the residue is stirred with 1M-sodium chloride the nuclear material dissolves to form a viscous solution of nucleoprotein. When the solution is diluted by pouring into six volumes of water, this complex precipitates out as thread-like material which readily coils round the stirring-rod. This material consists of three main components: nucleic acid (mainly DNA), histone, and a non-histone protein. They may be separated from each other by shaking a solution in 1M-sodium chloride with chloroform containing a little octanol (p. 58). When the mixture is centrifuged an upper aqueous layer forms, containing the nucleic acid in solution while the protein, which forms a gel at the chloroform-water interface, may be removed, washed with alcohol, and dried. From this mixed protein, histone may be separated by extraction with dilute acid, leaving a non-histone tryptophan-containing protein.

There are 15-20 major non-histone nuclear proteins which include actin and myosin. Although often referred to as acidic proteins, this is a misnomer since their pH values range from 3 to 9. These major proteins have molecular weights from 10 000 to 200 000 and are stable which suggests that they may play a structural role. In addition, there are many non-histone proteins present in the nucleus in small amounts, and some of these have been purified (e.g. DNA and RNA polymerases and histone phosphorylase). The nuclear proteins have been reviewed by several authors [106, 107, 129, 131-136, 274]. The role of the nucleus in protein synthesis is described in Chapter 13, and the possible function of some of the non-histone proteins is considered in Chapter 14.

2.5.2 Chromosomes and chromatin

The usual state of much of the nucleic acid and protein in the nucleus of a eukaryotic cell is a dispersed network of fibres known as chromatin. However, during the first stage of mitosis (*prophase*) thread-like structures, the *chromatids*, appear in the nucleus while the nucleolus disappears. These chromatids eventually, by a process of contraction and twisting, form the intensely basophilic compact *chromosomes*. At this stage the centrosome which lies outside the nucleus has divided into two halves which pass to opposite sides of the nucleus. The nuclear membrane and the nucleolus disappear and a structure known as the *spindle* makes its appearance, consisting of fine threads diverging from each centrosome to the equator of a spindle-shaped figure. At the stage of *metaphase* the chromosomes arrange themselves on the equator of the spindle and each divides into two equal portions which during *anaphase* pass to opposite poles of the spindle. This process ensures that each daughter cell obtains the same chromosome material as the mother cell has possessed. During the final stage or *telophase* a new nucleus forms round each group of daughter chromosomes which break up to form the chromatin of the new nucleus while a fresh nuclear membrane and nucleolus are formed. The chemical composition of the chromosomes was originally studied histochemically by Brachet [219], by the ultraviolet absorption technique by the Caspersson school, and chemically, after isolation in bulk, by Mirsky and Ris [220–223]. Recent reviews are recommended [60, 224, 225].

It can be shown by the standard procedures of cytology and genetics that the *genes*, the units of heredity, are lineally arranged in the chromatid string and pass thus to the chromosomes. As the chromatid thread forms and becomes more clearly visible it is seen to be a chain of nodules, rich in easily stainable nucleic acid, but separated by less deeply staining segments consisting mainly of proteins. During telophase certain parts of the chromatids do not disperse to more ill-defined material but persist as well-marked deeply staining nodules termed *heterochromatin*.

Giant (polytene) chromosomes are found in the salivary gland cells of certain dipteran larvae. These chromosomes are made up of many identical chromatids lined up side-by-side. The chromatids are partly condensed and, as the heterochromatic regions are aligned with each other, these chromosomes have a characteristic banded appearance [137, 276].

Some heterochromatic regions appear unchanged in preparations from different tissues. This is the so-called *constitutive heterochromatin*. *Facultative heterochromatin* is found only in specific tissues during a particular developmental stage. Heterochromatin is well marked in the neighbourhood of the nucleolus, where it forms the *nucleolus-associated chromatin*. The remaining chromatin material is termed the *euchromatin*. The two types of chromatin can be separated to some extent by differential centrifugation [138]. From studies using radioactive precursors it has become apparent that RNA is made in euchromatic regions while the heterochromatin represents regions where DNA is packed very densely into inactive regions of the chromosomes. This dense packaging has the effect of delaying replication of DNA in heterochromatic regions [139]. The diameter of the DNA double helix is about 2 nm while the finest chromosomal threads which can be seen in the ordinary light microscope have a diameter of 100–200 nm. Such threads must therefore contain either a large number of DNA strands organized into a complex nucleoprotein structure or a single long DNA helix coiled in a complex way [132]. Even the constituent threads of the lampbrush chromosomes in amphibian oocytes are about 20 nm in diameter, and are therefore considerably larger than DNA double helices. How the DNA helices are folded or coiled in chromosomal threads is not fully understood, although much useful information on chromosomal ultrastructure has been obtained by electron microscopy [7, 60, 140–143].

Recently our concept of chromatin structure has changed from an array of histone molecules arranged regularly alongside the DNA double helix to one where the DNA folds around a histone complex situated regularly along the chromatin fibre. Earlier X-ray work [144] suggested a repeating unit at intervals of 100 Å, and more recent work on nuclease digestion of chromatin showed the presence of units each about 200 nucleotide pairs long [145, 146]. It was on such data that Kornberg based his 'string of beads' model [147] where regions of DNA, protected by histones from nuclease digestion (the nucleosomes [148]), alternate with protein-free DNA (Plate VIII). Reconstitution experiments [148–150] show that histones H2a, H2b, H3, and H4 are all required for nucleosome formation but H1 is not. It appears that two molecules of each of the four histones are present in each nucleosome [272] (see also Chapter 14).

The closed circular DNA of SV40 and polyoma virus is also

associated with histones to form nucleosomes [151]. When the histones are removed the DNA assumes a superhelical conformation showing that the folding of the DNA around the histones produces constraints in the double helix. Indeed Kornberg's model suggests that the DNA when associated with histone is folded to about one-seventh of its normal length. Crick and Klug have suggested that within the nucleosome the DNA may be 'kinked' through 95–100° every 20 base pairs, giving 10 straight stretches of DNA in each 100 Å repeat [152]. Chambon has pointed out [151] that modification of the histone–DNA interaction may lead to unwinding of the extranucleosomal DNA, and Crick [153] has suggested that single-stranded regions of DNA may be vital for the binding of enzymes or regulatory factors involved in transcription of the genetic material. In such a structure there may be considerable lengths of DNA, in which the exact base sequence matters very little, interspersed with shorter stretches of specific sequences which are probably repeated at regular intervals (see p. 109). Histone H_1 may play a role in holding the nucleosomes together [268] and its phosphorylation may initiate condensation of the chromosomes at mitosis [269–271].

2.5.3 *The nucleolus*

In the resting nucleus a large part of the RNA is concentrated in the nucleolus which is particularly prominent in those cells in which a strongly basophilic cytoplasm indicates the presence of a high concentration of RNA. The nucleolus usually gives a negative Feulgen reaction, except perhaps in the peripheral regions adjacent to the nucleolus-associated chromatin. The central regions show a strong affinity for basic dyes, which is removed by ribonuclease, and a strong absorption of ultraviolet light at 257 nm, which is likewise abolished by ribonuclease [154]. This suggests that the central portion of the nucleolus at least is composed of ribonucleoprotein.

The DNA associated with the nucleolus is from a region of the chromosome known as the *nucleolar organizer* which codes for the synthesis of ribosomal RNA.

The biological significance of the nucleolus has been extensively studied and reviewed [61, 67, 155–163]. Its role in ribosome formation has been deduced from several lines of evidence [91]: (i) Its intense basophilia, indicating a high concentration of RNA, is particularly obvious in cells in which protein synthesis is very active.

THE OCCURRENCE OF NUCLEIC ACIDS

(ii) Accumulation of RNA in the nucleolus can be selectively prevented by low doses of actinomycin D (p. 300). (iii) The accumulation of newly formed RNA in the cytoplasm of growing cells can be prevented by irradiation of the nucleolus with a microbeam of ultraviolet light [164]. (iv) Mutants of the toad *Xenopus* in which the nucleolus is lacking do not survive beyond the gastrula stage and are unable to produce ribosomal RNA [165]. They lack virtually all the DNA complementary to 28S and 18S RNA (p. 53). (v) Kinetic studies in cells labelled with radioactive uridine have indicated that the nucleolus is the site of synthesis of 45S RNA which is the precursor of ribosomal RNA. This matter is discussed further in Chapter 12.

2.6 The mitochondria

It has been known since the early work of Warburg [30] in 1913 that cytoplasmic particles such as the mitochondria possess respiratory activity but their vital importance in the economy of the cell has only recently become apparent [8, 166].

The *mitochondria*, of which there are about 400 in a liver cell, account for about 17 per cent of the total cellular nitrogen and less than 4 per cent of the total RNA. They contain the cytochrome oxidase and other enzymes of the terminal oxidative pathway in electron transport and are characterized by their power of transferring the energy released by substrate oxidation into the high-energy phosphate compound adenosine triphosphate (ATP). As the seat of oxidative phosphorylation they are sometimes referred to as the 'power house of the cell'.

Mitochondria contain DNA which is quite distinct from the DNA of the nucleus [82, 167–171]. This DNA, which is in the form of a small cyclic molecule, carries the genetic information for the synthesis of some of the mitochondrial proteins and for the ribosomal RNA present in the mitochondria [82, 166, 172–175, 273].

2.7 Ribosomes and polysomes

The name 'ribosome' was introduced in 1957 to distinguish the particulate material of the microsomes from the membrane material [91, 176]. The ribosomes are electron-dense particles of diameter 20 nm containing about 40 per cent protein and 60 per cent RNA which are found in all types of living cell, both free and attached to the endoplasmic reticulum (p. 9). They play a vital part in the process of protein synthesis (Chapter 13) during which they become attached to a strand of messenger RNA to form complexes known as polysomes. Their properties and biological role have been extensively reviewed [20, 21, 79, 91, 176–184].

It is customary to characterize macromolecules and small particles (e.g. ribosomes) by their sedimentation coefficients expressed in Svedberg units (S). The sedimentation coefficient of a particle or molecule depends on both its molecular weight and its shape and is proportional to its rate of sedimentation in a centrifugal field.

The *Esch. coli* ribosome (sedimentation value 70S, mass 2.7×10^6) is composed of a 30S subunit (mass 0.9×10^6) and a 50S subunit (mass 1.8×10^6) (Fig. 2.3). When the magnesium concentration is

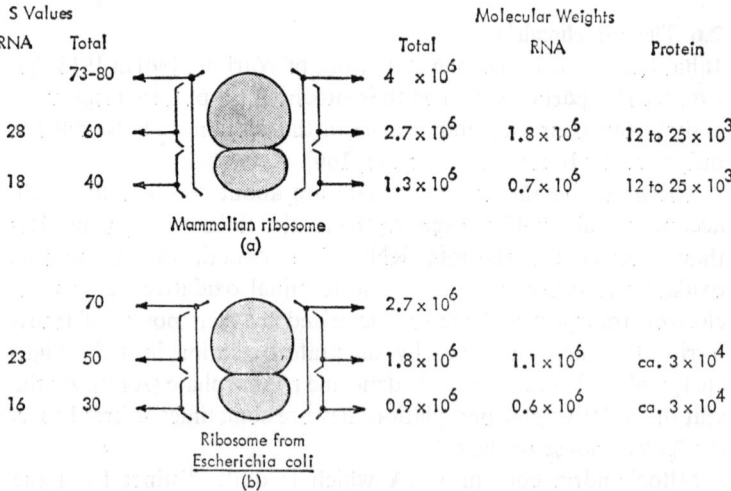

Fig. 2.3 *The S values and molecular weights of the components of* (a) *a mammalian ribosome and* (b) *a ribosome from* Esch. coli.

reduced below 0.5mM the ribosomal particle dissociates into the subunits; they reunite when the magnesium concentration is raised again [185]. The larger subunit contains two binding sites for tRNA (Chapter 13).

In the mammalian cell the ribosome has a sedimentation value of about 80S with subunits of about 40S and 60S [91, 186–190]. The ribosomes inside mitochondria are smaller, with a sedimentation of 55S, and they can be dissociated to subunits of about 29S and 39S [191, 192].

Ribosomal RNA comprises almost 80 per cent of the total RNA of the cell and is of high molecular weight (p. 53). It can be isolated by standard methods [83] and is of two kinds. The smaller subunits contain 16S RNA in *Esch. coli* and 18S RNA in mammalian cells, while the larger subunits contain 23S RNA in *Esch. coli* and 28S RNA in mammalian cells. Each ribosomal subunit contains one of these large RNA molecules while, in addition, the larger subunit contains one molecule of the 5S RNA mentioned on p. 53.

The bacterial ribosomes have been analysed in detail [184]. When ammonium chloride washed preparations are extracted with 4M-LiCl in 8M-urea the smaller subunit releases 21 proteins and the larger subunit 34 proteins [193]. Unwashed particles release many more proteins which are only transiently associated with the ribosome. The ribosomal proteins have been shown to be different from each other by sequencing studies [194] and immunological studies [195]. About 70 per cent of them are very basic and only three have isoelectric points as low as pH 5 [194].

By treating *Esch. coli* ribosomes with increasing concentrations of lithium chloride the proteins may be removed stepwise, 50S subunits yielding successively products of 40S, 36S, 28S, and 25S, and the 30S subunits products of 25S, 23S, 21S, and 19S [172, 196–200]. Alternatively, when 50S and 30S subunits from *Esch. coli* are treated with 5M-caesium chloride, proteins are removed to yield 'cores' of 40S and 23S. These cores are biologically inactive but can be restored to full activity by addition of the abstracted proteins [201, 202].

This partial reconstruction reaction is very rapid, being complete within a few minutes at 37°, which is in marked contrast to reconstruction reactions involving separated 16S RNA and the 21 proteins of the smaller ribosomal subunit [203]. This total reconstruction is extremely sensitive to ionic strength and proceeds most rapidly at 40–50°C [202]. It has been shown that only seven of the 30S proteins bind individually to 16S RNA, the remainder binding only in the presence of the initial seven. The reconstituted 30S particles are fully active by most criteria. However, omission of two of the proteins (S1 and S6) does not result in a significant reduction in activity [204].

It has not yet been possible totally to reconstruct the 50S subunit of the *Esch. coli* ribosome, possibly because of kinetic barriers. However, the 50S subunit from *Bacillus stearothermophilus* (a bacterium that lives in hot springs) has been totally reconstructed following incubation of purified RNA and protein components at 60° for 2 hours [205].

In the living cell ribosomes may be strung along a strand of messenger RNA (p. 375) to form the equivalent of a string of beads. Such structures are known as *polyribosomes* or *polysomes* [206] and are illustrated in Plate II and in Fig. 2.4. A polysome may contain some five or six ribosomes or as many as forty. In the process of protein synthesis, the ribosomes pass along the strand of messenger RNA from one end to the other, each ribosome spinning out a polypeptide chain as it goes (p. 377).

The ribosomes are spaced out on the messenger RNA strand with a gap of about 5–10 nm between them. As might be expected the

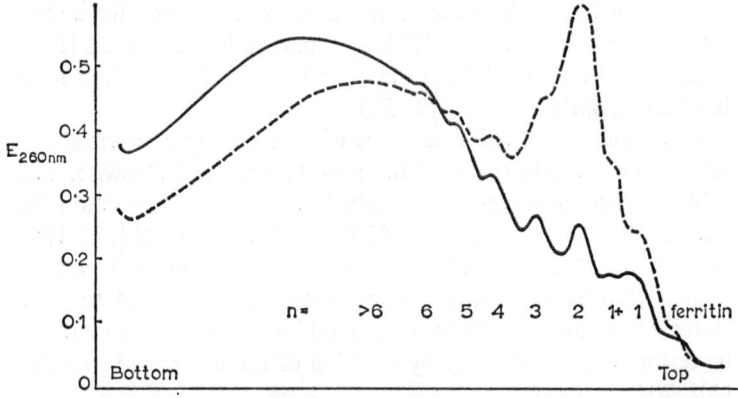

Fig. 2.4 *Distribution of polysomes from rat liver on a linear sucrose gradient. Solid line: from rat on normal diet. Broken line: from rat on tryptophan-deficient diet.* n = *ribosomes per polysome thread*

(*By courtesy of Dr W. H. Wunner*)

polysomes are very delicate structures and the greatest care must be taken during the process of preparation and isolation [207–209] in order to prevent mechanical breakage or degradation by RNase. Sedimentation on a sucrose gradient separates polysomes containing different numbers of ribosomes (Fig. 2.4).

The pattern of polysomes in liver cells is influenced by nutritional changes [210].

2.8 The cell sap and lysosomes

The cell sap containing the soluble proteins of the cytoplasm accounts for about 40 per cent of the total nitrogen of the cell. It also contains the transfer RNA (tRNA) which accounts for some 10–15 per cent of the total RNA of the cell.

THE OCCURRENCE OF NUCLEIC ACIDS

Also in the cytoplasm are particles known as *lysosomes* [211] first described in rat liver cells in 1955 [212] and now known to occur in most animal cells. They are particularly abundant in cells such as macrophages which are called upon to perform important digestive tasks.

Lysosomes (diameter 200–600 nm) are smaller than mitochondria from which they may be separated by sucrose gradient centrifugation [80, 211, 213–215]. They serve as storehouses of powerful digestive enzymes, including ribonuclease, deoxyribonuclease, phosphatases, cathepsins, glycosidases, and sulphatases.

Other related subcellular particles include the *peroxisomes* [216–218] which contain the enzymes uricase, catalase, and D-amino acid oxidase.

2.9 Histochemical identification of nucleic acids

Three main methods [226] may be employed to determine the localization of nucleic acids in individual cells or in tissue sections:

1. Staining by one of the few methods known to be specific for RNA or DNA. A few techniques have been devised which stain RNA or DNA selectively, e.g. the Feulgen stain.
2. Photography in ultraviolet light. The absorption of ultraviolet light at 260 nm by nucleic acids is so intense that if cells are photographed at this wavelength the structures containing nucleic acids can readily be identified.
3. Staining after treatment with specific enzymes. Tissue sections may be examined before and after treatment with such enzymes as ribonuclease and deoxyribonuclease.

The first two techniques may be employed in the measurement of the approximate amount of nucleic acid in a particular part of the cell, e.g. the nucleus.

2.9.1 *The Feulgen reaction*

This reaction was first described by Feulgen and Rossenbeck [227] in 1924 and has been extensively employed by histologists and cytologists for staining cell nuclei and chromosomes [228–230]. It depends upon the fact that the products of partial hydrolysis of DNA will restore the colour of basic fuchsin which has been decolourized with sulphurous acid (Schiff's reagent). It is usually applied to tissue sections which are taken down to water, plunged in 1N-HCl at

50–60° for 4–40 minutes (usually about 10 minutes), and then immersed in the fuchsin–sulphurous acid reagent for 15–90 minutes before washing and mounting. Nuclear material is stained a brilliant purple colour. (Full directions are given in many papers, e.g. that of de Tomasi [231].)

The classical interpretation of the Feulgen reaction is that the DNA in the nuclear material in the tissue is partially hydrolysed by the warm hydrochloric acid to give products which remain at the site of production and cause the development of the purple colour. Areas rich in DNA are therefore deeply stained. Treatment with acid is said to break the sugar linkages attached glycosidically to purine bases. This exposes 2-deoxyribose components which are still attached through phosphate linkages at C-3 and C-5 in the main nucleic acid chain. They are therefore firmly held in the furanose form which immediately becomes converted into a significant proportion of the aldehyde form, which in turn reacts with the Schiff's reagent to produce an insoluble coloured macromolecule [232–234].

The time of hydrolysis is a critical factor. If Carnoy's fluid has been used as fixative, the optimum hydrolysis time of 10 minutes corresponds to the removal of half the total base, presumably the purines [235]. Prolongation of the hydrolysis time results in progressive removal of the DNA from the preparation with consequent reduction in Feulgen staining.

Photometric measurements of the intensity of colour developed in tissue sections submitted to the Feulgen procedure have been made a basis for the quantitative estimation of DNA [236–240, 280].

2.9.2 *Ultraviolet microscopy*

The first description of a microscope suitable for use with ultraviolet light was given in 1904 by Köhler [241], who used optical parts made of fused quartz. The light source was a condensed high-tension metallic spark but the instrument is not easy to use on account of difficulties in focusing and in the location of fields.

A simplified quartz microscope outfit is described by Lavin [242], who uses the Köhler microscope with a quartz spiral mercury resonance lamp and focuses by means of a fluorescent plate.

One advantage of the use of ultraviolet light for microscope work is the greater degree of resolution which can be obtained – about twice that possible with visible light. The main use of ultraviolet microscopy, however, is to reveal the presence of nucleic acid in a tissue. When a section mounted on a quartz slide, usually in glycerol, is photographed in ultraviolet light, those parts which are rich in nucleic acid appear in a positive print as dark areas against a lighter background [243].

Quantitative ultraviolet cytophotometry has been developed to a high pitch of perfection by Caspersson and his associates in Stockholm [244-246].

The method does not distinguish between free purines or pyrimidines, nucleosides, nucleotides, and polynucleotides, nor can it differentiate between RNA and DNA, which both absorb ultraviolet light to an equal extent. Caspersson and his colleagues have therefore used the absorption method in conjunction with the Feulgen staining reaction (Section 2.9.1) in order to distinguish between RNA and DNA.

2.9.3 *The ribonuclease and deoxyribonuclease tests*

The use of the enzyme ribonuclease in conjunction with basophilic dyes such as the Unna–Pappenheim mixture to determine the localization of nucleic acids in tissue sections was first described in 1940 by Brachet who showed that, when a tissue section is incubated with an aqueous solution of ribonuclease, basophilic granules in the cell cytoplasm lose their capacity to take up pyronin, whereas the staining of the chromatin of the cell nucleus by the methyl green component in unaffected.

The analogous test employing deoxyribonuclease in place of ribonuclease has been used by only a few authors, but it has been employed to demonstrate the removal of DNA from chromosomes [247, 248] and from the nuclei of the cells of nervous tissue [249] and liver tissue [250, 251].

2.9.4 *Staining with basic dyes*

Since the nucleic acids are strongly acidic in nature, they have an affinity for basic dyes such as toluidine blue, celestine blue, pyronin, and methyl green, and those areas in a tissue section which readily take up such basic dyes are said to be *basophilic*.

While some basic dyestuffs such as toluidine blue stain both RNA and DNA, others are more selective, and the mixture of pyronin and methyl green [252], known as Unna–Pappenheim stain, has been shown by Brachet [253] to stain RNA red with pyronin and DNA green with methyl green [254].

Staining with methyl green is dependent upon the DNA being in a highly polymerized state. It has been made the basis of a quantitative microspectrophotometric estimation of DNA [255].

Fluorochrome dyes may also be used [256].

2.9.5 Chromosome banding

Chromosomes vary in size, arm length, and position of their centromeres. However, until recently it was not possible to identify individually more than a few chromosomes in a given karyotype (a preparation of metaphase chromosomes). If cells were labelled with tritiated thymidine towards the end of the S-period and autoradiography was performed on the metaphase chromosome preparation, certain chromosomes could be identified by characteristic labelling patterns [257].

With the application of the fluorochromes such as quinacrine mustard and quinacrine dihydrochloride by Caspersson [258, 259] it soon became apparent that each chromosome shows a characteristic banding pattern (Q-banding) [277, 278]. This allows the individual chromosomes to be recognized and any abnormalities to be readily mapped [260]. The methodology is simple, requiring only a 20-minute room temperature exposure to a dilute aqueous solution of the fluorochrome. The bands are visible as bright areas under a fluorescent microscope.

Soon after the discovery of Q-banding it became apparent that a variety of treatments brought about characteristic variable condensation in different regions of metaphase chromosomes and that the bands could be visualized by staining with giemsa. The staining is done under alkaline conditions [261] or following brief treatment with trypsin [262]. This G-banding technique produces dark bands in the light microscope corresponding to the bright fluorescent Q-bands. Although the molecular basis for the banding patterns is not completely elucidated the G-band pattern probably reflects the distribution of non-histone proteins along the chromosome [263, 264]. There appears to be no correlation between the stainability of a band and its time of replication [265].

A different procedure (C-banding) stains only certain bands, usually at the kinetochore, which are composed of constitutive heterochromatin [266, 267]. These bands are interbands in the G-banding procedure.

THE OCCURRENCE OF NUCLEIC ACIDS

REFERENCES

[1] Caspersson, T. and Schultz, J. (1938) *Nature*, **142**, 294; (1939) *ibid.*, **143**, 602
[2] Stenram, U. (1958) *Acta Histochem.*, **5**, 156
[3] Porter, K. R. (1955–6) *Harvey Lectures*, Series 51, p. 175
[4] Palade, G. E. (1953) *J. Histochem. Cytochem.*, **1**, 188
[5] Sjöstrand, F. S. and Hanzon, V. (1954) *Exp. Cell Res.*, **7**, 393
[6] Finean, J. B. (1968) *Chemical Ultrastructure in Living Tissues* (2nd Ed.). Springfield: Thomas
[7] Ris, H. (1961) *Canad. J. Genet. Cytol*, **3**, 95
[8] Lehninger, A. L. (1964) *The Mitochondrion*. New York: Benjamin
[9] Coult, D. A. (1966) *Molecules and Cells*. London: Longmans
[10] McElroy, W. D. (1971) *Cell Physiology and Biochemistry* (3rd Ed.). New Jersey: Prentice-Hall
[11] Neutra, M. and Leblona, C. P. (1969) *Sci. Amer.*, **220**, 100
[12] Palade, G. E. and Siekevitz, P. (1956) *J. Biophys. Biochem. Cytol.*, **2**, 171, 671
[13] Rendi, R. (1959) *Exp. Cell Res.*, **17**, 585
[14] Swanson, C. P. (1965) *The Cell* (2nd Ed.). New Jersey: Prentice-Hall
[15] Loewy, A. G. and Siekevitz, P. (1969) *Cell Structure and Function* (2nd Ed.). New York: Holt, Rinehart and Winston
[16] Allen, J. M. (1967) *Molecular Organization and Biological Function*. New York: Harper and Row
[17] Robertson, J. D. (1959) Biochem. Soc. Symp. No. 16: *The Structure and Function of Subcellular Components*, p. 3. Cambridge: University Press
[18] Brachet, J. and Mirsky, A. E. (Eds.) (1959) *The Cell*. New York: Academic Press
[19] Mercer, E. H. (1961) *Cells and Cell Structure*. London: Hutchinson
[20] Paul, J. (1967) (reprinted 1970) *Cell Biology* (2nd Ed.). London: Heinemann
[21] Campbell, P. N. (1966) *The Structure and Function of Animal Cell Components*. Oxford: Pergamon
[22] Wilson, G. B. and Morrison, J. H. (1966) *Cytology* (2nd Ed.). New York: Reinhold
[23] Stanier, R. Y., Doudoroff, M. and Adelberg, E. A. (1971) *General Microbiology* (3rd Ed.). London: Macmillan
[24] Davis, B. D., Dulbecco, R., Eisen, H. N., Ginsberg, H. S. and Wood, W. B. (1973) *Microbiology* (2nd Ed.). New York: Harper and Row
[25] Bensley, R. R. (1937) *Anat. Rec.*, **69**, 341
[26] Claude, A. (1943) *Science*, **97**, 451; (1943) *Biol. Symp.*, **10**, 111
[27] Claude, A. (1939) *Science*, **90**, 213; (1940) *ibid.*, **91**, 77; (1941) *Cold Spring Harbor Symp. Quant. Biol.*, **9**, 263
[28] Claude, A. (1946) *J. Exp. Med.*, **84**, 51
[29] Ackermann, D. (1904) *Z. Physiol. Chem.*, **43**, 299
[30] Warburg, O. (1910) *Z. Physiol. Chem.*, **70**, 413
[31] Laskowski, M. (1942) *Proc. Soc. Exp. Biol. Med.*, **49**, 534
[32] Villela, G. G. (1947) *Proc. Soc. Exp. Biol. Med.*, **66**, 398
[33] Dounce, A. L. and Lan, T. H. (1943) *Science*, **97**, 584
[34] Dounce, A. L. (1943) *J. Biol. Chem.*, **147**, 685; (1943) *ibid.*, **151**, 221, 235
[35] Dounce, A. L. (1955) *The Nucleic Acids*, Vol. 2, p. 93 (E. Chargaff and J. N. Davidson, Eds.). New York: Academic Press
[36] Mirsky, A. E. and Pollister, A. W. (1943) *J. Gen. Physiol.*, **30**, 117

[37] Allfrey, V., Stern, H., Mirsky, A. E. and Saetren, H. (1952) *J. Gen. Physiol.*, **35**, 529
[38] Behrens, M. (1938) *Aberhalden's Handbuch der Biologische Arbeitsmethoden*, Sect. 5, Part 10, p. 1363
[39] Dounce, A. L., Tishkoff, G. H., Barnett, S. R. and Freer, R. M. (1950) *J. Gen. Physiol.*, **33**, 629
[40] Kay, E. R. M., Smellie, R. M. S., Humphrey, G. F. and Davidson, J. N. (1956) *Biochem. J.*, **62**, 160
[41] Kirsch, W. M., Leitner, J. W., Gainey, M., Schutz, D., Lasher, R. and Nakane, P. (1970) *Science*, **168**, 1592
[42] Busch, H. (1967) *Methods in Enzymology*, Vol. 12, Part A, p. 421 (L. Grossman and K. Moldave, Eds.). New York: Academic Press
[43] Wang, T. Y. (1967) *Methods in Enzymology*, Vol. 12, Part A, p. 417 (L. Grossman and K. Moldave, Eds.). New York: Academic Press
[44] Schnieder, R. M. and Peterman, R. (1950) *Cancer Res.*, **10**, 751
[45] Philpot, J. St. L. and Stanier, J. E. (1956) *Biochem. J.*, **63**, 214
[46] Widnell, C. C. and Tata, J. R. (1964) *Biochem. J.*, **92**, 313
[47] Hymer, W. C. and Kuff, E. L. (1964) *J. Histochem. Cytochem.*, **12**, 359
[48] Fisher, H. W. and Harris, H. (1962) *Proc. Roy. Soc.*, B, **156**, 521
[49] Penman, S. (1966) *J. Mol. Biol.*, **17**, 117
[50] Fisher, W. D. and Cline, G. B. (1963) *Biochim. Biophys. Acta*, **68**, 640
[51] Allfrey, V. G., Littau, V. C. and Mirsky, A. E. (1964) *J. Cell Biol.*, **21**, 213
[52] *Methods of Separation of Subcellular Structural Components* (1963). Biochem. Soc. Symp. No. 23
[53] Allfrey, V. (1959) *The Cell*, Vol. 1, p. 193 (A. E. Mirsky and J. Brachet, Eds.). New York: Academic Press.
[54] Siebert, G. and Smellie, R. M. S. (1957) *Internat. Rev. Cytol.*, **6**, 383
[55] Siebert, G. (1967) *Methods in Cancer Research*, **2**, 287 (H. Busch, Ed.). New York: Academic Press
[56] Siebert, G. (1967) *Methods in Cancer Research*, **3**, 47 (H. Busch, Ed.). New York: Academic Press
[57] Roodyn, D. B. (1972) *Subcellular Components*, p. 15 (G. D. Birnie, Ed.). London: Butterworths
[58] Davison, P. F. and Mercer, E. H. (1956) *Exp. Cell Res.*, **11**, 237
[59] Bouteille, M., Laval, M. and Dupuy-Coin, A. M. (1974) *The Cell Nucleus*, Vol. 1, p. 3 (H. Busch, Ed.). Academic Press
[60] Dupraw, E. J. (1970) *DNA and Chromosomes*. New York: Holt, Rinehart and Winston
[61] Smetana, K. and Busch, H. (1974) *The Cell Nucleus*, Vol. 1, p. 73 (H. Busch, Ed.). Academic Press
[62] Muramatsu, M. and Busch, H. (1967) *Methods in Cancer Research*, **2**, 303 (H. Busch, Ed.)
[63] Busch, H., Hodneth, J. L., Morris, H. P., Neogy, R. and Unuma, T. (1968) *Methods in Cancer Research*, **4**, 179 (H. Busch, Ed.). Academic Press
[64] Vincent, W. S. (1952) *Proc. Nat. Acad. Sci.*, **38**, 139
[65] Monty, K. J., Litt, M., Kay, E. R. M. and Dounce, A. L. (1956) *J. Biophys. Biochem. Cytol.*, **2**, 127
[66] Maggio, R., Siekevitz, P. and Palade, G. E. (1963) *J. Cell Biol.*, **18**, 293
[67] Busch, H. (1967) *Methods in Enzymology*, Vol. 12, Part A, p. 448 (L. Grossman and K. Moldave, Eds.). New York: Academic Press

THE OCCURRENCE OF NUCLEIC ACIDS

[68] Francke, W. W. and Scheer, U. (1974) *The Cell Nucleus*, Vol. 1, p. 220 (H. Busch, Ed.). Academic Press
[69] Kasper, C. B. (1974) *The Cell Nucleus*, Vol. 1, p. 349 (H. Busch, Ed.). Academic Press
[70] Hoffman, E. J. (1965) *J. Cell Biol.*, **25**, 217
[71] Sakai, H. (1968) *Internat. Rev. Cytol.*, **23**, 89
[72] Mendelsohn, J. (1974) *The Cell Nucleus*, Vol. 2, p. 123 (H. Busch, Ed.). Academic Press
[73] Hogeboom, G. and Schneider, W. C. (1955) *The Nucleic Acids*, Vol. 2, p. 199 (E. Chargaff and J. N. Davidson, Eds.). New York: Academic Press
[74] Schneider, W. C. and Hogeboom, G. (1956) *Ann. Rev. Biochem.*, **25**, 201
[75] Anderson, N. G. (1956) *Physical Training Techniques in Biological Research*, Vol. 3, p. 299 (G. Oster and A. W. Pollister, Eds.)
[76] Allfrey, V. (1959) *The Cell*, Vol. 1, p. 193 (J. Brachet and A. E. Mirsky, Eds.). New York: Academic Press
[77] Hogeboom, G., Kuff, E. I. and Schneider, W. C. (1957) *Internat. Rev. Cytol.*, Vol. 6, p. 425 (G. H. Bourne and J. F. Danielli, Eds.). New York: Academic Press
[78] Palade, G. E. (1958) *Microsomal Particles and Protein Synthesis*, p. 36 (R. B. Roberts, Ed.)
[79] Osawa, S. (1968) *Ann. Rev. Biochem.*, **37**, 109
[80] *Methods of Separation of Subcellular Structural Components* (1963). Biochem. Soc. Symp. No. 23
[81] Birnie, G. D. and Fox, S. M. (1969) *Subcellular Components: Preparation and Fractionation*. London: Butterworths
[82] Roodyn, D. B. and Wilkie, D. (1968) *The Biogenesis of Mitochondria*. London: Methuen
[83] Moldave, K. (1967) *Methods in Enzymology*, Vol. 12, Part A, p. 607 (L. Grossman and K. Moldave, Eds.). New York: Academic Press
[84] Moldave, K. and Skogerson, L. (1967) *Methods in Enzymology*, Vol. 12, Part A, p. 478 (L. Grossman and K. Moldave, Eds.). New York: Academic Press
[85] Murray, R. K., Suss, R. and Pitot, H. C. (1967) *Methods in Cancer Research*, **2**, 239 (H. Busch, Ed.)
[86] G. D. Birnie (Ed.) (1972) *Subcellular Components*. London: Butterworths
[87] Tata, J. R. (1972) *Subcellular Components*, p. 185 (G. D. Birnie, Ed.). London: Butterworths
[88] Peterman, M. L., Mizen, N. A. and Hamilton, M. G. (1953) *Cancer Res.*, **13**, 372
[89] Littlefield, J. W., Keller, E. B., Gross, J. and Zamecnik, P. C. (1955) *J. Biol. Chem.*, **217**, 111
[90] Peterman, M. L., Hamilton, M. G., Balis, M. E., Smarth, K. and Pecova, P. (1958) *Microsomal Particles and Protein Synthesis*, p. 70 (R. B. Roberts, Ed.)
[91] Maden, B. E. H. (1968) *Nature*, **219**, 685
[92] Peterman, M. (1964) *Physical and Chemical Properties of Ribonucleoproteins*. Amsterdam: Elsevier
[93] Bonanou-Tzedaki, S. A. and Arnstern, H. R. V. (1972) *Subcellular Components*, p. 215 (G. D. Birnie, Ed.). London: Butterworths
[94] Birnie, G. D., Fox, S. M. and Harvey, D. R. (1972) *Subcellular Components*, p. 235 (G. D. Birnie, Ed.). London: Butterworths

[95] Gillchriest, W. C. and Boch, R. M. (1958) *Microsomal Particles and Protein Synthesis*, p. 1 (R. B. Roberts, Ed.)
[96] Kurland, C. G. (1971) *Methods in Enzymology*, **20**, 379 (L. Grossman and K. Moldave, Eds.)
[97] Takanami, M. (1967) *Methods in Enzymology*, **12**, Part A, p. 491 (L. Grossman and K. Moldave, Eds.)
[98] Albright, J. F. and Anderson, N. G. (1958) *Exp. Cell Res.*, **15**, 271
[99] Bomstein, R. A. and Sterberl, E. A. (1959) *Biochim. Biophys. Acta*, **31**, 548
[100] Thomson, J. F. and Kilpfel, F. J. (1958) *Exp. Cell Res.*, **14**, 612
[101] Meselson, M., Stahl, F. W. and Vinograd, J. (1957) *Proc. Nat. Acad. Sci.*, **43**, 581
[102] Wagner, R. P. and Mitchell, H. K. (1964) *Genetics and Metabolism* (2nd Ed.). New York: Wiley
[103] Busch, H. (Ed.) (1974) *The Cell Nucleus*. Academic Press
[104] Watson, M. L. (1954) *Biochim. Biophys. Acta*, **15**, 475; (1955) *J. Biophys. Biochem. Cytol.*, **1**, 257
[105] Benjamin, W. and Gelhorn, A. (1968) *Proc. Nat. Acad. Sci.*, **59**, 262
[106] Hnilica, L. S. (1967) *Progress in Nucleic Acid Research and Molecular Biology*, Vol. 7, p. 25 (J. N. Davidson and W. E. Cohn, Eds.). New York: Academic Press
[107] Busch, H. and Mauritzen, C. M. (1967) *Methods in Cancer Research*, Vol. 3, p. 392 (H. Busch, Ed.). New York: Academic Press
[108] Gronow, M. and Griffiths, G. (1971) *FEBS Lett.*, **15**, 340
[109] Siebert, G. (1967) *Methods in Cancer Research*, Vol. 3, p. 47 (H. Busch, Ed.). New York: Academic Press
[110] McIndoe, W. M. and Davidson, J. N. (1952) *Brit. J. Cancer*, **6**, 200
[111] Gulick, A. (1944) *Adv. Enzymol.*, **4**, 1
[112] Stedman, E. and Stedman, E. (1947) *Symp. Soc. Exp. Biol.*, **1**, 232
[113] Mayer, D. T. and Gulick, A. (1942) *J. Biol. Chem.*, **146**, 433
[114] Caspersson, T. (1941) *Naturwiss.*, **29**, 33
[115] Felix, K., Frocher, H. and Krekels, A. (1956) *Progr. Biophys. Biophys. Chem.*, **6**, 2
[116] Stedman, E. and Stedman, E. (1943) *Nature*, **152**, 556
[117] Elgin, S. C. R. and Weintraub, H. (1975) *Ann. Rev. Biochem.*, **44**, 725
[118] DeLange, R. J., Hooper, J. A. and Smith, E. M. (1972) *Proc. Nat. Acad. Sci.*, **69**, 882
[119] Fambrough, D. M. and Bonner, J. (1968) *J. Biol. Chem.*, **243**, 4434
[120] Panyim, S., Sommer, K. R. and Chalkley, R. (1971) *Biochem.*, **10**, 3911
[121] Hooper, J. A., Smith, E. L., Sommer, K. R. and Chalkley, R. (1973) *J. Biol. Chem.*, **248**, 3275
[122] Yeoman, L. C., Olson, M. O. J., Sugano, N., Jordan, J. J., Taylor, C. W., Starbuck, W. C. and Busch, H. (1972) *J. Biol. Chem.*, **247**, 6018
[123] DeLange, R. J. and Smith, E. L. (1971) *Ann. Rev. Biochem.*, **40**, 279
[124] Robbins, E. and Borun, T. W. (1967) *Proc. Nat. Acad. Sci.*, **57**, 409
[125] Moav, B. and Nemer, M. (1971) *Biochem.*, **10**, 881
[126] Jacobs-Lorena, M., Baglioni, C. and Borun, T. W. (1972) *Proc. Nat. Acad. Sci.*, **69**, 2095
[127] Borun, T. W., Scharff, M. D. and Robbins, E. (1967) *Proc. Nat. Acad. Sci.*, **58**, 1977
[128] Wilkins, M. H. F. (1956) *Cold Spring Harbor Symp. Quant. Biol.*, **21**, 75
[129] Shooter, K. V. (1958) *Progr. Biophys. Biophys. Chem.*, **8**, 309
[130] Schachman, H. K. (1957) *J. Cell Comp. Physiol.*, **49**, 71

THE OCCURRENCE OF NUCLEIC ACIDS

[131] Busch, H. and Davis, J. R. (1958) *Cancer Res.*, **18**, 1291
[132] Haggis, G. H., Michie, D., Muir, A. R., Roberts, K. B. and Walker, P. B. M. (1976) 2nd Ed. *Introduction to Molecular Biology*. London: Longmans
[133] Butter, J., Johns, E. W. and Phillips, D. M. P. (1968) *Progress in Biophysics and Mol. Biol.*, **18**, 211 (J. A. V. Butler and D. Noble, Eds.)
[134] Delange, R. J. and Smith, E. L. (1971) *Ann. Rev. Biochem.*, **40**, 279
[135] Smith, E. L., DeLange, R. J. and Bonner, J. (1970) *Physiol. Rev.*, **50**, 159
[136] Phillips, D. M. P. (Ed.) (1971) *Histones and Nucleohistones*. New York: Plenum
[137] Edström, J. E. (1974) *The Cell Nucleus*, Vol. 2, p. 293 (H. Busch, Ed.). Academic Press
[138] Frenster, J. H. (1969) *Handbook of Molecular Cytology*, p. 251 (A. Lima-de-Faria, Ed.). North Holland
[139] Lima-de-Faria, A. (1969) *Handbook of Molecular Cytology*, p. 277 (A. Lima-de-Faria, Ed.). North Holland
[140] Ambrose, E. J. (1956) *Progr. Biophys. Biophys. Chem.*, **6**, 26
[141] Stern, H. (1962) *Phys. Rev.*, **42**, 271
[142] Davies, H. G. and Small, J. V. (1968) *Nature*, **217**, 1122
[143] Olins, A. L. and Olins, D. E. (1974) *Science*, **183**, 330
[144] Pardon, J. F., Richards, B. M. and Cotter, R. I. (1974) *Cold Spring Harbor Symp. Quant. Biol.*, **38**, 75
[145] Burgoyne, L. A., Hewish, D. R. and Mobbs, J. (1974) *Biochem. J.*, **143**, 67
[146] Noll, M. (1974) *Nature*, **251**, 249
[147] Kornberg, R. D. (1974) *Science*, **184**, 868
[148] Oudet, P., Gross-Bellard, M. and Chambon, P. (1975) *Cell*, **4**, 281
[149] Kornberg, R. D. and Thomas, J. O. (1974) *Science*, **184**, 865
[150] Richards, B. M. and Pardon, J. F. (1970) *Exp. Cell Res.*, **62**, 184
[151] Germond, J. E., Hirt, B., Oudet, P., Gross-Bellard, M. and Chambon, P. (1975) *Proc. Nat. Acad. Sci.*, **72**, 1843
[152] Crick, H. F. C. and Klug, A. (1975) *Nature*, **255**, 530
[153] Crick, F. (1971) *Nature*, **234**, 25
[154] Davidson, J. N. and Waymouth, C. (1946) *J. Physiol.*, **105**, 191
[155] Vincent, W. S. (1955) *Internat. Rev. Cytol.*, **4**, 269
[156] Stich, H. (1956) *Experientia*, **12**, 7
[157] Sirlin, J. L. (1961) *Endeavour*, **20**, 146
[158] Sirlin, J. L. (1962) *Progr. Biophys. Biophys. Chem.*, **12**, 27
[159] Sirlin, J. L. and Jacob, J. (1962) *Nature*, **195**, 114
[160] Chipchase, M. I. H. and Birnstiel, M. L. (1963) *Proc. Nat. Acad. Sci.*, **50**, 1101
[161] Busch, H., Byvoet, F. and Smetana, K. (1963) *Cancer Res.*, **23**, 313
[162] Miller, Jr., O. L. and Beatty, B. R. (1969) *Handbook of Molecular Cytology*, p. 605 (A. Lima-de-Faria, Ed.)
[163] Busch, H. and Smetana, K. (1970) *The Nucleolus*. New York: Academic Press
[164] Perry, R. P., Hall, A. and Errera, M. (1961) *Biochim. Biophys. Acta*, **49**, 47
[165] Brown, D. D. and Gurdon, J. P. (1964) *Proc. Nat. Acad. Sci.*, **51**, 139
[166] Ashwell, M. and Work, T. S. (1970) *Ann. Rev. Biochem.*, **39**, 251
[167] Luck, D. J. L. and Reich, E. (1964) *Proc. Nat. Acad. Sci.*, **52**, 931
[168] Kalf, G. F. (1964) *Biochem.*, **3**, 1762
[169] Nass, S. and Nass, M. M. K. (1964) *J. Nat. Cancer Inst.*, **33**, 777
[170] Schatz, G., Haslbrunner, E. and Tuppy, H. (1964) *Monatsh. Chem.*, **95**, 1135

[171] Kasamatsu, H. and Vinograd, J. (1974) *Ann. Rev. Biochem.*, **43**, 695
[172] Craven, G. R., Voynow, P., Hardy, S. J. S. and Kurland, C. G. (1969) *Biochem.*, **8**, 2906
[173] Kroon, A. M. and Saccone, C. (Eds.) (1974) *Biogenesis of Mitochondria*. Academic Press
[174] Munn, E. A. (1974) *The Structure of Mitochondria*. Academic Press
[175] Chappell, J. B. and Hansford, R. G. (1972) *Subcellular Components*, p. 77 (G. D. Birnie, Ed.), London: Butterworths
[176] Roberts, R. B., Britten, R. J. and McCarthy, B. J. (1973) *Molecular Genetics*, Part I, p. 292 (J. H. Taylor, Ed.). New York: Academic Press
[177] Kurland, C. G. (1970) *Science*, **169**, 1171
[178] Maden, B. E. H. (1971) *Progress in Biophysics and Mol. Biol.*, **22**, 127 (J. A. V. Butler and D. Noble, Eds.)
[179] Peterman, M. (1964) *The Physical and Chemical Properties of Ribosomes*. Amsterdam: Elsevier
[180] Spirin, A. S. (1964) *Macromolecular Study of Ribonucleic Acids*. London: Reinhold
[181] Cox, R. (1970) *Science J.*, **6**, 56
[182] De Man, J. C. H. and Noorduyn, N. J. A. (1969) *Handbook of Molecular Cytology*, p. 1079 (A. Lima-de-Faria, Ed.)
[183] Nomura, M. (1969) *Sci. Amer.*, **221**, 28
[184] Nomura, M., Tissières, A. and Lengyel, P. (Eds.) (1974) *Ribosomes*. Cold Spring Harbor Labs.
[185] Tissières, A., Watson, J. D., Schlessinger, D. and Hollingsworth, B. R. (1959) *J. Mol. Biol.*, **1**, 221
[186] Noll, H., Staehelin, T. and Wettstein, F. O. (1963) *Nature*, **198**, 632
[187] Gierer, A. (1963) *J. Mol. Biol.*, **6**, 148
[188] Zimmerman, E. F. (1963) *Biochem. Biophys. Res. Commun.*, **11**, 301
[189] Warner, J. R., Rich, A. and Hall, C. E. (1962) *Science*, **138**, 1399
[190] Wool, I. G. and Stöffler, G. (1975) *Ribosomes*, p. 417 (M. Nomura, A. Tissières and P. Lengyel, Eds.). Cold Spring Harbor Labs.
[191] O'Brian, T. W., Denslow, N. D. and Martin, G. R. (1974) *Biogenesis of Mitochondria*, p. 347 (A. M. Kroon and C. Saccone, Eds.). Academic Press
[192] Greco, M., Pepe, G. and Saccone, C. (1974) *Biogenesis of Mitochondria*, p. 367 (A. M. Kroon and C. Saccone, Eds.). Academic Press
[193] Wittmann, H. G. (1974) *Ribosomes*, p. 93 (M. Nomura, A. Tissières and P. Lengyel, Eds.). Cold Spring Harbor Labs.
[194] Wittmann, H. G. and Wittmann-Liebold, B. (1974) *Ribosomes*, p. 115 (M. Nomura, A. Tissières and P. Longyel, Eds.). Cold Spring Harbor Labs.
[195] Tischendorf, G. W., Zeichhardt, H. and Stöffler, G. (1975) *Proc. Nat. Acad. Sci.*, **72**, 4820
[196] Itoh, T., Otaka, E. and Osawa, S. (1968) *J. Mol. Biol.*, **33**, 109
[197] Hardy, S. J. S., Kurland, C. G., Voynow, P. and Mora, G. (1969) *Biochem.*, **8**, 2897
[198] Maglott, L. and Staehelin, T. (1971) *Methods in Enzymology*, **20**, 408 (L. Grossman and K. Moldave, Eds.)
[199] Hindennach, I., Kaltschmidt, E. and Wittman, H. G. (1971) *Eur. J. Biochem.*, **23**, 12
[200] Hindennach, I., Stöffler, G. and Wittman, H. G. (1971) *Eur. J. Biochem.*, **23**, 7
[201] Traub, P., Hosokawa, K., Craven, G. R. and Nomura, M. (1967) *Proc. Nat. Acad. Sci.*, **58**, 2430

[202] Nomura, M. and Held, W. A. (1974) *Ribosomes*, p. 193 (M. Nomura, A. Tissières and P. Lengyel, Eds.). Cold Spring Harbor Labs.
[203] Traub, P. and Nomura, M. (1968) *Proc. Nat. Acad. Sci.*, **59**, 777
[204] Held, W. A., Mizushima, S. and Nomura, M. (1973) *J. Biol. Chem.*, **248**, 5720
[205] Nomura, M. and Erdmann, V. A. (1970) *Nature*, **228**, 144
[206] Rich, A. (1963) *Sci. Amer.*, **209**(6), 44
[207] Rich, A. (1967) *Methods in Enzymology*, Vol. 12, Part A, p. 481 (L. Grossman and K. Moldave, Eds.). New York: Academic Press
[208] Haschemeyer, A. E. V. and Gross, J. (1967) *Biochim. Biophys. Acta*, **145**, 76
[209] Wettstein, F. O., Staehelin, T. and Noll, H. (1963) *Nature*, **197**, 430
[210] Staehelin, T., Verney, E. and Sidransky, H. (1967) *Biochim. Biophys. Acta*, **145**, 105
[211] De Duve, C. (1963) *Sci. Amer.*, **208**(5), 64
[212] De Duve, C., Pressman, B. C., Gianetto, R., Wattiaux, R. and Appelmans, F. (1955) *Biochem. J.*, **60**, 604
[213] Allison, A. C. (1965) *Science J.*, **1**, No. 9, 32
[214] Wattiaux, R. (1969) *Handbook of Molecular Cytology*, p. 1159 (A. Lima-de-Faria, Ed.)
[215] Reid, E. (1972) *Subcellular Components*, p. 93 (G. D. Birnie, Ed.). London: Butterworths
[216] De Duve, C. and Baudhuim, P. (1966) *Physiol. Rev.*, **46**, 323
[217] De Duve, C. (1969) *Proc. Roy. Soc.*, B, **173**, 71
[218] Baudhuin, P. (1969) *Handbook of Molecular Cytology*, p. 1179 (A. Lima-de-Faria, Ed.)
[219] Brachet, J. (1941) *Arch. Biol.*, **53**, 207
[220] Mirsky, A. E. and Ris, H. (1947) *J. Gen. Physiol.*, **31**, 1 and 7
[221] Mirsky, A. E. and Ris, H. (1949) *Nature*, **163**, 666
[222] Denues, A. R. T. (1953) *Exp. Cell Res.*, **4**, 333
[223] Lamb, W. G. P. (1949) *Nature*, **164**, 109; (1950) *Exp. Cell Res.*, **1**, 571
[224] Hearst, J. E. and Botchan, M. (1970) *Ann. Rev. Biochem.*, **39**, 151
[225] Stellwagen, R. H. and Cole, R. D. (1969) *Ann. Rev. Biochem.*, **38**, 951
[226] Smetana, K. (1967) *Methods in Cancer Research*, **2**, 361 (H. Busch, Ed.)
[227] Feulgen, R. and Rossenbeck, H. (1924) *Hoppe-Seyler's Ztschr.*, **135**, 203
[228] Lessler, M. A. (1953) *Internat. Rev. Cytol.*, **2**, 231
[229] Swift, H. (1955) *The Nucleic Acids*, Vol. 2, p. 51 (E. Chargaff and J. N. Davidson, Eds.). New York: Academic Press
[230] Kurnick, N. B. (1955) *Internat. Rev. Cytol.*, **4**, 221
[231] de Tomasi, J. A. (1936) *Stain Technol.*, **11**, 137
[232] Overend, W. G. and Stacey, M. (1949) *Nature*, **193**, 538
[233] Brachet, J. (1947) *Symp. Exp. Biol.*, **1**, 207
[234] Lessler, M. A. (1951) *Arch. Biochem. Biophys.*, **32**, 42
[235] Di Stefano, H. S. (1948) *Proc. Nat. Acad. Sci.*, **34**, 75; (1948) *Chromosoma*, **3**, 282
[236] Walker, P. M. B. and Richards, B. M. (1959) *The Cell*, Vol. 1, p. 91 (J. Brachet and A. E. Mirsky, Eds.). New York: Academic Press
[237] Glick, D. (1959) *The Cell*, Vol. 1, p. 139 (J. Brachet and A. E. Mirsky, Eds.). New York: Academic Press
[238] Leuchtenberger, C. (1958) *General Cytochemical Methods*, Vol. 1, p. 219 (J. F. Danielli, Ed.)
[239] Mirsky, A. E. and Ris, H. (1951) *J. Gen. Physiol.*, **36**, 451
[240] Walker, P. M. B. and Deeley, E. M. (1956) *Exp. Cell Res.*, **10**, 155

[241] Köhler, A. (1940) *Z. Wiss. Mikr.*, **21**, 129
[242] Lavin, G. I. (1943) *Rev. Sci. Instr.*, **14**, 375
[243] Stedman, E. and Stedman, E. (1947) *Symp. Soc. Exp. Biol.*, **1**, 232
[244] Caspersson, T. (1950) *Cell Growth and Cell Function*. New York: Norton
[245] Caspersson, T. (1961) *Fed. Proc.*, **20**, 858
[246] Caspersson, T. (1947) *Symp. Soc. Exp. Biol.*, **1**, 127
[247] Kaufmann, B. P., McDonald, M. and Gay, H. (1951) *J. Cell Comp. Physiol.*, **38**, Suppl. 1, 71
[248] Catcheside, D. G. and Holmes, B. (1947) *Symp. Soc. Exp. Biol.*, **1**, 225
[249] Sanders, F. K. (1946) *Quart. J. Micr. Sci.*, **87**, 203
[250] Davidson, J. N. (1947) *Cold Spring Harbor Symp. Quant. Biol.*, **12**, 50
[251] Korson, R. (1951) *Stain Technol.*, **26**, 265
[252] Michaelis, L. (1947) *Cold Spring Harbor Symp. Quant. Biol.*, **12**, 131
[253] Brachet, J. (1940) *Compt. Rend. Soc. Biol.*, **133**, 88
[254] Kurnick, N. B. (1952) *Stain Technol.*, **27**, 233
[255] Kurnick, N. B. (1950) *Exp. Cell Res.*, **1**, 151; (1950) *J. Gen. Physiol.*, **33**, 243
[256] Armstrong, J. A. (1956) *Exp. Cell Res.*, **11**, 640
[257] Miller, O. J. (1970) *Adv. Human Genet.*, **1**, 35
[258] Caspersson, T., Huttén, M., Lindsten, J. and Zech, L. (1970) *Exp. Cell Res.*, **63**, 240
[259] Caspersson, T., Lomak, K. G. and Zech, L. (1971) *Hereditas*, **67**, 89
[260] Hecht, F., Wyandt, H. E. and Magenis, R. E. H. (1974) *The Cell Nucleus*, Vol. 2, p. 33 (H. Busch, Ed.). Academic Press
[261] Patil, S., Merrick, S. and Lubs, H. A. (1971) *Science*, **173**, 821
[262] Seabright, M. (1971) *Lancet*, **2**, 971
[263] Comings, D. E., Avelino, E., Okada, T. A. and Wyandt, H. E. (1973) *Exp. Cell Res.*, **77**, 469
[264] Stubblefield, E. (1973) *Internat. Rev. Cytol.*, **35**, 1
[265] Stubblefield, E. (1974) *The Cell Nucleus*, Vol. 2, p. 149 (H. Busch, Ed.). Academic Press
[266] Pardue, M. L. and Gall, J. G. (1970) *Science*, **168**, 1356
[267] Hsu, T. C. (1971) *J. Hered.*, **62**, 285
[268] Langmore, J. P. and Wooley, J. C. (1975) *Proc. Nat. Acad. Sci.*, **72**, 2691
[269] Bradbury, E. M., Inglis, R. J. and Matthews, H. R. (1974) *Nature*, **247**, 257
[270] Bradbury, E. M., Inglis, R. J. and Matthews, H. R. (1974) *Nature*, **249**, 553
[271] Gorovsky, M. A. and Keevert, J. B. (1975) *Proc. Nat. Acad. Sci.*, **72**, 2672
[272] Thomas, J. O. and Kornberg, R. D. (1975) *Proc. Nat. Acad. Sci.*, **72**, 2626
[273] Borst, P. (1972) *Ann. Rev. Biochem.*, **41**, 333
[274] Johnson, J. D., Douvas, A. S. and Bonner, J. (1974) *Internat. Rev. Cytol.*, Suppl. 4, 273
[275] Franke, W. W. (1974) *Internat. Rev. Cytol.*, Suppl. 4, 71
[276] Daneholt, B. (1974) *Internat. Rev. Cytol.*, Suppl. 4, 417
[277] Yunis, J. (Ed.) (1974) *Human Chromosome Methodology*. Academic Press.
[278] Schnedl, W. (1974) *Internat. Rev. Cytol.*, Suppl. 4, 237
[279] Mirsky, A. E. (1971) *Proc. Nat. Acad. Sci.*, **68**, 2945
[280] Davies, H. G. and Walker, P. M. B. (1953) *Progr. Biophys. Biophys. Chem.*, Vol. 3, p. 195 (J. A. V. Butler and J. T. Randall, Eds.)

CHAPTER 3

Chemical Constituents of Nucleic Acids

3.1 General

Before any account is given of the structure of the nucleic acids proper it is desirable to discuss the structure of the component parts that make up the nucleic acid molecule. Complete hydrolysis of the nucleic acids yields pyrimidine and purine bases, a sugar component, and phosphoric acid. Partial hydrolysis yields compounds known as nucleosides and nucleotides. Each of these component parts will be discussed in turn.

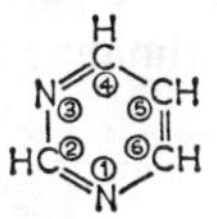

Pyrimidine

Chemical Abstracts system

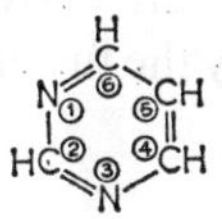

Beilstein system

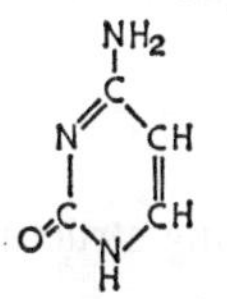

cytosine

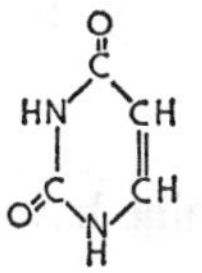

uracil

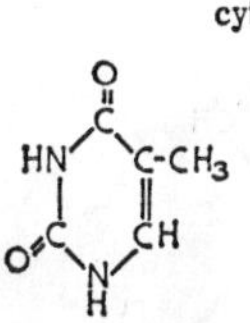

thymine
5-methyl-uracil

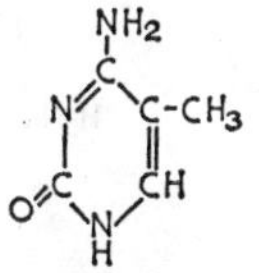

5-methyl cytosine

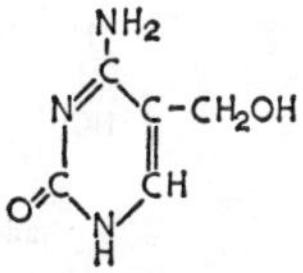

5-hydroxymethyl cytosine

3.2 Pyrimidine bases

The pyrimidine bases are all derivatives of the parent compound pyrimidine, and the derivatives found in the nucleic acids are cytosine found in both types of nucleic acid, uracil found in RNA, and thymine and 5-methylcytosine found in DNA. A fifth pyrimidine, 5-hydroxymethylcytosine, replaces cytosine in certain strains of coliphage (p. 144). The pyrimidine bases can undergo keto–enol tautomerism as shown below for uracil:

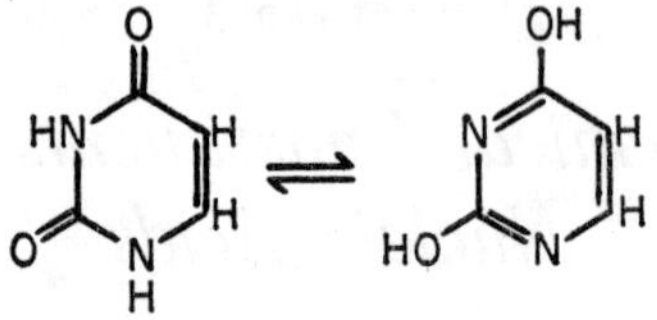

Lactam ⇌ Lactim

The numbering of the pyrimidine ring has caused considerable confusion since the official IUPAC system came into use. The old (Fischer or Beilstein) system shown on the preceding page is still used in Europe but the new system shown on the left predominates. It is used in this book.

3.3 Purine bases

Both types of nucleic acids contain the same purine bases, adenine and guanine. They are derivatives of the parent compound purine which is formed by the fusion of a pyrimidine ring and an iminazole ring.

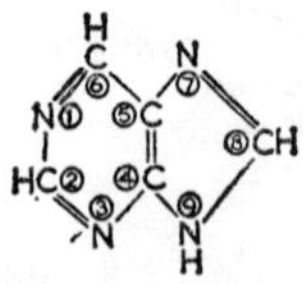

Purine

Adenine and guanine have the following structures:

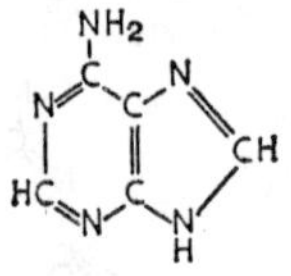

adenine 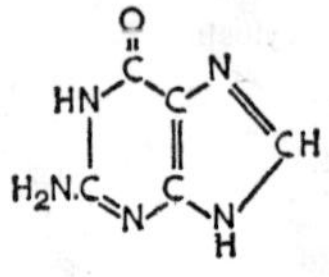   guanine

As in the pyrimidines so with the purines the bases can exist in two tautomeric forms as shown below for guanine:

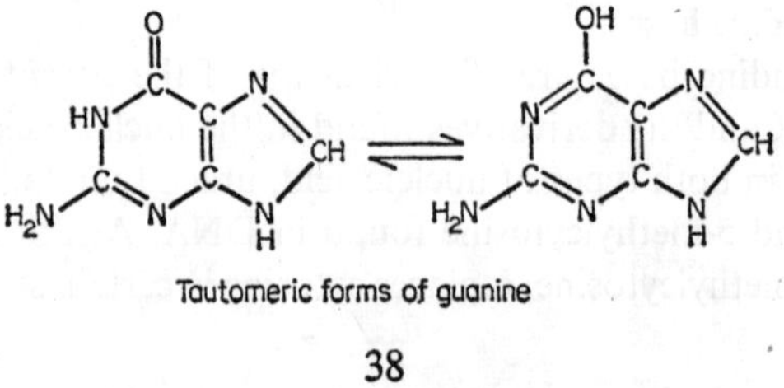

Tautomeric forms of guanine

CHEMICAL CONSTITUENTS OF NUCLEIC ACIDS

Other naturally occurring purine derivatives include hypoxanthine, xanthine, and uric acid.

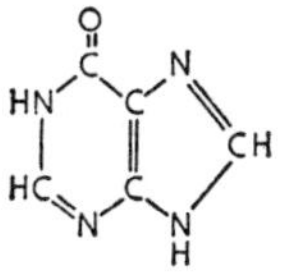

hypoxanthine 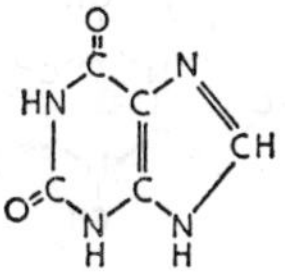   xanthine 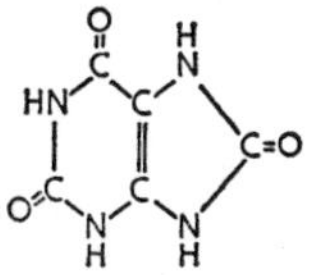   uric acid

Certain 'minor bases' are also found in small amounts in some nucleic acids [1–4]. For example, 'transfer' RNA (tRNA) which is discussed on pages 115 and 122 contains a wide variety of methylated bases, including thymine [3]. These unusual bases comprise less than 5 per cent of the total base content of the tRNA and vary in relative amounts from species to species. Some of the minor bases in RNA are listed in Table 3.1.

TABLE 3.1

Minor bases in RNA

1-methyladenine	dihydrouracil
2-methyladenine	5-hydroxyuracil
6-methyladenine	5-carboxymethyluracil
6,6-dimethyladenine	5-methyluracil (thymine)
6-isopentenyladenine	5-hydroxymethyluracil
2-methylthio-6-isopentenyladenine	2-thiouracil
6-hydroxymethylbutenyladenine	3-methyluracil
6-hydroxymethylbutenyl-2-methylthioadenine	4-thiouracil
	5-methylamino-2-thiouracil
1-methylguanine	5-methyl-2-thiouracil
2-methylguanine	5-uracil-5-hydroxyacetic acid
2,2-dimethylguanine	3-methylcytosine
7-methylguanine	4-methylcytosine
2,2,7-trimethylguanine	5-methylcytosine
hypoxanthine	5-hydroxymethylcytosine
1-methylhypoxanthine	2-thiocytosine
xanthine	4-acetylcytosine
6-aminoacyladenine	

The phenylalanine transfer RNA from yeast (p. 132) contains a most unusual base known as Wye base (Yt), the structure of which is shown below [5, 6, 49, 50].

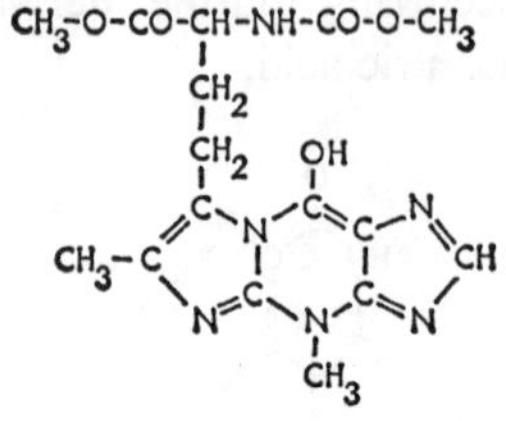

The chemistry of the pyrimidines and purines has been reviewed by Bendich [7] and by Ulbricht [8].

3.4 Pentose sugars

It has long been recognized that the nucleic acid originally prepared from yeast contained a pentose sugar which was identified as ribose by Levene [9] in 1909, using methods that were not absolutely conclusive. Subsequently Gulland [10] and his colleagues in 1943 proved without doubt that the pentose in yeast RNA is D-ribose by converting the aldonic acid from the sugar obtained on hydrolysis of the purine nucleotides of yeast RNA into the corresponding benziminazole which is easily identified.

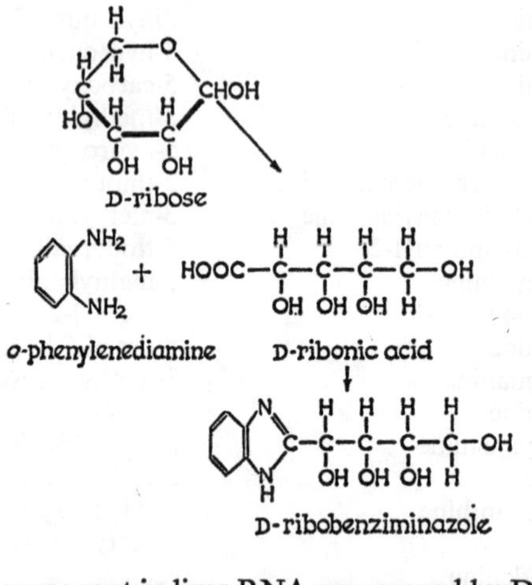

D-ribose

o-phenylenediamine   D-ribonic acid

D-ribobenziminazole

The sugar component in liver RNA was proved by Davidson and Waymouth in 1944 to be ribose by identification as the *p*-bromophenylhydrazone [11]. Since RNA derived from a number of sources including pancreas, liver, the tubercle bacillus, turnip yellow mosaic virus, and influenza virus yields in each case a pentose chromato-

CHEMICAL CONSTITUENTS OF NUCLEIC ACIDS

graphically identical with that from yeast RNA [12–14], it is widely assumed in the absence of evidence to the contrary that the pentose of RNA is always D-ribose. The pentose sugars from several strains of tobacco mosaic virus (TMV) have been identified as ribose by conversion into the di-n-propyl mercaptals [15].

β-D-ribopyranose   β-D-ribofuranose

β-D-2-deoxyribo-pyranose   β-D-2-deoxyribo-furanose

Some RNA's, notably ribosomal RNA's, contain very small amounts of $2'$-O-methylribose.

3.5 Deoxypentose sugars

Doubts surrounding the nature of the sugar present in thymus DNA were resolved when Levene and Mori [16] isolated the sugar from the guanine nucleoside of this nucleic acid and showed that it was a deoxypentose. Only two 2-deoxypentoses can exist, deoxyribose or ribodeose (arabinodeose) derived from ribose and arabinose, and deoxyxylose or xylodeose (lyxodeose) derived from xylose and lyxose.

Synthetic 1-deoxyribose (and its benzylphenylhydrazone) showed rotations of exactly the same value as the deoxypentose from thymus DNA, but of different direction. The carbohydrate in the guanine nucleoside component of thymus DNA is therefore 2-deoxyribose.

Since the sugars in the DNA's from several mammalian tissues, from various micro-organisms, and from fish sperm are chromatographically identical with that in thymus DNA [17–19] they are presumed to be also deoxyribose. The sugar component of the purine nucleosides of the DNA of *Mycobacterium phlei* has been shown to be identical with synthetic 2-deoxy-D-ribose [20], and the same sugar has been isolated from the DNA of herring and cod roe [21].

Glucose occurs glycosidically linked to hydroxymethylcytosine in the DNA from certain strains of bacteriophage (p. 145).

3.6 Nucleosides

A purine or a pyrimidine base may be condensed with a pentose or a deoxypentose sugar to form a *nucleoside*. Thus adenine condenses with ribose to form the nucleoside *adenosine*, guanine forms *guanosine*, cytosine forms *cytidine*, and uracil forms *uridine*. These *ribonucleosides* can be formed on partial hydrolysis of RNA. The ribonucleoside from hypoxanthine is named *inosine*. The nucleosides derived from 2-deoxyribose are known as deoxyribonucleosides – deoxyadenosine, deoxyguanosine, deoxycytidine, deoxythymidine and so on.

adenosine
9-β-D-ribofuranosyl adenine

guanosine
9-β-D-ribofuranosyl guanine

cytidine
1-β-D-ribofuranosyl cytosine

thymidine
1-β-D-2-deoxyribofuranosylthymine

The early investigations of the Levene group [22] showed that the nucleosides were ring N-glycosides which had no reducing properties

CHEMICAL CONSTITUENTS OF NUCLEIC ACIDS

until the sugar component was liberated by hydrolysis. In the nucleosides the sugar component has been shown, by application of the classical procedures of methylation and hydrolysis, to occur in the furanose condition [23], and this has been confirmed by periodate oxidation [24]. Since uridine can be converted into an N-methyl uridine which on hydrolysis yields N-3-methyluracil [25], it follows that the pyrimidine nucleosides are N-1-glycosides.

Spectroscopic evidence led Gulland [26] and his colleagues to the conclusion that in the purine nucleosides the sugar was attached in

TABLE 3.2

Minor nucleosides in RNA [51]
(see also Table 3.1)

1-ribosylthymine
5-ribosyluracil (pseudouridine)
2-ribosylguanine
2'-*O*-methyladenosine
2'-*O*-methylguanosine
2'-*O*-methyluridine
2'-*O*-methylcytidine
2'-*O*-methylpseudouridine
2'-*O*-methyl-4-methylcytidine

the N-9 position, and this conclusion was confirmed by the synthetic experiments of Todd and his associates, who also established the β configuration at the glycosidic linkage [27–30]. These structures have since been confirmed by X-ray crystallography. Cytidine, for example, has been shown by two-dimensional analysis to be cor-

pseudouridine (ψ)
5-β-D-ribofuranosyluracil

rectly represented as 1-β-D-ribofuranosylcytosine with the planar pyrimidine ring lying perpendicular to the non-planar ribofuranose ring [31].

The nucleoside 5-ribosyluracil has been obtained in small amounts from the digestion products of RNA and has been named pseudo-uridine [32–34]. It occurs particularly in transfer RNA (p. 116) and can be isolated in bulk from urine [32]. Other minor nucleosides found in small amounts in certain RNA's are listed in Table 3.2.

The corresponding nucleosides obtained from DNA are formulated as the 9-β-D-2'-deoxyribofuranosyls of guanine and adenine and the 1-β-D-2'-deoxyribofuranosyls of cytosine and thymine [33–36]. They are commonly referred to respectively as *deoxyguanosine*, *deoxyadenosine*, *deoxycytidine*, and *thymidine*. The last should, strictly speaking, be named *deoxythymidine* (or thymine deoxyribonucleoside). The rather rare ribonucleoside containing ribose and thymine is called *ribothymidine*.

3.7 Nucleotides

The nucleotides have been reviewed by Hutchinson [37], by Michelson [38], and by Ulbricht [8]. They are all phosphoric esters of the nucleosides. Those derived from ribonucleosides are usually referred to as *ribonucleotides* and those from deoxyribose nucleosides as

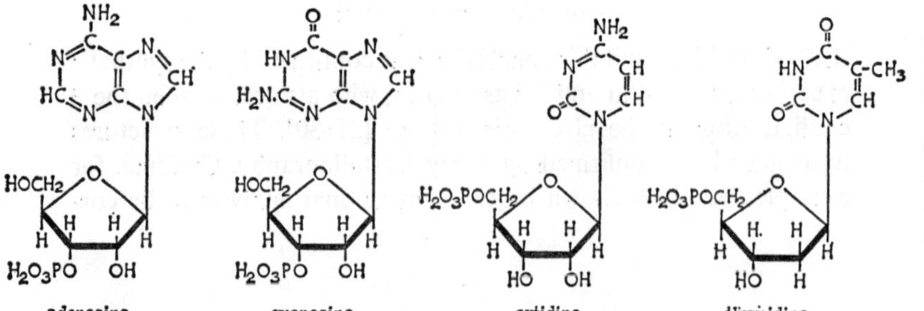

adenosine
3'-phosphate

guanosine
3'-phosphate

cytidine
5'-phosphate

thymidine
5'-phosphate

deoxyribonucleotides. These terms are sometimes abbreviated to *riboside, ribotide, deoxyriboside*, and *deoxyribotide*, but this usage is incorrect.

Since the ribonucleosides have three free hydroxyl groups on the sugar ring, three possible ribonucleoside monophosphates can be formed. Adenosine, for example, can give rise to three monophosphates (adenylic acids), adenosine 5'-phosphate, adenosine 3'-phosphate, and adenosine 2'-phosphate. The first of these was originally discovered in the free state in muscle and used to be referred to as muscle adenylic acid, while the second was originally obtained from

CHEMICAL CONSTITUENTS OF NUCLEIC ACIDS

alkaline hydrolysates of yeast RNA and used to be called yeast adenylic acid.

In the same way guanosine, cytidine, and uridine can give rise to three guanosine monophosphates (guanylic acids), three cytidine

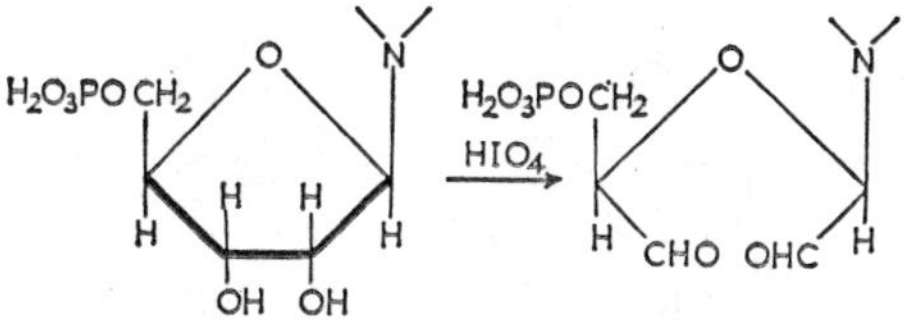

monophosphates (cytidylic acids), and three uridine monophosphates (uridylic acids) respectively. They are frequently referred to by the abbreviations shown in the table at the beginning of the book.

The ribonucleoside 5'-phosphates may be distinguished from the 2'- or 3'-phosphates by the fact that they are susceptible to periodate

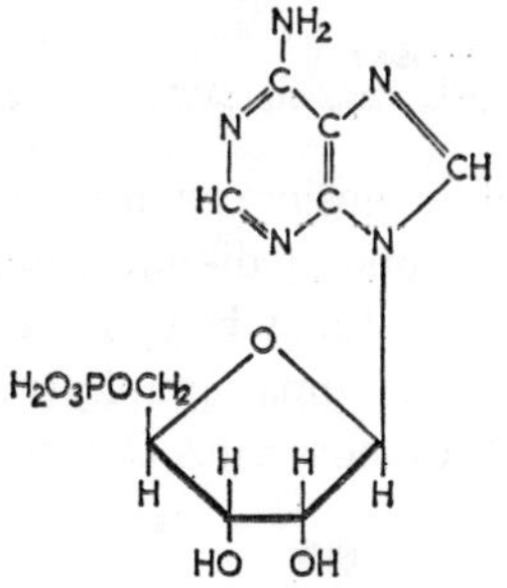

adenosine 5'-phosphate

oxidation owing to the presence of two *cis*-hydroxyl groups. This technique will not distinguish between deoxyribonucleoside 3'- and 5'-phosphates since in this instance there is no hydroxyl group on C-2.

Adenosine 5'-phosphate has the structure shown below which has been confirmed by synthesis [39]. On deamination it yields inosine 5'-phosphate (inosinic acid), a ribonucleotide containing hypoxanthine.

The ribonucleoside 5'-phosphates may be further phosphorylated at position 5' to yield 5'-di- and -tri-phosphates. Thus adenosine 5'-phosphate (AMP) yields adenosine 5'-diphosphate (ADP) and adenosine 5'-triphosphate (ATP). Adenosine 5'- and guanosine 5'-tetraphosphate have also been described.

THE BIOCHEMISTRY OF THE NUCLEIC ACIDS

The structures of these compounds have been confirmed by periodate oxidation [24] and by synthesis [39].

Similarly the other ribonucleoside 5'-phosphates yield such di- and tri-phosphates as GDP, CDP, UDP, GTP, CTP, and UTP. The

adenosine 3':5'-cyclic phosphate (cAMP)   adenosine 2':3'-cyclic phosphate

5'-monophosphates of adenosine, guanosine, cytidine, and uridine together with the corresponding di- and tri-phosphates all occur in the free state in the cell and may be extracted with dilute acid [40]. They may be separated by chromatography (p. 73) or electrophoresis.

Ribonucleoside 3',5'-diphosphates can be formed on hydrolysis

adenosinediphosphate (ADP)

adenosinetriphosphate (ATP).

CHEMICAL CONSTITUENTS OF NUCLEIC ACIDS

of RNA molecules and will be discussed later (p. 125) as will the ribonucleoside 2':3'-cyclic monophosphates (p. 119) and the 3':5'-cyclic monophosphates [40, 41].

Mono-, di-, and tri-phosphates of both purine and pyrimidine deoxyribonucleosides occur in small amounts in tissue extracts [42, 43] and can be formed from the corresponding deoxyribonucleosides by deoxyribonucleoside and deoxyribonucleotide phosphokinases [44, 45].

3.8 Methods of preparation

Details of practical methods are given in the volume edited by Colowick and Kaplan [46] and in Michelson's book [38].

When RNA is boiled with dilute acid the purine nucleotides are hydrolysed with liberation of adenine and guanine (which are easily isolated), ribose, and phosphoric acid. The pyrimidine nucleotides are more resistant to hydrolysis and must be broken down by heating with acid in an autoclave or in a sealed tube in order that cytosine and uracil may be liberated [47]. During this process there is a tendency for cytosine to be deaminated to uracil.

Hydrolysis of nucleic acids to the constituent bases may also be carried out by heating for 1 hour at 100° with 12N-perchloric acid. The bases may conveniently be separated and isolated by the use of a column of an ion-exchanger such as Dowex 50.

The ribonucleoside 2'- and 3'-phosphates may conveniently be prepared by subjecting RNA to alkaline hydrolysis and separating the resultant nucleotides by ion exchange chromatography. When RNA is hydrolysed under the influence of snake venom diesterase (p. 172) nucleoside 5'-phosphates are obtained and may be separated chromatographically on a preparative scale. The nucleoside 5'-phosphates may also be isolated from the acid-soluble fraction of animal tissues. In most tissues nucleotides of adenine are much more abundant than those of the other bases.

When DNA is acted on by appropriate enzymes (p. 174) a mixture of deoxyribonucleoside 5'- or 3'-phosphates is obtained and may be separated by chromatography on columns of Dowex 1 or 2 [48].

Details of the chromatographic separation of nucleotides and other derivatives are to be found in Colowick and Kaplan [46].

Nucleosides are conveniently prepared by dephosphorylation of nucleotides with the aid of a suitable monoesterase preparation (p. 192).

REFERENCES

[1] Adler, M., Weissman, B. and Gutman, A. B. (1958) *J. Biol. Chem.*, **230**, 717
[2] Littlefield, J. W. and Dunn, D. B. (1958) *Biochem. J.*, **70**, 642
[3] Smith, J. D. and Dunn, D. B. (1959) *Biochem. J.*, **72**, 294
[4] Davis, F. F., Carlucci, A. F. and Roubein, I. F. (1959) *J. Biol. Chem.*, **234**, 1525
[5] Nakanishi, K., Blobstein, S., Funamizu, M., Furutachi, N., Van Lear, G., Grunberger, D., Lanks, K. W. and Weinstein, I. B. (1971) *Nature New Biol.*, **234**, 107
[6] Thiebe, R., Zachau, H. G., Baczymskyj, L., Biemann, K. and Sonnenbichler, J. (1971) *Biochim. Biophys. Acta*, **240**, 163
[7] Bendich, A. (1955) *The Nucleic Acids,* Vol. 1, p. 81 (E. Chargaff and J. N. Davidson, Eds.). New York: Academic Press
[8] Ulbricht, T. L. V. (1964) *Purines, Pyrimidines and Nucleotides.* London: Pergamon
[9] Levene, P. A. and Jacobs, W. A. (1909) *Ber. Deutsch. Chem. Ges.*, **42**, 2102, 2469, 2474, 2703
[10] Barker, G. R. and Gulland, J. M. (1943) *J. Chem. Soc.*, 625; Barker, G. R., Farrar, K. R. and Gulland, J. M. (1947) *J. Chem. Soc.*, 21
[11] Davidson, J. N. and Waymouth, C. (1944) *Biochem. J.*, **38**, 375
[12] Vischer, E. and Chargaff, E. (1948) *J. Biol. Chem.*, **176**, 715
[13] Schwerdt, C. E. and Loring, H. S. (1947) *J. Biol. Chem.*, **167**, 593
[14] Ada, G. L. and Gottschalk, A., (1956) *Biochem. J.*, **62**, 686
[15] MacDonald, D. L. and Knight, C. A. (1953) *J. Biol. Chem.*, **202**, 45
[16] Levene, P. A. and Mori, T. (1929) *J. Biol. Chem.*, **83**, 803
[17] Vischer, E., Zamenhof, S. and Chargaff, E. (1949) *J. Biol. Chem.*, **177**, 429
[18] Chargaff, E., Vischer, E., Doniger, R., Green, C. and Misani, F. (1949) *J. Biol. Chem.*, **177**, 405
[19] Chargaff, E. and Lipshitz, R. (1953) *J. Amer. Chem. Soc.*, **75**, 3658
[20] Jones, A. S. and Laland, S. G. (1954) *Acta Chem. Scand.*, **8**, 603
[21] Laland, S. G. and Overend, W. G. (1954) *Acta Chem. Scand.*, **8**, 192
[22] Levene, P. A. and Bass, L. W. (1931) *Nucleic Acids.* New York: Chemical Catalog Company
[23] Levene, P. A. and Tipson, R. S. (1932) *J. Biol. Chem.*, **97**, 491
(1933) *ibid.*, **101**, 529
[24] Todd, A. R. (1946) *J. Chem. Soc.*, 647
[25] Levene, P. A. and Tipson, R. S. (1934) *J. Biol. Chem.*, **104**, 385
[26] Gulland, J. M. (1938) *J. Chem. Soc.*, 1722
[27] Lythgoe, B., Smith, H. and Todd, A. R. (1947) *J. Chem. Soc.*, 355
[28] Howard, G. A., Kenner, G. W., Lythgoe, B. and Todd, A. R. (1946) *J. Chem. Soc.*, 861
[29] Howard, G. A., Lythgoe, B. and Todd, A. R. (1945) *J. Chem. Soc.*, 556, 1052
[30] Davoll, J., Lythgoe, B. and Todd, A. R. (1946) *J. Chem. Soc.*, 833
(1948) *ibid.*, 967
[31] Furberg, S. (1950) *Acta Chem. Scand.*, **4**, 751
[32] *Biochemical Preparations* (1963) **10**, 135
[33] Gulland, J. M. (1947) *Symp. Soc. Exp. Biol.*, **1**, 1
(1947) *Cold Spring Harbor Symp. Quant. Biol.*, **12**, 95
[34] Lythgoe, B. and Todd, A. R. (1947) *Symp. Soc. Exp. Biol.*, **1**, 15
[35] Brown, D. M. and Lythgoe, B. (1950) *J. Chem. Soc.*, 1990

CHEMICAL CONSTITUENTS OF NUCLEIC ACIDS

[36] Manson, L. A. and Lampen, J. P. (1951) *J. Biol. Chem.*, **191**, 87
[37] Hutchinson, D. W. (1964) *Nucleosides and Coenzymes*. London: Methuen
[38] Michelson, A. M. (1963) *The Chemistry of Nucleosides and Nucleotides*. New York: Academic Press
[39] Baddiley, J. and Todd, A. R. (1947) *J. Chem. Soc.*, 648
[40] Jost, J.-P. and Rickenberg, H. V. (1971) *Ann. Rev. Biochem.*, **40**, 741
[41] Robison, G. A., Butcher, R. W. and Sutherland, E. W. (1971) *Cyclic AMP*. New York: Academic Press
[42] Potter, R. L. and Schlesinger, S. (1955) *J. Amer. Chem. Soc.*, **77**, 6714
[43] Schneider, W. C. (1957) *J. Nat. Cancer Inst.*, **18**, 569
[44] Keir, H. M. and Smellie, R. M. S. (1959) *Biochim. Biophys. Acta*, **35**, 405
[45] Grav, H. J. (1967) *Methods in Cancer Research*, **3**, 243. New York: Academic Press
[46] Colowick, S. P. and Kaplan, N. O. (Eds.) (1967) *Methods in Enzymology*, Vol. XII, Part A
[47] Hunter, A. and Hlynka, I. (1937) *Biochem. J.*, **31**, 486
[48] Cohn, W. E., Volkin, E. and Khym, J. X. (1957) *Biochemical Preparations*, **5**, 49
[49] Kasai, H., Goto, M., Ikeda, K., Zama, M., Mizuno, Y., Takemura S., Matsuura, S., Sugimoto, T., and Goto, T. (1976) *Biochemistry*, **15**, 898
[50] Blobstein, S. H., Gebert, R., Grunberger, D., Nakanishi, K., and Weinstein, I. B. (1975) *Arch. Biochem. Biophys.*, **167**, 668
[51] Nishimura, S. (1972) in *Progress in Nucleic Acid Research and Molecular Biology*, Vol. 12, p. 49. Eds. J. N. Davidson and W. E. Cohn, New York, Academic Press

CHAPTER 4

Isolation and Characterization of Nucleic Acids

4.1 General procedures for nucleic acid isolation

Detailed experimental information about laboratory methods for the isolation and purification of nucleic acids from various sources will be found in the handbooks edited by Cantoni and Davies [1] and by Grossman and Moldave [2]. Only general outlines of the principles involved will be given here.

4.1.1 *Disruption*

The first step in the isolation of a nucleic acid is its dissociation from the other cellular constituents and its separation from associated protein. Some methods for the disruption of animal cells have already been mentioned in Chapter 2. They include freezing and thawing, bursting by osmotic shock, or the use of lytic chemical agents. Mechanical methods may be employed such as grinding with a mortar and pestle with sand, alumina, or fine glass beads, or by a piston-type homogenizer such as that devised by Potter which can disrupt cells without serious damage to subcellular organelles. Devices such as the Waring blendor which rely on the use of cutting blades driven at very high speed are liable to be more disruptive to the organelles. Exposure to sonic or ultrasonic vibrations is a useful method for small amounts of material.

Bacterial cells may be lysed with surface-active agents such as sodium lauryl sulphate (sodium dodecyl sulphate, SDS) or, in suitable organisms, by the use of the enzyme lysozyme. Grinding with glass beads or alumina has been a popular method, and disruption by means of ultrasonic vibrations is also useful.

4.1.2 *Separation*

After cell disruption and removal of cell debris by centrifugation, the nucleic acid must be separated from associated protein. This may be achieved, as described below, by the use of phenol, or by

ISOLATION AND CHARACTERIZATION OF NUCLEIC ACIDS

repeated treatment with chloroform–isoamyl alcohol or with chloroform–octanol. Protein may also be removed by digestion with enzymes such as *pronase*, a broad-spectrum protease from *Streptomyces griseus* which digests almost any protein to amino acids.

In the isolation of RNA, care must be taken to avoid degradation by contaminating ribonucleases either from the experimental materials themselves or from the operator's fingers. Suitable ribonuclease inhibitors may have to be employed (Chapter 8). In the preparation of DNA, degradation by deoxyribonucleases (Chapter 8), which are activated by divalent metal ions, can be prevented by the presence of chelating agents and of SDS. In special cases, where even the slightest trace of nuclease must be avoided, autoclaved glassware must be used and the operator should wear gloves.

Conversely, purified ribonuclease may be used to remove contaminating RNA from preparations of DNA, and purified deoxyribonuclease to remove contaminating DNA from preparations of RNA.

4.1.3 *Purification*

The purification of nucleic acids is based on principles depending on molecular size, secondary structure, and base composition.

Separation of nucleic acid molecules according to size may be achieved by gel filtration on columns of dextran derivatives in the form of one of the many varieties of Sephadex in which smaller molecules penetrate the gel particles to a varying extent whereas large molecules are not retained and pass through readily. Molecules are therefore eluted from a column in the order of decreasing molecular size. Polyacrylamide gels are also commonly employed as described below.

Various methods of centrifugation are also commonly used. For example, when cells of *Esch. coli* are lysed with SDS and submitted to equilibrium centrifugation in a gradient of caesium chloride as described in page 84, the DNA separates as a discrete band separate from all other cell constituents.

Purification of nucleic acids may also be achieved on the basis of secondary structure using columns of methylated albumin on kieselguhr (MAK) which bind single-stranded DNA more firmly than double-stranded DNA, or of hydroxyapatite, a crystalline form of calcium phosphate, which retains double-stranded DNA while allowing single-stranded DNA to pass through.

Purification may also be carried out by making use of differences in base composition. For example, different transfer RNA's (p. 122) which vary in the proportions of bases can be separated by countercurrent distribution while DNA's of different base compositions have different buoyant densities and may be resolved by centrifugation to equilibrium in CsCl density gradients.

4.2 Isolation of RNA

The method of choice may vary according to the type of tissue employed and the particular RNA species to be isolated. For animal tissues it is sometimes convenient to carry out a preliminary fractionation of cytoplasmic and nuclear material (e.g. ribosomes, mitochondria, nucleoli, etc.). Probably the most commonly employed method for the preparation of undegraded RNA in good yields is based on treatment at elevated temperatures (63°) with a detergent (e.g. sodium dodecyl sulphate) to release and denature protein together with a solvent such as phenol for the denatured protein. The aqueous layer obtained upon centrifugation contains RNA and polysaccharides. Both are precipitated by ethanol but RNA may be extracted from the precipitate with 2-methoxyethanol from phosphate buffer. After dialysis the RNA is precipitated with ethanol. There are several modifications of the phenol method [3]; they have been discussed in detail by Kirby [4-6] and by Georgiev [7]. Traces of DNA can be removed by treatment with pancreatic DNase (purified free of RNase activity).

While RNA may be extracted in bulk from cells by methods such as these and subsequently fractionated by chromatographic or other methods, it is obviously desirable in certain circumstances to fractionate the cellular material before extracting the RNA. Ribosomes may be separated from cells as described in Chapter 2, and used in preparation of rRNA [8] or messenger RNA may be extracted from polysomes [9, 10]. Recently native messenger RNA has been isolated from purified polysomes simply by treatment with the broad-spectrum proteolytic enzyme *proteinase K* (from the fungus *Tritirachium album*) [11, 12]. Transfer RNA (tRNA) occurs in the soluble cytoplasmic fraction of cells which is obtained when a homogenate of a tissue is centrifuged to remove nuclei, mitochondria, microsomes, and ribosomes. This fraction may be used directly for the preparation of tRNA or it may be treated with acid to yield the 'pH 5 precipitate' from which the tRNA may be extracted by phenol.

ISOLATION AND CHARACTERIZATION OF NUCLEIC ACIDS

4.3 Types of RNA

RNA is a polymer, the monomer units of which are ribonucleoside monophosphates.

The living cell, whether from mammalian, bacterial, or other sources, contains three main kinds of RNA:

(a) The bulk of the RNA, about 80 per cent, is contained in the minute cytoplasmic particles known as ribosomes which are discussed in Chapter 2. Ribosomal RNA (rRNA) is of high molecular weight and is metabolically stable.

Ribosomal RNA falls into two main categories. In *Esch. coli* the 50S and 30S ribosomal subunits (p. 22) yield RNA's of molecular weights $1 \cdot 1 \times 10^6$ (23S RNA) and $0 \cdot 6 \times 10^6$ (16S RNA) respectively [13]. The two RNA's contain different base ratios, differ in base sequences [14, 15], and can hybridize with different sites on the bacterial genome (p. 331).

The 70S to 80S ribosomes from mammalian cells also yield two rRNA components corresponding to the 30–40S and 50–60S ribosomal subunits. These components vary in molecular weight according to the material of origin and method of separation and analysis [16], but the smaller component generally has a molecular weight of $0 \cdot 7 \times 10^6$ (18S) while the larger has a molecular weight of $1 \cdot 8 \times 10^6$ (28S).

Associated with the ribosomes are two RNA's of low molecular weight: (i) 5S RNA containing about 120 nucleotides, which is associated with the larger ribosomal subunit (p. 331) and has been obtained from bacterial [17, 18], animal [19–22], and plant cells [23, 24]; (ii) 5·8S RNA, of chain length about 130 nucleotides, which is associated with the 28S RNA [25, 26] in the large subunits of animal and plant ribosomes (p. 330).

(b) The next most abundant type, about 15 per cent, is transfer RNA (tRNA), a class of molecules which function as adaptors for amino acids in the course of protein synthesis (Chapter 13). They have a much lower molecular weight (23 000–28 000) [27] and sediment in the 4S region on zone centrifugation. Many different tRNA's exist, each being specific for one amino acid.

Transfer and ribosomal RNA's from mitochondria distinct from those present in cytoplasm have been described [30–32].

(c) The remaining 5 per cent or less is RNA with a base composition corresponding very closely to that of DNA. It is sometimes referred to as DNA-like RNA and includes 'messenger' RNA (mRNA) which is described in Chapter 12. The figures in the literature for the molecular weight of mRNA show wide variations [29]. In general, values of 0.5×10^6 or higher are quoted.

In addition to these RNA species, a number of other RNA molecules are to be found in cell nuclei and cytoplasm. Generally their existence is ephemeral and in some cases they appear to serve as precursors to the types of RNA already discussed, i.e. rRNA, tRNA, and mRNA [33]. These labile RNA's vary considerably in size and stability [34–36] and will be discussed in greater detail in Chapter 12.

(d) A normal constituent of *Esch. coli*, present in rather low amounts, is 6S RNA. It has been isolated from the supernatant fraction after sedimentation of ribosomes, and its nucleotide sequence has been determined [37]. It has no known function.

(e) The nuclei of mammalian cells also normally contain a number of low molecular weight monodisperse RNA's. Their sizes range from 100 to 180 nucleotide residues [38]. However, as is the case for the *Esch. coli* 6S RNA mentioned above, their cellular role is unknown at present.

4.4 Separation of RNA species

As isolated by any of the methods just mentioned, RNA tends to be a complex mixture of polynucleotides of various chain lengths together with breakdown products or *oligonucleotides*. (Oligonucleotides are usually defined as small polynucleotides containing fewer than about 20 nucleotide units.) In many cases further separation of the constituents of this complex mixture is required. Bentonite [8, 39], Macloid gel [40], or some other ribonuclease inhibitor [41] is sometimes added to prevent degradation by nucleases during the purification process.

4.4.1 *Gradient centrifugation*

These three main RNA fractions can also be readily separated in the process of *zone centrifugation* through sucrose density gradients [42, 43]. In this process an ultracentrifuge tube is prepared containing sucrose solution increasing in concentration from 5 per cent at the top to 25 per cent at the bottom. The solution of RNA's is carefully layered on the top and the tube is centrifuged at high speed for several hours (Fig. 4.1). The bottom of the tube is then punctured with a hypodermic needle and a series of samples of a few drops each is collected. The nucleic acid content of each sample is estimated by measurement of ultraviolet absorption (p. 68). An example of the results obtained by this method is shown in Fig. 4.3.

For large-scale work centrifugation in zonal rotors may be employed.

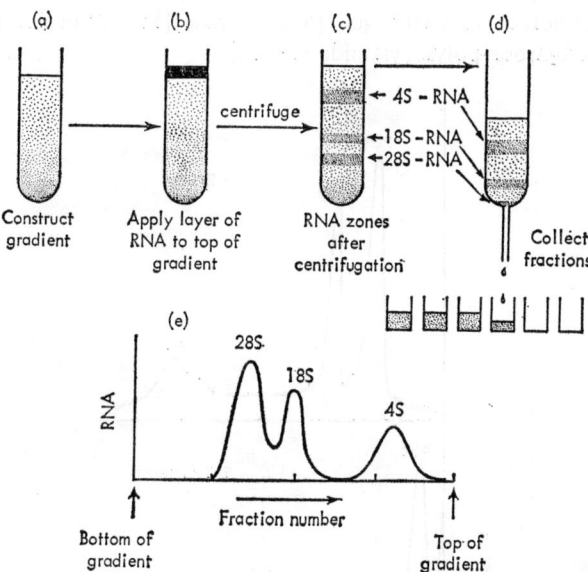

Fig. 4.1 *The process of gradient centrifugation. A sucrose gradient is constructed in a centrifuge tube at* (a), *the RNA solution is then applied carefully as a layer on top* (b), *and, on centrifuging, the RNA separates out into its main components according to molecular weight and shape of the molecules* (c). *When the tube is punctured* (d), *the gradient is collected a few drops at a time in tubes to give sequential fractions. The amount of RNA in each fraction is then determined by ultraviolet absorption measurements so as to give the pattern shown at* (e)

4.4.2 Gel electrophoresis

One of the most delicate methods for fractionating small amounts of ribonucleic acids is zone electrophoresis through polyacrylamide gels [44–52]. The fractions separate as discrete bands which may be located by scanning in ultraviolet light. The type of separation obtained by this method is illustrated in Fig. 4.2. It can be used on a preparative scale for the fractionation of RNA [53], and also in two dimensions to effect the separation of low molecular weight RNA's (e.g. tRNA's) [54, 55].

Normally the mobility of RNA's through the gels depends not

only on molecular size but also on secondary structure. However, the technique can be modified using formamide as electrophoretic solvent to determine RNA molecular weight [56]. The formamide destroys secondary structure and renders the RNA's conformationally homogeneous. Allied methods of separation include zone electrophoresis in starch gels [57], agarose [58], or in composite gels of agarose–polyacrylamide [59, 60].

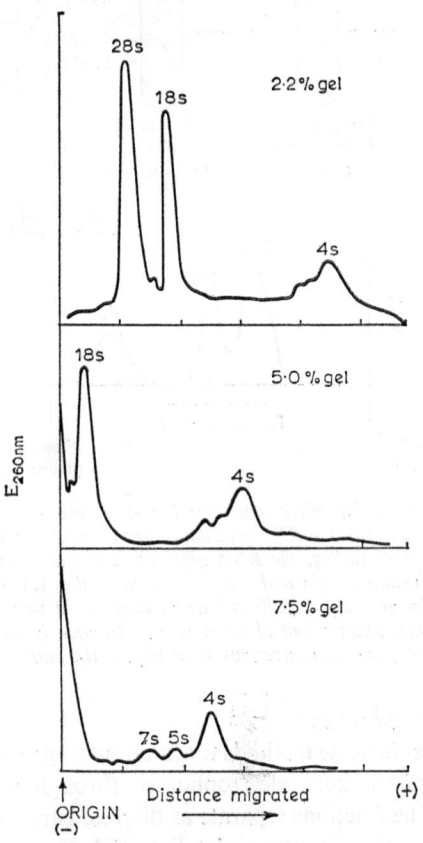

Fig. 4.2 *Schematic diagram illustrating the electrophoretic separation of RNA components (isolated from tumour cell cytoplasm using hot phenol and detergent) which can be achieved with the aid of polyacrylamide gels of various concentrations*

ISOLATION AND CHARACTERIZATION OF NUCLEIC ACIDS

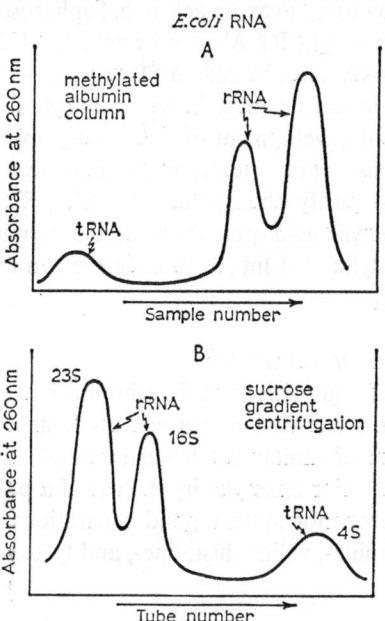

Fig. 4.3 (Above) *Separation of the two ribosomal RNA's and tRNA from Esch. coli on a column of methylated albumin.* (Below) *Separation of the same RNA's by sucrose gradient centrifugation and collection of the samples in tubes as shown in Fig. 4.1*

4.4.3 Chromatographic methods

Many systems have been described for the chromatographic separations of mixtures of RNA's. They have been extensively reviewed [4, 61].

Useful is the column of kieselguhr coated with methylated albumin (MAK) [62–64] which, on elution with increasing concentrations of sodium chloride, yields 3 peaks of RNA, corresponding to tRNA and the other two to the two major rRNA species (Fig. 4.3).

Columns of calcium phosphate [65–67] or of hydroxyapatite [68] have also been employed, and DEAE-cellulose (diethylaminoethyl-cellulose) [69, 70] using gradient elution with urea is particularly effective in separating oligonucleotides containing two to seven nucleotides [70].

Gel filtration using columns of Sephadex G-100 has been used [71] to separate various low molecular weight RNA's (tRNA and

5S RNA), and by using agarose gels (e.g. Sepharose) separation of higher molecular weight RNA's can be achieved [72, 73].

Columns of DNA immobilized on an inert support (e.g. cellulose) have been used in the purification of messenger RNA [74–77]. A recent and elegant development of this affinity approach is the use of synthetic chains of polyuridylic acid immobilized on cellulose or Sepharose 4B to purify eukaryotic mRNA's [78–80] which have tracts of polyadenylic acid (poly-A) at their 3'-termini (see Chapter 12). Oligothymidylic acid linked to cellulose can also be used for this purpose [81].

4.4.4 *Countercurrent distribution*
The technique of countercurrent distribution has been extensively employed in the separation of various fractions of tRNA [27, 4, 82–85] which are of similar molecular size and shape but vary in base composition. For example, by the use of a phosphate buffer–formamide–isopropanol system good separations have been obtained of the alanine-, valine, histidine-, and tyrosine-tRNA's from yeast [86].

4.5 The isolation of DNA
DNA exists in nature in the eukaryotic cell nucleus as deoxyribonucleoprotein [87], and it may be isolated by extraction from the cell followed by separation from the associated protein. The nucleoprotein complex may be extracted in 1M-sodium chloride [88, 89], and when the viscous solution is shaken with chloroform containing a little octyl or amyl alcohol [91] the protein forms a gel at the chloroform–water interface while the sodium salt of the nucleic acid remains the aqueous phase. A more convenient method for dissociating the DNA from the protein is with the aid of detergents such as sodium dodecyl sulphate [92] or sodium xylenesulphonate [93] but an alternative method using phenol is commonly employed. Full details of the phenol method for DNA isolation are described by Kirby [4, 6, 94–96].

The methods employed in isolating DNA vary according to the nature of biological material involved. Methods for animal, plant, and bacterial sources have been fully described [97–102]. For microorganisms in general one of the most satisfactory procedures is that of Marmur [103] which involves disruption of the cells, denaturation of cell debris, and removal of RNA by ribonuclease followed by selective precipitation of DNA with isopropanol. Chelating agents and sodium dodecyl sulphate are added to prevent bivalent metal ion contamination and degradation by deoxyribonuclease which requires bivalent metal ions for its hydrolytic action.

Methods for the isolation of DNA from mitochondria [104, 105] and from chloroplasts [106] have been described.

The most serious problems in any attempt to isolate DNA from natural sources are the avoidance of nuclease degradation and of shear degradation. The long thin threads which constitute the DNA molecules are very easily broken, even by shaking in solution. However, DNA may be isolated from *Esch. coli* by very gentle lysis of the cells in a layer on the surface of a caesium chloride solution and separation by gradient centrifugation of the entire DNA content of the bacterial chromosome as one molecule about 1 mm long with 3×10^6 base pairs corresponding to a mol. wt. of 2×10^9 [107]. High molecular weight DNA has been obtained from mammalian chromatin simply by treatment with the broad-spectrum proteolytic enzyme proteinase K, mentioned earlier [90].

Homogeneous preparations of DNA of monomolecular species are exceedingly difficult to obtain. The most suitable sources are the DNA-containing viruses discussed in Chapter 7.

Detailed experimental information about laboratory methods for the isolation and purification of DNA's from various sources will be found in the handbooks edited by Cantoni and Davies [1] and by Grossman and Moldave [2].

4.6 Types of DNA
DNA usually exists in the form of a double-stranded helix which may be linear in the form of a long flexible rod or ribbon, or cyclic with the two ends of the rod joined together to form an annular structure. Less commonly DNA is found in single-stranded form either linear or cyclic. These forms are discussed in detail in Chapter 5.

4.7 Separation of DNA species
4.7.1 *Zone centrifugation through sucrose gradients*
This separates DNA molecules according to size and shape. Several empirical relationships connect sedimentation coefficient (s) and molecular weight (M) for linear double-stranded DNA, e.g.

$$s = 2 \cdot 7 + 0 \cdot 1517 M^{0 \cdot 445} \qquad [108, 109]$$

Zone centrifugation also separates DNA molecules of the same molecular weight but of different shape, e.g. supercoiled open cyclic and linear forms of polyoma virua DNA (p. 141).

Zone centrifugation through alkaline sucrose gradients has proved useful in the separation of single-stranded cyclic and linear DNA. The method is rapid and gives good yields on a fairly large scale.

4.7.2 *Equilibrium centrifugation*

Equilibrium centrifugation in CsCl gradients [110] may be used to separate DNA molecules of different buoyant densities, e.g. single-stranded from double-stranded DNA, and is particularly useful in the separation of DNA's of different G+C contents (p. 84) (Chapter 5). Resolution can be improved [112] by the addition of actinomycin D [111] which differentially binds to G+C rich segments of DNA Netrospin is a drug which binds to DNA with A+T specificity and has been used to fractionate various DNA's in CsCl gradients [113].

Cs_2SO_4 gradients can also be used to fractionate DNA's but are mainly used to separate DNA complexes with the metal ions Ag^+ or Hg^{2+}. Hg^{2+} is considered to be an A+T specific reagent [114] whereas some evidence indicates Ag^+ to be more G+C specific [115]. NaI [116] and KI [117] gradients have also been recently introduced and can separate DNA of different base compositions.

4.7.3 *Equilibrium centrifugation in CsCl density gradients containing intercalating agents*

The dye ethidium bromide intercalates into DNA, that is, it becomes inserted between adjacent base pairs in the helix. This results in a slight untwisting of the helix and a decrease in buoyant density, but the physical constraints of a double-stranded supercoiled DNA molecule limit the amount of dye which can be bound per unit length. Consequently the density decrease is less in a supercoiled molecule than in a linear or open-cyclic molecule, and supercoiled forms can readily be separated from other DNA on a CsCl–ethidium bromide gradient. The related dye propidium iodide may be used in a similar way to give an improved separation.

4.7.4 *The use of column chromatography*

Native and denatured DNA may be fractionated by chromatography on columns of hydroxyapatite [118, 119]. Columns of methylated albumin on kieselguhr (MAK column) can also be used to separate

ISOLATION AND CHARACTERIZATION OF NUCLEIC ACIDS

denatured from native DNA and even DNA's of different G+C content [63, 120]. Columns of benzoylated [121] or benzoylated naphthoylated DEAE [122] have also been used. Molecular sieve chromatography, on the other hand, can be used to separate DNA's on the basis of size [123].

4.7.5 Electrophoresis

A linear relationship has been demonstrated between electrophoretic mobility and G+C content for a series of microbial DNA's [124].

4.7.6 Solvent extraction

Native DNA's differ in their partition coefficients in polyethylene glycol–dextran two-phase systems [125]. The system can also be used to separate single-stranded DNA from duplex DNA.

4.7.7 Interaction with basic proteins

At about 1M-NaCl polylysine interacts specifically with A+T rich DNA and can precipitate it from a mixture with G+C rich DNA [126]. Polyarginine, on the other hand, preferentially interacts with high G+C DNA [126].

4.8 Separation of RNA–DNA hybrids from RNA and DNA

RNA–DNA hybrids are formed by allowing the complementary single-stranded or denatured DNA and RNA species to come together in high salt concentrations at elevated temperatures and 'annealing' for a time. In early experiments hybrid formation was measured by equilibrium centrifugation in CsCl. This is time-consuming, and Nygaard and Hall [127] made a very important contribution when they found that RNA–DNA hybrids and denatured DNA would bind to nitrocellulose membranes but free RNA would not. A further development was the attachment of the DNA to nitrocellulose membranes prior to 'annealing' it with the RNA [128]. Hybrids may also be recovered by binding to hydroxyapatite [129] or by virtue of their resistance to single-stranded specific nuclease like the S1 nuclease from *Aspergillus* [130] (see Chapter 8).

The technique of hybrid formation between two complementary strands of DNA or between a strand of DNA and a complementary strand of RNA has proved of great value and is discussed in relation to specific examples on pages 104 and 319. The principles involved have been reviewed by several authors [131–136], and details of the experimental procedures are given by Gillespie [137].

4.9 Approaches to gene isolation

The formation and isolation of specific RNA–DNA hybrids can be used as a means of obtaining single-stranded DNA's complementary to specific cellular RNA's such as tRNA, 5S RNA, and ribosomal RNA. However, the isolation of the genes for these species as DNA duplexes is more complex and various approaches involving a number of the DNA fractionation techniques mentioned in Section 4.7 are discussed in a recent and extensive review [138].

With regard to protein specifying genes, although the fibroin genes from *Bombyx mori* have been studied using actinomycin–CsCl gradients [112], a most striking success has been the isolation by Beckwith and his colleagues [139] of the piece of DNA corresponding to the *lac* operon of *Esch. coli* [140], a single unit of transcription consisting of three structural genes specifying the three enzymes involved, together with their control genes, which is induced when lactose is supplied as carbon source (p. 399). The *lac* operon can be incorporated with opposite orientation into the relatively small genomes of the related bacteriophages λ and $\phi 80$. In these two phages the two complementary strands of the double-stranded DNA have different buoyant densities and can readily be separated into light and heavy strands, the two heavy strands containing opposing strands of the *lac* operon DNA in the same relative location. When the heavy strands from the two bacteriophages are mixed, a partially double-stranded DNA is produced in which the double-stranded region is formed by the complementary strands of the *lac* operon DNA (Fig. 4.4). The four protruding ends of the hetero-duplex are single-stranded portions of bacteriophage DNA which cannot form base pairs since they have similar and not complementary sequences. These ends are trimmed off by a nuclease which attacks only single-stranded DNA leaving the double-stranded DNA of the *lac* operon.

A recently developed approach to gene isolation capitalizes on the fact that DNA complexed with protein can be purified since it is protected from DNase digestion. The *lac* operator–repressor complex so treated resulted in the purification of the *lac* repressor binding site (operator) comprising 21 base pairs of DNA [141].

ISOLATION AND CHARACTERIZATION OF NUCLEIC ACIDS

Promoter sequences may be similarly purified by virtue of protection by RNA polymerase [142] (See also Chapter 5).

4.10 The chemical determination of nucleic acids in tissues

It is not the purpose of this book to give experimental details for the estimation of nucleic acids in tissues. For these the reader is referred

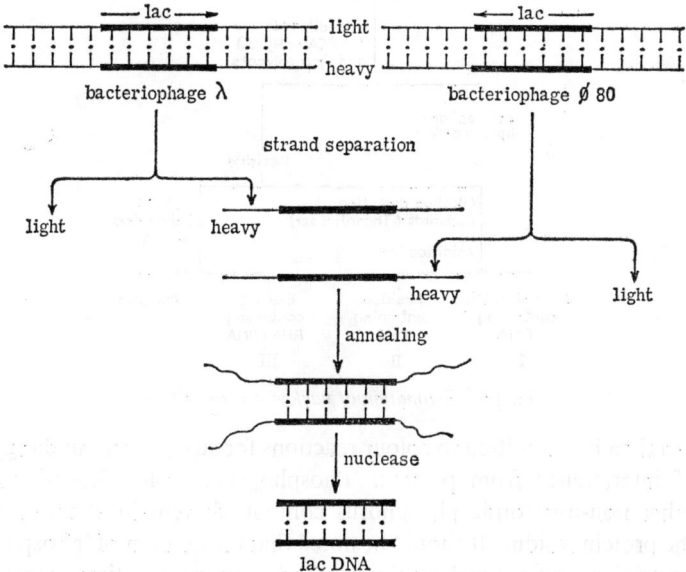

Fig. 4.4 *Isolation of the lactose operon. For details see text*

to the original papers or to recent reviews [143–147], particularly the extensive reviews by Munro and colleagues [148–150]. Only the general principles will be considered here.

All chemical methods for the estimation of nucleic acids are based on determination of (a) phosphorus, (b) total or reactive ribose or deoxyribose, or (c) purine and pyrimidine, and their accuracy is therefore limited by variations in percentage content of phosphorus, total sugar, and so on, between nucleic acids from different sources. In nearly all methods for the estimation of nucleic acids the finely divided tissue is extracted with acid, usually trichloroacetic acid (TCA), perchloric acid (PCA), or sulphuric acid, followed by lipid solvents (occasionally lipid solvents followed by acid). Special precautions must be taken to avoid losses of RNA and protein in the lipid solvents [151].

4.10.1 *The Schneider procedure*

In the procedure of Schneider [152] the nucleic acids are extracted by treatment of the acid-insoluble non-lipid fraction with dilute TCA at 90° for 15 minutes whereby the nucleic acids are split off as soluble products in the acid extract (fraction III in Fig. 4.5) which

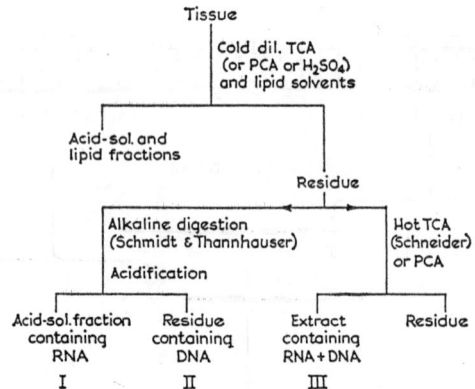

Fig. 4.5 *Estimation of nucleic acids in tissues*

can then be submitted to colour reactions for sugars without danger of interference from protein. 'Phosphoprotein phosphorus' and other non-nucleotide phosphorus compounds remain attached to the protein residue. It should be noted that the amount of 'phosphoprotein' as determined by this method is nearly ten times greater than that determined in the Schmidt–Thannhauser procedure.

A modification of the Schneider procedure involves the use of perchloric acid in place of trichloroacetic acid [153, 154].

The advantages of the Schneider method are its speed and simplicity. The main disadvantage is that DNA and RNA are not separated and can therefore only be estimated by the specific colour reactions for pentose and deoxypentose, which are very susceptible to interference by a variety of factors.

Modifications of the Schneider principle have been described to make it applicable to a few milligrams [155, 156] or even a few micrograms [157] of tissue.

4.10.2 *The Schmidt and Thannhauser procedure*

In the procedure of Schmidt and Thannhauser [158] the extracted tissue residue is incubated overnight with warm dilute alkali which breaks down RNA into acid-soluble nucleotides without affecting DNA in the same way. Some phosphoprotein phosphorus is split off at this stage as inorganic phosphate. When the alkaline digest is acidified, DNA is precipitated along with a large amount of degraded protein (fraction II in Fig. 4.5) and may be determined as DNA-P. The acid-soluble supernatant fluid may therefore be determined as RNA-P.

The main advantage of this method is that RNA and DNA are separated and can therefore be estimated by determination of phosphorus or sugar or purine and pyrimidine. Its main defect is that the 'RNA' fraction I contains phosphorus compounds other than nucleotides [159, 160]. Hence estimations based on P analysis tend to give values which are too high. The non-nucleotide P compounds in I may account for 25 per cent of the total P of the fraction in liver [159] or 80 per cent in nervous tissue [161] and include phosphopeptides and inositol phosphates [162]. The nucleotides may conveniently be separated in the pure state from these 'concomitants' by paper electrophoresis [159] and be estimated separately.

Another disadvantage of the Schmidt–Thannhauser principle is the tendency *in certain tissues only* for some DNA to break down and appear in the RNA fraction [163, 164]. This tendency varies according to the tissue and the species, and is for example much more evident in calf thymus than in rabbit thymus [165]. It may be controlled to some extent, but not completely, by adjusting the conditions of digestion to suit the tissue in question and by reducing the concentration of alkali from 1N to 0·5N or even 0·3N.

The phosphorus estimations themselves may be carried out by any of the standard methods for phosphorus, such as those of Fiske and Subbarow [166], Allen [167], Griswold *et al.* [168], or Berenblum and Chain [169]. Much more sensitive methods for phosphorus estimation have been described by Norberg [170] (lower limit 0·5 ng P) and by Engström [171], which can estimate the total phosphorus in a single cell, but these methods have not been combined with the Schmidt–Thannhauser separation procedure.

4.10.3 *Colour reactions of the sugar component*
These methods are colorimetric procedures more or less specific for pentose sugars or for deoxy-sugars and have been reviewed by Dische [172]. Pentose-containing compounds do not in general interfere in the methods for deoxypentose estimation and *vice versa*; at the same

time, most of the colorimetric sugar reactions are liable to interference by proteins, and pentose estimations may be upset by furfural derived from mucopolysaccharides. It is desirable therefore to free the nucleic acids as completely as is possible from the protein residue of the tissue before applying the colorimetric tests.

A major difficulty in all such methods is the calculation of the amount of DNA or RNA present from the amount of reactive ribose or deoxyribose found. In this connection it is important to remember that the sugars in purine nucleotides are more reactive than those in pyrimidine nucleotides. This difficulty is frequently evaded by calibrating the pentose or deoxypentose estimation with a purified specimen of yeast RNA or thymus DNA of known phosphorus content so that when the nucleic acid content of an unknown tissue is determined by such an estimation the result may be expressed in terms of nucleic acid phosphorus. If this is done, the accuracy of the result obviously depends on the purity and composition of the standard used.

It is also possible to use as standards pure ribose or deoxyribose or purified specimens of purine ribonucleotides and deoxyribonucleotides.

4.10.4 *Determination of ribose*

Most methods of ribose estimation depend upon the liberation of furfural when the ribose-containing material is heated with hydrochloric acid.

The most sensitive and most commonly employed test for pentoses [173], however, is the orcinol procedure which is frequently employed for the estimation of RNA. Details of suitable modifications are given by Schneider [145] and by Ceriotti [155].

These methods estimate only the purine-bound pentose of RNA, and their reliability is therefore limited by variations in the purine/pyrimidine ratio of RNA's from different sources.

4.10.5 *Determination of deoxyribose*

A commonly employed method for deoxyribose estimation is the Dische [174] diphenylamine reaction. It depends upon the formation of a blue colour when DNA is heated with diphenylamine in acid solution. The reacting sugar is found in the purine nucleotides rather than the pyrimidine nucleotides.

By the addition of acetaldehyde and by allowing the reaction to take place for several hours at 30° instead of for a few minutes at 100°, the sensitivity of the reaction is greatly increased and its susceptibility to interference by other compounds is greatly reduced [175, 176].

A useful and more sensitive quantitative colour reaction with indole has also been described by Ceriotti [155] and has been adopted by other authors [148, 177].

A very delicate fluorimetric method has been described for the estimation of as little as 3 ng DNA [179]. A fluorimetric method for the estimation of thymine in amounts of DNA of the order of 50 ng is also available [180]. Thymine in DNA can also be determined by techniques of gas chromatography [178].

4.10.6 *Determination of purines and pyrimidines by ultraviolet absorption measurements*

When RNA and DNA have been separated by the method of Schmidt and Thannhauser, the amount of purine and pyrimidine and hence of nucleic acid in each fraction may readily be estimated by measuring the extinction of the solution in the ultraviolet region [157, 161, 181] provided that certain precautions are taken [182, 183]. When such measurements are to be made, perchloric acid is a useful extracting and precipitating reagent, since it has a lower extinction in the ultraviolet than trichloroacetic acid, which absorbs powerfully at 260 nm.

The intense ultraviolet absorption in the wavelength range 250–290 nm exhibited by purine and pyrimidine derivatives is due to the conjugated double bond systems which they contain. This property is of practical importance in several respects:

(1) It provides a method by which small amounts of purine and pyrimidine derivatives in solution can be estimated provided that the molar extinction coefficient for each derivative is known.

(2) It enables the position of purine and pyrimidine derivatives to be detected on paper or thin-layer chromatograms and ionophoretograms for example. When exposed to ultraviolet light of wavelength 250–290 nm, filter paper fluoresces a light blue colour. The presence of a purine or pyrimidine derivative, by absorbing the incident radiation, 'quenches' this fluorescence, so that a spot of a purine or pyrimidine derivative on paper appears as a dark area when the paper is viewed in ultraviolet light of this wavelength.

(3) It provides a useful means of identifying different purine and pyrimidine derivatives. In aqueous solution each of the purine and pyrimidine bases and nucleosides has an individual spectrum which varies characteristically with pH (see Fig. 4.6). Full details of the spectra of nucleic acid derivatives will be found elsewhere [144, 184–186].

(4) It is the basis for the use of ultraviolet photography and ultraviolet spectrophotometry (Chapter 2).

The nucleic acids themselves have an ultraviolet spectrum of the form shown in Fig. 4.7.

It should be noted that the extinction coefficient of a nucleic acid

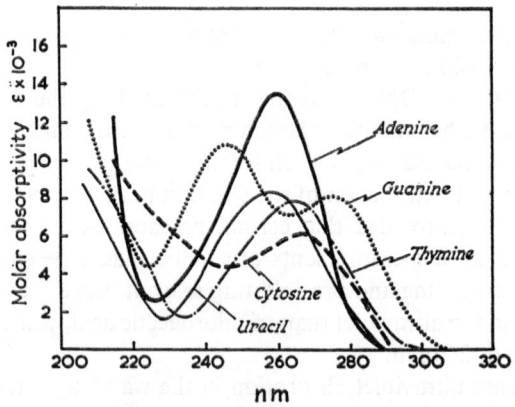

Fig. 4.6 *Ultraviolet absorption curves for purine and pyrimidine bases at pH7*

increases significantly on degradation or hydrolysis, since the sum of the extinctions of the constituent nucleotides is greater than the extinction of the polynucleotide. This 'hyperchromic effect' is due to an alteration in the resonance behaviour of the bases when they are present in high molecular weight polynucleotides (Chapter 5, p. 100).

4.10.7 *The nucleic acid content of tissues*

The nucleic acid content of different tissues as determined by chemical methods is generally in accordance with the values which might be expected on histological grounds. Tables showing the *concentrations* of both types of nucleic acid in different tissues are given in several reviews [143, 187–191] and a few figures obtained by the method of Schmidt and Thannhauser [158] are shown in

ISOLATION AND CHARACTERIZATION OF NUCLEIC ACIDS

Table 4.1, in which the results have been expressed in terms of nucleic acid phosphorus. As might be expected, the highly cellular organs, such as spleen, thymus, and pancreas, are rich in nucleic acid, whereas organs like brain or muscle have a much lower concentration. In tissues with a high nuclear/cytoplasmic ratio where the histological appearance reveals an abundance of nuclear material

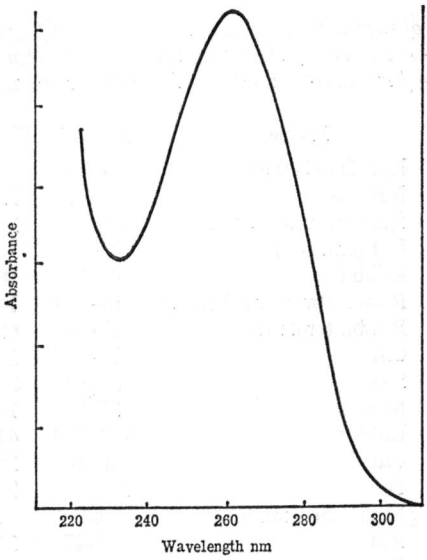

Fig. 4.7 *Ultraviolet absorption curve of a solution of the sodium salt of a specimen of yeast RNA showing the absorption maximum at 260 nm and the minimum at 230 nm*

the DNA concentration is high, but in tissues with a high cytoplasmic volume and a plentiful supply of cytoplasmic granules the RNA value is high. There is, in fact, a good correlation between the basophilia of the cytoplasm and the RNA concentration in the tissue.

4.11 Determination of nucleic acid base composition

Before the base composition of a sample nucleic acid can be investigated the nucleic acid must first be subjected to some form of hydrolysis. Both RNA and DNA may be hydrolysed to their constituent bases by treatment with 98 per cent formic acid at 175° for 30 minutes, or with 12M-perchloric acid at 100° for 1 hour [192, 193]. Neither method is absolutely quantitative, since formic acid gives low yields of uracil and perchloric acid causes some destruction of thymine. DNA can also be hydrolysed by treatment with 6M-HCl at 120° for 2 hours, but this causes the loss of some purines [194]. Satisfactory hydrolysis of RNA to a mixture of purine bases and pyrimidine nucleotides can be obtained by treatment with

TABLE 4.1

Concentrations of nucleic acids in tissues, expressed as mg ribonucleic acid phosphorus (RNA-P) and mg deoxyribonucleic acid phosphorus (DNA-P) per 100 g fresh tissue (method of Schmidt and Thannhauser)

Tissue	Species	RNA-P	DNA-P	Ratio R/D
Liver	Rat (200–240 g)	77–110	21–25	4·0
	Rat (69–80 g)	106–122	28–37	3·6
	Rat (pregnant female)	110–118	21–23	5·2
	Rat (embryo)	87–134	35–65	2·2
	Rabbit	44–76	16–29	2·7
	Rabbit (pregnant female)	67–138	14–17	6·8
	Rabbit (embryo)	87–105	61–84	1·3
	Cat	72–85	25–43	2·3
	Sheep	55–84	23–33	2·5
	Man	37–74	16–25	3·0
Pancreas	Rabbit	108–130	44–61	2·3
	Cat	130–165	38–49	3·4
	Ox	170–186	21–22	8·1
	Man (one sample)	42	31	3·6
Kidney	Rat	25–47	33–43	0·7
Brain	Rat	20–33	15–19	1·5
Spleen	Rat (200–240 g)	63–86	76–85	0·9
	Rat (60–80 g)	70–82	68–78	1·0
	Rabbit	67–79	81–96	0·8
	Cat	84–151	73–94	1·4
	Man (one sample)	36	77	0·5
Thymus	Rat (200–240 g)	87–116	181–242	0·5
	Rat (60–80 g)	114–135	181–261	0·6
	Rabbit	89–99	181–250	0·4
	Calf	80–100	224–250	0·4
Adrenal gland	Man	13–36	9–20	1·3–2·4

1M-hydrochloric acid at 100° for 1 hour [195]. RNA can easily be hydrolysed quantitatively to nucleoside 3'-phosphates by treatment with 0·3M-NaOH at 37° for 16 hours. Care must be taken to avoid the use of too strong alkali which can cause some deamination of cytidylic acid. Unfortunately the corresponding hydrolysis of DNA to nucleotides can be achieved only by enzymic methods. For example, digestion with micrococcal nuclease followed by spleen phosphodiesterase (see Chapter 8) will yield the deoxyribonucleoside 3'-monophosphates.

The hydrolysis products can be separated by chromatography or ionophoresis and estimated spectrophotometrically.

The relative proportions of nucleotides can also be determined by labelling a nucleic acid with ^{32}P, separating the nucleotides, and using the radioactivity of each as a measure of its relative abundance. This method, however, depends on an even distribution of label throughout the polynucleotide chain and can sometimes give rise to misleading results

An alternative procedure permitting accurate base composition analyses by tritium labelling has recently been developed [196]. Non-radioactive RNA is first enzymically degraded to nucleosides, using ribonuclease, snake venom phosphodiesterase (see Chapter 8), and alkaline phosphatase, and then treated successively with periodate and ^3H-labelled borohydride. The ^3H-labelled nucleoside derivatives are then separated and assayed for radioactivity to determine the RNA composition.

4.12 Separation and estimation of bases, nucleosides, and nucleotides

One of the principal factors in the rapid growth of nucleic acid biochemistry since 1945 has been the development of suitable chromatographic methods for separating bases, nucleotides, and oligonucleotides. These methods have found wide application not only in estimating the nucleotide composition of the nucleic acids but also in the investigation of many problems of nucleic acid metabolism [195–199]. They may be divided into four groups:

(1) Paper chromatography, which is particularly valuable for separating small quantities of purine and pyrimidine bases [197–199].

(2) Paper electrophoresis, which is chiefly useful for separating small quantities of nucleotides [201, 202].

(3) Thin-layer chromatography, which can be used for separation of bases, nucleosides, and nucleotides, and has the advantages of speed and simplicity together with considerable resolving power [200].

(4) Column chromatography on ion exchange resins, which can

be used for separating relatively large quantities of bases, nucleosides, and, more especially, nucleotides [199, 203–205].

Detailed experimental information about laboratory methods for the chromatography of nucleic acids and their derivations will be found in the handbook edited by Grossman and Moldave [2].

4.12.1 *Paper chromatography*

Full technical details of paper chromatography as applied to purine and pyrimidine derivatives may be found in the numerous reviews of the subject [193, 206–208].

One-dimensional or, preferably, two-dimensional paper chromatography is particularly suitable for separating the mixture of free bases obtained by strong acid hydrolysis of RNA or DNA. Among the most useful solvents for this purpose are isopropanol–concentrated HCl–water [193] and n-butanol–concentrated NH_4OH–water [209]. Similar methods can be used to separate the mixture of purine bases and pyrimidine nucleotides obtained by mild acid hydrolysis of RNA.

Whilst some paper chromatographic solvents permit a separation in one dimension of, for example, the mixture of ribonucleoside 3'-phosphates obtained on mild alkaline hydrolysis [210], two-dimensional paper chromatography has also been used [211]. Two-dimensional paper chromatography can also be conveniently used to separate the acid-soluble nucleotides in tissue extracts [29]; prior to the chromatography, the acid extract is purified by adsorption of the nucleotides on charcoal followed by elution with a suitable solvent.

The method of paper chromatography can be made quantitative. When the spots have been located, the paper is cut into sections, each corresponding to a different spot, and the base or nucleotide is eluted with acid. By determining the optical density of the acid extract at the appropriate region in the ultraviolet in a suitable spectrophotometer, the amount of each case can be determined provided that the molar extinction coefficients are known. The quantities involved are of the order of 5–40 μg.

4.12.2 *Paper electrophoresis*

One of the limitations of the paper chromatographic methods at present available is that they do not provide a rapid and easy method of separating the mixture of ribonucleoside 3'-phosphates obtained on mild alkaline hydrolysis of RNA. This can, however, be achieved by making use of the mobility of the nucleotides in an electric field in the procedure of electrophoresis [212, 201, 202, 213].

The method of electrophoresis has been scaled down to such an extent by Edström [214, 215] that it is possible to separate on a cellulose fibre the nucleotides obtained from the nucleic acids in portions of single cells.

4.12.3 *Thin-layer chromatography*

The main advantages of thin-layer methods are sharpness of resolution, great sensitivity, simplicity, and speed. Whereas partition chromatography on cellulose layers appears to be the thin-layer method of choice for bases and nucleosides [216–218], chromatography on anion-exchange layers, e.g. poly(ethyleneimine)-cellulose, is preferable for nucleotides [212, 219]. Two-dimensional anion-exchange thin-layer chromatography permits the resolution of complex mixtures of nucleotides that are difficult to separate on paper or on a single column [216].

4.12.4 *Column chromatography*

Paper chromatography and electrophoresis are suitable for very small amounts of material. For the separation of larger amounts of bases, nucleosides, and nucleotides, chromatography on columns of ion-exchange resins can be employed. Such methods were originally developed chiefly by W. E. Cohn and his colleagues [199, 203] using the cation exchanger Dowex-50 and the anion exchangers Dowex-1 and -2. The mixture of nucleotides is adsorbed on to the column through which an eluting solution containing a competing ion is passed. The eluate is collected in small fractions and the concentration of base or nucleotide in each is determined by measurements of absorbance in the ultraviolet spectrophotometer.

Column chromatography is useful not only for separating different bases or their derivatives but for separating closely similar derivatives such as the 2'-, 3'-, and 5'-phosphates of the same nucleoside [203].

In most systems of column chromatography of nucleic acid derivatives the concentration of the eluant is increased stepwise; for some purposes, such as the separation of a complex mixture of nucleotides, the technique of gradient elution on an anion exchange

THE BIOCHEMISTRY OF THE NUCLEIC ACIDS

resin is more suitable. At the pH of application to the column (pH 7–8), the nucleotides possess a net negative charge which diminishes as the pH is lowered so that the attraction of the nucleotide to the resin is reduced. This is done by passing through the resin a steadily increasing concentration of the appropriate acid or salt to elute the

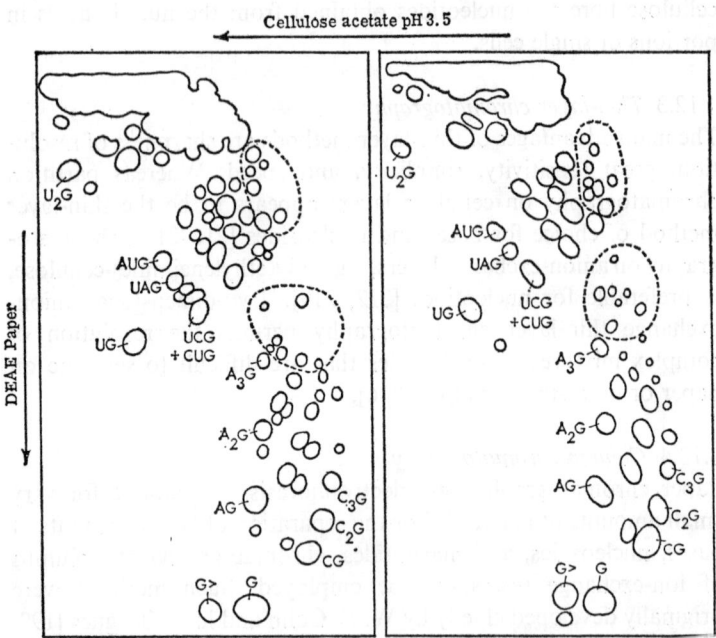

Fig. 4.8 *Diagram illustrating autoradiogram of fingerprinting of digests of ribosomal RNA labelled with ^{32}P from HeLa cells. Left, 18S RNA. Right, 28S RNA. First run, right to left, electrophoresis on cellulose acetate at pH 3·5. Second run, down, electrophoresis on DEAE paper with 7% formic acid. A few of the oligonucleotides shown have been named on the diagram. There are several qualitative or quantitative differences between the 18S and 28S RNA patterns. Regions showing conspicuous differences are indicated by circles*

(*By courtesy of M. Salim and Dr B. E. H. Maden*)

nucleotides successively. Those that are more firmly held on the resin are then eluted by the addition of steadily increasing amounts of salt to the acid solution entering the column. For example, Hurlbert, Schautz, Brumm, and Potter [204] separated the acid-soluble nucleotides of animal tissues on columns of Dowex-1 by eluting with steadily increasing concentrations of formic acid to which

ISOLATION AND CHARACTERIZATION OF NUCLEIC ACIDS

ammonium formate is gradually added. The nucleoside monophosphates come out first followed by the di- and finally the tri-phosphates.

New methods of high-speed chromatography on ion-exchange columns of narrow bore now permit very rapid analysis of nucleic acid constituents at the subnanomole level [220].

4.13 Separation of oligonucleotides

A large variety of methods exists for separation of oligonucleotides. Separation can be achieved by chromatography on paper, on DEAE-paper, on Dowex-1 columns, on DEAE-cellulose columns (in pre-

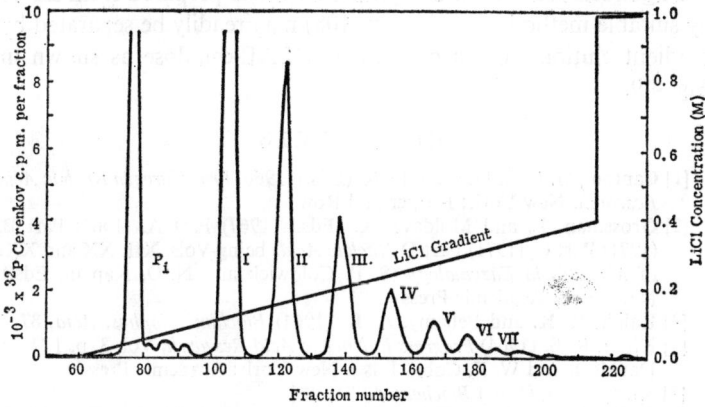

Fig. 4.9 Fractionation of pyrimidine runs into isostichs *An acid-digest of DNA was loaded on to a DEAE-cellulose column at pH 5·3. Purine bases were first removed by washing with buffer, and the pyrimidine runs were then eluted with a salt gradient: only this latter part of the procedure is shown here. This figure shows results for a calf thymus DNA digest*

(*By courtesy of Dr D. J. McGeoch*)

sence or absence of urea), and by polyacrylamide gel electrophoresis. Detailed information regarding these procedures can be obtained from the handbook edited by Grossman and Moldave [2].

Sanger and his colleagues [221–225] have developed a very sensitive two-dimensional 'fingerprinting' method for separating the oligonucleotides obtained by partial digestion of RNA highly labelled with ^{32}P. Fractionation in one dimension is by high-voltage ionophoresis on cellulose acetate at pH 3·5 and in the second dimension by ionophoresis with 7 per cent formic acid on DEAE-paper, an ion-exchange paper of opposite charge to that of the nucleotide. Alternatively thin-layer chromatography on DEAE-cellulose (diethylaminoethylcellulose) may be employed, and in the method known as 'homochromatography' a mixture of non-radioactive nucleotides is used to develop the chromatogram. The nucleotides are detected by autoradiography and are estimated by counting techniques. These methods have proved very successful in the sequence studies to be described in Chapter 5 and 6. They are illustrated in Fig. 4.8. With experience, the probable composition of a nucleotide may be determined from its position on the chromatogram grid. The technique has recently been adapted to 'fingerprint' RNA's labelled with high levels of ^{3}H (tritium) [226].

Oligonucleotides such as pyrimidine tracts prepared from DNA by suitable methods (isostichs, p. 108) may readily be separated by gradient elution from a column of DEAE-cellulose as shown in Fig. 4.9.

REFERENCES

[1] Cantoni, G. L. and Davies, D. R. (Eds.) (1966) *Procedures in Nucleic Acid Research*. New York: Harper and Row
[2] Grossman, L. and Moldave, K. (Eds.) (1967) Part A, (1968) Part B, (1971) Part C, (1971) Part D *Nucleic Acids* being Vols. XII, XX and XXI of *Methods in Enzymology* (S. P. Colowick and N. O. Kaplan, Eds.). New York: Academic Press
[3] Ralph, R. K. and Bellamy, A. R. (1964) *Biochim. Biophys. Acta*, **87**, 9
[4] Kirby, K. S. (1964) *Progress in Nucleic Acid Research*, Vol. 3, p. 1 (J. N. Davidson and W. E. Cohn, Eds.). New York: Academic Press
[5] Kirby, K. S. (1965) *Biochem. J.*, **96**, 266
[6] Kirby, K. S. (1968) *Methods in Enzymology*, Vol. 12, Part B, p. 87 (L. Grossman and K. Moldave, Eds.). New York: Academic Press
[7] Georgiev, G. P. (1967) *Progr. Nucleic Acid Res. Mol. Biol.*, **6**, 259
[8] Peterman, M. L. and Pavlovec, A. (1963) *J. Biol. Chem.*, **238**, 3717
[9] Henshaw, E. C. (1968) *J. Mol. Biol.*, **36**, 401
[10] Lockard, R. E. and Lingrel, J. B. (1969) *Biochem. Biophys. Res. Commun.*, **37**, 204
[11] Mach, B., Faust, C. and Vassalli, P. (1973) *Proc. Nat. Acad. Sci.*, **70**, 451
[12] Weigers, U., Kramer, G. and Hilz, H. (1973) *Biochem. Biophys. Res. Commun.*, **50**, 1039
[13] Kurland, C. G. (1960) *J. Mol. Biol.*, **2**, 83
[14] Aronson, A. I. (1962) *J. Mol. Biol.*, **5**, 453
[15] Aronson, A. I. (1963) *Biochim. Biophys. Acta*, **72**, 176
[16] Spirin, A. S. (1963) *Progress in Nucleic Acid Research*, Vol. 1, p. 301 (J. N. Davidson and W. E. Cohn, Eds.). New York: Academic Press
[17] Brownlee, G. G., Sanger, F. and Barrell, B. G. (1968) *J. Mol. Biol.*, **34**, 379
[18] Morell, P. and Marmur, J. (1968) *Biochem.*, **7**, 1141
[19] Knight, E. J. R. and Darnell, J. E. (1967) *J. Mol. Biol.*, **28**, 491
[20] Forget, B. G. and Weissman, S. M. (1969) *J. Biol. Chem.*, **244**, 3148
[21] Watson, J. D. and Ralph, R. K. (1967) *J. Mol. Biol.*, **26**, 541
[22] Ford, P. S. (1973) *Biochem. Soc. Symp.*, **37**, 69

ISOLATION AND CHARACTERIZATION OF NUCLEIC ACIDS

[23] Payne, P. I. and Dyer, T. A. (1971) *Biochem. J.*, **124**, 87
[24] Soave, C., Galante, E. and Torti, G. (1970) *Bull. Soc. Chim. Biol.*, **52**, 857
[25] Rubin, G. M. (1974) *Eur. J. Biochem.*, **41**, 197
[26] Pene, J. J., Knight, E., Jr. and Darnell, J. E. (1968) *J. Mol. Biol.*, **33**, 609
[27] Brown, G. L. (1963) *Progress in Nucleic Acid Research*, Vol. 2, p. 259 (J. N. Davidson and W. E. Cohn, Eds.). New York: Academic Press
[28] Greenman, D. L., Kenney, F. T. and Wicks, W. D. (1964) *Biochem. Biophys. Res. Commun.*, **17**, 449
[29] Singer, M. F. and Leder, P. (1966) *Ann. Rev. Biochem.*, **35**, 195
[30] Vesco, C. and Penman, S. (1969) *Proc. Nat. Acad. Sci.*, **62**, 218
[31] Lizzardi, P. M. and Luck, D. J. L. (1971) *Nature*, **229**, 140
[32] Epler, J. L. (1969) *Biochem.*, **8**, 2285
[33] Burdon, R. H. (1971) *Prog. Nucleic Acid Res. Mol. Biol.*, Vol. 11, p. 33 (J. N. Davidson and W. E. Cohn, Eds.). New York: Academic Press
[34] Scherrer, K., Latham, H. and Darnell, J. E. (1963) *Proc. Nat. Acad. Sci.*, **49**, 240
[35] Penman, S., Rosebash, M. and Penman, M. (1970) *Proc. Nat. Acad. Sci.*, **67**, 1878
[36] El-Khatib, S. M., Ro-Choi, T. S., Choi, Y. C. and Busch, H. (1970) *J. Biol. Chem.*, **245**, 3416
[37] Brownlee, G. G. (1971) *Nature New Biol.*, **229**, 147
[38] Weinberg, R. A. and Penman, S. (1968) *J. Mol. Biol.*, **38**, 289
[39] Fraenkel-Conrat, H., Singer, B. and Tsugita, A. (1961) *Virology*, **14**, 54
[40] Marcus, L., Bretthauer, R. K., Bock, R. M. and Halvorsen, H. O. (1963) *Proc. Nat. Acad. Sci.*, **50**, 782
[41] Payne, P. I. and Loening, U. E. (1970) *Biochim. Biophys. Acta*, **224**, 128
[42] McConkey, E. H. (1967) *Methods in Enzymology*, Vol. 12, Part A, p. 620 (L. Grossman and K. Moldave, Eds.). New York: Academic Press
[43] Vinograd, J. and Hearst, J. E. (1962) *Progress in Chemistry of Organic Natural Products*, **20**, 372
[44] Loening, U. E. (1967) *Biochem. J.*, **102**, 251
[45] Grossbach, U. and Weinstein, I. B. (1968) *Analyt. Biochem.*, **22**(2), 311
[46] Burdon, R. H. and Clason, A. E. (1969) *J. Mol. Biol.*, **39**, 113
[47] Caton, J. E. and Goldstein, G. (1971) *Analyt. Biochem.*, **42**, 14
[48] Loening, U. E. (1968) *Chromatographic and Electrophoretic Techniques* (2nd Ed.), Vol. 2, p. 437 (I. Smith, Ed.). London: Heinemann
[49] Richards, E. G. and Gratzer, W. B. (1968) *Chromatographic and Electrophoretic Techniques*, Vol. 2, p. 419 (I. Smith, Ed.). London: Heinemann
[50] Richards, E. G. and Lecanidou, R. (1971) *Analyt. Biochem.*, **41**, 43
[51] Richards, E. G. and Temple, C. J. (1971) *Nature*, **230**, 92
[52] De Wachter, R. and Fiers, W. (1971) *Methods in Enzymology*, Vol. 21, p. 167 (L. Grossman and K. Moldave, Eds.). New York: Academic Press
[53] Lanyon, W. G., Paul, J. and Williamson, R. (1968) *FEBS Lett.*, **1**, 279
[54] Stein, M. and Varrichio, F. (1974) *Analyt. Biochem.*, **61**, 112
[55] Reddy, R., Sitz, T. O., Ro-Choi, T. S. and Busch, H. (1974) *Biochem. Biophys. Res. Commun.*, **56**, 1017
[56] Pinder, J. C., Staynor, D. Z. and Gratzer, W. B. (1974) *Biochem.*, **13**, 5373
[57] Goldthwait, D. A. (1959) *J. Biol. Chem.*, **234**, 3245
[58] McIndoe, W. and Munro, H. N. (1967) *Biochem. Biophys. Acta*, **134**, 458

[59] Ringborg, U., Egyhezi, E., Daneholt, B. and Lambert, B. (1968) *Nature*, **220**, 1037
[60] Floyd, R. W., Stone, M. P. and Joklik, W. K. (1974) *Analyt. Biochem.*, **59**, 599
[61] Staehelin, M. (1963) *Progress in Nucleic Acid Research*, Vol. 2, p. 170 (J. N. Davidson and W. E. Cohn, Eds.). New York: Academic Press
[62] Mandell, J. D. and Hershey, A. D. (1960) *Analyt. Biochem.*, **1**, 66
[63] Sueoka, N. and Cheng, T. (1962) *J. Mol. Biol.*, **4**, 161
[64] Murakami, W. T. (1967) *Methods in Enzymology*, Vol. 12, Part A, p. 634 (L. Grossman and K. Moldave, Eds.). New York: Academic Press
[65] Main, R. K., Wilkins, M. J. and Cole, L. J. (1959) *J. Amer. Chem. Soc.*, **81**, 6490
[66] Bernardi, G. (1961) *Biochem. Biophys. Res. Commun.*, **6**, 54
[67] Brown, F., Newman, J. F. E. and Stewart, D. L. (1963) *Nature*, **197**, 590
[68] Bernardi, G. (1971) *Methods in Enzymology*, Vol. 21, p. 95 (L. Grossman and K. Moldave, Eds.)
[69] Kit, S. (1960) *Arch. Biochem. Biophys.*, **87**, 318
[70] Tomlinson, R. V. and Tener, G. M. (1962) *J. Amer. Chem. Soc.*, **84**, 2644
[71] Burdon, R. H., Martin, B. T. and Lal, B. M. (1967) *J. Mol. Biol.*, **28**, 357
[72] Erikson, R. L. and Gordon, J. A. (1966) *Biochem. Biophys. Res. Commun.*, **23**, 422
[73] Novakovic, M. B. and Petrovic, S. L. (1972) *Analyt. Biochem.*, **49**, 367
[74] Bautz, E. K. F. and Hall, B. D. (1962) *Proc. Nat. Acad. Sci.*, **48**, 400
[75] Bolton, E. T. and McCarthy, B. J. (1962) *Proc. Nat. Acad. Sci.*, **48**, 1390
[76] Bolton, E. T. and McCarthy, B. J. (1964) *J. Mol. Biol.*, **8**, 201
[77] McCarthy, B. J. and Bolton, E. T. (1964) *J. Mol. Biol.*, **8**, 184
[78] Lindberg, U. and Persson, T. (1972) *Eur. J. Biochem.*, **31**, 246
[79] Philipson, L., Wall, R., Glickman, G. and Darnell, J. E. (1971) *Proc. Nat. Acad. Sci.*, **68**, 2806
[80] Sheldon, R., Jurale, C. and Kates, J. (1972) *Proc. Nat. Acad. Sci.*, **69**, 417
[81] Aviv, H. and Leder, P. (1972) *Proc. Nat. Acad. Sci.*, **69**, 1408
[82] Goldstein, J., Bennett, T. P. and Craig, L. C. (1964) *Proc. Nat. Acad. Sci.*, **51**, 119
[83] Doctor, B. P., Connelly, C. M., Rushizky, G. W. and Sober, H. A. (1963) *J. Biol. Chem.*, **238**, 3985
[84] Holley, R. W. and Merrill, S. H. (1959) *Fed. Proc.*, **18**, 249
[85] Albertsson, P.-A. (1962) *Arch. Biochem. Biophys.*, Suppl. 1, 264
[86] Apgar, J., Holley, R. W. and Merrill, S. H. (1962) *J. Biol. Chem.*, **237**, 796
[87] Mirsky, A. E. (1971) *Proc. Nat. Acad. Sci.*, **68**, 2945
[88] Davidson, J. N. and Waymouth, C. (1944) *Biochem. J.*, **38**, 375
[89] Mirsky, A. E. and Pollister, A. W. (1946) *J. Gen. Physiol.*, **30**, 117
[90] Gross-Bellard, M., Oudet, P. and Chambon, P. (1973) *Eur. J. Biochem.*, **36**, 32
[91] Sevag, M. G., Lackman, D. B. and Smolens, J. (1938) *J. Biol. Chem.*, **124**, 425
[92] Kay, E. R. M., Simmons, N. S. and Dounce, A. L. (1952) *J. Amer. Chem. Soc.*, **74**, 1724
[93] Simmons, N. S., Chavers, S. and Orbach, H. K. (1952) *Fed. Proc.*, **11**, 390
[94] Kirby, K. S. (1956) *Biochem. J.*, **64**, 405
[95] Kirby, K. S. (1958) *Biochemical Preparations*, **6**, 79

ISOLATION AND CHARACTERIZATION OF NUCLEIC ACIDS

[96] Kirby, K. S. (1967) *Methods in Cancer Research*, Vol. 3, p. 1 (H. Busch, Ed.). New York: Academic Press
[97] Colowick, S. P. and Kaplan, N. O. (1957) *Methods in Enzymology*, Vol. III, Section V, Articles 100 to 105
[98] Smith, M. G. (1967) *Methods in Enzymology*, **12**, Part A, p. 545 (L. Grossman and K. Moldave, Eds.). New York: Academic Press
[99] Thomas, C. A., Berns, K. I. and Kelly, T. J. (1966) *Procedures in Nucleic Acid Research*, p. 535 (G. L. Cantoni and D. R. Davies, Eds.). New York: Harper and Row
[100] Hotchkiss, R. D. (1966) *Procedures in Nucleic Acid Research*, p. 541 (G. L. Cantoni and D. R. Davies, Eds.). New York: Harper and Row
[101] Thomas, C. A. and Abelson, J. (1966) *Procedures in Nucleic Acid Research*, p. 553 (G. L. Cantoni and D. R. Davies, Eds.). New York: Harper and Row
[102] Sinsheimer, R. L. (1966) *Procedures in Nucleic Acid Research*, p. 569 (G. L. Cantoni and D. R. Davies, Eds.). New York: Harper and Row
[103] Marmur, J. (1961) *J. Mol. Biol.*, **3**, 208
[104] Kalf, G. F. and Grece, M. A. (1967) *Methods in Enzymology*, **12**, Part A, p. 533 (L. Grossman and K. Moldave, Eds.). New York: Academic Press
[105] Borst, P., Van Bruggen, E. F. J., Ruttenberg, G. J. C. M. and Kroon, A. M. (1967) *Biochim. Biophys. Acta*, **149**, 156
[106] Eisenstadt, J. M. and Brawerman, G. (1967) *Methods in Enzymology*, Vol. 12, Part A, p. 541 (L. Grossman and K. Moldave, Eds.). New York: Academic Press
[107] Davern, C. I. (1966) *Proc. Nat. Acad. Sci.*, **55**, 792
[108] Crothers, D. M. and Zimm, B. H. (1965) *J. Mol. Biol.*, **12**, 525
[109] Hudson, B., Clayton, D. A. and Vinograd, J. (1968) *Cold Spring Harbor Symp. Quant. Biol.*, **33**, 435
[110] Szybalski, W. (1968), *Fractions*, **1**, 1
[111] Kersten, W., Kersten, H. and Szybalski, W. (1966) *Biochem.*, **5**, 236
[112] Lizardi, P. and Brown, D. D. (1974) *Cold Spring Harbor Symp. Quant. Biol.*, **38**, 701
[113] Peacock, W. J., Brutlag, D., Goldring, E., Appels, R., Hinton, C. W. and Lindsley, D. L. (1974) *Cold Spring Harbor Symp. Quant. Biol.*, **38**, 405
[114] Nandi, U. S., Wang, J. C. and Davidson, N. (1965) *Biochem.*, **4**, 1687
[115] Jensen, R. H. and Davidson, N. (1966) *Biopolymers*, **4**, 17
[116] Anet, R. and Strayer, D. R. (1969) *Biochem. Biophys. Res. Commun.*, **37**, 52
[117] DeKloet, S. R. and Andrean, B. A. G. (1971) *Biochim. Biophys. Acta*, **247**, 519
[118] Main, R. K., Wilkins, M. J. and Cole, L. J. (1959) *J. Amer. Chem. Soc.*, **81**, 6490
[119] Bernardi, G. (1961) *Biochem. Biophys. Res. Commun.*, **6**, 54
[120] Sueoka, N. and Cheng, Ts Ai-Ying (1967) *Methods in Enzymology*, Vol. 12, Part A, p. 562 (L. Grossman and K. Moldave, Eds.). New York: Academic Press
[121] Volvardy, A. and Venetianer, P. (1971) *Eur. J. Biochem.*, **20**, 513
[122] Iyer, V. N. and Rupp, W. D. (1971) *Biochim. Biophys. Acta*, **228**, 117
[123] Prunell, A. and Bernardi, G. (1973) *J. Biol. Chem.*, **248**, 3433
[124] Zeiger, R. S., Salomon, R., Dingman, C. W. and Peacock, A. C. (1972) *Nature New Biol.*, **238**, 65
[125] Albertsson, P. A. (1971) *Partition of Cell Particles and Macromolecules*, p. 323. New York: Wiley-Interscience.

[126] Leng, M. and Felsenfeld, G. (1966) *Proc. Nat. Acad. Sci.*, **56**, 1325
[127] Nygaard, A. P. and Hall, B. D. (1964) *J. Mol. Biol.*, **9**, 125
[128] Gillespie, D. and Spiegelman, S. (1965) *J. Mol. Biol.*, **12**, 829
[129] Kohne, D. E. (1968) *Biophys. J.*, **8**, 1104
[130] Ando, T. (1966) *Biochim. Biophys. Acta*, **114**, 158
[131] Walker, P. M. B. (1969) *Progr. Nucleic Acid Res. Mol. Biol.*, **9**, 301
[132] Kennell, D. E. (1971) *Progr. Nucleic Acid Res. Mol. Biol.*, **11**, 259
[133] McCarthy, B. J. and Church, R. B. (1970) *Ann. Rev. Biochem.*, **39**, 131
[134] Avery, R. J. and Midgley, J. E. M. (1969) *Biochem. J.*, **115**, 383
[135] Bishop, J. O. (1972) *Biochem. J.*, **126**, 171
[136] Birnstiel, M. L., Sells, B. H. and Purdom, I. F. (1972) *J. Mol. Biol.*, **63**, 21
[137] Gillespie, D. (1968) *Methods in Enzymology*, Vol. 12, Part B, p. 641 (L. Grossman and K. Moldave, Eds.). New York: Academic Press
[138] Brown, D. D. and Stern, R. (1974) *Ann. Rev. Biochem.*, **43**, 667
[139] Shapiro, J., Machattie, L., Evon, L., Ihler, G., Ippens, K. and Beckwith, J. (1969) *Nature*, **224**, 768
[140] Beckwith, J. R. and Zisper, D. (Eds.) (1970) *The Lactose Operon*. Cold Spring Harbor Laboratory Publication
[141] Gilbert, W., Maizels, N. and Maxam, A. (1974) *Cold Spring Harbor Symp. Quant. Biol.*, **38**, 845
[142] Chen, C.-Y., Hutchison, C. A. and Edgell, M. H. (1973) *Nature New Biol.*, **243**, 233
[143] Leslie, I. (1955) *The Nucleic Acids*, Vol. 2, p. 1 (E. Chargaff and J. N. Davidson, Eds.). New York: Academic Press
[144] Volkin, E. and Cohn, W. E. (1954) *Methods of Biochem. Analysis*, Vol. 1, p. 287 (D. Glick, Ed.). New York: Interscience
[145] Schneider, W. C. (1957) *Methods in Enzymology*, Vol. 3, p. 680 (S. P. Colowick and N. O. Kaplan, Eds.). New York: Academic Press
[146] Schmidt, G. (1957) ibid., pp. 671 and 775
[147] Webb, J. M. and Levy, H. B. (1958) *Methods of Biochemical Analysis*, Vol. 6, p. 1 (D. Glick, Ed.)
[148] Hutchison, W. C. and Munro, H. N. (1961) *Analyst*, **86**, 768
[149] Munro, H. N. and Fleck, A. (1966) *Methods of Biochemical Analysis*, Vol. 14, p. 113 (D. Glick, Ed.). New York: Interscience
[150] Munro, H. N. and Fleck, A. (1966) *Analyst*, **91**, 78
[151] Hallinan, T., Fleck, A. and Munro, H. N. (1963) *Biochim. Biophys. Acta*, **68**, 131
[152] Schneider, W. C. (1945) *J. Biol. Chem.*, **161**, 293
[153] Schneider, W. C., Hogeboom, G. H. and Ross, H. E. (1950) *J. Nat. Cancer Inst.*, **10**, 977
[154] Ogur, M., Minkler, S., Lindegren, G. and Lindegren, C. C. (1952) *Arch. Biochem. Biophys.*, **40**, 175
[155] Ceriotti, G. (1952) *J. Biol. Chem.*, **198**, 297; (1955) **214**, 59
[156] Steele, R., Sfortunato, T. and Ottolenghi, L. (1948) *J. Biol. Chem.*, **177**, 231
[157] Patterson, E. K. and Dackerman, M. E. (1952) *Arch. Biochem. Biophys.*, **36**, 97
[158] Schmidt, G. and Thannhauser, S. J. (1945) *J. Biol. Chem.*, **161**, 83
[159] Davidson, J. N. and Smellie, R. M. S. (1952) *Biochem. J.*, **52**, 594, 599
[160] Moulé, Y. (1956) *Bull. Soc. Chim. Biol.*, **38**, 175
[161] Logan, J. E., Mannell, W. A. and Rossiter, R. J. (1952) *Biochem. J.*, **51**, 470

[162] Hutchison, W. C., Crosbie, G. W., Mendes, C. B., McIndoe, W. M., Childs, M. and Davidson, J. N. (1956) *Biochim. Biophys. Acta*, **21**, 44
[163] Drasher, M. L. (1953) *Science*, **118**, 181
[164] McIndoe, W. M. and Davidson, J. N. (1952) *Brit. J. Cancer*, **6**, 200
[165] Logan, R. and Davidson, J. N. (unpublished results)
[166] Fiske, C. and Subbarow, Y. (1929) *J. Biol. Chem.*, **81**, 629
[167] Allen, R. J. L. (1940) *Biochem. J.*, **34**, 858
[168] Griswold, B. L., Humoller, F. L. and McIntyre, A. R. (1951) *Analyt. Chem.*, **23**, 192
[169] Berenblum, I. and Chain, E. (1938) *Biochem. J.*, **32**, 286
[170] Norberg, B. (1942) *Acta Physiol. Scand.*, **5** (Suppl. 14)
[171] Engström, A. (1964) *Acta Radiol.*, **43** (Suppl.)
[172] Dische, Z. (1955) *The Nucleic Acids*, Vol. 1, p. 285 (E. Chargaff and J. N. Davidson, Eds.). New York: Academic Press
[173] Mejbaum, W. (1939) *Hoppe-Seyler's Ztschr.*, **258**, 117
[174] Dische, Z. (1930) *Mikrochemie*, **8**, 4
[175] Burton, K. (1956) *Biochem. J.*, **62**, 315
[176] Burton, K. (1968) *Methods in Enzymology*, Vol. 12, Part B, p. 163 (L. Grossman and K. Moldave, Eds.). New York: Academic Press
[177] Keck, K. (1956) *Arch. Biochem. Biophys.*, **63**, 446
[178] Zahn, R. Z. (1970) *FEBS Lett.*, **6**, 141
[179] Kissane, J. M. and Robins, E. (1958) *J. Biol. Chem.*, **233**, 184
[180] Roberts, DeWayne and Friedkin, M. (1958) *J. Biol. Chem.*, **233**, 483
[181] Ogur, M. and Rosen, G. (1950) *Arch. Biochem.*, **25**, 262
[182] Fleck, A. and Munro, H. M. (1962) *Biochim. Biophys. Acta*, **55**, 571
[183] Fleck, A. and Begg, D. (1965) *Biochim. Biophys. Acta*, **108**, 333
[184] Beaven, G. H., Holiday, E. R. and Johnson, E. A. (1955) *The Nucleic Acids*, Vol. I, p. 493 (E. Chargaff and J. N. Davidson, Eds.)
[185] Pabst Laboratories, Milwaukee, Wisconsin. *Circular OR-10*
[186] California Corporation for Biochemical Research (1958) *Catalogue of Biochemical Data*
[187] Davidson, J. N. (1947) *Symp. Soc. Exp. Biol.*, **1**, 77
[188] Davidson, J. N. (1947) *Cold Spring Harbor Symp. Quant. Biol.*, **12**, 50
[189] Schneider, W. C. (1946) *J. Biol. Chem.*, **164**, 747
[190] Euler, H. V. and Hahn, L. (1948) *Arch. Biochem.*, **17**, 285
[191] Schneider, W. C. and Klug, H. L. (1946) *Cancer Res.*, **6**, 691
[192] Marshak, A. and Vogel, H. G. (1951) *J. Biol. Chem.*, **189**, 597
[193] Wyatt, G. R. (1951) *Biochem. J.*, **48**, 584; (1955) *The Nucleic Acids*, Vol. I, p. 243 (E. Chargaff and J. N. Davidson, Eds.). New York: Academic Press
[194] Daly, M. M., Allfrey, V. G. and Mirsky, A. E. (1950) *J. Gen. Physiol.*, **33**, 497
[195] Markham, R. and Smith, J. D. (1949) *Biochem. J.*, **45**, 216; (1950) *ibid.*, **46**, 509; (1951) *ibid.*, **49**, 401
[196] Randerath, K. and Randerath, E. (1973) *Methods in Cancer Research*, Vol. 9, p. 3 (H. Busch, Ed.). New York: Academic Press
[197] Vischer, E. and Chargaff, E. (1948) *J. Biol. Chem.*, **176**, 703, 715
[198] Chargaff, E., Levine, C. and Green, C. (1948) *J. Biol. Chem.*, **175**, 67
[199] Cohn, W. E. (1957) *Methods in Enzymology*, Vol. 3, p. 724 (S. P. Colowick and N. O. Kaplan, Eds.). New York: Academic Press
[200] Randerath, K. and Randerath, E. (1967) *Methods in Enzymology*, Vol. 12A, p. 323 (L. Grossman and K. Moldave, Eds.). New York: Academic Press

[201] Davidson, J. N. and Smellie, R. M. S. (1952) *Biochem. J.*, **52**, 594
[202] Markham, R. and Smith, J. D. (1952) *Biochem. J.*, **52**, 552
[203] Cohn, W. E. (1955) *The Nucleic Acids,* Vol. I, p. 211 (E. Chargaff and J. N. Davidson, Eds.). New York: Academic Press
[204] Hurlbert, R. B., Schautz, H., Brumm, A. F. and Potter, V. R. (1954) *J. Biol. Chem.*, **209**, 23
[205] Burtis, C. A. and Gere, D. R. (1970) *Nucleic Acid Constituents by Liquid Chromatography.* Walnut Creek, California: Varian Aerograph
[206] Bendich, A. (1957) *Methods in Enzymology,* Vol. 3, p. 715 (S. P. Colowick and N. O. Kaplan, Eds.). New York: Academic Press
[207] Markham, R. (1957) *Methods in Enzymology,* Vol. 3, p. 724 (S. P. Colowick and H. O. Kaplan, Eds.). New York: Academic Press
[208] Thomson, R. Y. (1960) *Chromatographic Techniques* (2nd Ed.), p. 295 (I. Smith, Ed.). London: Heinemann
[209] MacNutt, W. S. (1952) *Biochem. J.*, **50**, 384
[210] Nestle, M. and Roberts, W. K. (1968) *Analyt. Biochem.*, **22**, 349
[211] Hayashi, Y., Osawa, S. and Miura, K. (1966) *Biochim. Biophys. Acta*, **129**, 519
[212] Randerath, K. (1964) *J. Chromatog.*, **16**, 111, 126
[213] Smith, J. D. (1967) *Methods in Enzymology,* Vol. 12A, p. 350 (L. Grossman and K. Moldave, Eds.)
[214] Edström, J. E. (1964) *Biochim. Biophys. Acta*, **80**, 399
[215] Edström, J. E. (1964) *Methods in Cell Physiology,* Vol. 1, p. 417 (D. M. Prescott, Ed.). New York: Academic Press
[216] Randerath, K. and Randerath, E. (1967) *Methods in Enzymology,* Vol. 12A, p. 323 (L. Grossman and K. Moldave, Eds.)
[217] Holdgate, D. P. and Goodwin, T. W. (1964) *Biochim. Biophys. Acta*, **91**, 328
[218] Chmielewicz, Z. F. and Acara, M. (1964) *Analyt. Biochem.*, **9**, 94
[219] Randerath, K. (1963) *Biochim. Biophys. Acta*, **76**, 622
[220] Horvath, C. (1973) *Methods of Biochemical Analysis*, **21**, 79
[221] Brownlee, G. G. (1971) *Nature New Biol.*, **229**, 147
[222] Sanger, F., Brownlee, G. G. and Barrell, B. G. (1965) *J. Mol. Biol.*, **13**, 373
[223] Brownlee, G. G. and Sanger, F. (1969) *Eur. J. Biochem.*, **11**, 395
[224] Sanger, F. and Brownlee, G. G. (1970) *Biochem. Soc. Symp.*, **30**, 183
[225] Sanger, F. (1971) *Biochem. J.*, **124**, 833
[226] Shine, J., Dalgarno, L. and Hunt, J. A. (1974) *Analyt. Biochem.*, **59**, 360

CHAPTER 5

The Structure of DNA

5.1 The chemical composition of DNA

The common monomeric units of DNA are the four deoxyribonucleotides containing the bases adenine, cytosine, guanine, and thymine. Many DNA's, however, contain small amounts of other bases, e.g. 5-methylcytosine, which is particularly abundant in wheat germ DNA (see Table 5.1). In a few bacteriophages, one of the common pyrimidine bases is completely replaced by a different pyrimidine base, e.g. in T2, T4, and T6 5-hydroxymethylcytosine completely replaces cytosine (see p. 144), and in PBS 1 (a bacteriophage which attacks *Bacillus subtilis*) uracil replaces thymine.

TABLE 5.1

Molar proportions of bases (as moles of nitrogenous constituents per 100 g-atoms P) in DNA's from various sources (data from various authors)

Source of DNA	Adenine	Guanine	Cytosine	Thymine	5-Methyl-cytosine
Bovine thymus	28·2	21·5	21·2	27·8	1·3
,, spleen	27·9	22·7	20·8	27·3	1·3
,, sperm	28·7	22·2	20·7	27·2	1·3
Rat bone marrow	28·6	21·4	20·4	28·4	1·1
Herring testes	27·9	19·5	21·5	28·2	2·8
Paracentrotus lividus	32·8	17·7	17·3	32·1	1·1
Wheat germ	27·3	22·7	16·8	27·1	6·0
Yeast	31·3	18·7	17·1	32·9	—
Esch. coli	26·0	24·9	25·2	23·9	—
Mb. tuberculosis	15·1	34·9	35·4	14·6	—
ØX174	24·3	24·5	18·2	32·3	—

Methods for determining the molar proportions of bases by hydrolysis and chromatography are discussed in detail by Bendich [1].

The results of the analysis of a number of DNA's as shown in Table 5.1 reveal wide variations in the molar proportions of bases in DNA's from different species although the DNA's from the different organs and tissues of any one species are essentially the same. Extensive tables showing the molar proportions of bases have been published [2].

It was Chargaff [3] who first drew attention to certain regularities in the composition of DNA. The sum of the purines is equal to the sum of the pyrimidines; the sum of the amino bases (adenine and cytosine) is equal to the sum of the keto (oxo) bases (guanine and thymidine); adenine and thymine are present in equimolar amounts, and guanine and cytosine are also found in equimolar amounts. This equivalence of A and T and of G and C is of the utmost importance in relation to the formation of the DNA helix (p. 89) and may be referred to as Chargaff's rule. There are two major deviations from the rule in Table 5.1. (a) In wheat germ DNA guanosine and cytosine are not present in equimolar amounts but this is explained by the scarcity of cytosine being compensated by the presence of 5-methylcytosine. (b) In ØX174 DNA and in the DNA of several similar small coliphages the adenine is not equimolar with thymine nor is guanine with cytosine. This is because ØX174 DNA is single-stranded (see page 100).

DNA's fall into two main classes, the 'A-T rich type' in which adenine and thymine are in excess, and the much rarer 'G-C rich type' in which guanine and cytosine predominate.

Doty [4] has pointed out that the physical properties of DNA are strongly influenced by the percentage of G+C in the molecule, and the buoyant density of DNA in concentrated CsCl solutions (see p. 60) is no exception. G+C rich DNA has a higher buoyant density than A+T rich DNA [2] and there is a linear relationship between the buoyant densities (ρ) of different DNA's and their G+C contents (see Fig. 5.1). This can be expressed by the relationship:

$$\rho = 1\cdot 660 + 0\cdot 098 \, (GC)$$

where GC is the mole fraction of (G+C). The relative (G+C) content can also be determined from the thermal denaturation temperature which is discussed later (p. 105) [5], and from the ultraviolet spectrum of the DNA [6].

On the basis of measurements of this sort the relative (G+C) contents of the DNA's from a wide variety of sources have been determined and are shown in Table 5.2. While mammalian DNA's show a (G+C) content between 40 and 45 per cent, the range of bacterial DNA's is much wider (30–75 per cent). The significance

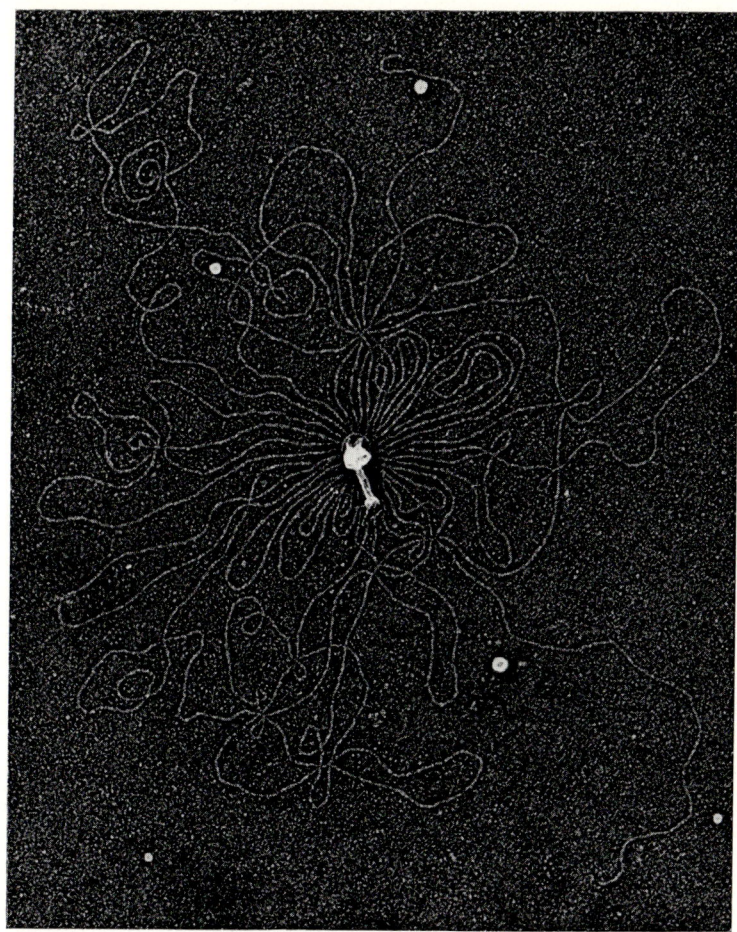

PLATE I *Electron micrograph showing the molecule of DNA emerging from an osmotically ruptured bacteriophage particle as a single thread, the two ends of which are visible*
(*By courtesy of the authors, Professor D. Lang, and Elsevier Publishing Company, from 'Biochim. Biophys, Acta'*, **61 (1962)**)

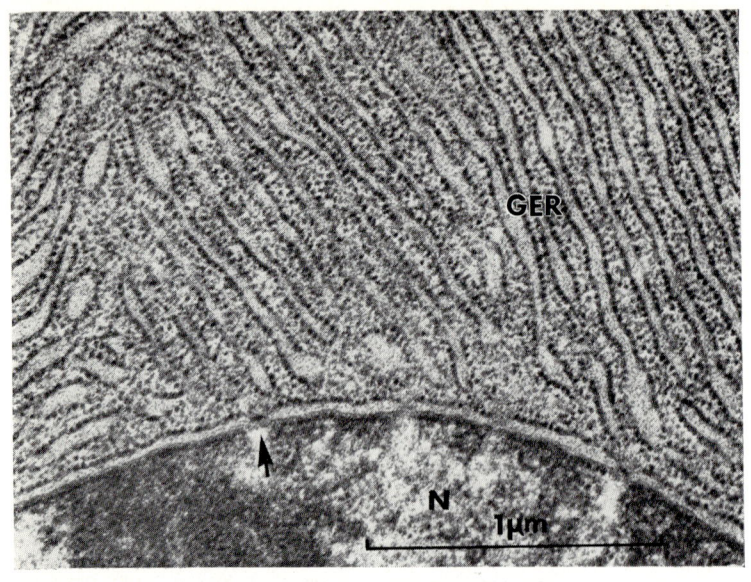

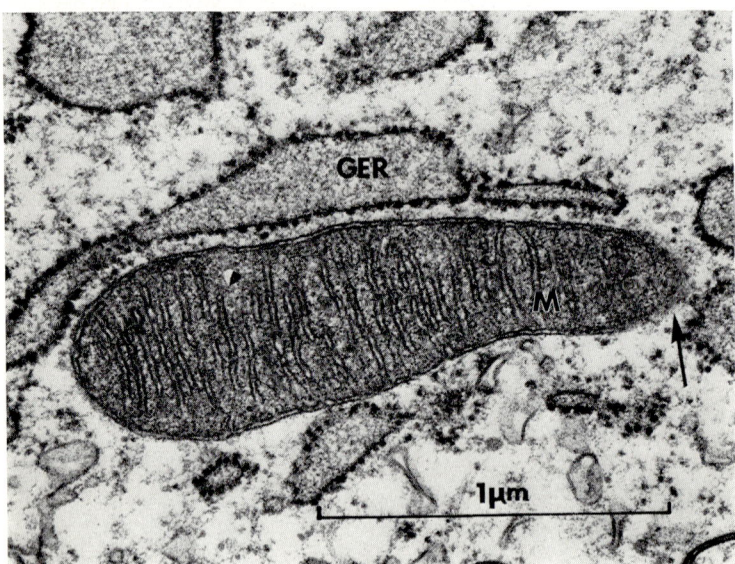

PLATE II
Above. *Zymogenic cell of mouse exocrine pancreas*
The granular endoplasmic reticulum (GER) and part of the nucleus (N) are shown. The membranes of the complex array of cisternal of the endoplasmic reticulum are studded with numerous ribosomes. The arrow points to a nuclear pore, which forms a discontinuity in the nuclear envelope
Below. *Submucosal gland cell of hen proventriculus*
A single mitochondrion (M) and adjacent cisternal of the granular endoplasmic reticulum (GER) are shown. The transverse shelves or cristae of the mitochondrion are clearly seen. The blurring of the mitochondrial membranes at the region arrowed is accounted for by obliquity of section and does not represent mitochondrial rupture

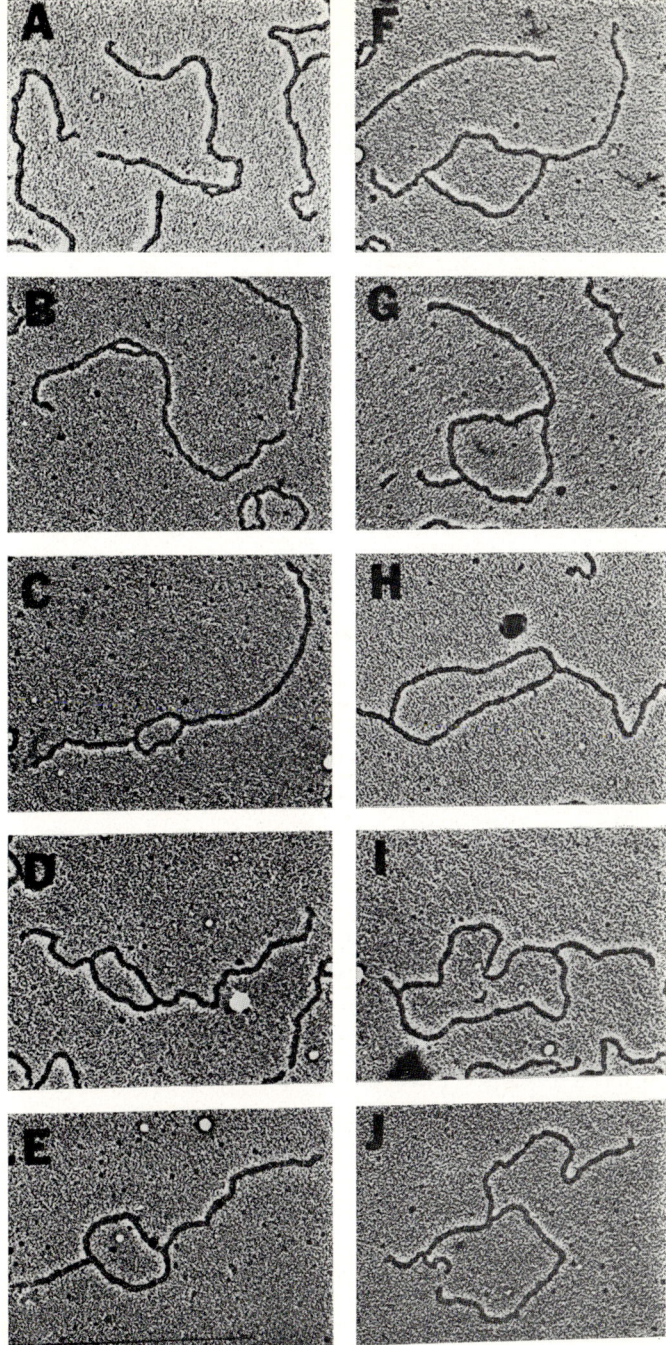

PLATE III *Replicating SV40 DNA molecules. The molecules have been cut at a unique site with the restriction endonuclease EcoRI and have been arranged in increasing degree of replication (A through J) and oriented with the short branch at the left*
(From Fareed, Garon and Salzman (see Chapter 11 ref. 17) with the authors' kind permission

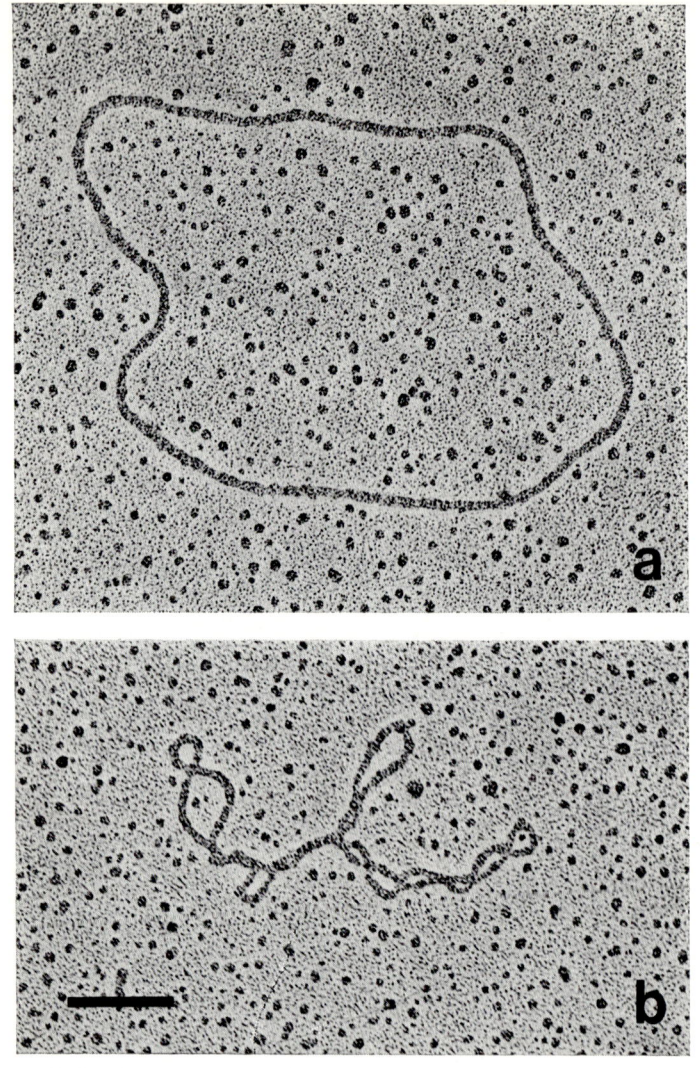

PLATE IV Open-circular (a) and supercoiled (b) forms of PM2 virus DNA. Bar represents 0·2 μm.

(Courtesy of Dr Lesley Coggins)

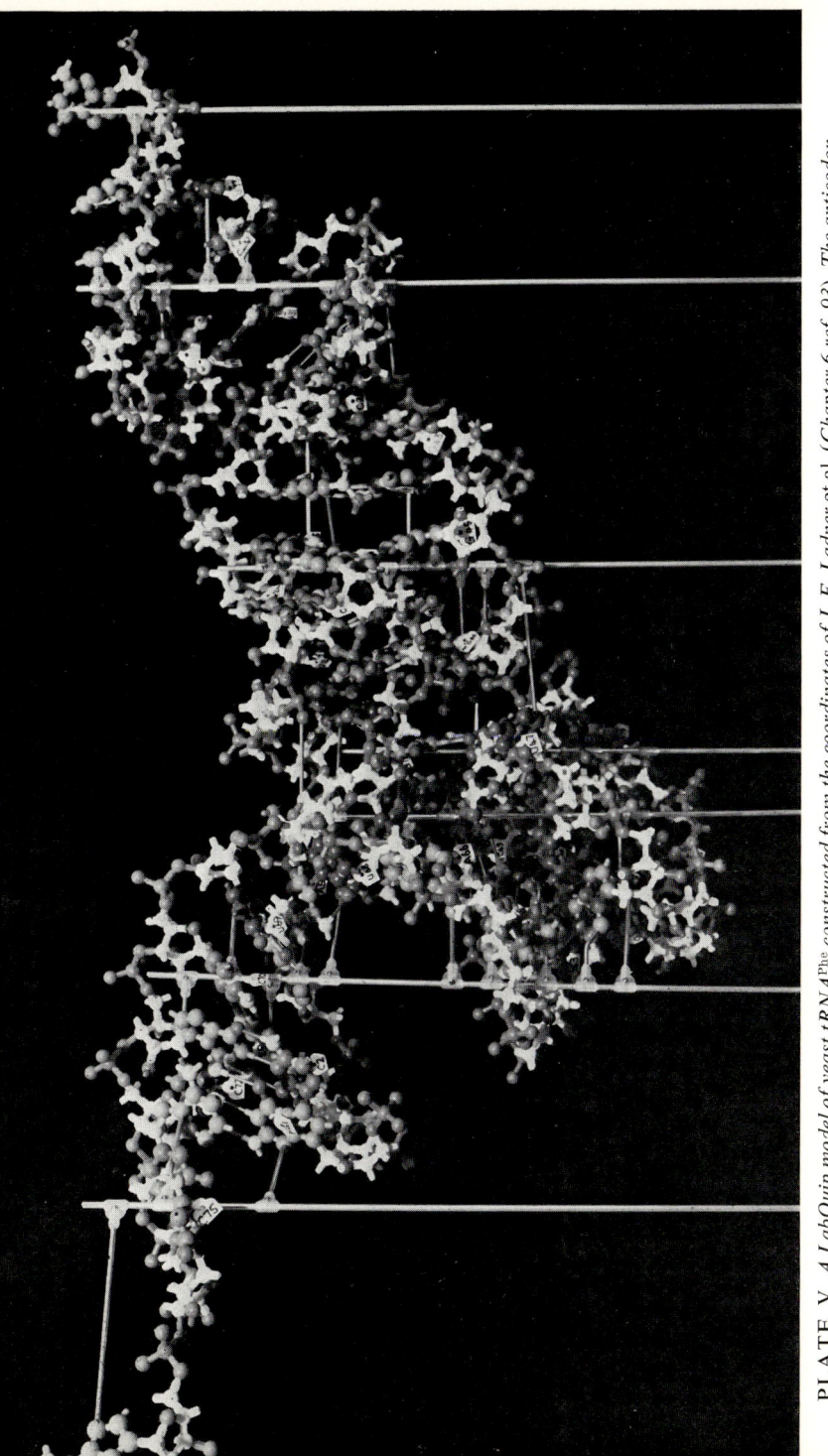

PLATE V *A LabQuip model of yeast tRNA^{Phe} constructed from the coordinates of J. E. Ladner et al. (Chapter 6 ref. 93). The anticodon and the –CCA terminus are at the right and left extremities of the model and are 77 Å apart. The GTΨC (IV) and dihydrouridine loops (I) are towards the back*

(Model by courtesy of Dr. J. P. Goddard)

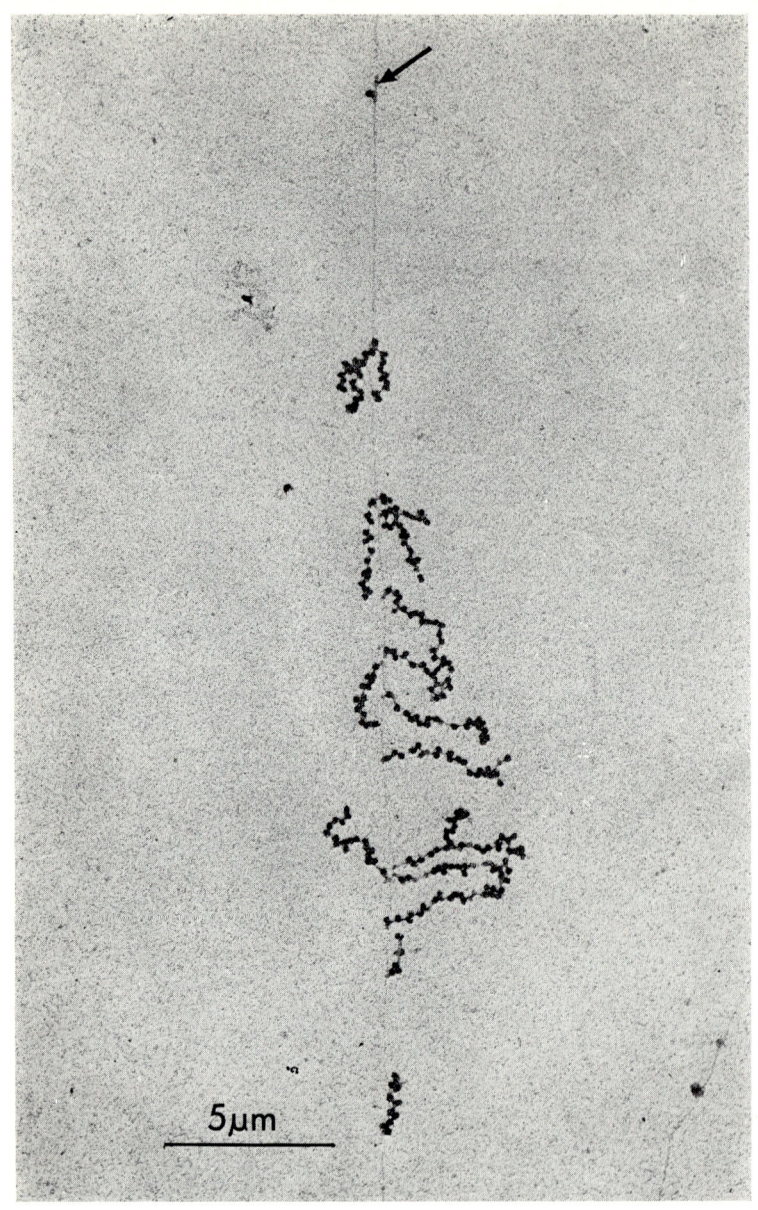

PLATE VI *Electron micrograph showing genetically active and inactive portions of* Esch. coli *chromosome. The polyribosomes attached to the active segments show imperfect gradients of increasing lengths. The arrow indicates a putative RNA polymerase molecule near an initiation site on the DNA strand.*

The absence of free polysomes suggests that all mRNA molecules in Esch. coli *in vivo are associated with the genome and that transcription and translation are intimately coordinated.*

(*By courtesy of Dr O. L. Miller, Jr*)

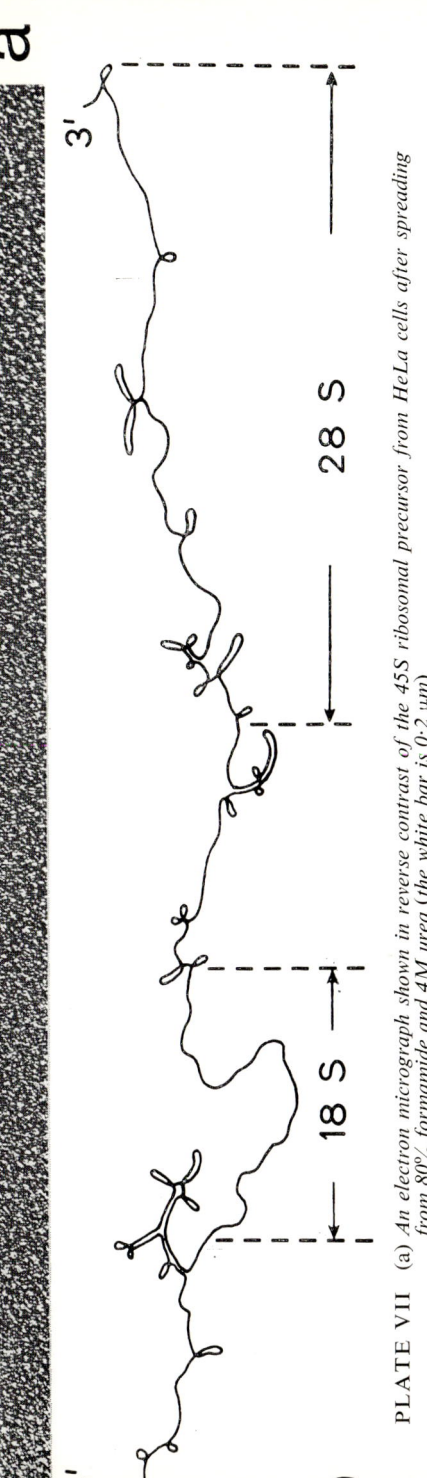

PLATE VII (a) *An electron micrograph shown in reverse contrast of the 45S ribosomal precursor from HeLa cells after spreading from 80% formamide and 4M urea (the white bar is 0·2 μm)*
(b) *A tracing of the same molecule showing secondary structure regions as hairpin loops (28S and 18S rRNA regions are indicated)*

(By courtesy of Dr P. K. Wellauer)

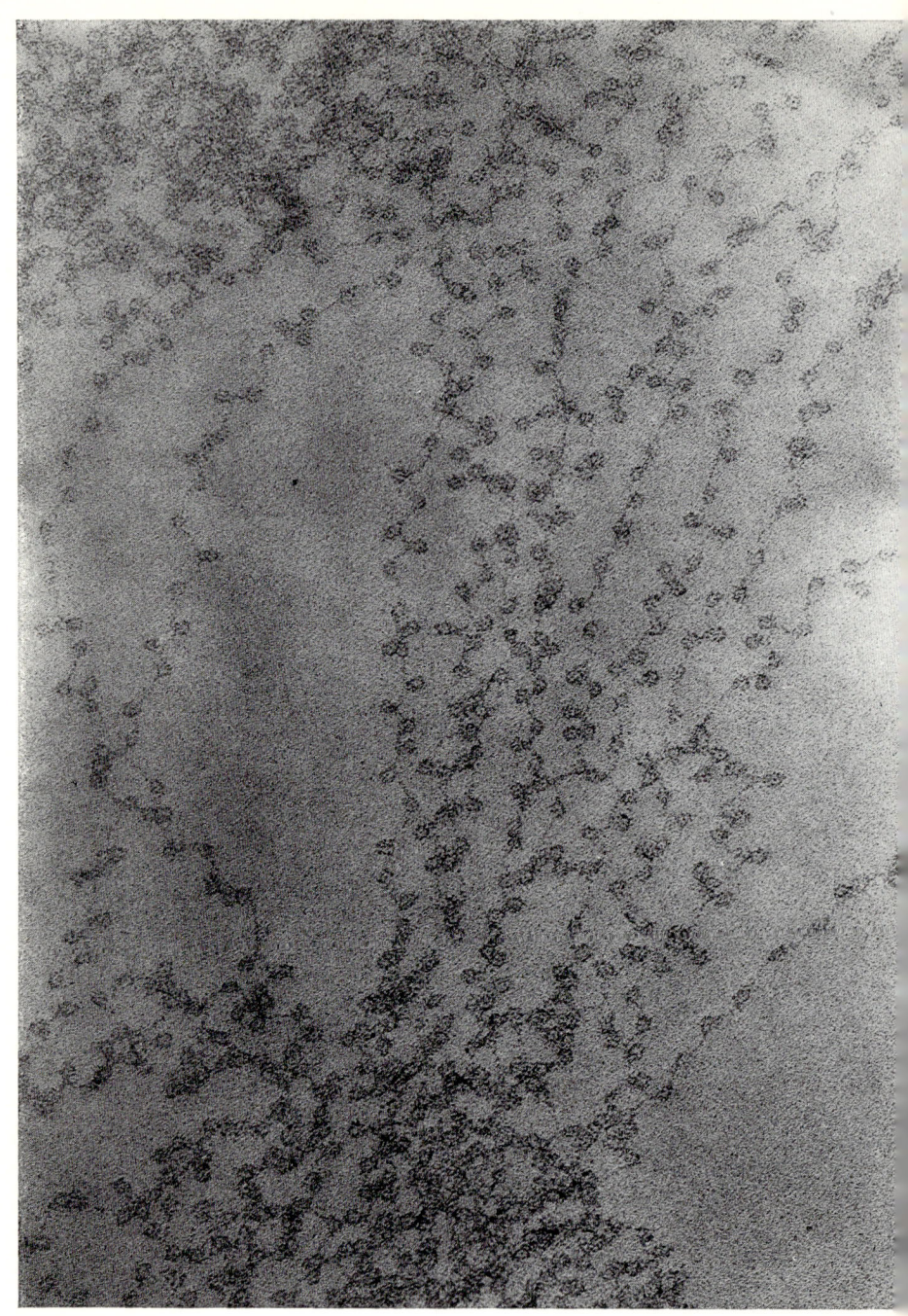

PLATE VIII *Chromatin fibres steaming out of a chicken erythrocyte nucleus. The bead-like structu... now termed nucleosomes or nu-bodies are about 70 Å in diam. The connecting strand is around 14(... in length. The sample was negatively strained with 5 mM uranyl acetate in water and the magnificat... 285 000 ×*

(By courtesy of Drs D. E. Olins and A. L. O...

of these variations in base content has been discussed in relation to the taxonomy of bacteria [7, 8] and protozoa [9] and to the evolution of various organisms [10].

As will be evident later (p. 275), biosynthetic polydeoxyribonucleotides containing only A and T or only G and C can be prepared enzymically. Poly(dA-dT) occurs naturally as a satellite component in the DNA of the crab (*Cancer*) where it comprises 10–30 per cent of the total DNA. The land crab (*Gercarcinus*) has

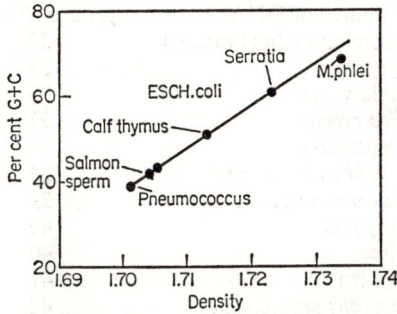

Fig. 5.1 *Relationship of density to content of guanine plus cytosine in DNA's from various sources* [4]

two satellite DNA's, poly(dA-dT) comprising 18 per cent of the total DNA and poly(dG) · poly(dC) comprising 3 per cent. Similar materials are found in other Crustacea [13].

5.2 Molecular weight of DNA

The molecular weights of DNA molecules are very difficult to determine accurately by the methods of classical chemistry since they range from 10^6 to more than 10^{10} (Table 5.3). Conventional analytical equilibrium ultracentrifugation is unsatisfactory since the available instruments are not stable at speeds low enough to balance centrifugal forces against diffusion forces. The best absolute methods of DNA molecular weight determination are light scattering on low-angle instruments [14–16], equilibrium analytical ultracentrifugation in caesium chloride gradients [17], and viscoelastic relaxation [18]. Most other methods have an empirical basis and rely on the few light scattering experiments which have been performed for their calibration. These more empirical methods include the measurement of intrinsic viscosity [19], sedimentation rate [19], electron microscopy [20], and autoradiography [21]. Generally a combination of two or more of these methods can be used [22]. The advent of the laser has led to the development of a new technique for determining diffusion coefficients of large molecules from the Doppler shift in the wavelength of the scattered laser light. This technique together with conventional analytical velocity centrifugation to determine sedimentation coefficients should lead to reliable absolute molecular

TABLE 5.2

The relative $(G+C)$ content of DNA's from various sources [11, 12]

Source of DNA	per cent $(G+C)$
Dictyostelium (slime mould)	22
M. pyogenes	34
Vaccinia virus	36
Bacillus cereus	37
B. megaterium	38
Haemophilus influenzae	39
Saccharomyces cerevisiae	39
Calf thymus	40
Rat liver	40
Bull sperm	41
Diplococcus pneumoniae	42
Wheat germ	43
Chicken liver	43
Mouse spleen	44
Salmon sperm	44
B. subtilis	44
T1 bacteriophage	46
Esch. coli	51
T7 bacteriophage	51
T3 bacteriophage	53
Neurospora crassa	54
Pseudomonas aeruginosa	68
Sarcina lutea	72
Micrococcus lysodeikticus	72
Herpes simplex virus	72
Mycobacterium phlei	73

weight determinations [23]. The determination of the molecular weight of DNA has also been complicated by the difficulties experienced in the preparation of whole DNA molecules since high molecular weight DNA is very susceptible to hydrodynamic shearing forces and to the action of contaminating nucleases. Thus all molecular weight determinations on an unknown DNA must be carried out in nuclease-free (heat sterilized) conditions, without pipetting or any other manipulation which puts shear stress on the

DNA; any methods which might themselves shear the DNA, such as viscosity determinations, must be used with caution. The size of large DNA molecules from individual chromosomes of eukaryotic cells has recently been determined by viscoelastic relaxation, and it appears that in *Drosophila* at least the DNA is contained in one

TABLE 5.3
DNA molecular weights

Source	Mol. wt.	Length	Number of nucleotide pairs	Conformation
Esch. coli chromosome	2×10^9	1 mm	3×10^6	Cyclic, duplex
H. influenzae chromosome	8×10^8	300 μm	12×10^5	—
Mycoplasma PPLO strain H-39	4×10^8	150 μm	6×10^5	—
Bacteriophage T2 or T4	1.3×10^8	50 μm	2×10^5	Linear, duplex
Bacteriophage λ	33×10^6	13 μm	0.5×10^5	Linear, duplex
Bacteriophage ØX174	1.6×10^6	0.6 μm	—	Cyclic, single-stranded
Polyoma virus	3×10^6	1.1 μm	4.6×10^3	Cyclic, duplex
Mouse mitochondria [92]	9.5×10^6	5 μm	14×10^3	Cyclic, duplex
Drosophila melanogaster chromosome	43×10^9	20 mm	65×10^6	—

complete molecule per chromosome [18]. The molecular weights of DNA's from a variety of sources are shown in Table 5.3.

5.3 The primary structure of DNA

As with RNA, the internucleotide bond in DNA is the phosphodiester linkage. Since C-4' in the sugar is occupied in ring formation and C-2' carries no hydroxyl group, only the hydroxyl groups at positions 3' and 5' in the sugar residue are available for internucleotide linkages. The absence of a hydroxyl group at C-2' makes cyclic phosphate formation impossible, and DNA is therefore not hydrolysed by alkali in the same way as is RNA. This property has been made use of in the separation of DNA from RNA (p. 64). The primary structure of the polynucleotide chain in DNA is shown in Fig. 5.2.

THE BIOCHEMISTRY OF THE NUCLEIC ACIDS

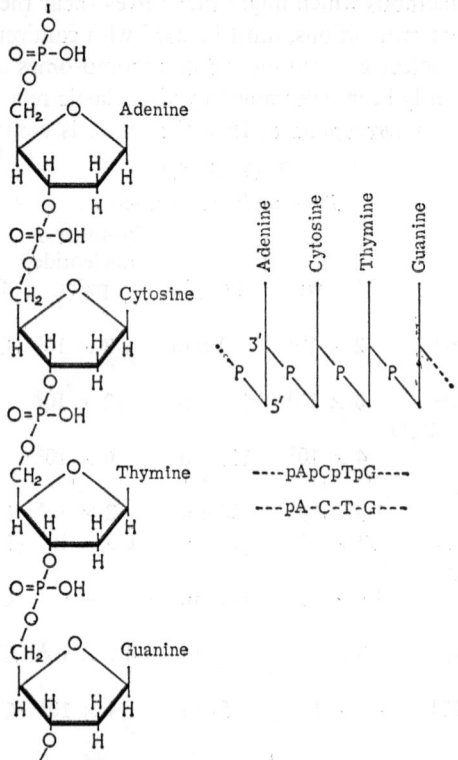

Fig. 5.2 *Part of the polynucleotide chain in DNA*

5.4 The secondary structure of DNA

X-ray diffraction has been extensively employed in the study of the molecular architecture of DNA by Astbury [24] and later by Franklin and Gosling [25] and on a very extensive scale by Wilkins and his colleagues [26–29].

Using early information obtained by this technique, and chemical observations, Watson and Crick [30–32] put forward the view in 1953 that the DNA molecule is double-stranded and in the form of a right-handed helix with the two polynucleotide chains wound round the same axis and held together by hydrogen bonds between their bases.

By making scale models they were able to show that the bases could fit in if they were arranged in pairs of one purine and one pyrimidine, and when the formation of hydrogen bonds between the

THE STRUCTURE OF DNA

bases was considered in detail it became evident that the most likely pairing arrangements were adenine with thymine and guanine with cytosine. The arrangement of hydrogen bonds is shown in Fig. 5.3. Other base pairing arrangements have since been suggested but have not been supported by any positive evidence [33].

It has already been mentioned (p. 3) that Chargaff [3] originally emphasized the equivalence of adenine and thymine and of guanine

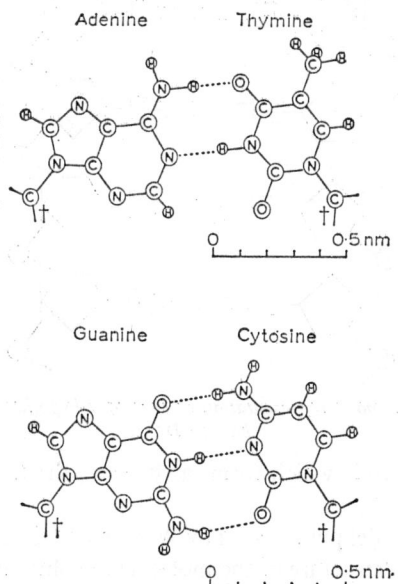

Fig. 5.3 *The pairing of adenine and thymine and of guanine and cytosine. Dotted lines indicate the hydrogen bonds. The carbon atoms marked † belong to sugar rings*

and cytosine in DNA's from many sources, and this remains one of the most important pieces of evidence for the double helical structure. The way in which the bases might be arranged in such a structure is shown in Fig. 5.4, which shows the linkage by hydrogen bonds of adenine in one chain to thymine in the other, or *vice versa*, and of guanine in the first chain to cytosine in the second, or *vice versa*. Accordingly, the order in which the bases occur in one chain automatically determines the order in the other chain, which is its complement. Apart from this essential condition, there are no restrictions on the sequence of pairs of bases along the chains. The pairs of bases are flat and may be stacked one above the other like a pile of plates.

A symbolic diagram of the Watson and Crick model is shown in Fig. 5.5 in which the two ribbons represent the phosphate sugar chains and the pairs of bases holding them together are shown by

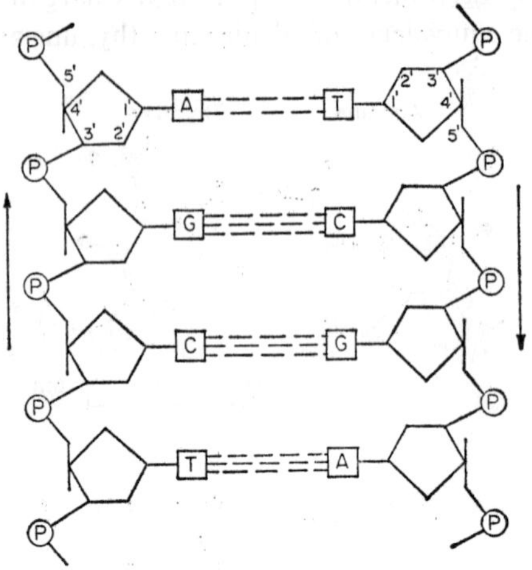

Fig. 5.4 *Diagrammatic representation of part of a hypothetical polynucleotide chain in DNA*

the horizontal rods which form, as it were, the treads of a spiral staircase.

This basic model put forward by Watson and Crick remains close to the accepted structure of the molecule in solution. However, the more refined X-ray diffraction studies of Wilkins and his colleagues [26–29] have shown that DNA fibres can have three possible structures (Table 5.4), and other techniques for the determination of secondary structure, such as ORD and CD [34] and low-angle X-ray scattering [35, 36], have been used to relate the structures of the fibres to the structure of the DNA molecule in solutions of various types.

The B structure of DNA is found in fibres of 92 per cent relative humidity and is believed to correspond most closely to the structure of DNA in solutions of low ionic strength [34]. This structure is very close to the original Watson–Crick model and is shown in Figs. 5.5 and 5.6. The two chains are of opposite polarity in the sense that the internucleotide linkage in one strand is $3' \rightarrow 5'$ while in the other it is $5' \rightarrow 3'$ (Fig. 5.4). The two helices are both right-

THE STRUCTURE OF DNA

handed and cannot be separated without unwinding. The chains are held together by hydrogen bonding between the base pairs and are complementary, the sequence of bases in one thus being determined by the sequences of bases in the other.

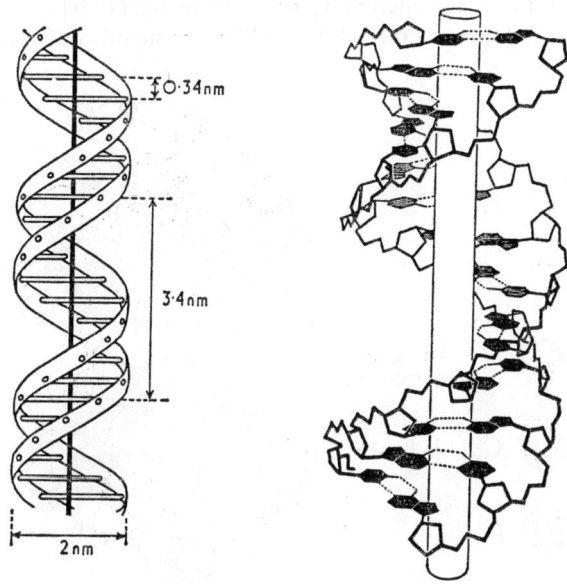

Fig. 5.5 (Left) *Diagrammatic representation of the DNA molecule as proposed by Watson and Crick* [30]. *The two phosphate-sugar chains are represented by ribbons and the pairs of bases holding the chains together are shown as horizontal rods.* (Right) *Drawing of a model of the DNA helix. The dotted lines represent the hydrogen bonds binding the bases*

The original belief that hydrogen bonds alone were responsible for the stability of the helix is now less widely held, and it is thought that a considerable part of the stability of the DNA double helix is maintained by base stacking forces or 'apolar' bonds. These bonds are consequent upon the hydrophobic nature of the bases themselves. This makes contact with an aqueous medium entropically unfavourable owing to the increase in water hydrogen bonding creating by such an interaction. Contact with the medium is minimized by vertical stacking of the bases above each other in the case of a single-stranded molecule, and by both stacking and combination of strands to give further reduction of the surface area exposed in double-stranded DNA [37]. Hydrogen bonds confer additional stability and also specificity.

The A structure of DNA is found in fibres of 75 per cent relative humidity when the counterion for the DNA is sodium, potassium, or caesium. While this structure is also in the form of a right-handed double helix, it differs from the B structure in that the bases are not flat but are tilted; the pitch of the helix and number of bases per turn is altered, and the structure of the backbone is slightly different. The

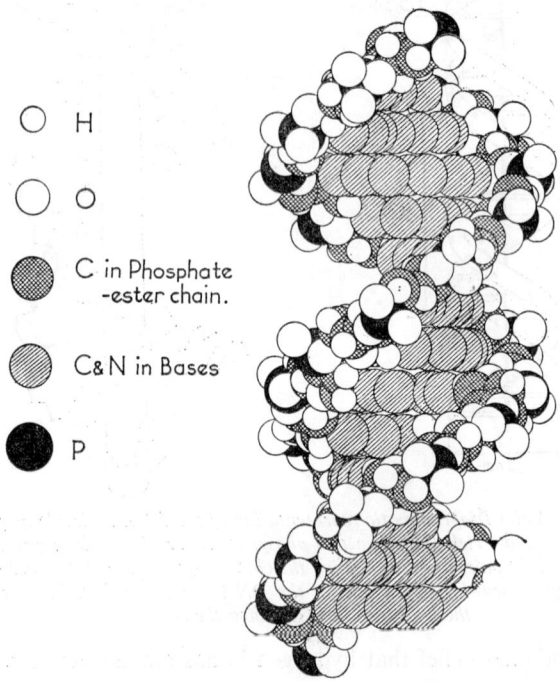

Fig. 5.6 *Model of proposed structure of DNA* [44]

biological interest in the A structure lies in the fact that it is believed to be very close to the conformation adopted by double-stranded RNA and by DNA–RNA hybrids in solution [38]. Because of the presence of the extra 2′-hydroxyl group, RNA appears to be unable to adopt the B conformation. Thus when it is engaged as a template for making RNA (Chapter 12) the DNA molecule must presumably adopt the A conformation. Various organic solvents [39] and proteins [40] have been suggested to force the DNA from the B form into the A conformation.

The C form of DNA is found in fibres of lithium DNA at 66 per cent relative humidity. It is also believed to occur in concentrated salt solutions [34] and in ethylene glycol [41]. Under these conditions the base stacking forces will be much reduced and the relative power of the hydrogen bond strengthened. Again the basic structure is a double right-handed helix with small differences in the pitch, number of residues per turn, base tilt, and backbone conformations relative to the B structure. It has been suggested that this type of structure occurs in chromatin [42] and in some viruses [43]. Bram

TABLE 5.4

The different forms of DNA

Form	Pitch (nm)	Residues per turn	Inclination of base pair from horizontal
A Na salt 75 per cent relative humidity	2·8	11	20°
B Na salt 92 per cent relative humidity	3·4	10	0°
C Li salt 66 per cent relative humidity	3·1	9·3	6°
DNA–RNA hybrid	2·8	11	20°

[35, 36] has studied the A, B, and C structures in solution together with two metastable structures which he has called the T and P structures. He has concluded that much greater scope for structural diversity lies within AT rich regions of the DNA molecule.

5.4.1 *Accepted abbreviations and short-hand notations for DNA*
Part of a deoxyribopolynucleotide strand as found in DNA is shown on the left in Fig. 5.2. For simplicity this may be abbreviated to the form shown on the upper right of the figure or even further to the form shown on the lower right in the same manner as abbreviated versions of ribopolynucleotides may be depicted (p. 120), provided that it is made clear in any accompanying description that the polynucleotide chain is of the deoxyribose type.

Double-stranded DNA may be shown schematically thus:

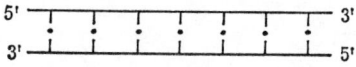

where the centre dots represent the hydrogen bonds between complementary bases.

Another method of illustrating double-stranded DNA in abbreviated schematic form is shown in Fig. 5.4.

5.4.2 *The dynamic secondary structure of DNA*

Despite the essential stability of the DNA molecule which befits its role as a carrier of genetic information, it is evident that the double helix is not a totally fixed and rigid structure but undergoes considerable internal deformation in a continuous manner. This dynamic aspect of DNA secondary structure is best shown by tritium exchange experiments [45, 46] which show that small segments of the double helix can swing apart and protrude into the external medium in a manner closely dependent on the environment of the DNA molecule. Thus, whereas X-ray diffraction studies and circular dichroism show the average conformations that the DNA molecule as a whole can assume, tritium exchange experiments show the amount of deformation and twisting of these structures that can occur in localized areas. Fluorescence experiments have also confirmed that internal distortions of the DNA molecule occur to a hitherto unsuspected degree [47].

5.5 The tertiary structure of DNA

The DNA from many sources, particularly cellular chromosomal DNA, cannot be isolated in the form of intact molecules because of its large size and consequent sensitivity to shear and enzymic degradation. However, the DNA from many viruses, from mitochondria [48] and chloroplasts [49], and from a few other sources (e.g. bacterial plasmid DNA) can be isolated in an undegraded form. The examination of such undegraded DNA has shown that the double helical secondary structure of DNA can in some instances be further constrained giving rise to tertiary structures such as supercoiled and open-cyclic molecules. Linear DNA occurs naturally and is also a product of the degradation of cyclic DNA. Single-stranded DNA, both in linear and cyclic forms, is found in some viruses, and can be formed artificially by denaturing double-stranded DNA (see p. 100). The properties of these different forms of DNA will now be described.

THE STRUCTURE OF DNA

5.5.1 Double-stranded linear DNA

Intact linear double-stranded DNA molecules are found in viruses (see Chapter 7) in the molecular weight region 120×10^6 and in chromosomes in even larger molecules [18] although some of the larger molecules may have been cyclic molecules sheared open by isolation procedures. Linear molecules of lower molecular weight, which are the result of degradation during isolation, are also found in commercial heterogeneous preparations such as calf thymus DNA. Although the conventional DNA models in Figs. 5.5 and 5.6 would suggest that such molecules are rigid about their long axis, it is evident that they must possess some degree of flexibility in order to be accommodated in virus particles and cells (Plate I). From a physicochemical point of view, short lengths of DNA below 500 000 in molecular weight may be treated as slightly flexible rods [50] whereas the longer DNA molecules behave more like stiff worm-like coils [51]. The stiffness of the double-helical structure gives to DNA solutions their characteristic viscosity, especially when the DNA is of high molecular weight.

5.5.2 Double-stranded cyclic DNA

Double-stranded cyclic DNA has been isolated from animal viruses such as polyoma and SV40 and from bacterial viruses such as PM2 (Chapter 7). It also occurs as an intermediate in the replication of other bacterial viruses such as λ and ØX174 and among bacterial plasmid DNA's. RNA tumour viruses have recently been shown to form a circular duplex intermediate during animal cell infection [52]. The entire chromosome of *Esch. coli* which was known for many years to be circular but was too large for conventional DNA isolation procedures has now been isolated and its physical and biological properties have been studied *in vitro* [53, 54]. In eukaryotic cells double-stranded circular DNA has been isolated from both mitochondria and chloroplasts. The occurrence and properties of circular DNA molecules has been reviewed [55]. All circular DNA molecules have two possible structural forms which differ considerably in physical, chemical, and biological properties.

Covalently closed cyclic DNA molecules have no breaks at all in either strand, with the consequence that any change in the secondary structure of the DNA which affects the number of residues per turn of the helix must have concomitant effects on the tertiary structure of the molecule. Such DNA can best be envisaged as a

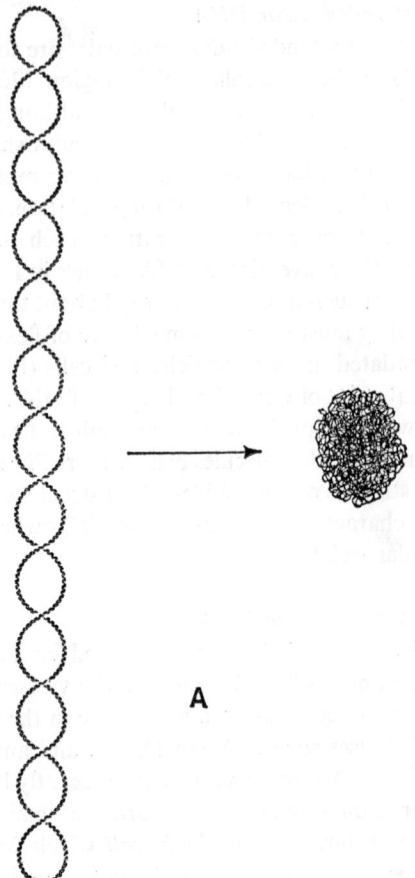

Fig. 5.7 *Some typical cyclic DNA molecules*

A. *Covalently closed cyclic duplex DNA; a typical example of molecular weight about 3×10^6 containing 12 superhelical turns. At pH values above 12·6 and at temperatures above 100° this collapses into a compact form as shown on the right*

B. *An open cyclic DNA molecule of the same size. This molecule (left) has only one-strand breakage (at X) and so is denatured to one single-stranded linear molecule and one single-stranded closed cyclic molecule (right). Denaturing conditions are much milder than for A and are similar to those for linear DNA*

C. *A typical catenane containing a closed cyclic duplex DNA molecule as in A, linked to an open one as in B. Other combinations such as AA and BB and triple linkages can also occur*

The contour length in all molecules shown is approximately 1700 nm

THE STRUCTURE OF DNA

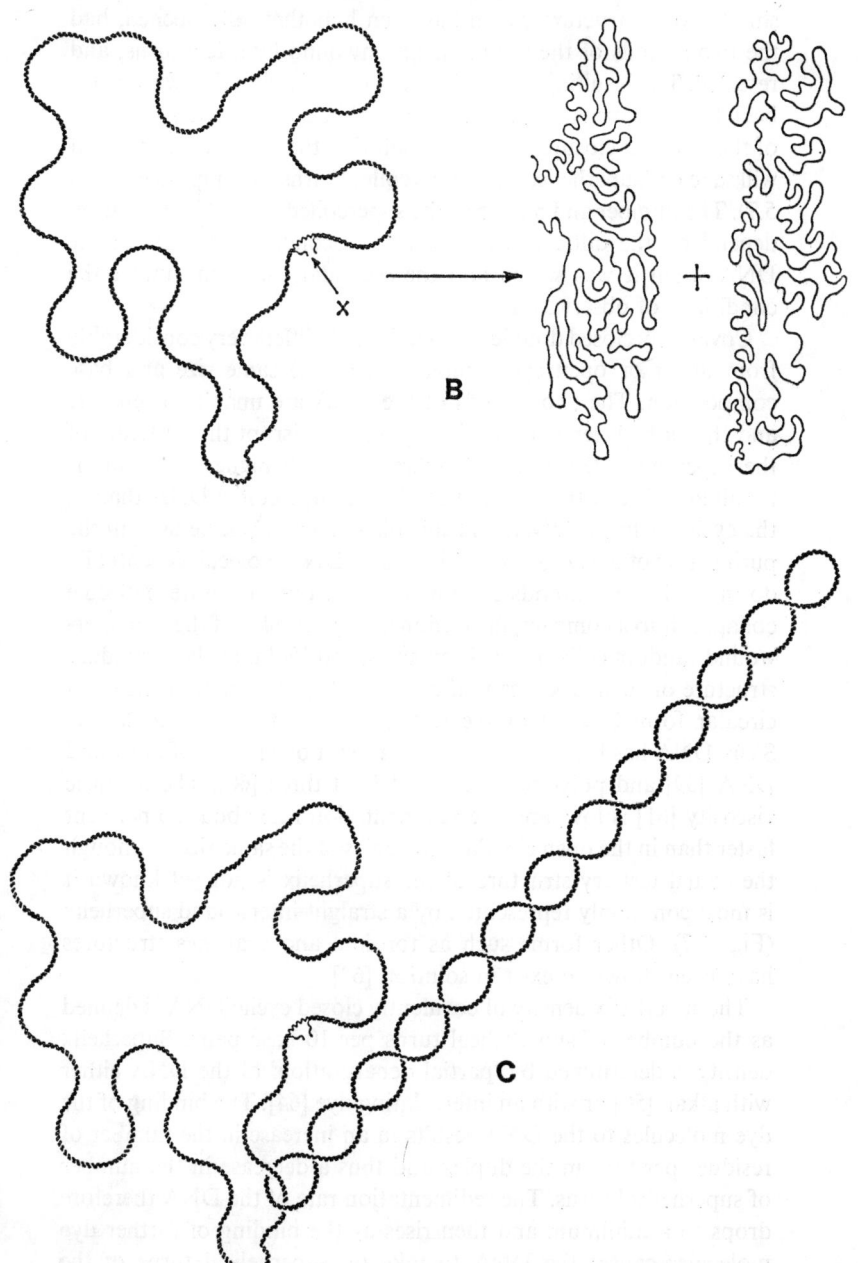

simple cyclic structure which has been hypothetically opened, had the two strands of the double helix unwound by a few turns, and resealed. The resulting molecule could try to rewind the two strands back to their normal structure but would be unable to do so because of the covalent closure, with the result that the circular duplex itself will take on 'superhelical' or 'supercoiled' turns to compensate (Fig. 5.7). The number and nature of the supercoiled turns will, of course, depend on the difference between the secondary structure of the DNA when it was sealed and the secondary structure under the conditions of observation.

Covalently closed double-stranded DNA differs very considerably from linear or open cyclic molecules of the same size and base composition. The two strands of the DNA are unable to separate and thus a high temperature is required to disrupt the structure of the supercoiled DNA [56]. Similarly a much higher pH value is required to break the base-pairing in the supercoiled DNA than in the cyclic or linear DNA, and this phenomenon can be used in the purification of the DNA [57]. When these DNA molecules eventually do melt, the two strands cannot separate, but the entire molecule collapses into a compact, fast-sedimenting complex of the two interwound random coils. It has been suggested [58] that the secondary structure of such DNA can differ detectably from that of the open circular form because of the tertiary restraints on the molecule. SV40 DNA has been shown to have at least one region of unwound DNA [59] and polyoma to have at least three [60]. The intrinsic viscosity [61] is low, and the sedimentation rate about 20 per cent faster than in the open circular molecules of the same size. Although the actual tertiary structure of the superhelix is not yet known it is most commonly represented by a straight interwound superhelix (Fig. 5.7). Other forms such as toroidal and branches structures have been shown to exist in solution [62].

The superhelix density of covalently closed cyclic DNA is defined as the number of superhelical turns per 10 base pairs. Superhelix density is determined by 'partial denaturation' of the DNA either with alkali [56] or with an intercalating dye [64]. The binding of the dye molecules to the DNA results in an increase in the number of residues per turn in the duplex and thus a decrease in the number of superhelical turns. The sedimentation rate of the DNA therefore drops to a minimum and then rises as the binding of further dye molecules causes the DNA to take on superhelical turns of the opposite sense. The superhelix density of most covalently closed DNA molecules was until recently thought to be about $-0\cdot 03$ in neutral caesium chloride but recent estimates have resulted in a doubling of this value [63, 64]. Superhelix density is affected by both temperature and ionic strength [65]. Electron micrographs of various supercoiled molecules are shown in Plate IV.

Open double-stranded cyclic DNA can be described as linear DNA which has been joined at the ends to form a circle but in which one or other of the two strands has a break such that it can rotate round the non-broken strand. Denaturation of such a molecule would be expected to lead to one single-stranded circular DNA molecule and one single-stranded linear DNA molecule (Fig. 5.7B). The extent to which such molecules are present in the cell is still difficult to assess (Chapter 7). Open cyclic DNA can also occur where there is a break on both of the single strands of the duplex but the two breaks are far apart. This is the case with DNA molecules with cohesive ends (p. 142).

The secondary structure of such cyclic molecules is believed to be close to that of equivalent linear molecules and the density and T_m are the same as that of linear molecules with the same (G+C) content. The more compact nature of the tertiary structure, however, leads to a lower intrinsic viscosity than is found in the equivalent linear DNA molecules [66] and to slightly faster sedimentation in the analytical ultracentrifuge. In general, however, there are no dramatic differences in physicochemical behaviour from linear double-stranded DNA of the same size.

5.5.3 *Catenanes*

Some of the open and closed cyclic DNA molecules isolated from cells infected with SV40 [67] or ØX174 [68] and from mitochondria of human leukaemic leucocytes [69] have been shown to form linked dimers or trimers, i.e. systems of interlocking rings (Fig. 5.7). These cells also contain complex DNA of double the size of the circular DNA of the infecting virus. Owing to their scarcity, the physical properties of these molecules are not yet well described although they can be prepared *in vitro* [70]. Such molecules are thought to occur through errors in the DNA replication process. Their frequency is greatly increased in the presence of inhibitors of protein synthesis [71].

5.6 Denaturation and renaturation of DNA

5.6.1 *Single-stranded DNA*

Single-stranded DNA molecules rarely occur naturally. The best known example is the DNA molecule from the small spherical bacteriophage ØX174 first isolated by Sinsheimer in 1959 [72] but it is also found in the filamentous bacteriophages such as fd. Among the animal viruses single-stranded DNA is found in the parvoviruses. Some of these naturally occurring single-stranded DNA's are circular and others linear. The former can be distinguished by their insensitivity to exonucleases. In general they are comparatively small molecules with less than 10 000 nucleotides. The molar proportions of the bases do not show the usual equivalence of A and T, and G and C, required for double helix formation and the bases react readily with formaldehyde since the amino groups are not protected by hydrogen bondings as they are in double-stranded DNA. Single-stranded DNA molecules are very much more flexible than double-stranded molecules of the same size since they lack the rigid double helical structure. In general they behave in solution in a manner similar to RNA molecules.

5.6.2 *The helix–coil transition*

When double-stranded DNA molecules are subjected to extremes of temperature or pH, the hydrogen bonds in the double helix are ruptured and the DNA collapses into two single-stranded molecules. If heat is used as the denaturant, the temperature at which this collapse occurs is known as the T_m or transition temperature. The absorption at 260 nm of any polynucleotide is due to that of its component bases. However, this absorption tends to be suppressed in the double-stranded DNA molecule where the bases are stacked above one another and inhibited from swinging out freely in solution by hydrogen bonds, and consequently it is very much lower than that of equimolar amounts of the component bases or nucleotides free in solution. This inhibition of absorption is to some extent relieved when the DNA molecule goes through the helix–coil transition and the bases are no longer so rigidly stacked although they still interact to some extent. As a consequence of this the absorption of DNA solutions rises by about 20–30 per cent as the DNA undergoes the melting process, and the transition process is usually measured by ultraviolet spectroscopy (Fig. 5.8). The increase in absorbance on melting is known as the *hyperchromic effect*.

THE STRUCTURE OF DNA

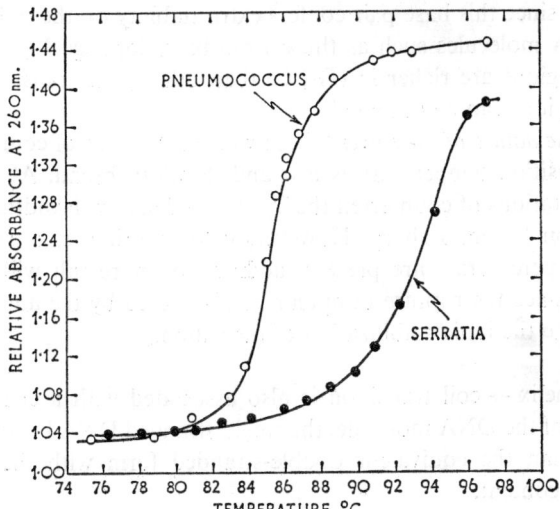

Fig. 5.8 *Increase in absorbance at 260 nm with rising temperature for DNA from two different sources* [4]

Such a transition can be characterized by several factors:

(1) *The nature of the DNA.* Homogeneous DNA, such as viral DNA, melts over a short temperature range but heterogeneous DNA melts over a longer range.

(2) *The (G+C) content of the DNA.* Figure 5.9 shows that the melting temperature of any DNA can be related to its (G+C)

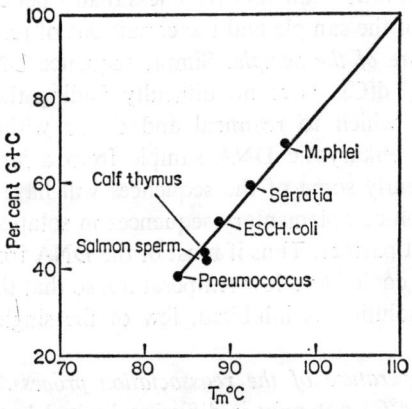

Fig. 5.9 *Relationship between the content of guanine plus cytosine and melting temperature* (T_m) *in DNA's from various sources* [4]

content since this base pair confers extra stability on the molecule. In DNA molecules such as those from bacteriophage λ, in which some regions are richer in (G+C) than others, the transitions of both regions can be observed.

(3) *The nature of the solvent.* In low concentrations of counterion the transition temperature is low and its width broad. At higher concentrations of counterion the T_m is raised and with width of the transition becomes sharp. However, if 'denaturing' salts such as sodium perchlorate are present, addition of more salt will lower the T_m since the rupture of apolar bonds caused by the anion will overcome the ionic stabilization of the cation.

The helix → coil transition is also associated with a change in density of the DNA molecule, the single-stranded DNA being more dense than the equivalent double-stranded form with the same (G+C) content.

5.6.3 *The renaturation of DNA*
When two DNA strands are returned from the extreme conditions which caused them to melt to their original state they may reassociate to form a double helix again. However, whether or not they do so depends on a variety of factors. In principle the process should follow simple second-order kinetics but, because of the high probability of two sequences which were not previously matched forming an imperfect union which is then difficult to dissociate, the degree of renaturation can vary from less than 1 per cent according to the nature of the sample and the conditions of renaturation.

(1) *The nature of the sample.* Simple sequence DNA molecules such as $d(G)_n \cdot d(C)_n$ have no difficulty finding the appropriate sequence with which to reanneal and do so without difficulty. However, if a eukaryotic DNA sample from a large genome is reannealed, clearly some of the sequences will have to encounter many other non-complementary sequences in solution before they find the correct partner. Thus if most of the DNA from eukaryotic cells is quickly cooled to a low temperature, so that the diffusion of the DNA in solution is inhibited, few of the single strands will reassociate.

(2) *The temperature of the reassociation process.* At very low temperatures (4°C) not only is diffusion limited but, for a DNA molecule which has become mismatched with a strand having only

a few complementary bases, the opportunities to break away and continue its search for the correct complementary strands are reduced. Consequently DNA which is heated beyond the T_m and then quickly cooled to a low temperature will be denatured whereas solutions maintained at high temperatures below the T_m may renature.

(3) *The size of the DNA fragments.* Large string-like fragments of single-stranded DNA encounter diffusion problems and frequently cannot reanneal correctly since this effectively reduces their opportunities for finding complementary strands. For this reason DNA is often sheared for reannealing experiments.

(4) *The ionic strength of the solution.* Two highly charged DNA molecules are likely to repel one another, and the presence of salt is necessary to mask this repulsion in renaturation.

(5) *The concentration of the DNA.* Clearly, at higher concentrations the probability of two complementary strands encountering each other is raised.

(6) *The time allowed for reannealing experiments.* If renaturation is allowed under ideal conditions, two DNA samples of identical concentration should take different times to reanneal according to the genome size [see (1)]. For this reason the term *Cot value* has been defined for the study of reannealing of DNA and also for the study of the formations of DNA-RNA hybrids. *Co* represents the DNA concentration and *t* represents time in seconds. This is again monitored by ultraviolet spectroscopy or, more frequently, by hydroxyapatite chromatography since this can be used to separate double- and single-stranded DNA molecules and can be used at higher concentrations and hence for shorter times. Figure 5.10 shows the reannealing process and Fig. 5.11 the representation of *Cot* values for a variety of DNA molecules [73, 74].

5.7 Nucleotide sequence analysis of DNA

The techniques of DNA sequence analysis were developed later than those of RNA sequence analysis (Chapter 6) for two main reasons. Firstly the base-specific ribonucleases used for controlled degradation of RNA have no counterpart among the deoxyribonucleases, and secondly the smallest homogeneous DNA molecules which could be readily isolated in large amounts were for many years at least 5000 nucleotides long. This situation has changed dramatically in recent years as new techniques have been developed to allow the

THE BIOCHEMISTRY OF THE NUCLEIC ACIDS

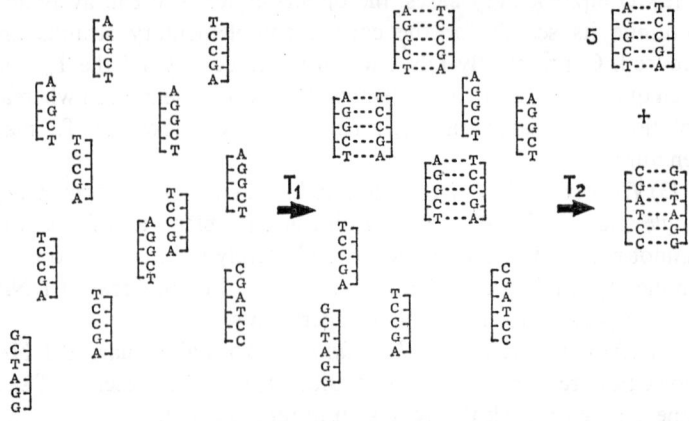

Fig. 5.10 *Renaturation of complementary strands of DNA with time. After time T_1 most of the strands which are frequently repeated have found complementary partners but the unique strand has not. After a longer time, T_2, all strands have found complementary sequences*

isolation of substantial quantities of small discrete pieces of DNA amenable to sequence analysis. The major tools now employed for this task are the restriction nucleases (Chapter 8) which hydrolyse DNA molecules at a limited number of sites because of their high sequence specificity. Since each one of these enzymes is specific for a different sequence, the simultaneous use of two or three can result in the production, from the initial larger DNA molecule, of a large

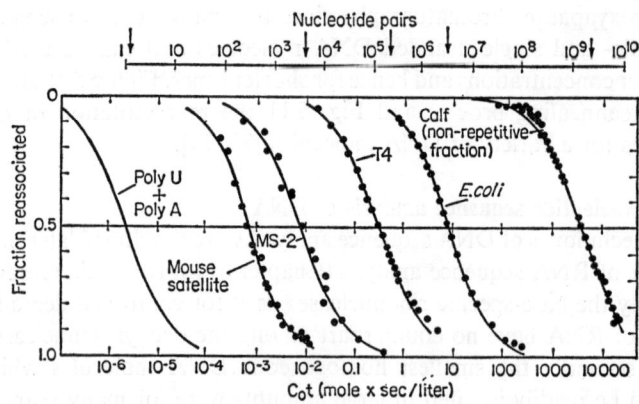

Fig. 5.11 *The rate of reassociation of double-stranded polynucleotides from various sources showing how the rate decreases with the complexity of the organism and its genome (from [73])*

number of homogeneous DNA fragments of a size suitable for sequence analysis [77, 78]. These fragments can be separated by gel electrophoresis and sequenced individually. A second method of isolation of small pieces of DNA is particularly useful in the study of DNA protein interactions. Many specific proteins involved in gene expression bind very tightly indeed to their sites on the DNA molecule so that the remainder of the DNA may be digested away by deoxyribonucleases leaving the binding site protected. The protein can then be removed and the binding site sequenced [79-81].

Once a homogeneous DNA molecule of a size amenable to analysis has been prepared there are two basic approaches to sequencing. The first is called the direct method. Radioactive single-stranded DNA molecules are digested with endonuclease IV from *Esch. coli* infected with bacteriophage T4. This enzyme has a strong preference (but not an absolute specificity) for the production of fragments with a cytidine 5'-monophosphate terminus. If it is used at high ionic strength it degrades the DNA molecule at a limited number of sites as the single strands tend to fold back on themselves leaving a limited number of sites exposed. The radioactive fragments can then be separated and treated again with the same enzyme at lower ionic strength, and the smaller fragments produced can again be separated for sequence analysis [82, 83]. This procedure leads to the production of sequences of average length 10-20 residues which can be analysed by a variety of methods involving two-dimensional homochromatography of partial exonuclease digests [84, 85]. The indirect method of DNA sequence analysis involves the use of the DNA fragment as a template and analysis of the structure of the complementary product. If the enzyme RNA polymerase is used, the sequence of the RNA product can be determined by the well established techniques of RNA sequencing. However, RNA polymerase is not always suitable for this method since it requires specific initiation points to be present on the DNA fragment. Consequently, DNA polymerase which requires a primer sequence for the complementary strand can be used in conjunction with a primer molecule long enough to specify a single clear initiation point on the template. If small sequences are being determined, it is possible by use of a variety of radioactive precursors to determine the order in which they are incorporated and hence the sequence of the complementary strand [86, 87]. Longer sequences can be analysed by the use of DNA polymerase under conditions

THE BIOCHEMISTRY OF THE NUCLEIC ACIDS

which permit one of the four bases in the product to be attached to a ribonucleotide sugar residue rather than a deoxyribonucleotide one. The mixed sugar complementary strand can then be hydrolysed with ribonuclease or alkali to yield short sequences of radioactive

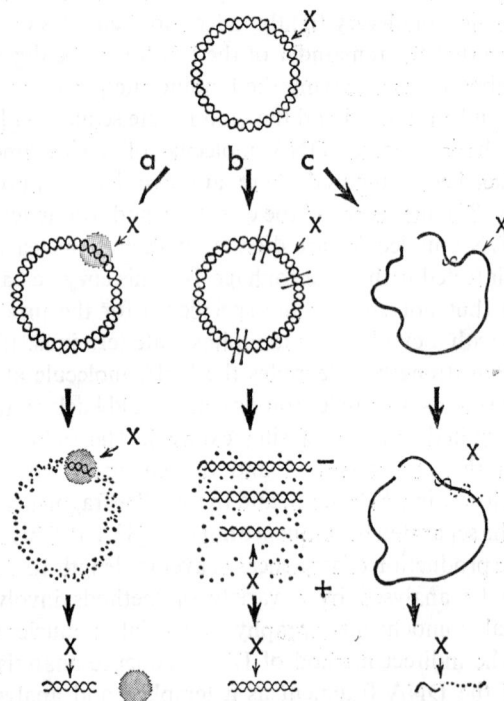

Fig. 5.12 *Methods of isolation of a specific genomic region for sequencing*

a. *If a specific protein which is known to bind to a particular region of the DNA is available, then the DNA protein complex can be digested with nucleases under tight binding conditions and then dissociated. This method is limited by the lack of availability of suitable proteins in many cases*

b. *Restriction nuclease digestion of the whole genome yields a variety of fragments which can be separated by gel electrophoresis. The fragment obtained by this procedure may still contain more material than is desirable. More than one restriction nuclease can be used*

c. *If the sequence of a short segment of DNA is already known through techniques such as* a, *then neighbouring sequences can be determined by attachment of the specific primer sequence to the denatured template DNA molecule. The complementary sequence is then synthesized using DNA polymerase III. This method has the advantage of yielding radioactive fragments of varying length to separate and analyse if samples are taken at various time intervals during the synthetic reaction. A total picture can then be built up starting with the shorter molecules. However, a suitable primer sequence must be available*

THE STRUCTURE OF DNA

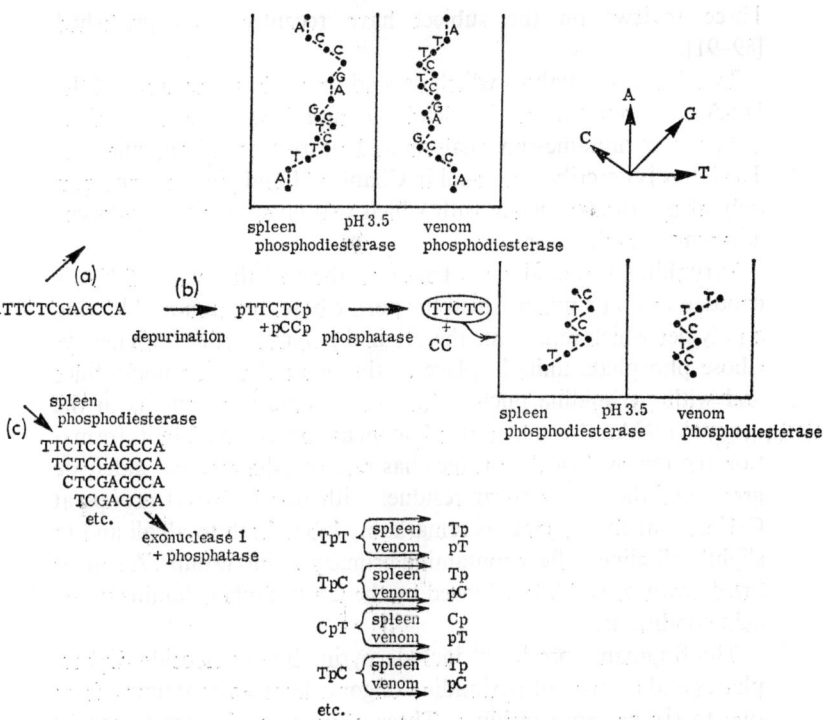

Fig. 5.13 *Various methods of sequence determination for a small oligodeoxyribonucleotide*

a. *Electrophoresis and homochromatography of the spleen and venom phosphodiesterase partial digests of the entire fragment. The sign and magnitude of the vector between each polynucleotide spot in the digest gives an indication of which base has been removed. For short fragments this method can give total sequence*

b. *Depurination gives a simplification for large fragments by allowing the sequences of the pyrimidine clusters to be determined. This sequence can then be fitted into the large oligodeoxyribonucleotide*

c. *Spleen phosphodiesterase partial digests treated with exonuclease and phosphatase leading to the hydrolysis of the 5'-dinucleotide. Subsequent hydrolysis of this with spleen and venom phosphodiesterase gives the sequence of the dinucleotide. A progressive picture of the polynucleotide can thus be built up*

deoxyribonucleotides which can be sequenced by homochromatography of partial exonuclease digests as in the direct method. Since the base attached to the ribonucleotide sugar residue can be varied, there is considerable potential in this technique [88]. Sequences of up to 1000 nucleotides have now been determined by use of these methods, and the complete sequences of some of the small viral DNA molecules are likely to be determined in the near future.

107

THE BIOCHEMISTRY OF THE NUCLEIC ACIDS

Three reviews on the subject have recently been published [89–91].

Two forms of analysis which depend on the base sequence of the DNA, but do not themselves lead to complete sequence determination, are pyrimidine-run analysis and nearest-neighbour analysis. The latter is described in detail in Chapter 11 and gives information only about the frequency with which two given nucleotides occur adjacent to each other.

Pyrimidine-run analysis is based on the fact that when DNA is exposed to dilute mineral acid the purine bases are removed leaving a polymer which represents the original polynucleotide with deoxyribose phosphate units in place of the original purine nucleotides and with pyrimidine nucleotides in the same positions as in the original DNA. This material is known as *apurinic acid*. In its formation the removal of the purines has released the reactive aldehyde groups of the deoxy sugar residues with free hydroxyl groups at C-1' so that the polymer is remarkably labile both to alkali and to slightly alkaline buffers containing primary amino groups. A similar breakdown of DNA is achieved by the use of diphenylamine under acid conditions.

The fragments produced include pyrimidine nucleoside diphosphates and a series of pyrimidine oligonucleotides containing from two to six or more residues. These oligonucleotide tracts can be separated by chromatography [84]. In this way the frequencies of occurrence of pyrimidine nucleotides in DNA's from several sources have been determined, and it is well established that pyrimidines frequently occur in clusters. Such tracts or clusters have the general structure Py_nP_{n+1} and have been named *isostichs*. For example, the dinucleotides pTpTp and pCpCp are isostichs but are not isomers, whereas pTpCp and pCpTp are both isomers and isostichs.

Treatment of DNA with permanganate yields a product (O-DNA) in which all the guanine, cytosine, and thymine residues are oxidized to ureido groups. Treatment of O-DNA with 1N-alkali at 100° for 1 hour yields components of the type $(adenine)_n(deoxyribose\ phosphate)_{n+1}$. Results obtained by this procedure suggest that more adenine residues occur in tracts of three, four, or five than would be expected on the basis of random distribution [92]. Such purine isostichs have also been obtained from the counterpart of apurinic acid, the less well known *apyrimidinic acid* which is prepared by treating DNA with hydrazine to remove the pyrimidines leaving the purines *in situ* [93]. It has been used to study purine isostichs and in the chemical proof that the two strands of DNA are antiparallel.

A limited amount of information about DNA sequences can be obtained by the technique of nearest neighbour sequence analysis described in Chapter 11 which reveals the frequency with which any one base is found next to another base in DNA [94]. This method has shown that the CpG doublet is very rare in mammalian DNA.

5.8 The structure of the DNA in the eukaryotic chromosome

The DNA of most eukaryotic cells is composed partly of unique sequences of DNA, partly of 'moderately' repeated sequences, and partly of highly repeated sequences. Highly repeated sequences are present in all eukaryotic cells although they vary in amount from 1 per cent of the genome to more than 50 per cent and average about 15 per cent. They are found in blocks of several million repeat units of a sequence which is between 6 and 8 base pairs long. These DNA sequences frequently have a different $G+C$ content from the bulk of the cell DNA and consequently band at a different or 'satellite' position on caesium chloride gradients (page 60). Guinea pig and *Drosophila* satellite DNA's are of at least three types, the latter group differing from each other in only one base pair [95, 96]. Originally the satellites were thought (from renaturation kinetics) to be several hundred base pairs long but sequencing has shown them to be very much smaller. A typical example of a satellite DNA sequence is shown in Fig. 5.14. No satisfactory biological role has yet been proposed for such DNA. Satellite regions are usually clustered near the centromere region of the chromosome and may be responsible for the maintenance of its structural integrity in some way [97]. Because of their highly repeated sequences such fragments of DNA often form circles (Fig. 5.14).

Moderately repetitive DNA includes the genes coding for histones (Chapter 14) and ribosomal RNA. About 1000 copies of such genes may be present per genome although the number of ribosomal genes is variable [98]. The sequences of the two major species of ribosomal RNA are arranged in tandem with spacer DNA in between.

Unique DNA sequences are more difficult to detect but appear to be present in the genes which code for proteins of more specialized functions such as haemoglobin [76] or fibroin [99]. The arrangement of unique and repetitive regions of DNA along the chromosome is not fully established but over a substantial portion of the genome

THE BIOCHEMISTRY OF THE NUCLEIC ACIDS

unique and moderately repetitive sequences appear to alternate. The distance between moderately repetitive sequences has been reported as 2000 to 3000 bases in most metazoa [100] but greater in *Drosophila* [101].

Another structural feature of eukaryotic DNA is the high proportion of 'palindromic regions' or regions of inverted repetition [103]. These are sequences of DNA arranged in two-fold symmetry such

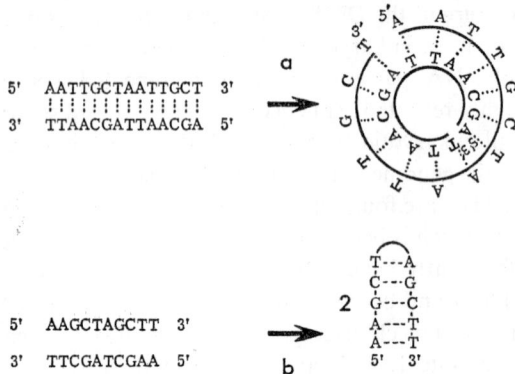

Fig. 5.14 *Two examples of unusual structures formed by eukaryotic chromosomal DNA on denaturation and reannealing*
a. *Repeated DNA's such as satellite DNA can form circular molecules*
b. *DNA's with inverted repetition form hairpin loops or palindromes*

that the molecule can open up into a cruciform structure as first suggested by Gierer [104]. The majority of those present in eukaryotic DNA are 300–1200 nucleotides long and comprise up to 30 per cent of the genome in some species. Prokaryotic DNA's have neither this quantity nor length of palindromic regions but many areas of the bacterial genome which have been sequenced because of their functional interest, such as repressor binding sites [79, 105] and RNA polymerase binding sites [80], show extensive two-fold symmetry, and it is likely that the interaction of the protein with the DNA involves the stabilization of some cruciform type of structure (Fig. 5.14).

The DNA in the eukaryotic chromosome is extensively folded. Various methods of packaging the stiff double helix, including supercoiling [106] and kinking [107], have been suggested to account for this.

REFERENCES

[1] Bendich, A. (1957) *Methods in Enzymology*, Vol. III, Sect. V, p. 715 (S. P. Colowick and N. O. Kaplan, Eds.)
[2] Sober, H. A. (Ed.) (1968) *Handbook of Biochemistry*, pp. H-11, H-30
[3] Chargaff, E. (1963) *Essays on Nucleic Acids*. Amsterdam: Elsevier
[4] Doty, P. (1961) *Harvey Lectures*, **55**, 103
[5] Mandel, M. and Marmur, J. (1968) *Methods in Enzymology*, Vol. 12, Part B, p. 195 (L. Grossman and E. Moldave, Eds.). New York: Academic Press
[6] Felsenfeld, G. (1968) *Methods in Enzymology*, Vol. 12, Part B, p. 247 (L. Grossman and K. Moldave, Eds.). New York: Academic Press
[7] Marmur, J. (1963) *Ann. Rev. Microbiol.*, **17**, 329
[8] Sueoka, N. (1964) *The Bacteria*, p. 419 (I. C. Gunsalus and R. Y. Stanier, Eds.)
[9] Schildkraut, C. L., Mandel, M., Levisohn, S., Smith-Sonneborn, J. E. and Marmur, J. (1962) *Nature*, **196**, 795
[10] Freese, E. (1962) *J. Theor. Biol.*, **3**, 82
[11] Schildkraut, C. L., Marmur, J. and Doty, P. (1962) *J. Mol. Biol.*, **4**, 430
[12] Kit, S. (1963) *Ann. Rev. Biochem.*, **32**, 43
[13] Laskowski, M. (1972) *Progr. Nucleic Acid Res. Mol. Biol.*, Vol. 12, p. 161 (J. N. Davidson and W. E. Cohn, Eds.). New York: Academic Press
[14] Harpst, J. A., Krasna, A. I. and Zimm, B. H. (1968) *Biopolymers*, **6**, 595
[15] Krasna, A. I., Dawson, J. R. and Harpst, J. A. (1970) *Biopolymers*, **9**, 1017
[16] Krasna, A. I. (1970) *Biopolymers*, **9**, 1029
[17] Schmidt, V. W. and Hearst, J. E. (1969) *J. Mol. Biol.*, **44**, 143
[18] Kavenoff, R. and Zimm, B. H. (1963) *Chromosoma*, **41**, 1
[19] Crothers, D. M. and Zimm, B. H. (1965) *J. Mol. Biol.*, **12**, 525
[20] Lang, D. (1970) *J. Mol. Biol.*, **54**, 557
[21] Leighton, S. B. and Rubenstein, I. (1969) *J. Mol. Biol.*, **46**, 313
[22] Freifelder, D. (1970) *J. Mol. Biol.*, **54**, 567
[23] Dubin, S. B., Benedek, G. B., Bancroft, F. C. and Freifelder, D. (1970) *J. Mol. Biol.*, **54**, 547
[24] Astbury, W. T. (1947) *Symp. Soc. Exp. Biol.*, **1**, 66
[25] Franklin, R. and Gosling, R. G. (1953) *Nature*, **171**, 740; **172**, 156
[26] Langridge, R., Wilson, H. R., Hooper, C. W., Wilkins, M. H. F. and Hamilton, L. D. (1960) *J. Mol. Biol.*, **3**, 547
[27] Fuller, W., Wilkins, M. H. F., Wilson, H. R. and Hamilton, L. D. (1965) *J. Mol. Biol.*, **12**, 60
[28] Marvin, D. A., Spencer, M., Wilkins, M. H. F. and Hamilton, L. D. (1961) *J. Mol. Biol.*, **3**, 547
[29] Davies, D. R. (1967) *Ann. Rev. Biochem.*, **36**, 321
[30] Watson, J. D. and Crick, F. H. C. (1953) *Nature*, **171**, 737, 964
[31] Watson, J. D. (1968) *The Double Helix*. New York: Atheneum
[32] Olby, R. (1964) *The Path to the Double Helix*. Macmillan
[33] Arnott, S. (1970) *Science*, **167**, 1694
[34] Tunis-Schneider, M. J. B. and Maestre, M. F. (1970) *J. Mol. Biol.*, **52**, 521
[35] Bram, S. (1971) *J. Mol. Biol.*, **58**, 277
[36] Bram, S. (1973) *Cold Spring Harbor Symp. Quant. Biol.*, **38**, 83
[37] Sinonaglu, O. (1968) *Molecular Associations in Biology*, p. 427 (B. Pullman, Ed.). New York: Academic Press
[38] Tunis, M. J. B. and Hearst, J. E. (1958) *Biopolymers*, **6**, 128

[39] Brahms, J. and Mommaerts, W. H. F. M. (1964) *J. Mol. Biol.*, **10**, 73
[40] Shih, T. Y. and Fasman, G. D. (1971) *Biochem.*, **10**, 1675
[41] Green, G. and Mahler, H. R. (1971) *Biochem.*, **10**, 2200
[42] Fasman, G. D., Schaffrausen, B., Goldsmith, L. and Adler, A. (1970) *Biochem.*, **9**, 2814
[43] Yang, J. T. and Samejima, T. (1969) *Progr. Nucleic Acid Res. Mol. Biol.*, Vol. 9, p. 223 (J. N. Davidson and W. E. Cohn, Eds.). New York: Academic Press
[44] Feughelman, M., Langridge, R., Seeds, W. E., Stokes, A. R., Wilson, H. R., Hooper, C. W., Wilkins, M. H. F., Barclay, R. K. and Hamilton, L. D. (1955) *Nature*, **175**, 834
[45] McConnell, B. and Von Hippel, P. H. (1970) *J. Mol. Biol.*, **50**, 297
[46] Hanson, C. V. (1971) *J. Mol. Biol.*, **58**, 847
[47] Wahl, P., Paoletti, J. and Le Pecq, J.-B. (1970) *Proc. Nat. Acad. Sci.*, **65**, 417
[48] Nass, M. M. K. (1969) *Science*, **165**, 25
[49] Kirk, J. T. O. (1971) *Autonomy and Biogenesis of Mitochondria and Chloroplasts*, p. 267. C.S.I.R.O.
[50] Cohen, G. and Eisenberg, H. (1966) *Biopolymers*, **4**, 429
[51] Kratky, O. and Porod, G. (1949) *Rec. Trav. Chim.*, **68**, 1106
[52] Guntaka, R. V., Mahy, B. W. J., Bishop, J. M. and Varmus, H. E. (1975) *Nature*, **253**, 507
[53] Worcel, A. and Burgi, E. (1972) *J. Mol. Biol.*, **71**, 127
[54] Pettijon, D. E. and Hecht, R. (1973) *Cold Spring Harbor Symp. Quant. Biol.*, **38**, 31
[55] Helinski, D. R. and Clewell, D. B. (1971) *Ann. Rev. Biochem.*, **40**, 899
[56] Vinograd, J., Lebowitz, J. and Watson, R. (1968) *J. Mol. Biol.*, **33**, 173
[57] Jansz, H. S., Pouwels, P. H. and Schiphorst, J. (1967) *Biochim. Biophys. Acta*, **123**, 626
[58] Campbell, A. M. and Lochhead, D. S. (1971) *Biochem. J.*, **123**, 661
[59] Beard, P., Morrow, J. F. and Berg, P. (1973) *J. Virol.*, **12**, 1303
[60] Monjardino, J. and James, A. W. (1975) *Nature*, **225**, 249
[61] Revet, B. M. J., Schmir, M. and Vinograd, J. (1971) *Nature New Biol.*, **229**, 10
[62] Campbell, A. M. and Jolly, D. J. (1973) *Biochem. J.*, **133**, 209
[63] Wang, J. C. (1974) *J. Mol. Biol.*, **89**, 783
[64] Pulleyblank, D. E. and Morgan, A. R. (1975) *J. Mol. Biol.*, **91**, 1
[65] Wang, J. C. (1968) *J. Mol. Biol.*, **43**, 25
[66] Douthart, R. J. and Bloomfield, V. A. (1968) *Biopolymers*, **6**, 1297
[67] Rush, M. G., Eason, R. and Vinograd, J. (1971) *Biochim. Biophys. Acta*, **228**, 585
[68] Rush, M. G., Kleinschmidt, A. K., Hellman, W. and Warner, R. C. (1967) *Proc. Nat. Acad. Sci.*, **58**, 1676
[69] Hudson, B. and Vinograd, J. (1969) *Nature*, **221**, 332
[70] Wang, J. C. (1970) *Biopolymers*, **9**, 489
[71] Bourgaux, P. (1973) *J. Mol. Biol.*, **77**, 197
[72] Sinsheimer, R. L. (1970) *Harvey Lectures*, **64**, 69
[73] Britten, R. J. and Kohne, D. E. (1968) *Science*, **161**, 529
[74] Britten, R. J., Graham, D. E. and Neufeld, B. R. (1974) *Methods in Enzymology*, Vol. XXIX, Part E, p. 363 (L. Grossman and K. Moldave, Eds.). New York: Academic Press
[75] Walker, P. M. B. (1971) *Progr. Biophys. Mol. Biol.*, **23**, 145

[76] Leder, P., Ross, J., Gielen, J., Packman, S., Ikawa, Y., Aviv, H. and Swan, D. (1973) *Cold Spring Harbor Symp. Quant. Biol.*, **38,** 753
[77] Danna, K. and Nathans, D. (1971) *Proc. Nat. Acad. Sci.*, **68,** 2913
[78] Danna, K., Sack, G. and Nathans, D. (1973) *J. Mol. Biol.*, **78,** 363
[79] Gilbert, W. and Maxam, A. (1973) *Proc. Nat. Acad. Sci.*, **70,** 3581
[80] Sugimoto, K., Okamoto, T., Sugisaki, H. and Takanami, M. (1975) *Nature* **253,** 410
[81] Robertson, H. D. (1975) *J. Mol. Biol.*, **92,** 363
[82] Ziff, E. B., Sedat, J. W. and Galibert, F. (1973) *Nature New Biol.*, **241,** 34
[83] Barrell, B. G., Weith, H. L., Donelson, J. E. and Robertson, H. D. (1975) *J. Mol. Biol.*, **92,** 377
[84] Ling, V. (1972) *J. Mol. Biol.*, **64,** 87
[85] Jay, E., Bambara, R., Padmanabhan, R. and Wu, R. (1974) *Nucleic Acids Res.*, **1** 331
[86] Wu, R. and Taylor, E. (1971) *J. Mol. Biol.*, **57,** 491
[87] Hedgpeth, J., Goodman, H. M. and Boyer, H. W. (1973) *Proc. Nat. Acad. Sci.*, **69,** 3448
[88] Sanger, F., Donelson, J. E., Coulson, A. R., Kossel, H. and Fischer, D. (1974) *J. Mol. Biol.*, **90,** 315
[89] Salser, W. A. (1974) *Ann. Rev. Biochem.*, **43,** 923
[90] Murray, K. and Old, R. W. (1974) *Progr. Nucleic Acid Res. Mol. Biol.*, Vol. 14 (W. E. Cohn, Ed.). New York: Academic Press
[91] Murray, K. (1974) MTP International Review of Science, Vol. 6, *Biochemistry of Nucleic Acids,* p. 1. London: Butterworths
[92] Jones, A. S. and Walker, R. T. (1964) *Nature,* **202,** 24, 1108
[93] Türler, H. and Chargaff, E. (1969) *Biochim. Biophys. Acta,* **195,** 446
[94] Subak-Sharpe, J. H., Bürk, R. R., Crawford, L. V., Morrison, J. M., Hay, J. and Keir, H. M. (1966) *Cold Spring Harbor Symp. Quant. Biol.*, **31,** 737
[95] Southern, E. M. (1970) *Nature,* **227,** 794
[96] Gall, J. G., Cohen, E. H. and Atherton, D. D. (1973) *Cold Spring Harbor Symp. Quant. Biol.*, **38,** 417
[97] Hearst, J. E., Cech, T. R., Marx, K. A., Rosenfeld, A. and Aleen, J. R. (1973) *Cold Spring Harbor Symp. Quant. Biol.*, **38,** 329
[98] Crippa, M. and Ticchini-Valentini, G. P. (1971) *Proc. Nat. Acad. Sci.*, **68,** 2769
[99] Suzuki, Y. L. P., Gage, L. P. and Brown, D. D. (1972) *J. Mol. Biol.*, **70,** 637
[100] Davidson, E. H., Galau, G. A., Angerer, R. C. and Britten, R. J. *Chromosoma,* **51,** 253
[101] Manning, J. E., Schmid, C. W. and Davidson, N. (1975) *Cell,* **4,** 141
[102] Lee, C. S. and Thomas, C. A. (1973) *J. Mol. Biol.*, **77,** 25
[103] Wilson, D. A. and Thomas, C. A. (1973) *J. Mol. Biol.*, **84,** 115
[104] Gierer, A. (1966) *Nature,* **212,** 1480
[105] Maniatis, T., Ptashne, M., Barrell, B. G. and Donelson, J. (1974) *Nature,* **250,** 394
[106] Pardon, J. F., Richards, B. M., Skinner, L. G. and Ockey, C. H. (1973) *J. Mol. Biol.*, **76,** 267
[107] Crick, F. H. C. and Klug, A. (1975) *Nature,* **255,** 530

CHAPTER 6

The Structure of RNA

6.1 The molar proportions of bases

The various types of RNA found in living cells have already been discussed in Chapter 2. All forms of RNA – rRNA, tRNA, mRNA, and 5S RNA – have structural features in common which will be discussed in this chapter.

It used to be thought that the nucleic acid molecule contained equimolar amounts of the four constituent bases which were linked together as nucleotides to form a *tetra*-nucleotide structure; it was not until about 1950 that the tetranucleotide hypothesis was finally abandoned as the result of accurate chromatographic analysis by Chargaff and his colleagues, by Markham, Smith, Wyatt, and others who showed clearly that nucleic acids vary widely in the molar proportions of bases according to the material of origin [1].

The relative molar proportions of bases in different RNA's from several sources are shown in Table 6.1. The figures vary from one RNA to another, and although the striking complementarity shown by the bases of DNA is not evident, Elson and Chargaff [2] were the first to point out that the number of nucleotides carrying an amino group in the 6-position (in adenine and cytosine) is, in general, roughly equal to the number having a 6-oxo (6-keto) group (in guanine and uracil).

Pseudouridine (ψ) and the methylated bases (p. 43) are amongst the so-called 'minor' or 'modified' bases of RNA [3], found to be particularly abundant in the tRNA fraction (Tables 6.2 and 6.3). The relative base compositions of different RNA fractions from several types of cell are also illustrated in Table 6.1. Most ribosomal RNA's (rRNA's) have relatively high levels of guanosine plus cytosine (usually 50–60 per cent) regardless of the base composition of the cellular DNA (see preceding chapter). On the other hand, the messenger RNA class as a whole appears to have a guanine plus cytosine composition more closely corresponding to that of the cellular DNA. In some RNA's such as those in reovirus and wound

THE STRUCTURE OF RNA

TABLE 6.1

Molar proportions of bases (as moles per 100 moles nucleotide) in RNA's from various sources [17, 76, 78–87]

Source	Type	Adenine	Guanine	Cytosine	Uracil
Human (HeLa)	18S rRNA	21·0	29·6	27·5	21·7
	28S rRNA	16·2	35·6	31·7	16·5
	5S rRNA	18·3	30·3	29·0	22·3
	mRNA (bulk)	32·5	21·0	21·6	24·8
Rat liver	18S rRNA	20·1	31·2	28·1	22·5
	28S rRNA	15·5	36·3	30·3	17·9
	5S rRNA	18·2	33·9	25·7	22·2
Rabbit reticulocytes	mRNA (globin)	21·2	20·7	34·5	23·7
Esch. coli	16S rRNA	24·8	31·0	22·7	21·5
	23S rRNA	25·4	33·5	21·5	19·6
	mRNA (bulk)	25·1	27·1	24·1	23·7
P. aeruginosa	16S rRNA	25·1	32·8	21·6	20·5
	23S rRNA	26·3	31·2	21·2	21·3
B. subtilis	16S rRNA	26·5	29·6	22·3	21·6
	23S rRNA	26·5	32·0	22·5	19·3

tumour virus (and also in tRNA) there is close equivalence between A and U and between C and G (Table 6.4). The importance of this in relation to base-pairing is discussed on page 127.

6.2 The primary structure of RNA

This subject has been reviewed in detail by Markham [6] and others.

The most important preliminary consideration is the nature of the internucleotide link. Alkaline fission of RNA results in neutraliza-

TABLE 6.2

Relative proportions of additional components in rat liver RNA as moles/100 moles uridine [5]

	Ribosomal RNA	Transfer RNA
Pseudouridine	7·5	25
5-Methylcytosine	0·4	10
6-Methylaminopurine	0·5	8·1
6-Dimethylaminopurine	0·1	0·1
1-Methylguanine	0·1	3·3
2-Methylamino-6-hydroxypurine	0·1	2·3
2-Dimethylamino-6-hydroxypurine	0·1	3·0

tion of alkali; it is clear therefore that some or all of the phosphoric acid groups are involved in the internucleotide linkages. Since the intact nucleic acid may be deaminated by nitrous acid, the amino groups do not take part in the linkages; neither do the oxo groups

TABLE 6.3
Molar proportions of bases in the RNA's of various transfer RNA's

Source	Adenine	Guanine	Cytosine	Uracil	Uracil as ψ	Methyl bases	Ref.
Yeast tRNA	19.4	26.6	25.1	20.1	4.6	3.1	[4]
Rabbit liver tRNA	16.6	31.1	27.8	15.9	4.3	3.5	[4]
Esch. coli tRNA	18.3	30.3	30.3	15.9	2.4	2.2	[4]

of guanine or uracil, for electrometric titration reveals that they are unsubstituted. The early work of Gulland and Jackson suggested the involvement of C-5' and this was confirmed by Cohn and Volkin [7], who treated RNA with phosphodiesterase from snake venom and obtained a mixture of 5'-phosphates of all four nucleosides, which were separated chromatographically.

On the other hand, digestion with alkali yields a mixture of

TABLE 6.4
Molar proportions of bases in the RNA's of cetain viruses

Type of virus	Adenine	Guanine	Cytosine	Uracil	Ref.
Tobacco mosaic	29.8	25.4	18.5	26.3	[68]
Turnip yellow mosaic	22.6	17.2	38.0	22.2	[69]
Poliomyelitis	28.6	24.0	22.0	25.4	[70]
Mouse encephalomyocarditis	27.3	23.5	23.2	25.9	[71]
Reovirus type 3	28.0	22.3	22.0	27.9	[61]
Wound tumour	31.1	18.6	19.1	31.3	[61]

nucleoside 2'- and 3'-phosphates as was shown by Cohn and Carter in 1949 [8–10]. Consequently, it would appear that the main internucleotide linkages are phosphodiester groups connecting C-5' in one nucleotide with C-2' or C-3' in the next nucleotide. A dinucleotide containing such a (C-3')-O-P-O-(C-5') linkage is shown in Fig. 6.1.

The possibility that the linkage is (C-2')-(C-5') rather than

THE STRUCTURE OF RNA

(C-3')-(C-5') is excluded by the observation that hydrolysis with diesterase from spleen yields nucleoside 3'-phosphates [11]. The enzymic digestion of RNA is discussed in greater detail in Chapter 8.

Part of a polynucleotide chain built up as described above is shown in Fig. 6.2. It is *theoretically* possible that such a structure could carry side chains attached either at C-2' or at triply esterified phosphate groups, but this has not been observed.

The molecular weights of RNA may be determined by light-scattering measurements, by sedimentation velocity measurements,

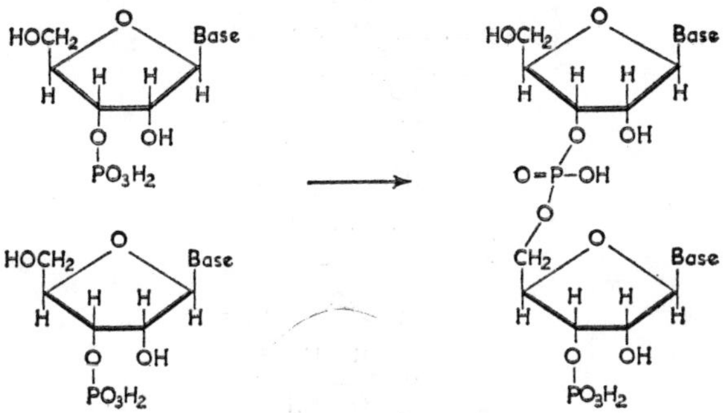

Fig. 6.1

by viscosity measurements, by equilibrium ultracentrifugation [14], or by polyacrylamide gel electrophoresis in formamide [15].

Bacteria, *actinomycetes*, blue-green algae, and higher plant chloroplasts all have rRNA's of molecular weight $1 \cdot 1 \times 10^6$ (23S) and $0 \cdot 55 \times 10^6$ (16S) whereas the corresponding values for higher plants, ferns, algae, fungi, and some protozoa are $1 \cdot 3 \times 10^6$ (26S) and $0 \cdot 7 \times 10^6$ (18S) [16]. The 18S component is common to all animals, but the large 28S component has evolved from $1 \cdot 4 \times 10^6$ daltons in sea urchins to $1 \cdot 75 \times 10^6$ daltons in mammals [16]. (In mitochondria, e.g. mammalian, the rRNA's are somewhat smaller; the 21S species is $0 \cdot 56 \times 10^6$ daltons and the 12S species $0 \cdot 36 \times 10^6$ daltons [73].) Whilst the ribosomal RNA's from most organisms possess broadly similar base compositions with G plus C contents between 50 and 60 per cent, there are some exceptions to this, e.g. *Drosophila* (40 per cent) and *Tetrahymena* (43 per cent) [17].

The RNA's of certain viruses show special features which will be discussed in Chapter 8.

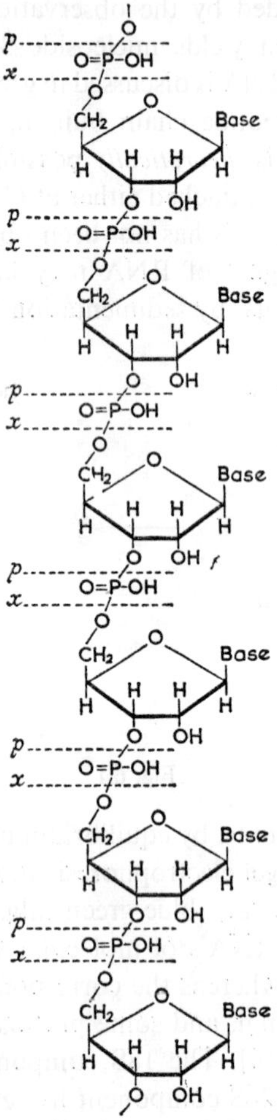

Fig. 6.2 *A section of the polynucleotide chain in the RNA molecule. Hydrolysis with alkali or with spleen phosphodiesterase breaks the linkages at x; hydrolysis with snake venom phosphodiesterase breaks the linkages at p*

6.3 Alkaline degradation

The mechanism of alkaline digestion of RNA has been explained by Todd and his collaborators [18, 19] on the basis that alkaline hydrolysis yields two isomeric forms of each nucleotide, originally termed the *a* and *b* nucleotides upon their discovery by Cohn and Carter [9, 10] but subsequently proved to be the nucleoside 2'- and 3'-phosphates respectively [9, 20–23]. These isomers are readily con-

THE STRUCTURE OF RNA

verted into a mixture of both under acid conditions but are stable without interconversion in alkaline solution. Interconversion involves the formation of a cyclic intermediate, the nucleoside 2':3'-phosphate (II in Fig. 6.3), which yields on hydrolysis a mixture of the 2'- and 3'-phosphates. For example, the tri-nucleotide I shown in Fig. 6.3 yields the cyclic nucleoside 2':3'-phosphates (II) which

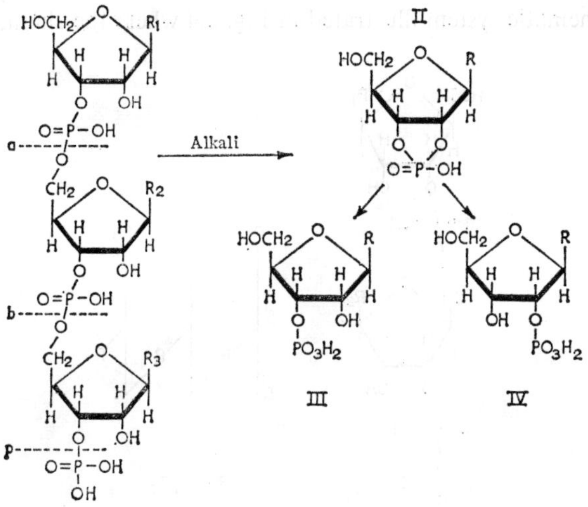

Fig. 6.3 *Hydrolysis of a trinucleotide by alkali. R represents a purine or pyrimidine base*

then give rise to a mixture of nucleoside 2'- and 3'-phosphates (III and IV).

In a similar manner, treatment with alkali of the polynucleotide shown in Fig. 6.2 would bring about hydrolysis to a mixture of nucleoside 2'- and 3'-phosphates by fission at x after cyclization as described above. Hydrolysis by pancreatic ribonuclease also proceeds via the formation of cyclic phosphates (Chapter 8). On the other hand, treatment with venom phosphodiesterase would yield nucleoside 5'-phosphates by rupture at p of the (C-3')-O-P linkages (see Chapter 8).

2'-O-Methylribose (p. 43) is found in rRNA and tRNA [24]. The internucleotide bond adjacent to a ribose residue methylated at the 2-position is of course resistant to hydrolysis by alkali and by pancreatic ribonuclease since cyclization between the 2'- and 3'-positions is not possible. Moreover, pancreatic RNase cannot degrade such internucleotide bonds which would otherwise be susceptible to its nucleolytic action.

6.4 Shorthand notation

The representation of polynucleotide chains by formulae such as are shown in Fig. 6.2 is clumsy and it has become customary to use the schematic system illustrated in Fig. 6.4 where the chain, shown

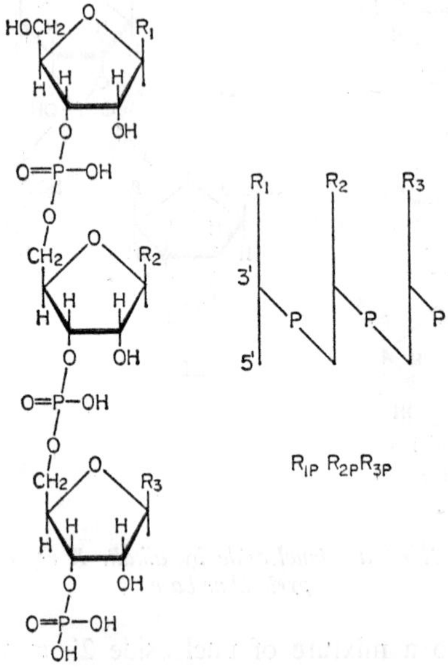

Fig. 6.4 *A trinucleotide containing the bases R_1, R_2, and R_3 shown in full on the left and in short-hand forms on the right*

in full on the left, is abbreviated as on the right. The vertical line denotes the carbon chain of the sugar with the base attached at C-1'. The diagonal line from the middle of the vertical line indicates the phosphate link at C-3', while that at the end of the vertical line remote from the base denotes the phosphate link at C-5'. This system may be used for either RNA or DNA.

To simplify further the representation of specific polynucleotides in shorthand notation, the following system originally suggested by Heppel, Ortiz, and Ochoa [25] and now embodied in the Rules of the CBN (see Preface) is commonly employed. A phosphate group

THE STRUCTURE OF RNA

is denoted by p; when placed to the right of the nucleoside symbol, the phosphate is esterified at C-3' of the ribose moiety; when placed to the left of the nucleoside symbol, the phosphate is esterified at C-5' of the ribose moiety. Thus, UpUp or U-Up is a dinucleotide with one phosphate monoesterified at C-3' of a uridine residue and a phosphodiester bond between C-5' of that same uridine residue and C-3' of the other uridine group. UpU or U-U would be the dinucleoside monophosphate, uridylyl(3'→5')uridine. The letter p *between* nucleoside residues may be replaced by a hyphen.

The following examples illustrate the method:

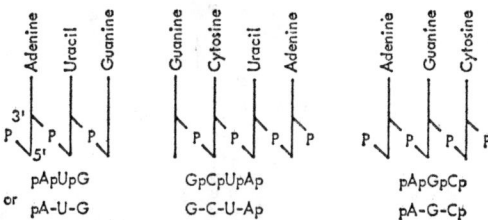

The letters A, G, C, U, and T represent adenosine, guanosine, cytidine, uridine, and ribothymidine respectively. The prefix d (e.g. dA) may be used to indicate a deoxyribonucleoside.

There is now agreement on nomenclature for cyclic-terminal nucleotides. They may be represented by using the symbol -cyclic-p to indicate a 2':3'-phosphoryl group or by means of the symbol >p. Thus, U-cyclic-p or U>p is uridine 2':3'-phosphate and UpU-cyclic-p or UpU>p is the cyclic-terminal dinucleotide.

6.5 The determination of nucleotide sequences

One of the most important problems in nucleic acid chemistry is the determination of the primary structure of RNA's, that is the order in which the nucleotide units are arranged along the polynucleotide chain. Spectacular progress recently made in this field has resulted in the determination of nucleotide sequences of some of the transfer RNA's and 5S RNA's which have the double advantage of consisting of relatively short polynucleotide chains and of lending themselves to relatively easy isolation in the pure state as a monomolecular species.

The methods employed in sequence determinations consist essentially in controlled degradation of the RNA with enzymes and separation of the products by chromatography or electrophoresis

[26–34]. By examining the products of hydrolysis at various stages and by using appropriate combinations of enzymes, it is possible to determine the composition of each fragment and to work out how the fragments may be pieced together to establish the sequence in the complete RNA chain [35].

In 1965 Holley and his colleagues [27, 36] worked out the complete sequence of nucleotides in the alanine transfer RNA (tRNA^{Ala}) of yeast (Fig. 6.5). This work, which took several years to complete, was recognized by the award of a Nobel Prize three years later. This was followed shortly by the primary structures for

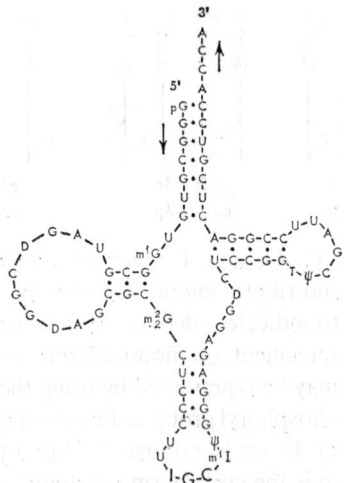

Fig. 6.5 *The structure of* tRNA^{Ala} *from yeast* [27]. *The anticodon is shown in heavy type. For an explanation of the symbols denoting the minor bases, the list of abbreviations in the Preface should be consulted*

years tRNA^{Ser} which exists in two forms (1 and 2) differing only in three nucleotides (Fig. 6.6) [37].

Initially the oligonucleotides that were liberated by enzymic digestion were normally detected by their absorption of ultraviolet light. This limited both the scale and the rate at which sequence determination could progress. To reduce the scale, Sanger and his colleagues [31] prepared ^{32}P-labelled RNA's by *in vivo* labelling and studied enzymic digests of these, detecting and estimating the oligonucleotides after two-dimensional electrophoretic separation by autoradiographic and counting techniques. This elegant approach is now known as 'fingerprinting' (see also Chapter 4,

THE STRUCTURE OF RNA

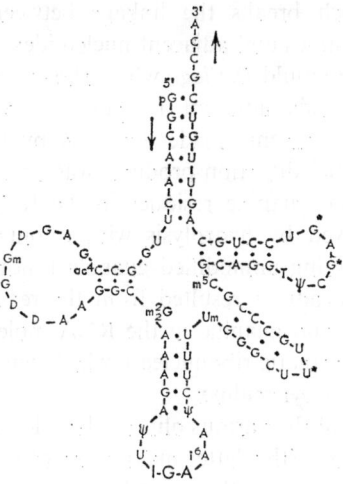

Fig. 6.6 *The structure of tRNA$_2^{Ser}$ from yeast* [37]. *The anticodon is shown in heavy type. In the isoacceptor tRNA$_1^{Ser}$ the three nucleotides indicated with asterisks are different, U being replaced by C and G by A*

Fig. 4.8) and not only is it suitable for very small amounts (0·5 mg) of ^{32}P-labelled RNA but it is also extremely rapid. Historically it was first used to determine the primary structure of the small, 120 nucleotide long, 5S ribosomal RNA from *Esch. coli* rather than a tRNA [31, 32] (Fig. 6.7).

The determination of the complete primary structure of the 5S RNA molecule involved the study of the products of hydrolysis, either partial or complete, by a number of enzymes (see also Chapter 8). These were ribonuclease T$_1$ from takadiastase which specifically breaks the internucleotide bonds between guanosine 3′-monophosphate and the 5′-hydroxyl groups of adjacent nucleotides, pancreatic

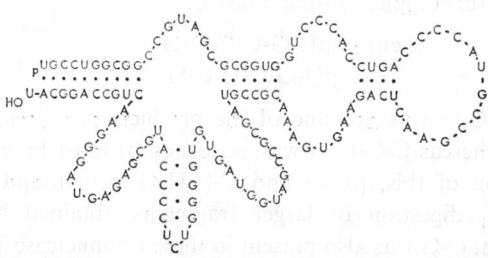

Fig. 6.7 *A possible structure of* Esch. coli *5S rRNA* [32].

ribonuclease which breaks the linkage between 3′-phosphorylpyrimidine nucleotides and adjacent nucleotides, and ribonuclease U_2 from the slime mould *Ustilago* which shows a specificity for the internucleotide bonds adjacent to purine nucleotides [34]. In addition to the fragments made available by these methods, a further set of partial digestion products was obtained after partial methylation of the guanine residues in the RNA with dimethyl sulphoxide followed by hydrolysis with ribonuclease T_1 (acting only at the remaining unmodified guanine residues). Yet another set of digestion products resulted from the reaction of a carboimide with uracil and guanine in the RNA molecule followed by digestion with pancreatic ribonuclease which acts only on linkages next to unmodified pyrimidines.

The separation of the various oligonucleotides by electrophoresis and chromatography (this latter making use of nucleotides in solution as displacing ions, in the procedure known as homochromatography) (see p. 75) was followed by identification of the components mainly by electrophoretic techniques. The determination of many partial sequences made it possible to deduce the complete sequence. We here describe the evidence for the elucidation of the structure of the 5′-terminal sequence.

One of the end products of digestion by ribonuclease T_1 was:

pU-G

which was clearly the terminal dinucleotide.

An oligonucleotide product of ribonuclease T_1 digestion of the chemically methylated 5S RNA could be further hydrolysed with pancreatic ribonuclease to yield:

pU, C, U, methyl-G-C, G

On the basis of its origin and of the structure of the terminal dinucleotide, this oligonucleotide must be:

either pU-G-C-C-U-G (i)
or pU-G-C-U-C-G (ii)

Since C-U-C-G was not one of the products of ribonuclease T_1 digestion whereas C-C-U-G was, sequence (i) must be correct. In confirmation of this, pU-G and C-C-U-G were found on ribonuclease T_1 digestion of larger fragments obtained by partial degradation. C-G was also present in these ribonuclease T_1 digests, and treatment of this terminal region with pancreatic ribonuclease

THE STRUCTURE OF RNA

gave two molecules of G-G-C. In digests of the whole 5S molecule, only two G-G-C sequences were obtained; consequently both must lie within this terminal fragment.

Hydrolysis by pancreatic ribonuclease of carbo-imide treated RNA gave U-G-G-C and C-U-G-G-C, so allowing the addition of G-C-G-G-C to sequence (i) giving:

$$\text{pU-G-C-C-U-G-G-C-G-G-C} \qquad \text{(iii)}$$

which was confirmed by the presence of G-C-G-G in a partial digest by ribonuclease T_1, a sequence which must have been derived from G-G-C-G-G.

Partial digestion of 5S RNA with pancreatic ribonuclease gave (iii) with the addition of one C while digestion with ribonuclease T_1 gave an added C-C-G so that the sequence could be extended to:

$$\text{pU-G-C-C-U-G-G-C-G-G-C-C-G} \qquad \text{(iv)}$$

The discovery of a corresponding T_1 fragment containing not C-C-G but C-C-U-U-A-G indicated the presence of two distinct 5S species of partial sequence as follows:

$$\text{pU-G-C-C-U-G-G-C-G-G-C-C-G} \qquad \text{(iv)}$$
$$\text{pU-G-C-C-U-G-G-C-G-G-C-C-U-U-A-G} \qquad \text{(v)}$$

Pancreatic ribonuclease gave (iv) with an added U which by comparison with (v) would be expected to continue with A-G. This was shown to be true by the presence in the digest of the partially methylated 5S RNA of the sequence:

$$\text{C-C-G-U-A-G}$$

Thus the 16 residues at the 5'-terminal could be arranged in order. In a similar manner the structure of the remainder of the molecule was deduced (see Fig. 6.7).

It now turns out that the 5'-ends of all naturally occurring molecules are characterized by the presence of at least one phosphate group (they can have three) and the 3'-ends by the presence of free hydroxyl groups, i.e. they have the form $(pp)pN(pN)_n pN_{OH}$ where N is any nucleoside and n is a variable but large number. The 5'-terminal nucleotide of a radioactive RNA is released by alkaline hydrolysis in the form (pp)pNp whereas all the other internal nucleotides are liberated as Np. This terminal group can be separated from the four mononucleotides by two-dimensional electrophoresis [34]. A method used to determine the 3'-terminal sequence depends on the fact that in a complete ribonuclease T_1 digest of an RNA (see Chapter 8) the only product which is not susceptible to degradation with alkaline phosphatase is the oligonucleotide derived from the 3'-end. All other nucleotides have a 3'-phosphate and are therefore susceptible. Thus, if an enzyme digest of an RNA is fractionated by electrophoresis in one dimension followed by treatment with alkaline phosphatase before electrophoresis in a second dimension at right-angles to the first dimension, the only product remaining on the 'diagonal' should be the 3'-terminal oligonucleotide [39].

Another method for the isolation and sequence determination of the 3'-terminal polynucleotides of isotopically labelled RNA molecules involves their selective isolation on columns of cellulose derivatives containing covalently bound dihydroxyboryl groups [88, 89].

An alternative to labelling RNA's *in vivo* by growth in media containing high levels of ^{32}P has recently been developed. A ribonuclease digest of the unlabelled RNA in question is prepared. The oligonucleotides produced are subsequently labelled at their free 5'-OH groups with a ^{32}P-ester group, by using polynucleotide kinase (induced in *Esch. coli* by T4 bacteriophage) and γ-^{32}P-ATP as the ^{32}P-phosphate donor (see also Chapter 8). The resulting 5'-terminally ^{32}P-labelled oligonucleotides can then be fractionated by the standard 'fingerprinting' methods developed for studying uniformly ^{32}P-labelled oligonucleotides [40]. This approach has been useful in preparing fingerprints of unlabelled mammalian messenger RNA's [41].

Since Sanger and his colleagues [38] elucidated the structure of *Esch. coli* 5S RNA (Fig. 6.7), nine other 5S rRNA's of both eukaryotic and prokaryotic origin have been sequenced [42]. From an evolutionary point of view it appears that the structure has been fairly rigorously conserved since all are 120 nucleotides long. Nevertheless there are some slight variations in the sequences observed [42].

The 'fingerprinting' technique has also been applied to many tRNA's. In fact, the primary sequences of around forty are now known [43]. $tRNA_f^{Met}$ from *Esch. coli*, which acts as a chain initiator in protein synthesis (p. 379), has been shown to have a different sequence [44] from that of $tRNA^{Met}$ which functions in a different way (p. 379). $tRNA^{Tyr}$ of *Esch. coli* differs from su_3^{+Tyr};

mutation to amber suppression results in a single base change in the anticodon (see page 372) from GUA and CUA [45].

Attention has now turned to longer RNA molecules such as those found in ribosomes and in bacteriophages. So far only the regions adjacent to the 46 methylated nucleotides in the 18S and the 71 methylated nucleotides in the 28S rRNA's of mammalian cells have been sequenced [46]. However, in the case of the *Esch. coli* 16S rRNA, 95 per cent of the primary structure has been elucidated recently by Ebel and his collaborators [47]. Since 1520 nucleotides are involved, only a small portion of the sequence is shown in Fig. 6.10.

Progress with the primary structure of the RNA of a number of viruses, whilst initially focused on a number of specific regions, has nonetheless yielded considerable insight into the general sequence arrangement of these RNA's (e.g. from R17, Qβ, and MS2) in relation to their genetic function [48]. Perhaps the most spectacular progress has recently been made by Fiers and his colleagues with their elucidation of the sequence of MS2 RNA [49, 50]. Starting from the 5'-end this so far includes the leader sequence, the A-protein cistron, the first intercistronic region, the coat protein cistron, the second intercistronic region, and the beginning of the replicase cistron. Some of this [50] is shown in Fig. 6.11.

6.6 The secondary structure of RNA

While DNA (see previous Chapter) has a firmly established helical structure, the nature of the secondary and tertiary structure of RNA is less well defined but has already been partially established [14, 51].

In solutions of low ionic strength RNA molecules behave like typical highly swollen polyelectrolyte chains but an increase in the ionic strength causes the chains to contract upon themselves so as to display relatively low intrinsic viscosities and high sedimentation rates. This suggests the existence of base pairing in certain regions of the RNA chain such as is known to occur in some of the biosynthetic polyribonucleotides.

Under the influence of the enzyme polynucleotide phosphorylase which is discussed on page 349, biosynthetic polynucleotides may be obtained with some rather interesting features. If the substrate employed is adenosine diphosphate (ADP), the polymer formed is a polyribonucleotide containing adenine as its only base. It is usually referred to as poly(A). Poly(U) may be formed similarly from UDP as substrate, and if the substrate is an equimolar mixture of ADP and UDP the product is poly(A,U).

These biosynthetic polymers behave towards alkali and hydrolytic enzymes in the same way as do naturally occurring polyribonucleotides.

When equimolar amounts of poly(A) and poly(U) are mixed in dilute aqueous solution they form a complex known as poly(A) · poly(U) in which the adenine moieties of one strand are linked by hydrogen bonds to the uracils of the complementary strand [52, 53] (Fig. 6.8). The X-ray diffraction pattern of this complex indicates a

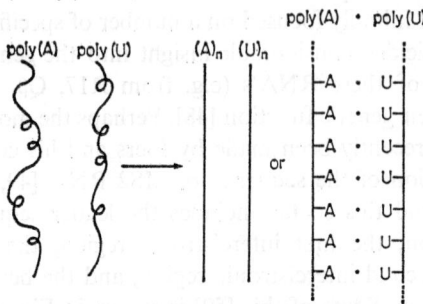

Fig. 6.8 *The association of single strands of poly(A) and poly(U) to form the poly(A) · poly(U) complex in which the two strands are linked by hydrogen bonding between A and U. The complex exists in three dimensions in the form of a helix*

double helical structure, as in DNA, with ten base pairs per turn of the helix, the pitch of which is 3·4 nm.

The formation of the helical complex is accompanied by a 34 per cent depression in the absorbance at 260 nm below the value for the sum of the two constituents (the *hypochromic effect*).

The helical complex behaves in many ways like DNA. For example, it shows the phenomenon of 'molecular melting' or 'helix–coil transition'. When it is heated in 0·15M-NaCl at neutral pH, the absorbance at 260 nm rises sharply (the *hyperchromic effect*) by 34 per cent at a temperature of about 60°, the so-called 'melting temperature' T_m. At the same time the specific optical rotation at 589 nm decreases markedly. These changes are due to the separation of the two component strands of the helix on heating; they are reversed when the helix is re-formed on cooling. This phenomenon is described in greater detail in Chapter 5 in relation to DNA.

THE STRUCTURE OF RNA

When solutions of naturally occurring ribosomal RNA or the RNA from certain viruses such as TMV (p. 139) are heated, similar but less pronounced changes occur. This initially suggested that the RNA chain is folded back on itself in a number of places so that pairs of bases may come together and become linked by hydrogen bonds, e.g. between adenine and uracil and between guanine and cytosine (Fig. 6.9). Since the segments of chain brought

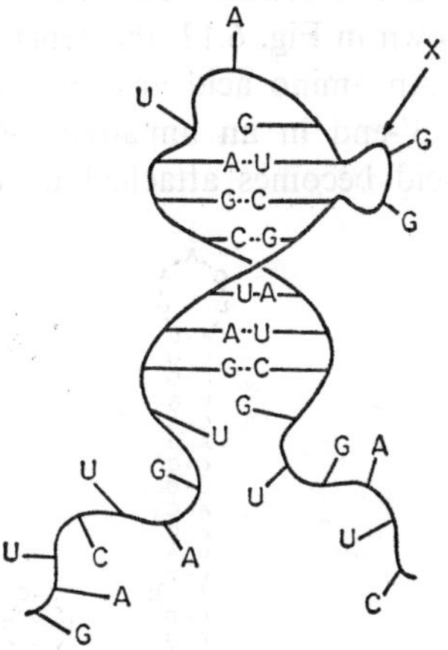

Fig. 6.9 *A possible secondary structure of RNA illustrating a helical region with complementary base-pairing* [54]. *A looped-out portion of the helix is shown at* X

into apposition in this way may not be exactly complementary, looping out of non-bonding residues at X as shown in Fig. 6.9 was suggested [54]. Such loops could be stabilized through 'base-stacking' interactions known to operate in single-stranded polynucleotides (e.g. polyA). The regions in which the RNA chain is folded back on itself were shown by X-ray analysis to be helical in structure (the bases are titled relative to the helix axis similar to the A rather than the B conformation in DNA as described in Chapter 5). The RNA molecule was therefore thought to consist of a polynucleotide chain containing short imperfect helical regions in which base pairing occurs, between 40 and 70 per cent of the nucleotides being involved in such helix formation [54–59]. Specific examples of possible helical regions in the RNA of bacteriophage MS2 and in *Esch. coli* 16S rRNA are shown in Figs. 6.11 and 6.10.

Early views on the occurrence of base-paired regions in tRNA were founded on several pieces of evidence: (i) the equivalence of

THE BIOCHEMISTRY OF THE NUCLEIC ACIDS

A and U and of C and G; (ii) the degree of hyperchromicity, the shape of the melting curve, and the unexpectedly high T_m which suggests a very stable structure with all the theoretically possible $A \cdot U$ and $G \cdot C$ base pairs in fact present; (iii) the behaviour of tRNA towards certain enzymes.

As mentioned earlier, the primary structures around forty tRNA are now known. Just about all these sequences can be fitted to the same hydrogen bonded secondary structure of loops and short helical region as shown in Fig. 6.12, the general features of which are as follows. (1) An amino acid arm, a helix of 7 base pairs, terminating at the 3'-end in an unpaired-C-C-A_{OH} sequence to which an amino acid becomes attached under the influence of

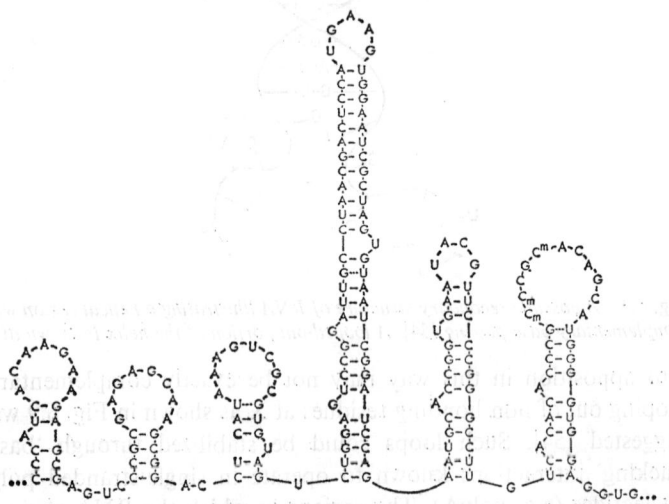

Fig. 6.10 *An example of the possible secondary structure of a small portion of the* Esch. coli *16S rRNA* [47] (*a letter m denotes a nucleotide residue which has been methylated*)

appropriate enzymes as described in Chapter 13. (2) A dihydrouracil loop (I) containing 8–11 nucleotides at the end of a helical stem of 3 or 4 base pairs. (3) An anticodon loop (II) containing 7 nucleotides at the end of a helical stem of 5 base pairs. The anticodon itself, a group of three nucleotides at the centre of the loop, is discussed in Chapter 13. (4) An 'extra arm' (III) which varies widely from species to species, containing 3–18 nucleotides. (5) A pseudouridine–ribothymidine loop (IV) containing 7 nucleotides,

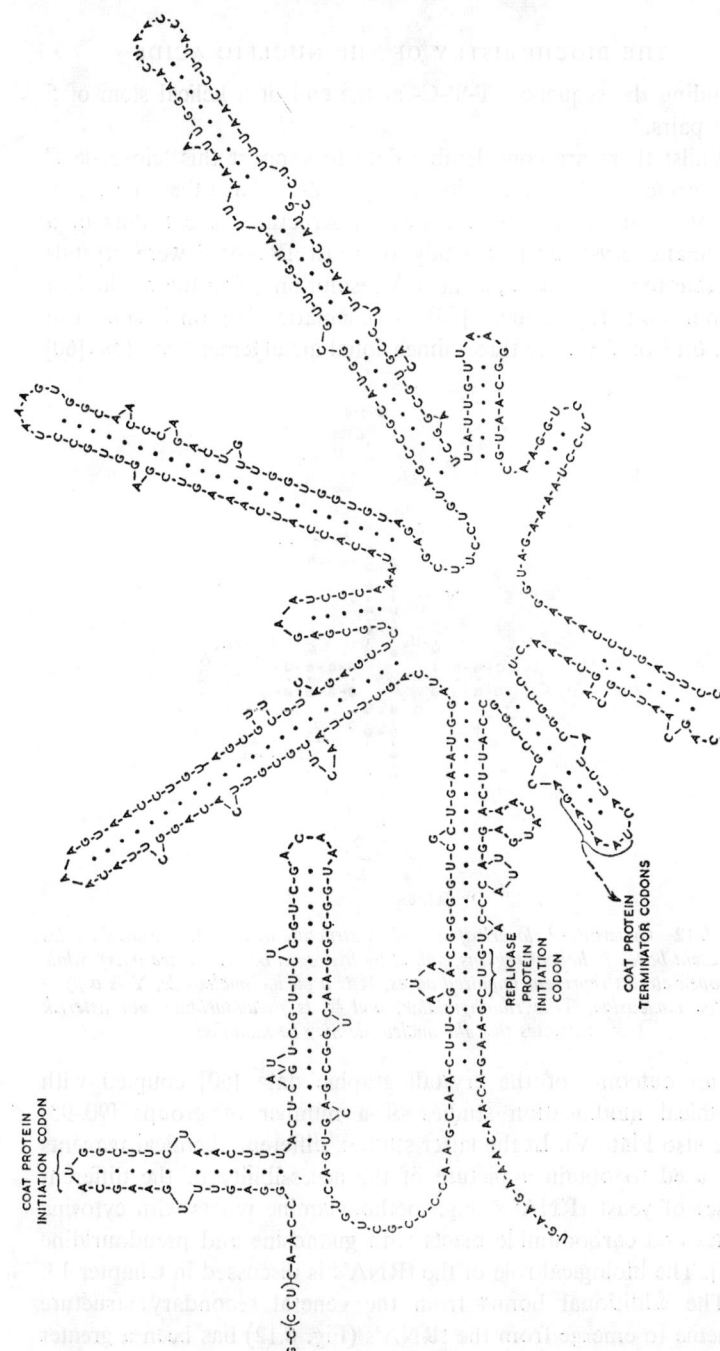

6.11 *The nucleotide sequence in MS2 RNA before the coat protein cistron (folded in the proposed 'flower' conformation), the intercistronic region preceding the replicase cistron, and the beginning of the replicase cistron* [50]

including the sequence -T-Ψ-C- at the end of a helical stem of 5 base pairs.

Whilst there are considerable data to support this 'clover-leaf' arrangement in two dimensions, only recently have there been any unambiguous data regarding tertiary structure. The results of a systematic crystallization study of yeast tRNA^{Phe} were crystals suitable for X-ray analysis at 3 Å resolution using the method of isomorphous replacement [60]. A schematic diagram is shown in Fig. 6.13 of the basic three-dimensional arrangement deduced [60]

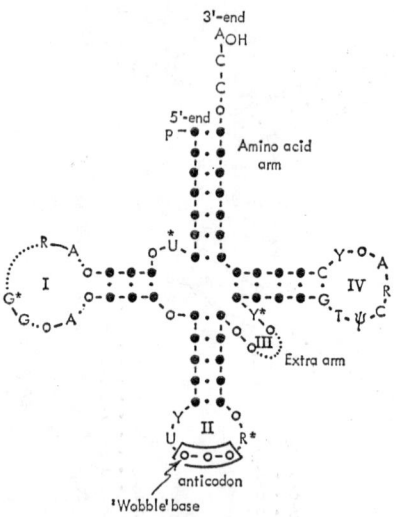

Fig. 6.12 *Generalized cloverleaf secondary structure of tRNA. The solid circles represent bases in helical regions, paired by hydrogen bonds (centre dots), while the open circles represent unpaired bases. R is a purine nucleoside, Y is a pyrimidine nucleoside, T is ribothymidine, and Ψ is pseudouridine. An asterisk indicates that the nucleoside may be modified*

as an outcome of the crystallographic data [60] coupled with chemical modification studies of a number of groups [90–93] (see also Plate V). In the latter studies, different chemical reagents are used to obtain a picture of the accessibility of the different bases of yeast tRNA^{Phe}, e.g. methoxylamine reacts with cytosine bases and carbodiimide reacts with guanosine and pseudouridine [92]. The biological role of the tRNA's is discussed in Chapter 13.

The additional bonus from the general secondary structure scheme to emerge from the tRNA's (Fig. 6.12) has been a greater appreciation of the 'rules' that appear to govern RNA secondary structure [74]. Only G · C, A · U, and *sometimes* G · U, pairs are allowed. All helices consist of at least three base pairs without G · U pairs, or four base pairs when G · U pairs are allowed. Looped out bases are non-existent, or very rare and 'hairpin' loops contain at least three nucleotides [74].

Numerous attempts have been made to deduce a secondary structure for 5S ribosomal RNA by applying these 'rules' and the diagonal procedure of Tinoco and his colleagues [74] and a possible secondary structure for this RNA can be deduced (Fig. 6.7) [42] which is consistent with phylogenetic [42] and physical data [75]. The same approach has led to the secondary structure proposals for the *Esch. coli* 16S rRNA and the MS2 RNA in Figs. 6.10 and 6.11. Messenger RNA's from both mammalian and

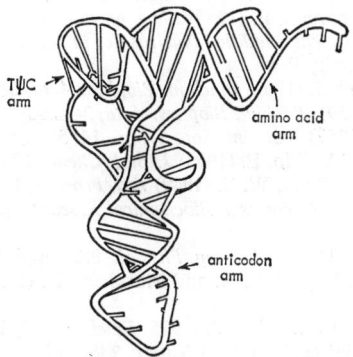

Fig. 6.13 *A schematic model of yeast tRNA^{Phe}; the ribose phosphate backbone is drawn as a continuous cylinder with bars to indicate hydrogen bonded base pairs* [60]

bacterial sources can also be shown to have considerable secondary structure [12, 13].

In some other RNA's the extent of helix formation appears to be complete. For example, double helical structure is found in the RNA of several viruses (Chapter 7) including reovirus and wound tumour virus, and for the replicative forms of polio virus [61–63], Sendai virus [64], and EMC virus [65], on the basis of complementary base ratios, thermal denaturation data, resistance to pancreatic ribonuclease, failure to react with formaldehyde, and X-ray analysis [66, 67].

The secondary structure features of RNA molecules such as

those just discussed can be abolished by treatment with formamide [15]. This not only destroys base-pairing but it also appears to eliminate nearest-neighbour 'base stacking' interactions in such single-stranded polymers as polyA. Another effective RNA denaturant is dimethyl sulphoxide [72]. This technique has also been used to decide whether large RNA molecules such as rRNA are covalently continuous. This is true for the *Esch. coli* ribosomal RNA's [76], and the same is true for mammalian rRNA's with one qualification. On treating 28S rRNA with such denaturants a small 5·8S fragment of RNA is released [77]. This fragment appears to be hydrogen bonded to the parent 28S molecule [77] (see p. 53).

REFERENCES

[1] Chargaff, E. (1963) *Essays on Nucleic Acids*. Amsterdam: Elsevier
[2] Elson, D. and Chargaff, E. (1955) *Biochim. Biophys. Acta,* **17,** 367
[3] Hall, R. H. (1970) *The Modified Nucleosides in Nucleic Acids*. New York: Columbia University Press
[4] Cantoni, G. L. *et al.* (1962) *Biochim. Biophys. Acta,* **61,** 354
[5] Dunn, D. B. (1959) *Biochim. Biophys. Acta,* **34,** 286
[6] Markham, R. (1957) *Biochem. Soc. Symp.,* **14,** 5
[7] Cohn, W. E. and Volkin, E. (1953) *J. Biol. Chem.,* **203,** 319
[8] Carter, C. E. and Cohn, W. E. (1949) *Fed. Proc.,* **8,** 190
[9] Cohn, W. E. (1956) *Currents in Biochemical Research,* p. 460 (D. E. Green, Ed.). New York: Interscience
[10] Cohn, W. E. (1951) *J. Cell Comp. Physiol.,* **38,** Suppl. 1, 21
[11] Whitfield, P. R., Heppel, L. A. and Markham, R. (1955) *Biochem. J.,* **60,** 15
[12] Holder, J. W. and Lingrel, J. B. (1975) *Biochem.,* **14,** 4209
[13] Ricard, B. and Salser, W. (1974) *Nature,* **248,** 314
[14] Boedtker, H. (1968) *Methods in Enzymology,* Vol. 12, Part B, p. 429 (L. Grossman and K. Moldave, Eds.). New York: Academic Press
[15] Pinder, J. C., Staynov, D. Z. and Gratzer, W. B. (1974) *Biochem.,* **13,** 5373
[16] Loening, U. E. (1968) *J. Mol. Biol.,* **38,** 355
[17] Maden, B. E. H. (1971) *Progr. Biophys. Mol. Biol.,* **22,** 127
[18] Brown, D. M. and Todd, A. R. (1952) *J. Chem. Soc.,* 44
[19] Brown, D. M. and Todd, A. R. (1955) *The Nucleic Acids,* Vol. 1, p. 409 (E. Chargaff and J. N. Davidson, Eds.). New York: Academic Press
[20] Brown, D. M., Fasman, G. D., Magrath, D. J., Todd, A. R., Cochran W. and Woolfson, M. M. (1953) *Nature,* **172,** 1184
[21] Cavalieri, L. F. (1953) *J. Amer. Chem. Soc.,* **75,** 5268
[22] Khym, J. X. and Cohn, W. E. (1954) *J. Amer. Chem. Soc.,* **76,** 1818
[23] Cohn, W. E. and Doherty, D. G. (1956) *J. Amer. Chem. Soc.,* **78,** 2863
[24] Hall, R. H. (1964) *Biochem.,* **3,** 876
[25] Heppel, L. A., Ortiz, P. J. and Ochoa, S. (1967) *J. Biol. Chem.,* **229,** 679
[26] Cantoni, G. L., Ishikura, H., Richards, H. H. and Tanaka, K. (1963) *Cold Spring Harbor Symp. Quant. Biol.,* 28, 123
[27] Holley, R. W. (1966) *Sci. Amer.,* **214**(2), 30

[28] Holley, R. W. (1968) *Progress in Nucleic Acid Research and Molecular Biology*, Vol. 8, p. 37 (J. N. Davidson and W. E. Cohn, Eds.). New York: Academic Press
[29] Gilham, P. T. (1970) *Ann. Rev. Biochem.*, **39**, 227
[30] Hindley, J. (1958) *Chromatographic and Electrophoretic Techniques*, Vol. 1, p. 979 (I. Smith, Ed.). London: Heinemann
[31] Sanger, F. (1967) *Methods in Enzymology*, Vol. 12, p. 361 (L. Grossman and K. Moldave, Eds.). New York: Academic Press
[32] Sanger, F. and Brownlee, G. G. (1970) *Biochem. Soc. Symp.*, **30**, 183
[33] Sanger, F. (1971) *Biochem. J.*, **124**, 833
[34] Brownlee, G. G. (1972) *Determination of Sequences in RNA*. Amsterdam, New York: North-Holland, American Elsevier
[35] Madison, J. T. (1968) *Ann. Rev. Biochem.*, **37**, 131
[36] Holley, R. W., Apgar, J., Everett, G. A., Madison, J. T., Marquisee, M., Merrill, S. H., Penswick, J. R. and Zamir, A. (1965) *Science*, **147**, 1462
[37] Zachau, H. G., Dutting, D., Feldman, H., Melchers, F. and Karau, W. (1966) *Cold Spring Harbor Symp. Quant. Biol.*, **31**, 417
[38] Brownlee, G. G., Sanger, F. and Barrell, B. G. (1968) *J. Mol. Biol.*, **34**, 379
[39] Dahlberg, J. E. (1968) *Nature*, **220**, 348
[40] Szekeley, M. and Sanger, F. (1969) *J. Mol. Biol.*, **43**, 607
[41] Labrie, F. and Sanger, F. (1969) *Biochem. J.*, **114**, 29P
[42] Fox, G. E. and Woese, C. R. (1975) *Nature*, **256**, 505
[43] Barrell, B. G. and Clark, B. F. C. (1974) *Handbook of Nucleic Acid Sequences*. Oxford: Joynson-Bruvvers
[44] Dube, S. K., Marcker, K. A., Clark, B. F. C. and Cory, S. (1968) *Nature*, **218**, 232
[45] Goodman, H. M., Abelson, J., Landy, A., Brenner, S. and Smith, J. D. (1968) *Nature*, **217**, 1919
[46] Maden, B. E. H. and Salim, M. (1974) *J. Mol. Biol.*, **88**, 133
[47] Ehresmann, C., Stiegler, P., Mackie, G. A., Zimmermann, R. A., Ebel, J. P. and Fellner, P. (1975) *Nucleic Acids Res.*, **2**, 265
[48] Weissman, C., Billiter, M. A., Goodman, H. M., Hindley, J. and Weber, H. (1973) *Ann. Rev. Biochem.*, **42**, 303
[49] Fiers, W. et al. (1975) *Nature*, **256**, 273
[50] Min Jou, W., Haegeman, G., Ysebart, M. and Fiers, W. (1972) *Nature*, **237**, 82
[51] Spirin, A. S. (1963) *Macromolecular Structure of Ribonucleic Acids*. London: Methuen
[52] Warner, R. C. (1957) *J. Biol. Chem.*, **229**, 711
[53] Rich, A. and Davies, D. R. (1956) *J. Amer. Chem. Soc.*, **78**, 3548
[54] Fresco, J. R. (1963) *Informational Macromolecules*, p. 121 (H. J. Vogel, V. Bryson and J. O. Lampen, Eds.). New York: Academic Press
[55] Spirin, A. S. (1963) *Progress in Nucleic Acid Research*, Vol. 1, p. 301, (J. N. Davidson and W. E. Cohn, Eds.). New York: Academic Press
[56] Fresco, J. R., Alberts, B. M. and Doty, P. (1960) *Nature*, **188**, 98
[57] Fresco, J. R., Klotz, L. C. and Richards, E. G. (1963) *Cold Spring Harbor Symp. Quant. Biol.*, **28**, 83
[58] Cox, R. A. (1966) *Biochem. J.*, **98**, 841
[59] Cox, R. A. and Katchalsky, A. (1972) *Biochem. J.*, **126**, 1039
[60] Robertus, J. D., Ladner, J. E., Finch, J. T., Rhodes, D., Brown, R. S. Clark, B. F. C. and Klug, A. (1974) *Nature*, **250**, 546
[61] Gomatos, P. J. and Tamm, I. (1963) *Proc. Nat. Acad. Sci.*, **49**, 707

[62] Gomatos, P. J. and Tamm, I. (1963) *Proc. Nat. Acad. Sci.*, **50**, 878
[63] Geiduschek, E. P., Moohr, J. W. and Weiss, S. B. (1962) *Proc. Nat. Acad. Sci.*, **48**, 1078
[64] Tikchonenko, T. I., Kisseljov, F. L., Bukrinskaha, A. G. and Smirnov, Y. (1964) *Nature*, **202**, 1363
[65] Montagnier, L. and Sanders, F. K. (1963) *Nature*, **199**, 664
[66] Langridge, R. and Gomatos, P. J. (1963) *Science*, **141**, 694
[67] Tomita, K. and Rich, A. (1964) *Nature*, **201**, 1160
[68] Wade, H. E. and Morgan, D. M. (1953) *Nature*, **171**, 529
[69] Markham, R. and Smith, J. D. (1951) *Biochem. J.*, **49**, 401
[70] Schaffer, F. L. et al. (1960) *Virology*, **10**, 530
[71] Faulkner, P. et al. (1961) *Biochem. J.*, **80**, 597
[72] Strauss, J. H., Kelly, R. B. and Sinsheimer, R. (1968) *Biopolymers*, **6**, 793
[73] Borst, P. (1972) *Ann. Rev. Biochem.*, **41**, 333
[74] Tinoco, J., Uhlenbeck, O. C. and Levine, M. D. (1971) *Nature*, **230**, 362
[75] Connors, P. G. and Beeman, W. W. (1972) *J. Mol. Biol.*, **71**, 31
[76] Stanley, W. M. and Bock, R. M. (1965) *Biochem.*, **4**, 1309
[77] Pene, J. J., Knight, E. and Darnell, J. E. (1968) *J. Mol. Biol.*, **33**, 609
[78] Soiero, R., Birnboim, H. C. and Darnell, J. E. (1966) *J. Mol. Biol.*, **19**, 362
[79] Warner, J. R., Soiero, R., Birnboim, H. C. and Darnell, J. E. (1966) *J. Mol. Biol.*, **19**, 349
[80] Jeanteur, P., Amaldi, F. and Attardi, G. (1968) *J. Mol. Biol.*, **33**, 757
[81] Mahler, H. R. and Cordes, E. H. (1971) *Biological Chemistry*. London, New York: Evanston, Harper and Row
[82] Fraser, N. W., Burdon, R. H. and Elton, R. W. (1975) *Nucleic Acids Res.*, **2**, 2131
[83] Ford, P. J. (1973) *Biochem. Soc. Symp.*, **37**, 69
[84] Laycock, D. G. and Hunt, J. A. (1969) *Nature*, **221**, 1118
[85] Hadjivassilou, A. and Brawerman, G. (1967) *Biochem.*, **6**, 1934
[86] Attardi, G. and Amaldi, F. (1970) *Ann. Rev. Biochem.*, **39**, 183
[87] Sueoka, N. (1964) *The Bacteria*, Vol. 5, p. 419 (I. C. Gunsalus and R. Y. Stanier, Eds.). New York: Academic Press
[88] Rosenberg, M. and Gilham, P. T. (1971) *Biochem. Biophys. Acta*, **246**, 337
[89] Rosenberg, M. (1974) *Nucleic Acids Res.*, **1**, 653
[90] Cramer, F. (1971) *Progr. Nucleic Acid Res. Mol. Biol.*, **11**, 391
[91] Levitt, M. (1969) *Nature*, **224**, 759
[92] Robertus, J. D., Ladner, J. E., Finch, J. T., Rhodes, D., Brown, R. S., Clark, B. F. C. and Klug, A. (1974) *Nucleic Acids Res.*, **1**, 927
[93] Ladner, J. E., Jack, A., Robertus, J. D., Brown, R. S., Rhodes, D. S., Clarke, B. F. C. and Klug, A. (1975) *Nucleic Acids Res.*, **2**, 1629

CHAPTER 7

Nucleic Acids in Viruses and Plasmids

7.1 Viruses

Viruses, which were once looked upon as curiosities – perhaps a missing link between the living and the non-living – have proved to be invaluable tools for the study of nucleic acids. The genetic information of viruses is carried either in DNA or in RNA enclosed in a protein coat which both protects the nucleic acid from damage and confers a specific host-range on the potential infectivity of the particle. Most cells, including those of plants, animals, and bacteria, are susceptible to infection by viruses. Viruses which infect bacteria are known as bacteriophages.

Viral DNA varies in mol. wt. from a little over 10^6 to more than 10^8 (cf. the value of $2 \cdot 2 \times 10^9$ for *Esch. coli* DNA) and can often be extracted from the virus without degradation. Such intact DNA molecules have revealed an unexpected variety of tertiary structure. Studies with cellular DNA have not proved so informative because the exceptional shear fragility of such very large molecules only allows the isolation of broken pieces.

The mol. wt. of viral RNA varies from about 10^6 to more than 10^7. In many cases the RNA can be extracted as complete genomes providing polycistronic mRNA of known function [1]. Uniform populations of altered mRNA can be selectively produced in large numbers by growing the appropriate virus mutants [2].

The information for progeny virus synthesis is carried by the viral nucleic acid. This was inferred from the early experiments of Hershey and Chase [3], who showed that, after infection of *Esch. coli* with bacteriophage T2, the viral DNA enters the cell while the coat remains on the surface (see Chapter 10). The inference was confirmed when viral nucleic acids were isolated which could productively infect cells to give mature active virus particles of the appropriate genotype.

Infective RNA was first isolated from tobacco mosaic virus

TABLE 7.1
Properties of some viral nucleic acids

	Host cell	Mol. wt. ×10⁻⁶	Single- or double-stranded	Shape	Terminal repetition	Circular permutation	Single-strand 'nicks'
DNA Bacteriophages							
T2	Esch. coli	130	Double	Linear	Yes	Permuted	No
T5	Esch. coli	85	Double	Linear	—	Unique	Yes
T7	Esch. coli	25	Double	Linear	Yes	Unique	No
λ	Esch. coli	32	Double	Linear	Yes (exposed)	Unique	No
ØX174	Esch. coli	1·7	Single	Cyclic	—	—	No
P22	Salmonella	26	Double	Linear	Yes	Permuted	No
RNA Bacteriophage							
MS2	Esch. coli (male)	1·1	Single	Linear	—	—	—
DNA Animal viruses							
polyoma	Mammals	3	Double	Cyclic	—	—	No
herpes	Man	68	Double	—	—	—	—
RNA Animal viruses							
poliovirus	Man	2·2	Single	Linear	—	—	—
reovirus	Mammals	12	Double	Linear (in several pieces)	—	—	—
Plant virus							
TMV	Tobacco plant	2	Single	Linear	—	—	—

(TMV) [4]. Unlike that of the intact virus, the infectivity was sensitive to RNase and insensitive to TMV antiserum. Infective RNA has also been isolated from other plant viruses (e.g. turnip yellow mosaic virus), animal viruses (e.g. poliomyelitis virus [5]), and bacteriophages (e.g. MS2 [6]). Infective DNA has been isolated from animal viruses (e.g. polyoma virus [7]) and from bacteriophages (e.g. ØX174 [8] and λ [9]). Plant and animal cells are sensitive to the direct application of an appropriate preparation of infective viral nucleic acid, but bacteria often are not. *Esch. coli* can be made sensitive by pretreatment with lysozyme which removes part of the cell wall to form spheroplasts [8]. Successful infection requires complete nucleic acid molecules, unless the cell is infected concurrently with a second 'helper' virus (which must be genetically different from the infecting nucleic acid, so that the progeny can be differentiated). 'Helper' assays have been used for the DNA of bacteriophages T1 [10] and λ [11].

The contributions which viruses, and in particular bacteriophages, have made to the understanding of molecular biology are numerous and important. Several books [12–21], symposia [22–24], and reviews [25–28] deal with the many aspects.

7.2 Morphology of viruses

Chemical analysis of viruses reveals a nucleic acid content varying from 1 per cent to 50 per cent according to the virus, the remainder being almost entirely protein, and in some cases lipoprotein. This apparent simplicity of structure covers a wide range of complexity and sophistication [29], but this range is many orders of magnitude less than the complexity found at the cellular level.

Most plant viruses contain single-stranded RNA as in TMV, but a few contain double-stranded RNA as in wound tumour virus [30] or double-stranded DNA as in cauliflower mosaic virus [31]. The virus particles can be rod-shaped (e.g. TMV) or spherical (e.g. cowpea chlorotic mottle virus).

An extensive study of TMV (for reviews see [32–34]), which has a particle weight of 4×10^7 and is rod-shaped measuring 15×300 nm, has led to a detailed picture of its structure (Fig. 7.1). A helical array of about 2100 identical protein subunits of mol. wt. 17 400 surround a single-stranded RNA molecule of mol. wt. 2×10^6. Cowpea chlorotic mottle virus is spherical and is composed of 180 identical subunits of mol. wt. about 20 000 arranged on the surface of an icosahedron in 32 morphological units of 20 hexamers (on the faces) and 12 pentamers (on the vertices) [35, 36].

Animal viruses contain single-stranded RNA as in poliovirus, double-stranded RNA as in reovirus [30], single-stranded DNA as in minute virus of the mouse [37], or double-stranded DNA as in polyoma virus. The small viruses are spherical (e.g. poliovirus, polyoma) and the large viruses have both a virus-specified protein coat and a lipoprotein envelope similar to the cell cytoplasmic membrane (e.g. influenza virus, herpes virus [38]).

Bacteriophages contain single-stranded RNA (e.g. MS2), single-stranded DNA (e.g. ØX174 [39]), or double-stranded DNA (e.g. the T-phages). The single-stranded DNA bacteriophages are either

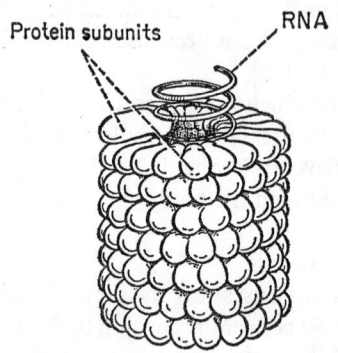

Fig. 7.1 *Segment of the tobacco mosaic virus particle showing the protein subunits forming a helical array* [32]. *The RNA lies in a helical groove in the protein subunits some of which have been omitted to show the top two turns of the RNA helix*

spherical (e.g. ØX174) or filamentous (e.g. fd or M13 [40]) where the circular DNA molecule is wrapped in a protein coat and then formed into a filament of two nucleoprotein strands by bringing the opposite sides of the circle together. All the RNA bacteriophages are spherical. MS2 consists of 180 protein subunits all identical except for one (the maturation protein).

Many of the double-stranded DNA bacteriophages have a more intricate structure. The T-even bacteriophages (Fig. 7.2) have a head (which contains the DNA) (Plate I), a tail (through which the DNA is injected into the host cell), and a base plate with 6 tail fibres which recognize, and attach to, sites on the surface of the host cell [41].

NUCLEIC ACIDS IN VIRUSES AND PLASMIDS

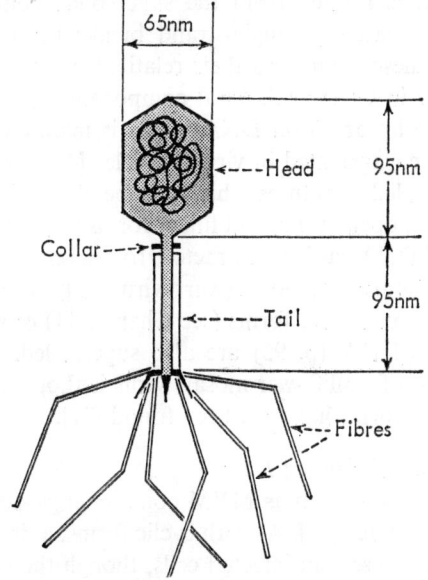

Fig. 7.2 *The structure of a T-even phage particle*

7.3 Structure of viral DNA
The DNA of many viruses has been extracted intact and the structure determined. Several types of structure have been found [27, 42].

7.3.1 *Single-stranded cyclic DNA*
The DNA of bacteriophage ØX174 has been shown by Sinsheimer to be single-stranded [39, 43, 44] (p. 100) and in the form of a continuous cyclic polynucleotide chain [44] of mol. wt. 1.7×10^6. Similar cyclic single-stranded DNA can be isolated from other spherical bacteriophages related to ØX174 such as S13, and the filamentous bacteriophages such as fd contains similar DNA molecules.

The structure and properties of single-stranded cyclic DNA are discussed in detail in Chapter 5.

7.3.2 *Double-stranded cyclic DNA*
Three forms of double-stranded DNA can be isolated from purified polyoma virus [45, 46]. A supercoiled form (21S, component I), an open cyclic form (16S, component II), and a linear form (14·5S, component III). All the forms have a molecular weight of 3×10^6.

Component II is formed from the supercoiled component I by introducing at least one single-strand break; the structures and properties of these forms and their relationship to each other are discussed fully in Chapter 5. Viral component III forms a minor fraction of the linear virion DNA which is mainly cellular DNA that has been encapsulated in virus particles [47]. Electron microscopy has revealed structures which can be clearly identified with the supercoiled, open cyclic, and linear forms (Plate IV).

Supercoiled DNA has been extracted from several viruses, including SV40 [48], human papilloma (wart) virus [49], and bacteriophage PM2 [50]. The replicative forms (see Chapter 11) of ØX174 DNA (p. 295) and λ DNA (p. 95) are also supercoiled. The double-stranded DNA of cauliflower mosaic virus is thought to be cyclic, but supercoiled forms have not been found [31].

7.3.3 *Single-stranded linear DNA*
Some animal viruses such as MVM contain single-stranded DNA of mol. wt. $1 \cdot 5 \times 10^6$ [37]. As with cyclic forms, a double-stranded intermediate is present in infected cells, though the role this plays has not yet been delineated [51].

7.3.4 *Double-stranded linear DNA*
Many viruses contain double-stranded linear DNA but most of those which have been studied in detail exhibit some additional feature such as cohesive ends, terminal repetition, circular permutation, or 'nicks'. These are discussed in turn.

7.3.5 *Cohesive ends*
When DNA extracted from bacteriophage λ (mol. wt. 30×10^6) is heated to 65° and cooled slowly, its sedimentation coefficient is 37S, but when it is quick-cooled its sedimentation coefficient is only 32S [52]. This behaviour is a result of the 5'-ends of the DNA projecting as single strands beyond the 3'-ends, the two single-stranded regions being complementary. These cohesive ends can base-pair and convert the DNA into a cyclic molecule which is disrupted at 65°. The complementary sequences of the cohesive ends, which are 12 nucleotides long, have been determined [53]. The hydrogen-bonded cyclic form can be converted with polynucleotide ligase (p. 285) into a cyclic form with both strands continuous. *Esch. coli* DNA polymerase, which adds on nucleotides to the 3'-ends, abolishes the ability to form cyclic molecules, but the

ability is regained after treatment with *Esch. coli* exonuclease III which removes the newly added bases from the 3'-ends [54]. Electron microscopy has shown the 37S form to be cyclic and the 32S form to be linear. Some other lysogenic bacteriophages contain DNA with a similar structure (e.g. Φ80 [55]).

7.3.6 *Terminal repetition*
The DNA of bacteriophage T7 (mol. wt. 25×10^6) is double-stranded and linear, and the sequence (about 0·7 per cent of the total) at the beginning is repeated at the end of each molecule [56]. Treatment of such molecules with *Esch. coli* exonuclease III results in the formation of cohesive ends which cause circle formation under suitable conditions. Terminal repetition has been detected in the DNA from several bacteriophages (e.g. T2, T4, T3, P22) and is believed to play a role in replication by enabling multiple length concatamers to be formed (see p. 293).

7.3.7 *Circular permutation*
If the linear double-stranded DNA of bacteriophage T2 (mol. wt. 130×10^6) is denatured and allowed to reanneal slowly, circular molecules are formed which can be detected with the electron microscope [57]. These are formed because the bacteriophage DNA molecules do not have a unique sequence, but the population is a collection of molecules with sequences which are circular permutations of each other (Fig. 7.3). The molecules also show terminal repetition of the sequences at the beginning. The DNA's from bacteriophages T4 and P22 also exhibit circular permutation and terminal repetition.

7.3.8 *Nicks*
Three specific breaks have been found by electron microscopy in one strand of the double-stranded linear DNA extracted from bacteriophage T5 [58]. Accordingly, when the DNA is denatured, five single-stranded pieces are produced (instead of two) from each T5 DNA molecule. Bacteriophages SP8 and SP50 may also contain specific breaks.

7.3.9 *The rule of the ring* [59]
Many of the features of double-stranded viral DNA can be related to a cyclic form. Some phage DNA's are converted into cyclic forms during replication (e.g. the supercoiled forms isolated from ØX174

[60] and λ [61] infected cells), and others have a structure which can most easily be explained if their synthesis involved at some stage a cyclic form [42, 62–64]. The replication of a cyclic DNA may give rise to concatenates [65] (long molecules made by continuously

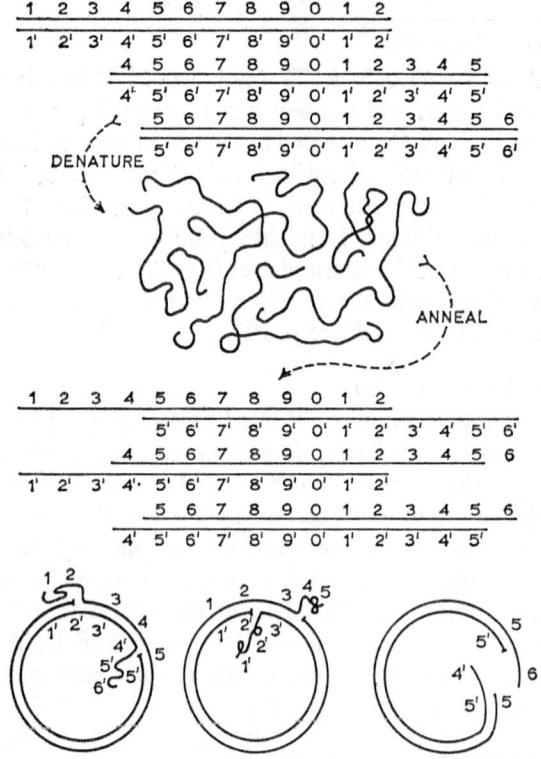

Fig. 7.3 *Formation of cyclic DNA by denaturing and annealing a permuted collection of duplexes* [22]. *Notice that each permutation is also terminally repetitious. One repetitious terminal from each strand cannot find a complementary partner and is left out of the circular duplex. Their separation depends on the relative permutation of the partner chains*
(reproduced from [22] by permission of Academic Press Inc.)

repeating the basic molecular unit) which may in turn be the precursors of molecules with cohesive ends, terminal repetition, and those which are circularly permuted [59].

7.3.10 *Modification of viral DNA*
The DNA's of the T-even bacteriophages contain 5-hydroxymethylcytosine in place of cytosine [66] and the hydroxyl group of this base

can be glucosylated (Tables 7.2 and 7.3) [67]. Growth of the bacteriophage in a UDP-glucose-deficient host produces bacteriophages with DNA which is not glucosylated, and such DNA is degraded when the bacteriophages infect the normal host [68].

TABLE 7·2

Molar proportions of bases in the DNA from certain strains of coliphage

Strain	Adenine	Guanine	Thymine	Cytosine	Hydroxymethylcytosine
T2	32·4	18·3	32·4	—	17·0
T4	32·4	18·3	32·4	—	17·0
T6	32·5	18·3	32·5	—	16·7
T5	30·3	19·5	30·8	19·5	—
T3	23·7	26·2	23·5	27·7	—

The DNA's of many bacteriophages are modified by host specific mechanisms when grown in certain strains of *Esch. coli* (*Esch. coli* K12 and *Esch. coli* B), but not when grown in other strains (e.g. *Esch. coli* C) [69, 70]. Bacteriophage λ modified by growth in K12 or B will only grow efficiently in the same strain, K12 or B respectively or in C. Unmodified phage will only grow efficiently in C.

TABLE 7.3

Per cent glucosylation of hydroxymethylcytosine residues in the DNA of T-even phages [67]

	T2	T4	T6
Unglucosylated	25	0	25
α-Glucosyl	70	70	3
β-Glucosyl	0	30	0
β-Glucosyl-α-glucosyl (Diglucosyl)	5	0	72

When the infecting bacteriophage fails to grow (i.e. is restricted), the bacteriophage DNA is degraded. The restriction/modification system is considered in detail in Chapter 8.

7.4 Structure of viral RNA

Like viral DNA, viral RNA can often be extracted as complete molecules. Single-stranded and double-stranded forms have been isolated and characterized.

7.4.1 *Single-stranded RNA*

The RNA bacteriophages such as MS2, the small animal RNA viruses such as poliomyelitis virus, and most plant viruses contain one molecule of single-stranded RNA (mol. wt. $1-3 \times 10^6$) per particle [42]. The replication of these viral RNA molecules appears to involve a double-stranded intermediate (see p. 346).

A second type of single-stranded RNA viruses includes the oncornaviruses or RNA tumour viruses such as the leukaemia viruses. These viruses contain RNA of molecular weight around 10^7 daltons (60–70S) but this can be dissociated to subunits of about 3×10^6 molecular weight. Replication of this RNA takes place by way of a DNA intermediate and involves the enzyme known as reverse transcriptase (see p. 296) [71].

7.4.2 *Double-stranded RNA*

Reovirus RNA is double-stranded [30]. Each virus particle contains about 12×10^6 daltons of RNA which, when extracted, is in pieces of three sizes (mol. wt. 2.3×10^6, 1.3×10^6, and 8×10^5), plus 50–100 single-stranded oligonucleotides rich in adenine [72]. Two plant viruses, would tumour virus and rice dwarf virus, have also been shown to contain double-stranded RNA.

7.5 Replication of viruses

The replication of a virus can be considered in stages: (1) adsorption of the virus on to the host cell; (2) penetration of the viral nucleic acid into the cell; (3) development of virus specific functions, alteration of cell functions, replication of the nucleic acid, and synthesis of other virus constituents; (4) assembly of the progeny virus particles; (5) release of virus particles from the cell [21].

(1) Viruses will only infect certain specific cells, i.e. they have a limited host-range because the coat (or tail) will only recognize and adsorb to specific sites on the appropriate cell walls. Polyoma virus adsorbs to neuraminidase-sensitive sites on mouse cells [73]. The host range of the T-phages is a property of their tail fibres. Some T-even bacteriophages are free to adsorb to the bacterial cell wall site only in the presence of tryptophan; in the absence of this amino acid the tail fibres are folded back and attached to the tail sheath [74]. The initial reversible interaction of the tail fibres with the cell wall is followed by the formation of a permanent attachment. The

NUCLEIC ACIDS IN VIRUSES AND PLASMIDS

small male-specific bacteriophages (e.g. MS2, R17) attach only to the f-pili of male *Esch. coli* cells [75] (see p. 150).

(2) The penetration of the viral nucleic acid into the cell involves a phage mechanism (e.g. T-phage), host-cell mechanism (e.g. MS2), or possibly the simple removal of the coat once the virus is in the cell (e.g. polyoma).

After attachment to the cell, the lysozyme present in the base of the bacteriophage T4 tail probably hydrolyses part of the cell wall. This allows the tail core to penetrate into the cell as the contractile tail sheath contracts and the small amount of ATP present in the phage tail is hydrolysed to ADP [76]. How the DNA passes from the head through the tail and into the cell (a process equivalent to passing a 10-metre long piece of string down a straw) is not understood.

The injection of bacteriophage T5 DNA takes place in two stages. Eight per cent of the DNA (the first step transfer DNA) enters the cell and directs mRNA and protein synthesis. One of these proteins is required to complete the injection of the remaining DNA [77].

After attachment of bacteriophage MS2 to the f-pilus the RNA leaves the phage and is then transported inside the length of the pilus to the cell. This last step requires cellular energy [78, 79]. An alternative mechanism proposed for entry of M13 into male *Esch. coli* involves retraction of the pilus with the filamentous bacteriophage attached [80]. Replication to the duplex form is necessary for the bacteriophage DNA to be drawn into the host, and when entry is effected a considerable fraction of the capsid protein is deposited in the inner cell membrane [21].

Polyoma virus, after attachment to the animal cell wall, is probably taken into the cell by natural pynocytosis.

(3) The metabolic processes of the cell are usually modified after virus infection [77, 81]. Viral mRNA directs the synthesis of specific enzymes, and the rates of host cell DNA, RNA, and protein synthesis are altered. The viral nucleic acid replicates and virus constituents are synthesized. These different viral functions are divided into two groups, the early and late functions, which appear to be controlled either at the level of transcription or translation. Early functions include the biosynthesis of enzymes required for the replication of the nucleic acid, and late functions include the formation of the virus coat and other constituents. T-even bacteriophages turn off the synthesis of host cell DNA and redirect the synthesis of

DNA precursors to fit their particular requirements (e.g. hydroxymethylcytosine triphosphate is produced; see Chapter 11). Polyoma virus, on the other hand, stimulates the synthesis of host cell DNA [82, 83], particularly if the cells are in a resting state before infection [84].

Many animal viruses stimulate the production of *interferon* in the host cell [85–88]. Interferon is a species specific protein (mol. wt. 30 000) which elicits the synthesis of another protein [89] which in turn disrupts the replication cycle of the virus probably by preventing attachment of the viral mRNA to the ribosomes [90]. Several other agents, including double-stranded RNA [91] and statolon, an antiviral agent from fungi, also stimulate interferon production. The active component of statolon has been shown to be double-stranded RNA derived from a virus present in the fungi [92]. The production of interferon can be stimulated by the artificial polynucleotide poly(I) · poly(C) [93].

The replication of virus nucleic acids is described in Chapters 11 and 12.

(4) The assembly of virus particles appears to be either spontaneous [36] or to involve a series of virus directed steps [94].

The reconstitution of the rod-shaped TMV has been studied extensively [95–97]. When the coat protein and viral RNA are mixed in the correct ionic environment, virus particles are formed which possess up to 80 per cent of the original infectivity. Attempts to reconstitute small spherical viruses have proved more difficult. MS2 RNA and coat protein form morphologically complete particles which are not infective, possible because they lack the maturation protein [98]. Cowpea chlorotic mottle virus can be partially degraded and reassembled to form particles indistinguishable from the original virus, and separated protein and nucleic acid have been mixed under conditions such that infectious particles are formed which have the same appearance, serological properties, and sedimentation coefficient as intact virus [99]. Such experiments are consistent with the suggestion that the nature of the virus coat subunits alone directs the size and shape (spherical or rod) of the completed virus particles [100].

More complex viruses do not assemble spontaneously from separate nucleic acid and protein, but are pieced together step by step. This process has been partially characterized for T4 (see p. 157) and λ [69, 101].

(5) Interference with the normal metabolic processes may lead to the eventual death of the infected cell, followed by natural lysis, but in some instances it has been shown that the virus actively causes cell lysis. Bacteriophage T4 codes for a lysozyme [102] which digests the host cell wall, causing the release of the progeny virus particles.

7.6 Lysogeny

When a virulent bacteriophage infects a cell, the virus replicates and the cell is killed. The temperate bacteriophages, however, can either kill the cell or lysogenize it [103]. A lysogenic cell usually carries the DNA of the bacteriophage integrated in the cell genome (as a prophage). The bacteriophage genes are transcribed (in a controlled way), replicated, and inherited along with the cell genes. Most of the bacteriophage functions are repressed (e.g. those involved in the lytic development), but others are not (e.g. those involved in the maintenance of lysogeny). One gene function which is not repressed in the lysogenic cell confers immunity against further infection of the cell by the same bacteriophage. This immunity is quite specific.

The stable lysogenic cell can be induced by various agents, most of which interfere with normal DNA synthesis [104]. This results in the bacteriophage DNA being cut out of the cell genome, probably by a single recombination event [105]. The DNA can then replicate and function in a virulent manner, resulting in cell lysis and liberation of a burst of progeny bacteriophage.

The survival of virulent bacteriophages depends on a continuous supply of susceptible bacteria (e.g. in sewage, a rich source of bacteriophages) while the temperate bacteriophages can survive and replicate with a limited population of cells which are protected from further infection.

Prophages are either integrated at a specific site on the bacterial chromosome (e.g. λ which normally attaches near the *gal* locus of *Esch. coli* [106]) or else at many sites [107] or at random (e.g. phage μ). The specific site can be identified on the genetic map of the bacteria. One temperate bacteriophage, P1, exists in the lysogenic cell as a prophage which is not integrated but is somehow strictly limited to one copy per cell [108].

7.7 Plasmids

Plasmids are similar to temperate bacteriophages like P1 but have no extracellular existence. They are stable genetic elements which exist in an extrachromosomal state. They differ from most temperate bacteriophages in that they enjoy a semi-autonomous, self replicating status without lowering host cell viability [109–111].

Some plasmids are recognized because they confer on their host resistance to certain drugs (R-factors) or cause their host to make colicins (colicinogenic factors). A colicin is a protein which is toxic to related bacterial cells but not to the host cell. Others (F-factors) cause the host to produce a 'pilus' or tube by means of which the plasmid may be transferred to another bacterium by conjugation (cf. a virus which causes the host to produce a protein coat to aid in transfer of viral nucleic acid). Combinations of F-factors with R-factors or colicinogenic factors are also known. Thus R222 has a molecular weight of 68×10^6 in *Esch. coli* and is made up of two parts (a 54×10^6 molecular weight transfer unit and a 12×10^6 molecular weight resistance determining unit) which exist separately in *Proteus mirabilis* [112].

Plasmids are duplex supercoiled DNA molecules which range in size from 1.5×10^6 to 1.5×10^8 daltons. The large ones are present in only 1 or 2 copies per cell while there may be up to 20 copies per cell of the smaller ones [21, 117]. They are transferred between cells either by viral mediated transduction or by conjugation via the sex pilus. The presence of a sex (F) factor in a bacterium excludes the entry of a second sex factor by conjugation which is similar to lysogeny, but more than one type of plasmid may be stably maintained in a cell if the two are not closely related. A bacterium possessing an F-factor (F^+) is male and transfer occurs to female (F^-) cells. Generally F^+ cells are rare but each usually contains several F-factors. When one of these is transferred the recipient cell also becomes F^+ and soon all the cells in a population will become F^+. As R-factors are also transferred the spread of drug resistance throughout the bacterial population is alarming. Conjugation crosses species barriers and dangerous pathogens such as *Shigella flexneri* and *Vibrio cholera* have become simultaneously resistant to as many as 8 of the antibacterial drugs in current use.

Replication of most plasmids in *Esch. coli* requires the normal RNA polymerase and is sensitive to rifampicin (see p. 322). Moreover, replication of the small plasmids (e.g. col E_1) appears to require DNA polymerase I, and DNA polymerase III is not required (see p. 280) [21, 117]. Normally replication is highly coupled to host chromosome replication (i.e. it is stringent [113]) and occurs by the formation of a replication bubble which expands in one or two directions – the Cairns model [114] (see Plate III). However, at conjugation transfer is asymmetric and probably involves DNA replication by a rolling circle mechanism. A nick is produced in one of the strands of the plasmid DNA and the 5′-end passes through the conjugation tube. The complementary strand is synthesized in the recipient.

In one in 10 000 F^+ cells the F-factor becomes integrated into the bacterial chromosome by reciprocal crossing over. A plasmid which can become integrated is known as an episome. For insertion the F-factor breaks at a unique site but this linear piece of DNA integrates at random with regard to position and orientation. Like the λ-prophage, an integrated plasmid loses its ability for independent replication [115, 118]. However, *Esch. coli* mutants defective in initiation of chromosomal replication can be rescued by integration of an F-factor which takes over control of initiation [115]. However, it still causes the formation of pili and can be transferred by conjugation, but now, between the leading 5′-end of the plasmid DNA and the trailing 3′-end, there is the entire bacterial chromosome which gets dragged along the conjugation tube, into the female cell. Since transfer of the whole chromosome takes about 60 minutes, breakage of the conjugation tube normally occurs before transfer is complete. When this occurs the second half of the F-factor is not transferred and so the recipient remains female. Some of the bacterial fragments transferred are integrated into the recipient's chromosome which thereby shows a *h*igh *f*requency of *r*ecombination of genetic markers. Hence cells with an integrated F-factor are called *Hfr*.

Because of the mode of integration of an F-factor into the bacterial chromosome, different Hfr strains transfer the genes of the bacterial chromosome in a different order and with different polarity. However, closer study shows the gene order to be circularly permuted, providing evidence for the linear arrangement of genes on a circular chromosome [116].

Sometimes an Hfr strain may revert to F^+, and when this occurs a section of the bacterial chromosome may also be excised and found in the plasmid (e.g. Flac contains the lac operon integrated into an F-factor which is thereby more than doubled in size). Such plasmids are known as F'-factors. This gene is now transferred to a recipient bacterium at conjugation (cf. transduction).

7.8 Tumour viruses

Some animal viruses (tumour viruses) can alter (transform) the infected cell, without killing it, so that it has new properties which are typically neoplastic [119, 120]. Uninfected hamster cells do not form tumours when injected into new-born hamsters, and do not grow when suspended in nutrient agar, but polyoma-transformed hamster cells form tumours and grow in agar [20].

Several DNA-containing tumour viruses have been identified, including polyoma, SV40, rabbit papilloma, human papilloma, and adeno viruses 7, 12, and 18.

Some of these viruses can interact with different cells in different ways. Polyoma virus will transform hamster cells but will replicate in, and kill, mouse cells. SV40, on the other hand, transforms mouse cells and kills green monkey cells. A temperature-sensitive mutant of polyoma [121] will transform mouse cells at the non-permissive temperature, 38·5°, and then replicate when the temperature of the transformed cell is reduced to 31° [122]. This suggests that at least one virus function necessary for replication is not required for transformation. It is not understood why the different interactions occur with the different cell types.

Several lines of evidence show that the viral genome is present in the transformed cells, in a stable inheritable state (cf. the lysogenic state of some bacteria). mRNA isolated from polyoma transformed cells will hybridize with polyoma virus DNA [123]. SV40-transformed mouse cells can be fused with green monkey cells using inactivated Sendai virus, and the hybrid cells liberate active SV40 particles [20, 124].

Although some SV40 transformed cells appear to contain only one copy of the viral genome per cell [125], others contain up to nine copies [126]. The loss of some of the neoplastic properties of polyoma transformed cells can be associated with the loss of a biochemically marked chromosome from artificial tetraploid cells, which is consistent with the idea that the polyoma genome is integrated into only one chromosome [127]. However, further evidence is required to define the site (or sites) of integration.

Two virus specific proteins have been identified in cells trans-

NUCLEIC ACIDS IN VIRUSES AND PLASMIDS

formed by polyoma, SV40, or the adenoviruses. One, the T antigen, is found in the nucleus and is also present in infected cells. Its detection therefore provides an easy method for determining whether or not a cell is transformed by a particular virus. The second virus specific protein is the (tumour specific) transplantation antigen (TSTA) which is present on the cell surface. It is this surface antigen which causes the rejection of transplantable tumours by animals previously immunized with the virus [20]. The function of these two proteins is not known; neither is it known what viral product gives rise to the neoplastic character of the transformed cell. However, a temperature sensitive mutant of polyoma virus (ts-3) will transform hamster cells at the permissive temperature, and the transformed state of such cells is temperature dependent, i.e. the cells behave normally at the non-permissive temperature and as transformed cells at the permissive temperature [128].

The leukaemia viruses and the sarcoma viruses are closely related RNA tumour viruses. The leukaemia viruses are a natural cause of leukaemia and the sarcoma viruses transform cells in tissue culture. Avian, murine, and feline RNA tumour viruses have been isolated; the feline sarcoma virus will transform human cells in tissue culture, but there is no evidence which relates human leukaemia to such viruses.

The RNA tumour viruses contain single-stranded RNA of molecular weight about 10^7 (70S) [129] along with a considerable amount of 4S material and smaller amounts of 28S, 18S, and 7S RNA [130] and DNA [131] which may be of cellular origin. The virus particles, which are enveloped, also contain an RNA-dependent DNA polymerase [132, 133] which, in the presence of the four deoxyribonucleoside triphosphates and an RNA template, catalyses the formation of first an RNA–DNA hybrid and then double-stranded DNA [134] (see Chapter 11). This ability of the virus to produce double-stranded DNA might explain how these viruses can permanently transform cells, if such DNA can be integrated into the cellular chromosome [135] (cf. the DNA tumour viruses, p. 152). This suggestion is substantiated by the observation that chick cells transformed with avian sarcoma virus contain DNA which will hybridize with viral RNA. However, *all* untransformed chick cells examined also contain such DNA sequences. This infers that all chicks carry the viral information, but that the expression is suppressed except in leukaemic chicks [20, 136].

7.9 Transduction

The transfer of genetic material from one bacterium to another, mediated by a bacteriophage vector, is termed transduction [14, 15] (p. 227). The bacterial genes transduced by a particular bacteriophage are either closely linked (specialized transduction, e.g. λ) or come from many places on the bacterial chromosome (generalized transduction, e.g. P22).

Bacteriophage λ can lysogenize bacteria and become integrated as a prophage in the host cell chromosome. Induction normally results in excision of the bacteriophage DNA by a single recombination event following pairing between terminally repeated sequences [105]. However, if an illegitimate pairing occurs between part of the bacteriophage genome and part of the cell genome, then the excised piece of DNA will contain both bacteriophage and host DNA. The heterologous DNA is incorporated into bacteriophage particles. These transducing bacteriophages can now infect another cell so that the DNA becomes integrated as a prophage, thus introducing new bacterial genes into the new host.

Transducing bacteriophages are invaluable tools for fine genetic analysis of bacteria [137], and have recently been used for partial purification of specific bacterial genes (specific because of the selection procedures used) in large quantity by growing the transducing bacteriophage under appropriate conditions and then isolating the DNA from the purified particles [138]. However, even at a bacteriophage DNA concentration of 3 mg/ml the gene concentration is only 10^{-7}M.

7.10 Virus genetics

The viruses provide a unique opportunity to characterize completely the information content of a functional genome.

A mutation in a viral genome can cause the inactivation of a gene function. If the missing function can be characterized (e.g. a missing enzyme) then the nature of the gene, and the mutant, is defined. However, if the function is essential, the mutation is lethal and the mutant cannot be propagated and studied. The discovery of conditional lethal mutants [139–141], therefore, which lack a gene function under one set of conditions but regain it under another set of conditions, has revolutionized the study of virus, and in particular, phage genetics. Stocks of conditional lethal mutants can be grown under permissive conditions and then the mutant gene function studied and identified under non-permissive conditions.

Two classes of conditional lethal mutation are particularly useful. Temperature sensitive mutants [140] are not viable at the non-permissive temperature (e.g. 42°) but grow at the permissive temperature (e.g. 30°). This is due to a single base change in the DNA causing the introduction of the wrong amino acid into the protein, and reducing the stability of its configuration at the higher temperature.

Amber mutants [139] of a bacteriophage will only grow in a permissive host (which contains a suppressor). This is caused by a single base change altering an amino acid coding triplet to UAG (p. 369), which is read as stop under the normal, non-permissive conditions, and the protein is terminated at that point. Permissive host bacteria, which suppress the mutation, have a species of tRNA which translates UAG as an amino acid and allows the protein to be completed. Different amino acids are added by different classes of permissive host.

When two virus mutants, with mutations in different genes, infect the same cell under non-permissive conditions, the missing function of each can be supplied by the other, i.e. complementation takes place, and progeny viruses are formed. If the two mutants have mutations in the same gene, complementation cannot occur and no progeny are formed. Thus complementation between mutants is used to determine the number of complementation groups, i.e. the number of different genes in which the mutations occur. If sufficient mutants have been isolated for each gene of the virus to be represented, then the total number of essential viral genes can be estimated.

The joint growth of two bacteriophages in the same cell can also result in the formation of a few progeny phages, recombinants, which carry genetic characters of both parental bacteriophages. A study of the frequency of formation of recombinants from parental bacteriophages carrying known mutations allows the construction of a genetic map; the frequency of recombinants with both genetic characters is related to the distance between the two mutation points on the genome. Such a map shows the relative positions of the mutations and therefore of the genes in which these mutations occur, and it can be linear or circular [141].

Although the smallest known plasmid (the 15T$^-$ minicircles of *Esch. coli*) can probably code for only two proteins, the most studied plasmids (the F-factors) can code for about 100 proteins yet hardly 20 have been accounted for and most of these are involved in conjugal transfer [110]. About one-third of the F genome is non-essential but is believed to be involved in recombination events when F integrates with the bacterial chromosome [147].

The small RNA bacteriophages (e.g. R17) contain all their genetic information in a single strand of RNA about 3300 nucleotides long (which can code for about 1100 amino acids or about 3–4 proteins of average size). Three complementation groups corresponding to three viral coded proteins have been identified (p. 370) and these may represent the total genetic potential of the virus [142]. In order they are: the coat protein (129 amino acids, mol. wt. 14 000), the maturation protein (mol. wt. about 37 000) [143], and the RNA-dependent RNA polymerase (p. 370) (synthetase; mol. wt. about 50 000). The coat protein contains no histidine, which however is present in the maturation protein, another structural component of the virus particle. It seems likely that there is only one maturation protein molecule per particle and that this protein is concerned with the assembly of the RNA into the bacteriophage. In the infected cell, the synthetase is an early protein made about 10 minutes after infection, while the coat and maturation proteins are late proteins made (in appropriate amounts) about 20 minutes later. The mechanism of this control is not understood, but an *in vitro* protein synthesizing system using viral RNA as messenger exhibits controlled translation and provides a model system for the study of control mechanisms [1, 2].

Small single-stranded DNA bacteriophages contain a ring of about 5500 nucleotides which could code for between 5 and 10 proteins. Conditional lethal mutants of ØX174 have been obtained in eight cistrons. Amber mutations early in seven of the eight cistrons have allowed the characterization of the corresponding proteins. Dual labelling of proteins from infected permissive cells (where the cistron product is synthesized) and from non-permissive cells (where only the small initial polypeptide, terminated by the amber mutation, is made) followed by acrylamide gel electrophoresis in SDS of the mixed extracts shows the position of the cistron product by a variation in the isotope ratio. The molecular weights of seven of the proteins have been determined in this way. Four of the proteins are structural components of the virus particle [144].

Polyoma virus and SV40 are tumour viruses. Since intense interest surrounds the nature of the viral coded function which induces the transformed state, because of its relevance to general problems of cancer, attempts are being made to characterize all the gene products. The 5000 base-pairs of the DNA could code for about 5–10 proteins, and temperature sensitive mutants (no suppressor system for amber-type mutations has been found in animal cells) fall into five complementation groups, but only the function of those associated with the formation of coat protein has been identified [119].

The extensive study of the genetic structure of bacteriophage T4 [141] has resulted in the isolation of conditional lethal mutants in nearly 70 genes. The genetic map of T4, like that of the host cell (*Esch. coli*), is circular. Many of the mutants have been characterized and provide sophisticated genetic 'reagents' for a wide variety of studies. Two examples will illustrate this. The mechanism of DNA replication is not completely understood, but in T4 infected cells the bacteriophage induced polynucleotide ligase (gene 30) is involved. Under non-permissive conditions (when the ligase is not made) short pieces of newly synthesized DNA accumulate, but when the conditions are changed and an active ligase is formed, the pieces are joined and DNA synthesis continues normally [21, 145].

The assembly of the complex structure of T4 is not understood either. However, Wood and Edgar [146] have shown that about 45 viral genes are involved in the process, and many of the steps have been characterized by *in vitro* complementation using the partially finished pieces (e.g. heads, tails, fibres) found in cells infected with different mutants under non-permissive conditions. Eight genes have been assigned to the component parts of the head, and a further 8 implicated in the assembly of these components. The head, which then appears to be morphologically complete, requires the action of two more gene products before it can interact spontaneously with assembled tails. The base plate components (12 genes) are assembled in 2 steps, then the core (3 genes) and the sheath (1 gene) are added by progressive polymerization from the base plate. Two gene products are required to finish the tails before they can be joined to the heads. The assembled heads and tails are modified (1 gene) before the tail fibres are added. This last step requires complete tail fibres (2 genes for components and 3 for assembly) and a labile factor (L) which has many properties of an enzyme. Phage particles assembled *in vitro* are active and possess characteristics which vary with the source of the parts (e.g. the genotype of the head and the host range of the tail fibres).

REFERENCES

[1] Nathans, D., Notani, G., Schwartz, J. and Zinder, N. (1962) *Proc. Nat. Acad. Sci.*, **48**, 1424
[2] Lodish, H. F. (1968) *J. Mol. Biol.*, **32**, 681
[3] Hershey, A. D. and Chase, M. (1952) *J. Gen. Physiol.*, **36**, 39
[4] Gierer, A. and Schramm, G. (1956) *Z. Naturforsch.*, **116**, 138
[5] Alexander, H. E., Koch, G., Mountain, I. M. and van Damme, O. (1958) *J. Exp. Med.*, **108**, 493
[6] Davies, J. E., Strauss, J. A. and Sinsheimer, R. L. (1961) *Science*, **134**, 1427
[7] Crawford, L., Dulbecco, R., Fried, M., Montagnier, L. and Stoker, M. (1964) *Proc. Nat. Acad. Sci.*, **52**, 148
[8] Guthrie, G. D. and Sinsheimer, R. L. (1963) *Biochim. Biophys. Acta*, **72**, 290
[9] Young, E. T. and Sinsheimer, R. L. (1967) *J. Mol. Biol.*, **30**, 147
[10] Van De Pol, J. H., Veldhuisen, G. and Cohen, J. A. (1961) *Biochim. Biophys. Acta*, **48**, 417
[11] Kaiser, A. D. and Hogness, D. S. (1960) *J. Mol. Biol.*, **2**, 392
[12] Luria, S. E. and Darnell, J. E. (1967) *General Virology*. New York: Wiley
[13] Horsfall, F. L. and Tamm, I. (Eds.) (1965) *Viral and Rickettsial Infections of Man*. London: Pitman
[14] Stent, G. S. (1963) *Molecular Biology of Bacterial Viruses*. San Francisco: Freeman
[15] Hayes, W. (1969) *The Genetics of Bacteria and Their Viruses*. 2nd Edn. Oxford: Blackwell
[16] Fenner, F. (1968) *The Biology of Animal Viruses*. New York: Academic Press
[17] Cairns, J., Stent, G. S. and Watson, J. D. (1966) *Phage and the Origins of Molecular Biology*. New York: Cold Spring Harbor Lab. Quant. Biol.
[18] Fraenkel-Conrat, H. (Ed.) (1968) *The Molecular Basis of Virology*. Amsterdam: Reinhold
[19] Mathews, C. K. (1971) *Bacteriophage Biochemistry*. New York: Van Nostrand
[20] Tooze, J. (1973) *The Molecular Biology of Tumour Viruses*. Cold Spring Harbor Lab.
[21] Kornberg, A. (1974) *DNA Synthesis*. San Francisco: Freeman
[22] Colter, J. S. and Parenchych, W. (Eds.) (1967) *The Molecular Biology of Viruses*. New York: Academic Press
[23] Crawford, L. V. and Stoker, M. G. P. (Eds.) (1968) *The Molecular Biology of Viruses*. Cambridge: University Press
[24] Fox, C. F. and Robinson, W. S. (Eds.) (1973) *Virus Research*. New York: Academic Press
[25] Cohen, S. S. (1963) *Ann. Rev. Biochem.*, **32**, 83
[26] Wittmann, H. G. and Scholtissek, C. (1966) *Ann. Rev. Biochem.*, **35**, 299
[27] Thomas, C. A. and MacHattie, L. A. (1967) *Ann. Rev. Biochem.*, **36**, 485
[28] Crawford, L. V. (1969) *Adv. Virus Res.*, **14**, 89

[29] Horne, R. W. (1963) *Sci. Amer.*, **208**, (1), 76
[30] Gomatos, P. J. and Tamm, I. (1963) *Proc. Nat. Acad. Sci.*, **50**, 878
[31] Russell, G. J., Follett, E. A. C. and Subak-Sharpe, J. H. (1971) *J. Gen. Virol.*, **11**, 129
[32] Klug, A. and Caspar, D. L. D. (1960) *Adv. Virus Res.*, **7**, 225
[33] Knight, C. A. (1963) *Chemistry of Viruses. Protoplasmatologia*, Vol. 5, No. 2
[34] Markham, R. (1963) *Progress in Nucleic Acid Research*, Vol. 2, p. 61 (J. N. Davidson and W. E. Cohn, Eds.). New York: Academic Press
[35] Bancroft, J., Hills, G. and Markham, R. (1967) *Virology*, **31**, 354
[36] Leberman, R. (1968) *Symposium No. 18, Society for General Microbiology*, p. 183. Cambridge: University Press
[37] Crawford, L. V., Follett, E. A., Burdon, M. G. and McGeoch, D. J. (1969) *J. Gen. Virol.*, **4**, 37
[38] Watson, D. H. (1968) *Symposium No. 18, Society for General Microbiology*, p. 208. Cambridge: University Press
[39] Sinsheimer, R. L. (1959) *J. Mol. Biol.*, **1**, 43
[40] Marvin, D. A. and Hoffmann-Berling, H. (1963) *Z. Naturforsch.*, **186**, 884
[41] Kellenberger, E. (1962) *Adv. Virus Res.*, **8**, 1
[42] Crawford, L. V. (1968) *Symposium No. 18, Society for General Microbiology*, p. 163. Cambridge: University Press
[43] Sinsheimer, R. L. (1962) *Sci. Amer.*, **201**(1), 109
[44] Sinsheimer, R. (1970) *Harvey Lectures*, **64**, 69
[45] Dulbecco, R. and Vogt, M. (1963) *Proc. Nat. Acad. Sci.*, **50**, 236
[46] Weil, R. and Vinograd, J. (1963) *Proc. Nat. Acad. Sci.*, **50**, 730
[47] Winocour, E. (1967) *The Molecular Biology of Viruses*, p. 577 (J. S. Colter and W. Paranchych, Eds.). New York: Academic Press
[48] Crawford, L. V. and Black, P. H. (1964) *Virology*, **24**, 388
[49] Crawford, L. V. (1965) *J. Mol. Biol.*, **13**, 362
[50] Espejo, R. T., Canelo, E. S. and Sinsheimer, R. L. (1969) *Proc. Nat. Acad. Sci.*, **63**, 1164
[51] Tattersall, P., Crawford, L. V. and Shatkin, A. J. (1973) *J. Virol.*, **12**, 1446
[52] Hershey, A. D., Burgi, E. and Ingraham, L. (1963) *Proc. Nat. Acad. Sci.*, **49**, 748
[53] Wu, R. and Taylor, E. (1971) *J. Mol. Biol.*, **57**, 491
[54] Strack, H. B. and Kaiser, A. D. (1965) *J. Mol. Biol.*, **12**, 36
[55] Yamagishi, H., Nakamura, K. and Ozeki, H. (1965) *Biochem. Biophys. Res. Commun.*, **20**, 727
[56] Ritchie, D. A., Thomas, C. A., MacHattie, L. A. and Wensinv, P. C. (1967) *J. Mol. Biol.*, **23**, 365
[57] Thomas, C. A. and MacHattie, L. A. (1964) *Proc. Nat. Acad. Sci.*, **52**, 1297
[58] Bujard, H. (1969) *Proc. Nat. Acad. Sci.*, **62**, 1167
[59] Thomas, C. A. (1967) *J. Cell Physiol.* (Suppl.) **70**(2), 13
[60] Sinsheimer, R. L., Starman, B., Nagler, C. and Guthrie, S. (1962) *J. Mol. Biol.*, **4**, 142
[61] Young, E. T. and Sinsheimer, R. L. (1967) *J. Mol. Biol.*, **30**, 165
[62] Sinsheimer, R. L. (1968) *Symposium No. 18, Society for General Microbiology*, p. 101. Cambridge: University Press
[63] Thomas, C. A., Ritchi, D. A. and MacHattie, L. A. (1967) *The Molecular Biology of Viruses*, p. 9 (J. S. Colter and W. Paranchych, Eds.). New York: Academic Press

[64] Gilbert, W. and Dressler, D. (1968) *Cold Spring Harbor Symp. Quant. Biol.*, **33**, 473
[65] Frankel, F. R. (1966) *J. Mol. Biol.*, **18**, 127
[66] Wyatt, G. R. and Cohen, S. S. (1950) *Biochem. J.*, **55**, 774
[67] Lehman, I. R. and Pratt, E. A. (1960) *J. Biol. Chem.*, **235**, 3254
[68] Hattman, S. and Fukasawa, T. (1963) *Proc. Nat. Acad. Sci.*, **50**, 297
[69] Hershey, A. D. (Ed.) (1971) *The Bacteriophage Lambda.* New York: Cold Spring Harbor
[70] Arber, W. (1968) *Symposium No. 18, Society for General Microbiology,* p. 295. Cambridge: University Press
[71] Green, M. and Gerard, G. F. (1974) *Progr. Nucleic Acid Res. Mol. Biol.*, **15**, 1 (W. E. Cohn, Ed.)
[72] Bellamy, A. R. and Joklik, W. K. (1967) *Proc. Nat. Acad. Sci.*, **58**, 1389
[73] Crawford, L. V. (1962) *Virology*, **18**, 177
[74] Stent, G. S. and Wollman, E. L. (1950) *Biochim. Biophys. Acta*, **6**, 307
[75] Crawford, E. M. and Gesteland, R. F. (1964) *Virology*, **22**, 165
[76] Kozloff, L. M. and Lute, M. (1959) *J. Biol. Chem.*, **234**, 534
[77] McCorquodale, D. J., Oleson, A. E. and Buchanan, J. M. (1967) *The Molecular Biology of Viruses*, p. 31 (J. S. Colter and W. Paranchych, Eds.). New York: Academic Press
[78] Paranchych, W. (1966) *Virology*, **28**, 90
[79] Brinton, C. C. and Beer, H. (1967) *The Molecular Biology of Viruses*, p. 251 (J. S. Colter and W. Paranchych, Eds.). New York: Academic Press
[80] Marvin, D. A. and Hohn, B. (1969) *Bact. Rev.*, **33**, 172
[81] Keir, H. M. (1968) *Symposium No. 18, Society for General Microbiology,* p. 67. Cambridge: University Press
[82] Dulbecco, R., Hartwell, L. H. and Vogt, M. (1965) *Proc. Nat. Acad. Sci.*, **53**, 403
[83] Winocour, E., Kaye, A. M. and Stollar, V. (1965) *Virology*, **27**, 156
[84] Fried, M. and Pitts, J. D. (1968) *Virology*, **34**, 761
[85] Isaacs, A. (1964) *Adv. Virus Res.*, **10**, 1
[86] Wolstenholme, G. E. W. and O'Connor, M. (1968) *Interferon*, Ciba Foundation Symp. London: Churchill
[87] Harris, M. (1970) *Science*, **170**, 1068
[88] Colby, C. (1971) *Progress in Nucleic Acid Research and Molecular Biology*, Vol. 11, p. 1 (J. N. Davidson and W. E. Cohn, Eds.). New York: Academic Press
[89] Friedman, R. M. and Sonnabend, J. A. (1964) *Nature*, **203**, 366
[90] Joklik, W., Jungwirth, C., Oda, K. and Woodson, B. (1967) *The Molecular Biology of Viruses*, p. 473 (J. S. Colter and W. Paranchych, Eds.). New York: Academic Press
[91] Merigan, T. C. (1970) *Nature*, **228**, 219
[92] Banks, G. T., Buck, K. W., Chain, E. B., Himmelweit, F., Marks, J. E., Tyler, J. M., Hollings, M., Last, F. T. and Stone, O. M. (1968) *Nature*, **218**, 542
[93] Schafter, T. W. and Lockart, R. Z. (1970) *Nature*, **226**, 449
[94] Wood, W. B. and Edgar, R. S. (1967) *Sci. Amer.*, **217**(1), 60
[95] Fraenkel-Conrat, H. and Williams, R. C. (1955) *Proc. Nat. Acad. Sci.*, **41**, 690
[96] Fraenkel-Conrat, H. and Singer, B. (1957) *Biochim. Biophys. Acta*, **24**, 540
[97] Fraenkel-Conrat, H. and Singer, B. (1964) *Virology*, **23**, 354
[98] Hohn, T. (1967) *Eur. J. Biochem.*, **2**, 152

[99] Bancroft, J. B. and Hiebert, E. (1967) *Virology*, **32**, 354
[100] Crick, F. H. C. and Watson, J. D. (1956) *Nature*, **177**, 473
[101] Weigle, J. J. (1966) *Proc. Nat. Acad. Sci.*, **55**, 1462
[102] Streisinger, G., Mukai, F., Dreyer, W. J., Miller, B. and Horiuchi, S. (1961) *Cold Spring Harbor Symp. Quant. Biol.*, **26**, 25
[103] Thomas, R. (1968) *Symposium No. 18, Society for General Microbiology*, p. 315. Cambridge: University Press
[104] Tomizawa, J. I. and Owaga, T. (1967) *J. Mol. Biol.*, **23**, 247
[105] Campbell, A. (1962) *Adv. Genetics*, **11**, 101
[106] Lederberg, E. M. and Lederberg, J. (1953) *Genetics*, **38**, 51
[107] Jacob, F. and Wollman, E. L. (1957) *The Chemical Basis of Heredity*, p. 468 (W. D. McElroy and B. Glass, Eds.). Baltimore: Johns Hopkins
[108] Ikeda, H. and Tomizawa, J. I. (1968) *Cold Spring Harbor Symp. Quant. Biol.*, **33**, 791
[109] Meynell, G. G. (1972) *Bacterial Plasmids*. London: Macmillan
[110] Sherratt, D. J. (1974) *Cell*, **3**, 189
[111] Campbell, A. M. (1969) *Episomes*. New York: Harper and Row
[112] Nishimura, Y., Ishibashi, M., Meynell, E. and Hirota, Y. (1967) *J. Gen. Microbiol.*, **49**, 89
[113] Willetts, N. (1972) *Ann. Rev. Gen.*, **6**, 257
[114] Lovett, M. A., Katz, L. and Helinski, D. R. (1974) *Nature*, **251**, 337
[115] Nishimura, Y., Caro, L., Berg, C. M. and Hirota, Y. (1971) *J. Mol. Biol.*, **55**, 441
[116] Jacob, F. and Wollman, E. L. (1961) *Sexuality and the Genetics of Bacteria*. New York: Academic Press
[117] Goebel, W. and Schremph, H. (1972) *Biochim. Biophys. Acta*, **262**, 32
[118] Caro, L. G. and Berg, C. M. (1969) *J. Mol. Biol.*, **45**, 325
[119] Dulbecco, R. (1967) *Sci. Amer.*, **216**(4), 28
[120] MacPherson, I. (1967) *Brit. Med. Bull.*, **23**, No. 2, 144
[121] Fried, M. (1965) *Proc. Nat. Acad. Sci.*, **53**, 486
[122] Cuzin, F., Vogt, M., Dieckmann, M. and Berg, P. (1970) *J. Mol. Biol.*, **47**, 317
[123] Benjamin, T. L. (1966) *J. Mol. Biol.*, **16**, 359
[124] Watkins, J. F. and Dulbecco, R. (1967) *Proc. Nat. Acad. Sci.*, **58**, 1396
[125] Gelb, L. D., Kohne, D. E. and Martin, M. A. (1971) *J. Mol. Biol.*, **57**, 129
[126] Ozanne, B., Vogel, A., Sharp, P., Keller, W. and Sambrook, J. (1973) *Lepetit Colloq. Biol. Med.*, **4**
[127] Marin, G. and Littlefield, J. W. (1968) *J. Virol.*, **2**, 69
[128] Dulbecco, R. and Eckhart, W. (1970) *Proc. Nat. Acad. Sci.*, **67**, 1775
[129] Duesberg, P. H. (1968) *Proc. Nat. Acad. Sci.*, **60**, 1511
[130] Duesberg, P. H. (1970) *Current Topics Microbiol. Immunol.*, **51**, 74
[131] Levinson, W. E., Varmus, H. E., Garapin, A. C. and Bishop, J. M. (1972) *Science*, **175**, 76
[132] Temin, H. M. and Mitzutani, S. (1970) *Nature*, **226**, 1211
[133] Baltimore, D. (1970) *Nature*, **226**, 1209
[134] Spiegelman, S., Burny, A., Das, M. R., Keydar, J., Schlom, J., Travnicek, M. and Watson, V. (1970) *Nature*, **227**, 563
[135] Temin, H. M. (1964) *Virology*, **23**, 486
[136] Duesberg, P. H. (1971) *Proc. Nat. Acad. Sci.*, **68**, 2505
[137] Demerec, M. and Demerec, Z. E. (1956) *Mutation*, Brookhaven Symp. Biol., No. 8, p. 75
[138] Gilbert, W. and Müller-Hill, B. (1967) *Proc. Nat. Acad. Sci.*, **58**, 2415

[139] Epstein, R. H., Bolle, A., Steinberg, C. M., Kellenberger, E., Boy De La Tour, E., Chevalley, R., Edgar, R. S., Sussman, R. S., Denhardt, G. H. and Lielausis, A. (1963) *Cold Spring Harbor Symp. Quant. Biol.*, **28**, 375
[140] Edgar, R. S. and Lielausis, I. (1964) *Genetics*, **49**, 649
[141] Edgar, R. S. and Epstein, R. H. (1965) *Sci. Amer.*, **212**(2), 71
[142] Tooze, J. and Weber, K. (1967) *J. Mol. Biol.*, **28**, 311
[143] Steitz, J. A. (1968) *J. Mol. Biol.*, **33**, 923
[144] Mayol, R. F. and Sinsheimer, R. L. (1970) *J. Virol.*, **6**, 310
[145] Okazaki, R., Okazaki, T., Sakable, K., Sugimoto, K., Kaenuma, R., Sugino, A. and Iwatsuki, N. (1968) *Cold Spring Harbor Symp. Quant. Biol.*, **33**, 129
[146] Wood, W. B. and Edgar, R. S. (1967) *Sci. Amer.*, **217**(1), 60
[147] Hsu, M. T. (1974) *J. Bact.*, **118**, 425

CHAPTER 8

Nucleases and Related Enzymes

Enzymes which catalyse the breakdown of nucleic acids by hydrolysis of phosphodiester bonds have been found in almost all biological systems [1–6]. Some, the *ribonucleases*, are quite specific RNA, others, the *deoxyribonucleases*, act only on DNA, while a third group of non-specific *nucleases* is active against either nucleic acid.

The *phosphorylases*, polynucleotide phosphorylase and pyrophosphorylase, are also capable of depolymerizing RNA, but their degradative role *in vivo* is uncertain and they are dealt with elsewhere in this volume (Chapter 12).

The *phosphomonoesterases* act on polynucleotides or oligonucleotides with a terminal phosphate group or on a mononucleotide to liberate inorganic phosphate. Their substrates will often be products of nuclease action.

In all nucleolytic enzymes tested, the P–O bond is cleaved, as shown by ^{18}O incorporation [7].

8.1 Classification of nucleases

Classification schemes for the nucleases have been discussed by Laskowski [1] and by Barnard [6]. Three main features of nuclease action can be used as a basis for classification.

The first of these is *substrate specificity*, i.e. action on RNA, DNA, or both, as discussed above. The second is *mode of attack*; polynucleotides can be attacked at points within the polymer chain *endolytically* or stepwise from one end of the chain *exolytically*. Thus we may have *endonucleases* which produce oligonucleotides and cause rapid changes in physical properties (e.g. in viscosity of DNA), and *exonucleases* which produce mononucleotides but with rather less drastic effects on nucleic acid physical properties. A few enzymes appear to act as both endo- and exo-nuclease, e.g. *micrococcal nuclease* [8]. The third feature is *mode of phosphodiester bond cleavage*. Most biological polymers can, like proteins and carbohydrates, be split in only one way; polynucleotides can be cleaved in two ways to give products bearing (i) 5′-*phosphoryl end groups* by hydrolysis of the bond between the 3′-OH and the phosphate group or (ii) 3′-*phosphoryl end groups* by hydrolysis of the bond between the 5′-OH and the phosphate group.

Additional criteria may be used to define further the action of a nuclease. These include *specificity towards secondary structure of substrate, direction of attack by exonuclease* ($3' \rightarrow 5'$ or $5' \rightarrow 3'$), and *preferential endonucleolytic bond cleavage*, e.g. GpX→Gp, by ribonuclease T_1. However, in only a few cases (e.g. the restriction endonucleases) are base specificities absolute, and relative differences in reaction rates with different bases are more common.

Experimental details for the preparation and handling of several of these enzymes are to be found in the handbooks edited by Cantoni and Davies [3] and by Grossman and Moldave [9].

8.2 Ribonucleases (RNases)

8.2.1 *Endonucleases forming 3′-phosphate groups*

(a) *Pancreatic ribonuclease* (E.C.2.7.7.16)* (for reviews see [1, 10–12]). In 1920 Jones [13] described a heat-stable enzyme present in the pancreas which was capable of digesting yeast RNA. The enzyme was purified by Dubos and Thompson [15] and was crystallized in 1940 by Kunitz [16] who named it *ribonuclease*.

Crystalline pancreatic RNase prepared by the method of Kunitz tends to be contaminated with traces of proteolytic enzymes which have on occasion given rise to misleading results. The crystallization of pancreatic RNase absolutely free from proteolytic contaminants has been described by McDonald [17].

Pancreatic RNase is a very small protein, mol. wt. 13 700, is stable over a wide pH range, and is remarkably resistant to heat in slightly acid solution, although it is readily inactivated by alkali. It has no action on DNA and is strongly antigenic. Its maximum activity is in the range pH 7·0–8·2, with the optimum at pH 7·7. Its optimum temperature is 65°.

As the result of the work of Moore, Stein, and their collaborators [18, 19], the sequence of amino acids in the pancreatic RNase molecule has been fully worked out and the active site and mechanism of action determined [20–23]. Recently, the complete structure of the

* These index numbers refer to the classification system in the Report of the Enzyme Commission [14].

protein was obtained by X-ray crystallography [24, 25] and the total chemical synthesis of pancreatic RNase A has been achieved [26–28].

Pancreatic RNase is a highly specific endonuclease which splits the bond between the phosphate residue at C-3' in a *pyrimidine* nucleotide to C-5' in the next nucleotide in sequence. The basic feature of its action is an intramolecular attack on the phosphodiester bond using the 2'-OH group to form an obligatory 2':3'-cyclic phosphate intermediate which is then hydrolysed by the enzyme to give pyrimidine 3'-phosphates either as free nucleotides or as a terminal nucleotide residue in an oligonucleotide (Fig. 8.1). The products of short

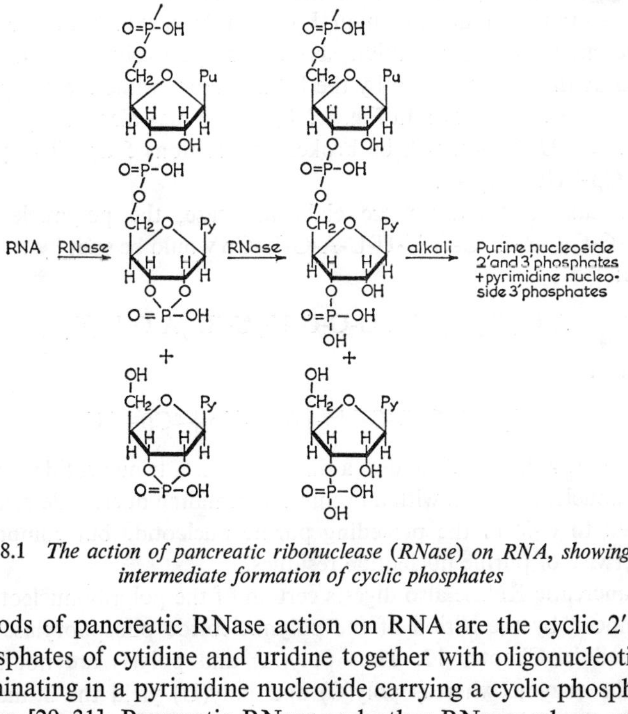

Fig. 8.1 *The action of pancreatic ribonuclease (RNase) on RNA, showing the intermediate formation of cyclic phosphates*

periods of pancreatic RNase action on RNA are the cyclic 2':3'-phosphates of cytidine and uridine together with oligonucleotides terminating in a pyrimidine nucleotide carrying a cyclic phosphate group [29–31]. Pancreatic RNase and other RNases whose mode of action is similar have been classified as *cyclizing* by the Standing Committee on Enzymes of the International Union of Biochemistry [14]. The cyclizing RNases have the possibility of forming both the 2'- and 3'-monoester by hydrolysis of the cyclic phosphate (Fig. 8.1); most, if not all, form the 3'-ester exclusively.

Subsequent treatment of oligonucleotides resulting from pancreatic ribonuclease digestion with alkali yields purine 2'- and 3'-nucleotide by cyclization and fission as described earlier, together with the 3'-isomers of the terminal nucleotides. These pyrimidine nucleotides are, of course, stable in alkaline solution; in acid they quickly yield an equilibrium mixture of the 2'- and 3'-phosphates.

Thus, pancreatic RNase may be regarded as a highly specific phosphodiesterase which will hydrolyse only secondary phosphate esters of pyrimidine nucleoside 3'-phosphates. It will therefore also hydrolyse the cyclic 2':3'-secondary phosphates of the pyrimidine nucleosides.

The action of pancreatic RNase may be illustrated as follows. The pentanucleotide shown in Fig. 8.2(a), in which Pu and Py represent purine and pyrimidine residues respectively, will be hydrolysed at the points shown by the broken lines, while the ribopolynucleotide chain shown in Fig. 8.2(b), which may also be expressed as pApCpUpGpAp, will be broken at positions 5 and 7 to yield pApCp+Up+GpAp.

To take a slightly more elaborate case, the polynucleotide A-C-C-C-C-A-G-G-G-U-U-U-A-G-U-Cp would be split by RNase thus:

$$\text{A-C-/C-/C-/-C-A-G-G-G-U-/U-/U-/A-G-U-/Cp}$$

to yield

$$\text{A-Cp+A-G-G-G-Up+A-G-Up+4Cp+2Up}$$

i.e. 6 pyrimidine nucleotides, a dinucleotide, a trinucleotide, and a pentanucleotide each with a terminal pyrimidine nucleotide residue linked to C-3' of the preceding purine nucleotide but composed otherwise of purine nucleotide residues.

Pancreatic RNase also digests certain of the polyribonucleotides produced by the action of the polynucleotide phosphorylase described in Chapter 12. Thus poly(A) and poly(I) are not split by the enzyme whereas poly(C) and poly(U) yield the 3'-mononucleotides. However, the specificity of pancreatic RNase for pyrimidines is not absolute, since Ap diester bonds in a polynucleotide are also attacked, albeit considerably less readily than Up diester bonds [32].

The action of the enzyme may be demonstrated by making use of the fact that a solution of uranyl acetate in dilute trichloroacetic acid completely precipitates RNA but not its split products [33]. Pancreatic RNase renders about half the phosphorus of the nucleic acid non-precipitable by the uranyl reagent. Enzyme activity may also be determined spectrophotometrically [34–36], manometrically [37], or by the action on RNA labelled with ^{32}P [38]. The methods available have been reviewed [23].

(b) *Ribonuclease* T_1 (E.C.3.1.4.8) [39] has been the subject of intensive study, and our knowledge of its chemistry approaches that of pancreatic RNase. It is obtained from *Aspergillus oryzae* and specifically hydrolyses the internucleotide bonds of RNA between 3'-GMP and the 5'-OH groups of adjacent nucleotides (Fig. 8.2). Ribonuclease T_1 is a small heat-stable and acid-stable endonuclease. A second enzyme from the same source, ribonuclease T_2, preferentially attacks Ap residues and will digest tRNA almost totally to 3'-monophosphates [40]. Enzymes similar to ribonuclease T_1 are common in fungi and bacteria, e.g. ribonuclease U_1 from *Ustilago sphaerogena* [41].

One important outcome of the purification and characterization of RNA endonucleases with particular specificities has been their use in the production of defined fragments of RNA. This has led to the complete sequence analysis of several RNA species. Among the RNases employed have been pancreatic and T_1 although T_2 and U_2 (from *U. sphaerogena*) are also used. This topic is illustrated in Chapter 6.

Many tissues and organisms contain RNA endonucleases which produce 3'-monophosphates, although none of these has been so closely studied as the two enzymes just described. Some of the more important of these enzymes deserve brief mention.

(c) *Rat liver* contains several ribonucleases [42–44]. Two are endonucleases forming 3'-monophosphates. One has a pH optimum at about 6·0 and hydrolyses all phosphodiester bonds in RNA with the production of nucleoside 2':3'-cyclic phosphates. The other has a pH optimum at about 8·0. It hydrolyses phosphodiester bonds only between adjacent pyrimidine nucleotides leaving purine-rich oligonucleotide tracts.

(d) *Mouse L cell nucleoli* [45] and probably also the nucleoli from *Novikoff hepatoma* [46] and *HeLa cells* [47] contain an endonucleolytic activity for single-stranded RNA with possible relevance to ribosomal precursor RNA processing. It is inhibited by Mg^{2+} and has a marked preference for cytidylate residues.

(e) *Ribonuclease NU from human KB cells*. This enzyme exhibits a high degree of specificity in that it can cleave the precursor to tyrosine tRNA in two places and the single-stranded RNA of phage Φ80 in four specific places [48].

(f) *Ribonuclease I* from *Esch. coli* normally occurs in the free state but on conversion of the cells to spheroplasts it is released into the medium. In broken cell preparations it is found in association with 30S ribosomal subunits [49]. Its intracellular role is unclear since strains lacking this enzyme are perfectly viable [50].

(g) *Ribonuclease III* from *Esch. coli* is also found in ribosome fractions [51–54]. It attacks only double-stranded RNA and requires Mg^{2+} or Mn^{2+} and K^+, Na^+, or NH_4^+. Considerable biochemical and genetic data implicate this nuclease in the post-transcriptional processing of ribosomal RNA precursors and of messenger RNA precursors (see Chapter 12). Analogous enzyme activities are found in animal cells [55–57].

8.2.2 *Endonucleases which form 5′-monophosphates*

The mechanism by which these nucleases cleave the phosphodiester bond is noncyclizing, i.e. a direct attack of water on a 3′,5′-phosphodiester is catalysed and thus a 2′-OH group is not required. For this reason, many 5′-monophosphate-forming endonucleases will attack RNA and DNA.

(a) *Ribonuclease IV* from *Esch. coli* [53, 58, 59]. This appears to be fairly specific in that it will cleave phage R17 RNA into a 15S fragment carrying the 5′-terminus of the original molecule and a 21S fragment lacking the 5′-end.

(b) *Ribonuclease P* of *Esch. coli* [60, 61]. This highly specific endonuclease is found in association with ribosomes and is responsible for the removal, as a single fragment, of the extra nucleotides from the 5′-ends of tRNA precursor molecules as shown in Chapter 12 (Fig. 12.10). Mutants deficient in this enzyme accumulate transfer RNA precursor molecules [75]. Its ionic requirements are very similar to those of ribonuclease III mentioned above. A similar activity can be detected in mammalian cells [62, 63].

(c) *Ribonuclease H*. This endonucleolytic activity was originally discovered in calf thymus tissue [64, 65] and specifically degrades the RNA strand of DNA–RNA hybrid to acid-soluble products with 5′-monophosphate ends. It does not degrade single-stranded RNA or DNA, double-stranded RNA or DNA, or the DNA strand of the DNA–RNA duplex. So far it has been purified from calf thymus tissue [71, 72], rat liver [66], chick embryo cells [67], KB cells [67], *Ustilago* [68], and *Esch. coli* [69, 70]. Cellular ribonuclease H has no associated DNA polymerase activity, and the enzyme from eukaryotic cells has a molecular weight of 70 000–90 000 [67, 72].

(d) *Rat liver alkaline ribonuclease I*. This enzyme cleaves RNA non-specifically to give products with a 5'-monophosphate terminus. The activity requires Mg^{2+} and operates maximally at pH 7·5 [49]. Similar activities have been reported in nuclei from pig heart and mouse ascites cells [78, 272].

8.2.3 *RNA exonucleases*

(a) *Ribonuclease II from Esch. coli*. RNase II is only loosely bound to the ribosomes and acts on single-stranded RNA's as an exonuclease in the 3'→5' direction. It requires both K^+ and Mg^{2+} ions [73, 74, 76, 77], and the products of its action are 5'-mononucleotides and a residual oligonucleotide fraction. Regarding a possible intracellular role, it may be involved in 'trimming' of tRNA, mRNA, and rRNA precursors after endonucleolytic cleavages by RNase P or RNase III [75].

(b) 3'-OH *specific exonucleases of mammalian nuclei*. Exonucleases specific for 3'-OH ends of single-stranded RNA's have been detected in ascites cell nuclei [78], rat liver nuclei [78], and mouse L-cell nuclei [79]. They all yield 5'-mononucleotides.

(c) *Oligoribonuclease from Esch. coli*. This recently discovered enzyme activity shows a marked preference for oligoribonucleotides. In fact the reaction rate is inversely proportional to the chain length of the substrate. For full activity Mn^{2+} is required and the exonucleolytic cleavage is in the 3'→5' direction yielding 5'-mononucleotides [80, 81]. Its cellular role may be to complete the digestion of oligoribonucleotides for example resistant to RNase II as described above.

(d) *Ribonuclease H of RNA tumour virus particles*. RNase H also turns out to be an ubiquitous activity of oncornavirions [82]. However, at least in the case of avian myeloblastosis virions [69], the nuclease activity appears to be an integral component of the virion RNA-dependent DNA polymerase (reverse transcriptase) [83]. Additionally the polymerase-associated RNase H is unlike the previously mentioned RNase H species of prokaryotic and eukaryotic origin in that it acts exonucleolytically in *both* 5'→3' and 3'→5' directions [69].

(e) One enzyme which acts, in a sense, like a limited RNA exonuclease is *CCA pyrophosphorylase*. This has been purified from *Esch. coli* [84] and, in the presence of inorganic phosphate, forms CTP and ATP by removal of the CpCpA specifically from the 3'-OH terminus of tRNA. The reaction is freely reversible and the action towards both bases appears to be catalysed by one protein [85].

(f) *Polynucleotide phosphorylase* can be regarded as a depolymerase of RNA. It catalyses the reversible reaction whereby a polyribonucleotide reacts with inorganic phosphate to yield ribonucleoside diphosphates. It is discussed in greater detail in Chapter 12. Most of the well-characterized exonucleases which degrade RNA are also active against DNA, and these enzymes will be dealt with as non-specific nucleases.

The properties of the ribonucleases known to occur in *Esch. coli* are summarized in Table 8.1.

8.2.4 *Ribonuclease inhibitors*

Rat liver contains a protein which acts as a powerful inhibitor of pancreatic RNase but does not affect RNase T_1 or plant RNases [86, 87]. Heparin also inhibits pancreatic RNase. The clay *bentonite* is a powerful inhibitor of RNase and is commonly employed to prevent degradation of RNA during isolation [88]. Polyvinyl sulphate and *Macaloid* have often been employed in a similar way, but, at least for sea urchin ribonuclease activity, they are less effective inhibitors than bentonite [89].

Diethyl pyrocarbonate (*Baycovin*) has been used as an RNase inhibitor [90], particularly in the extraction of nucleic acids. It has an advantage over several other inhibitors in that it is water-soluble, but there is evidence [91] that it reacts with RNA.

Ribonuclease from human skin [92] may seriously contaminate glassware, dialysis tubing, and other laboratory materials. A similar problem exists for deoxyribonuclease.

8.3 Non-specific nucleases
8.3.1 *Endonucleases*

(a) *Micrococcal nuclease* (E.C.3.1.4.7). This enzyme is found in cultures of *Staphylococcus* and degrades DNA to a mixture of nucleoside 3′-monophosphates and oligonucleotides with 3′-phosphate termini [93]. It attacks RNA and, preferentially, heat-denatured DNA. It requires Ca^{2+} for maximum activity. Its structure and chemical properties have been recently reviewed [94, 95]. 3′-Phosphate-forming endonucleases have been reported to be present in a variety of snake venoms [96].

TABLE 8.1

Ribonucleases of Esch. coli *(for references see text)*

	Type of activity	Substrate	Ionic requirement	Cleavage product(s)
RNase I	endo-(2′-, 3′-nucleotides as intermediates)	Single-stranded RNA's	—	3′-Mononucleotides
RNase II	exo (3′→5′ direction)	Single-stranded RNA's	K^+, Mg^{2+}	5′-Mononucleotides plus some resistant oligonucleotides
RNase III	endo-	Double-stranded RNA's	K^+, Mg^{2+}	3′-Phosphorylated oligonucleotides (10–25 units)
RNase IV	endo- (specific position)	R17 RNA		Two specific fragments (see text)
RNase P	endo- (specific) position)	tRNA precursor	Mg^{2+}, Mn^{2+} K^+, NH_4^+	tRNA plus fragment (see text)
RNase H	endo-	RNA strand of DNA–RNA hybrid	K^+, Mg^{2+}	5′-Phosphorylated oligonucleotides
OligoRNase	exo-(3′→5′ direction)	Short oligoribonucleotides	Mn^{2+}	5′-Mononucleotides
Polynucleotide phosphorylase	exo-(3′→5′ direction)	Single-stranded RNA's	Mg^{2+}	5′-Nucleoside diphosphates (see text)

(b) *Neurospora crassa nuclease* [97]. This enzyme has been considerably purified from conidia of *Neurospora* and attacks DNA or RNA to give oligonucleotides with a 5′-phosphate terminus. It exhibits a preference for guanosine or deoxyguanosine residues, but its most interesting property is an absolute requirement for denatured polynucleotide. It is active under a wide range of conditions and requires Ca^{2+} or Mg^{2+}.

(c) *Nuclease S1 from Aspergillul oryzae*. This enzyme is very similar to the *N. crassa* nuclease in that it hydrolyses phosphodiester bonds in single-stranded DNA or RNA [98, 99]. It shows a requirement for Zn^{2+}.

(d) Several 5′-phosphate-forming endonucleases have been reported from a variety of mammalian cells, in invertebrates, plants, and bacteria. A common feature is a requirement for Mg^{2+} [100].

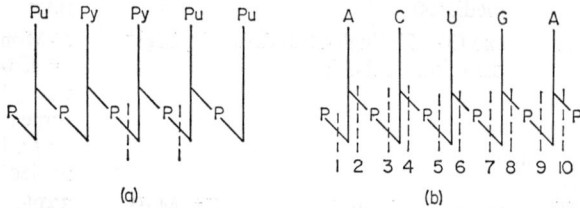

Fig. 8.2 (a) *The pentanucleotide containing 3 purine and 2 pyrimidine nucleotide units is split by ribonuclease at the broken lines.* (b) *A pentanucleotide containing 2 adenine nucleotide residues and one residue each of cytosine, uracil, and guanine nucleotides with monoesterified phosphate residues at each end is split by pancreatic ribonuclease at positions 5 and 7, by ribonuclease T_1 at position 9, by 5′-monoesterase at 1, by 3′-monoesterase at 10, by venom diesterase at 2, 4, 6, and 8, and by spleen diesterase at 3, 5, 7, and 9*

One of these, the *mung bean nuclease* of Laskowski [101] has been extensively purified.

8.3.2 *Non-specific exonucleases*

(a) *Venom phosphodiesterase* [102]. The venom of several species of snakes contains a phosphodiesterase which is commonly employed in the preparation of nucleoside 5′-phosphates. The enzyme occurs naturally in association with a high concentration of phosphomonoesterase from which it can be freed by chromatography and acetone fractionation [103, 104].

Venom diesterase hydrolyses RNA to nucleoside 5′-monophosphates (Fig. 8.2) starting at the 3′-hydroxyl end of the chain, and is also active in hydrolysing the oligonucleotides produced by the action of deoxyribonuclease I on DNA to deoxyribonucleoside

NUCLEASES AND RELATED ENZYMES

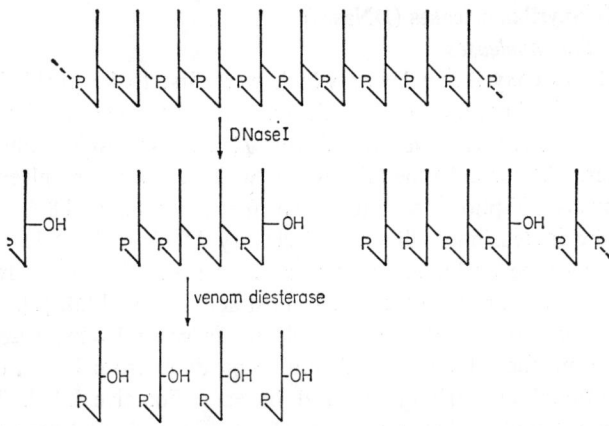

Fig. 8.3 *The digestion of DNA by DNase I followed by venom diesterase to yield deoxyribonucleoside 5'-monophosphates*

5'-phosphates (Fig. 8.3). The presence of 3'-phosphoryl terminal group confers resistance on the substrate.

(b) *Spleen phosphodiesterase* [105]. This enzyme hydrolyses RNA to nucleoside 3'-monophosphates (Fig. 8.2) starting at the 5'-hydroxyl end, and also acts on the mixture of oligonucleotides produced from DNA by spleen deoxyribonuclease II to yield deoxyribonucleoside 3'-phosphates (Fig. 8.4) [106]. It is inactive with oligonucleotides carrying a 5'-phosphomonoester end-group.

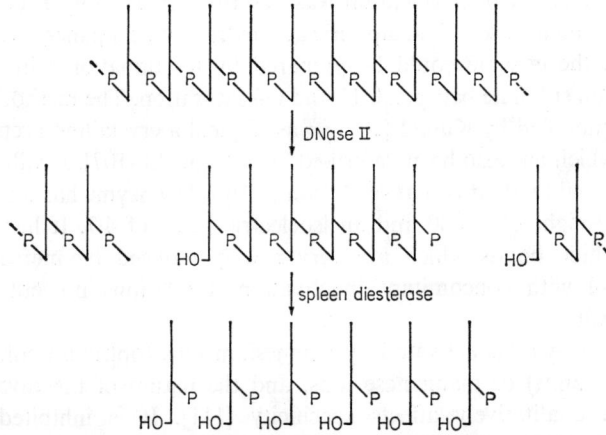

Fig. 8.4 *The digestion of DNA by DNase II followed by spleen diesterase to yield deoxyribonucleoside 3'-monophosphates*

173

8.4 Deoxyribonucleases (DNases)
8.4.1 *Endonucleases*

The two deoxyribonucleases which were first to be purified and characterized are both endonucleases. The first type exemplified by *pancreatic deoxyribonuclease* (*DNase I*) is a 5'-phosphomonoester former. The second type (*DNase II*) which is found in spleen and thymus is a 3'-phosphomonoester former (Figs. 8.3 and 8.4). It has become increasingly difficult to classify DNA endonucleases as DNase I-type or DNase II-type and we should probably regard these as extremes between which most activities will fall [1].

The activity of DNA endonucleases is generally measured by estimating the release of acid-soluble products from DNA, either as ultraviolet-absorbing material or as radioactive label. These methods are useful in the presence of extensive endonuclease action; where extreme sensitivity has been required, supercoiled circular viral DNA is used as a substrate. Only one phosphodiester bond cleavage is required to alter the physical properties of such molecules and allow separation of intact and cleaved molecules.

(a) *Pancreatic deoxyribonuclease* (*DNase I*) (E.C.3.1.4.5) (for reviews see [10, 93, 107]). This enzyme breaks down DNA into oligonucleotides of average chain length 4 units with a free hydroxyl group on position 3' and a phosphate group on position 5' (Fig. 8.3). It requires magnesium ions and has an alkaline optimum pH in the range 6·8–8·2.

The method of purification was described in 1946 by McCarty [108], who used a 0·25N-sulphuric acid extract of beef pancreas from which the enzyme could be prepared by fractionation with ammonium sulphate between 0·17 and 0·3 saturation. The method has been modified by Kunitz [109] so as to yield a crystalline preparation which has also been described by McDonald [107]. A valuable additional method is that of Polson [110]. The enzyme has a molecular weight of 31 000 and an isoelectric point of 4·7. It has two disulphide bonds which are very readily reduced by mercaptoethanol with concomitant inactivation. Ca^{2+} ions prevent this inactivation.

The enzyme is activated by magnesium ions (optimum concentration 3mM) or manganese ions, and the nature of the divalent cation qualitatively affects specificity [111]. It is inhibited by fluoride [112], and citrate at 0·01M inhibits completely the magnesium-activated but not the manganese-activated enzyme.

Citrate, borate, and fluoride exert their inhibitory action by removing the activating magnesium ions while other inhibitors such as sodium sulphide and thioglycollic acid appear to react with the functional groups of the enzyme protein [113].

Pancreatic DNase hydrolyses native DNA more rapidly than denatured DNA. As the early products of the reaction are worse substrates than the initial DNA the enzyme is autoretarding. In the early stages of the reaction single-stranded nicks are produced towards the centre of the DNA molecule [114] but later on the Pu-p-Py bond is preferentially cleaved leading to a final product of di- and oligo-nucleotides. Small oligodeoxyribonucleotides, apurinic acid [115], or deaminated single-stranded DNA [116] are not hydrolysed. The biosynthetic polymers poly(dA)·poly(dT), poly(dI)·poly(dC), poly(dG)·poly(dC) are degraded in part by pancreatic DNase. The resistance of the poly(dC) chain in the latter two copolymers to hydrolysis by the enzyme is overcome by adding Ca^{2+} to the Mg^{2+} or by replacing Mg^{2+} by Mn^{2+} [111]. The enzyme can be freed of ribonuclease contamination by electrophoresis [117] or by ion-exchange chromatography [118], and has been shown to exist in multiple forms [101].

(b) *Deoxyribonuclease II (DNase II)* (E.C.3.1.4.6) (for reviews see [10, 119]). A deoxyribonuclease of molecular weight 40 000, pH optimum in the range 4·5–5·5, and no requirement for magnesium ions has been isolated from spleen and thymus.

Double-stranded DNA is degraded by splenic DNase II, in part by a 'one-hit' process that hydrolyses both strands of the double helix at the same point [114, 120]. This initial phase of the reaction is followed by the slower release of oligonucleotides of chain length from 14 to 100 nucleotides. The final stage produces oligonucleotides of average chain length 6 units, which have a free 5′-hydroxyl group and a phosphate residue on position 3′ (Fig. 8.4).

The properties of the two main types of DNase are summarized in Table 8.2

(c) *Streptococcal deoxyribonuclease.* Streptococcal deoxyribonuclease (streptodornase) is a deoxyribonuclease of the endonuclease type, cleaving the 3′-phosphate bond, and producing 5′-phosphoryl-terminated fragments of various lengths [121]. Only traces of mononucleotides are produced, together with small amounts of dinucleotides, but the majority of the fragments are larger than dinucleotides. The preferential cleavage involves the pY-R bonds. The optimal pH is 7, and the optimal Mg^{2+} ion concentration is 0·02M [122]. The enzyme is inhibited by RNA.

At least two other distinct DNases have been characterized from group A *Streptococci* [123].

(d) *Endonuclease I from* Esch. coli [121, 124–126]. This enzyme is an endonuclease which attacks DNA producing scissions at many points along the DNA chain. At each scission an exonucleolytic activity removes about 400 nucleotides [127] which are released as a mixture of oligonucleotides of average chain length 7 units

TABLE 8.2

The properties DNase I and DNase II

	DNase I	DNase II
Substrate	DNA	DNA
pH optimum	7–8	4–5
Activators	Mg^{2+}, Mn^{2+}, Co^{2+}	0·3M-Na^+
Inhibitors	Citrate, EDTA	Mg^{2+}
Product	5'-Phosphoryl terminated oligonucleotides	3'-Phosphoryl terminated oligonucleotides

terminated by a 5'-phosphoryl group. It is highly specific for DNA, attacking native DNA seven times more readily than denatured DNA to give random double-stranded breaks [128]. RNA is an inhibitor of the action of this enzyme.

(e) *Endonuclease II from* Esch. coli [129, 130] produces single-strand breaks in double-stranded DNA which has been alkylated with monofunctional alkylating agents (e.g. methyl methanesulphonate). It can also slowly attack non-alkylated DNA. The role of endonuclease II in repair is discussed in Chapter 11. It has a pH optimum at pH 8–9 and is stimulated by Mg^{2+} or Mn^{2+}. It produces single-stranded nicks with 5'-phosphoryl end groups. (A similar enzyme which recognizes thymine dimers in duplex DNA is also discussed in Chapter 11.) It has been shown recently that endonuclease II and exonuclease III activities are present in the same enzyme molecule [131, 132].

(f) *ATP-dependent endonucleases.* DNA endonuclease activity dependent on the presence of ATP has recently been described in several organisms [133, 134]. The enzyme from *M. luteus* acts preferentially on native double-stranded DNA to give fragments with 5'-monophosphate termini [134]. Three molecules of ATP are hydrolysed for every phosphodiester bond cleaved [135]. An inter-

NUCLEASES AND RELATED ENZYMES

esting feature of the *Esch. coli* enzyme is its ability to nick single-stranded circular DNA of phage fd [136]. Associated with purified preparations is an exonuclease activity towards native DNA, also ATP-dependent. These activities have been implicated in the recombination process.

(g) *Phage-induced endonucleases.* Nucleases are induced following infection of bacteria with a variety of bacteriophages. Two of the best studied are endonuclease II and endonuclease IV induced after infection of *Esch. coli* with phage T4 [137–139].

(1) *Endonuclease II* makes single-strand breaks in double-stranded DNA other than that of T4 to give products (at least from phage λ) of about 10^3 nucleotides. These have 5′-phosphoryl and 3′-OH termini. It differs from *Esch. coli* endonuclease II in its inability to attack T4 DNA (glucosylated or not).

(2) *Endonuclease IV* hydrolyses single-stranded DNA to give considerably smaller products than endonuclease II, again with 5′-phosphoryl termini, but with dCMP exclusively in that position see Chapter 5. DNA containing hydroxymethylcytosine (i.e. T4 DNA) is inactive as a substrate.

A mechanism has been suggested [139] whereby these enzymes, with the help of a bacteriophage-induced exonuclease (see p. 178) may be involved in the degradation of host DNA after T4 infection (Fig. 8.5).

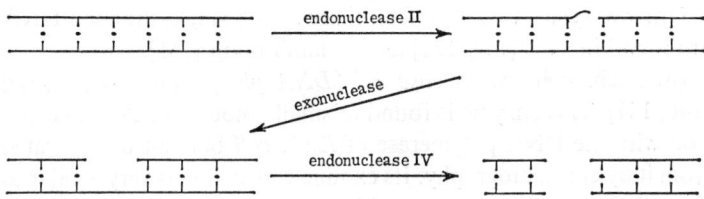

Fig. 8.5 *Hypothetical scheme for degradation of cell DNA after infection by bacteriophage T4*

(3) Another endonuclease which plays a specific role in the replication of bacteriophage DNA is that coded for by gene A of ØX174. This breaks a single phosphodiester bond to initiate phage DNA synthesis [140].

(h) *Mammalian virus endonucleases.* Much less is known about DNases induced by mammalian viruses than about phage-induced activities. However, deoxyribonuclease activities are increased after infection with several mammalian viruses. Some of these form part of the virus structure, and an example is the *endonuclease* associated with the penton protein of *adenovirus 2*. This enzyme preferentially attacks native DNA, cleaves both strands to give large fragments, and is inhibited by tRNA. It is active against all DNA's tested except glucosylated T4 DNA [141]. An endonuclease is also present in the core of *vaccinia virus* particles [142].

8.4.2 *DNA exonucleases*

The bacterial DNA-specific exonucleases (phosphodiesterases) from *Esch. coli* are of considerable general interest and their mode of action has been worked out in some detail [124, 143–147]. Their properties are summarized in Table 8.3 and [271].

(a) Esch. coli *exonuclease I* [143, 144, 148]. This enzyme hydrolyses heat-denatured single-stranded DNA and has hardly any effect on native double-stranded DNA. It is an exonuclease hydrolysing the DNA chain stepwise beginning at the 3'-hydroxyl end, and releasing deoxyribonucleoside 5'-monophosphates until only a dinucleotide is left.

The enzyme does not cleave free dinucleotides or the 5'-terminal dinucleotide portion of a polydeoxyribonucleotide chain, but it can degrade bacteriophage DNA's containing glucosylated hydroxymethylcytosine (p. 145) quantitatively to their constituent mononucleotides. It has no effect on polyribonucleotides.

Other enzymes which preferentially attack single-stranded DNA are found in liver [149, 150] and in lamb brain [151].

(b) Esch. coli *exonuclease III* (*DNA phosphatase-exonuclease*) [146, 147]. This enzyme is found in small amounts in close association with the DNA polymerase of *Esch. coli* but can be separated from it by chromatography. Its exonuclease action is very similar to that of *Esch. coli* exonuclease II but, in addition, it acts as a phosphatase highly specific for a phosphate residue esterified to the 3'-hydroxyl terminus of a DNA chain (Fig. 8.6). It does not release inorganic phosphate from deoxyribonucleoside 3'- or 5'-monophosphates from oligodeoxyribonucleotides of short chain length or from 3'-phosphoryl-terminated RNA, but it does attack DNA with a phosphoribonucleotide terminus.

As an exonuclease it carries out a stepwise attack on the 3'-hydroxyl end of the DNA chain releasing mononucleotides (Fig. 8.5) but it acts only on double-stranded DNA, degrading it until

35–45 per cent has been digested. If the enzyme begins its attack from both 3'-hydroxyl ends of the double-stranded molecule (Fig. 8.7), when nearly half has been degraded, the residual acid-insoluble DNA will be single-stranded and resistant to further attack although it is still susceptible to the action of exonuclease I.

Similar enzymes are found in *Diplococcus pneumonia* [152] and *B. subtilis* [153]. An enzyme of opposite polarity has been found in

TABLE 8.3

Properties of DNA exonucleases from Esch. coli [271]

Enzyme	Reqd. DNA structure	End group attacked	Extent of action	Products
Exonuclease I	Single-stranded	3'-OH	Up to terminal dinucleotide	Mono-and di-nucleoside 5'-monophosphates
Exonuclease II (3'→5' activity associated with DNA polymerase I)	Prefers single-stranded	3'-OH	Complete	Nucleoside 5'-monophosphates
Exonuclease III	Double-stranded	3'-OH or 3'-OP (initial attack removes terminal P_i)	To 40 per cent degradation	P_i nucleoside 5'-monophosphates and larger, single-stranded oligonucleotides
Exonuclease IV	Prefers oligonucleotides	3'-OH	Complete (cleaves dinucleotides)	Nucleoside 5'-monophosphates
Exonuclease V	Single- and double-stranded	3'-OH or 5'-OH		Oligonucleotides
Exonuclease VI (5→3' activity associated with DNA polymerase I)	Double-stranded	5'-OH or 3'-OP	Excises mismatched regions	Mostly mono- and di-nucleotides with 20–25% longer
Exonuclease VII	Single-stranded	3'-OH or 5'-OH		Oligonucleotides

bacterial cells infected with bacteriophage λ which produces deoxyribonucleoside 5′-monophosphates stepwise from the 5′-end of DNA chains [142]. This enzyme is involved in λ-recombination [154].

(c) Esch. coli *exonuclease IV*. This exonuclease shows little

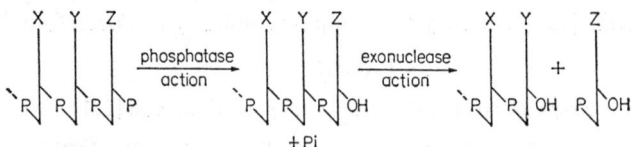

Fig. 8.6 *Sequential action of* Esch. coli *exonuclease III (DNA phosphatase-exonuclease) on a DNA chain terminated by a nucleotide carrying a 3′-phosphate group*

activity towards single- or double-stranded DNA, and exhibits a considerable preference (twenty-fold) for DNA predigested with pancreatic deoxyribonuclease. In this sense it could be termed *oligonucleotide diesterase*. It can be separated by DEAE-cellulose chromatography into two fractions (IVA and IVB) [155].

(d) Esch. coli *exonuclease V*. The role of this enzyme in recombination is discussed on p. 305. It is made of two subunits and one

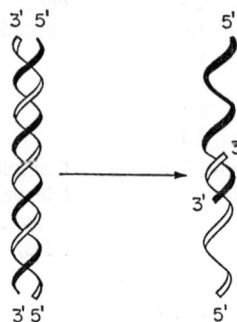

Fig. 8.7 *Mechanism of action of stepwise attack of* Esch. coli *exonuclease III on native DNA beginning at the 3′-hydroxyl terminus*

of them (β) is defective in mutants with altered *rec B* or *rec C* genes. The enzyme shows DNA-dependent ATPase activity as well as ATP-dependent DNase activity and may act as both an endo- and an exo-nuclease [156].

(e) *Exonucleases associated with* Esch. coli *DNA polymerase I.*

The terms exonuclease II and VI have been used to define the $3' \to 5'$ and the $5' \to 3'$ DNA exonuclease activities which form part of the protein of *Esch. coli* DNA polymerase I (Chapter 11) [157].

$3' \to 5'$ *activity* [145]. This activity resides in the same protein molecule that possesses DNA polymerase activity (p. 277). Like exonuclease I it commences attack at the 3'-hydroxyl terminus of a polydeoxyribonucleotide chain, with the stepwise release of deoxyribonucleoside 5'-monophosphates, but unlike exonuclease I it also attacks dinucleotides. It will, for example, hydrolyse the oligonucleotide pT-T-T-T-T to 5pT. The enzyme attacks denatured DNA in preference to native DNA [158] and has no effect on oligonucleotides bearing a 3'-phosphomonoester group or on RNA. Most bacterial and phage DNA polymerases so far investigated (with the possible exception of those from *B. subtilis*) have an associated $3' \to 5'$ exonuclease [157].

Evidence that the enzyme is in fact an exonuclease with these properties comes from several sources.

(1) Exhaustive digestion of ^{32}P-labelled d(A-T) copolymer results in the conversion of 99 per cent of the ^{32}P into an acid-soluble form which can be accounted for in terms of 5'-monophosphates.

(2) Partial digestion results in the release of a proportion of radioactivity which is the same as the proportion of monophosphates formed.

(3) When d(A-T) copolymer specifically labelled with ^{32}P-dTMP at the 3'-hydroxyl end is used as substrate, 90 per cent of the ^{32}P-labelled material is made acid-soluble when less than 10 per cent of the unlabelled nucleotides from the interior of the chain have been released. This indicates attack from the 3'-hydroxyl end of the chain.

(4) Treatment of DNA having transforming activity (p. 223) from *B. subtilis* results in a 36 per cent drop in viscosity with 46 per cent of the initial transforming activity still present. With an endonuclease (DNase I) a 36 per cent drop in viscosity is accompanied by a drop in transforming activity to 0·1 per cent of the original value owing to breakage of the chains at critical regions.

(5) With ^{32}P-labelled native DNA as substrate the decrease in viscosity is more rapid than the release of ^{32}P-mononucleotides. This suggests an exonucleolytic attack on a double helical polynucleotide of opposite polarity with initiation of the hydrolysis at the two 3'-

hydroxyl groups at opposite ends leaving the opposing strand in single-stranded form of low viscosity.

$5' \to 3'$ *activity*. This activity is specific towards native DNA and will function in the presence and absence of a 5'-phosphate group on the substrate. The products are mostly mononucleotides, with 20–25 per cent dinucleotides or longer [159].

The DNA polymerase I molecule is susceptible to protease action such that it specifically splits into two fragments. One of these (76 000 mol. wt.) has DNA polymerase I and $3' \to 5'$ exonuclease activity; the other (34 000 mol. wt.) shows $5' \to 3'$ exonuclease activity but only if the cleavage is carried out in the presence of DNA [160] (p. 277). *B. subtilis* DNA polymerase contains no $5' \to 3'$ exonuclease activity [153] but a separate enzyme exists with this function. $5' \to 3'$ exonuclease activity is also present in *E. coli* DNA polymerase III, but is absent from *E. coli* DNA polymerase II.

(f) *Exonuclease VII* degrades single-stranded DNA from either end, releasing oligonucleotides. It does not require ATP [271].

8.4.3 *Virus-induced DNA exonucleases*

Phage T4 induced DNA polymerase has a $3' \to 5'$ exonuclease activity similar to that of the host polymerase [158]. In addition Koerner [161] has described an *exonuclease A* from bacteriophage T4-infected cells which is distinct from the polymerase-associated activity. This is an *oligonucleotide diesterase* (as *Esch. coli* exonuclease IV) which liberates 5'-monophosphates quantitatively from the 3'-terminus [162]. It is possible that this exonuclease is involved in degradation of cell DNA (see Fig. 8.5).

Phage SP3 induces in *B. subtilis* SB19 an exonuclease which attacks the 5'-terminus of single-stranded DNA, but releases only dinucleotides (90 per cent) and trinucleotides [163]. At no time during digestion can larger fragments be detected.

Poxvirus particles contain an exonuclease activity and a second is induced in infected cells [164]. Exonuclease is induced after *herpes virus* infection [165]. The enzyme is a 5'-phosphate-former and acts on native or denatured DNA.

8.4.4 *Mammalian DNA exonucleases*

Activities corresponding to two exonucleases are found in *rabbit liver* [166, 167]. *DNase III* preferentially attacks denatured DNA from the 3'-terminus to give dinucleotides in addition to monomers.

NUCLEASES AND RELATED ENZYMES

DNase IV is specific for native DNA and attacks from the 5'-terminus.

8.4.5 *Deoxyribonuclease inhibitors*

Most animal tissues contain enzymes similar to DNase I and DNase II. Many tissues also contain inhibitors of both, of which the best known are the protein inhibitors of DNase I, which are particularly abundant in the crop gland of pigeons [93] and in calf spleen [168].

8.5 DNA methylation

8.5.1 *Cellular DNA*

Two classes of methylated base occur in the DNA of all cellular organisms so far examined. The first is derived from dTTP directly during DNA synthesis and the second from methylation of a preformed polydeoxynucleotide. This second class of methylated bases contains only 5-methylcytosine, first discovered by Hotchkiss in calf thymus DNA [85] and 6-methyladenine which appears to be restricted to the DNA from lower organisms. 6-Methyladenine may be present as the sole product of polynucleotide methylation, e.g. in *Esch. coli* 15T$^-$ which contains 1 mole per cent base as 6-methyladenine [169].

5-Methylcytosine, which is widespread in nature, may also occur alone and to greatly differing extents, from an almost negligible level in insect DNA [170, 171] through 1·0–1·5 per cent in mammalian DNA [170] to 5–6 per cent in plant DNA [170, 172]. Variation in DNA methylation occurs between organelles in certain cell types. Main band DNA from *Euglena gracilis* is methylated but satellite DNA is not [173]; in tobacco leaves nuclear DNA is methylated but not chloroplast DNA [172]. On the other hand, the level of methylation of mouse nuclear satellite DNA is twice that of the main band [174]. The chromosomal genes for ribosomal RNA are methylated in *Xenopus* but the amplified ribosomal RNA genes present in the oocyte are not [175].

8.5.2 *Viral DNA*

5-Methylcytosine and 6-methyladenine also occur in bacteriophage DNA, in addition to many other modifications, e.g. glucosylated 5-hydroxymethylcytosine [176]. Bacteriophage T2, T4, T7, and P1 [177, 178] contain 6-methyladenine while T3 and T5 have a complete lack of methylated bases other than thymine [179]. Polyoma, herpes simplex type I, and pseudorabies virus DNA's are not methylated at the polynucleotide level [180–182].

8.5.3 *The methylation reaction*

Methylation of DNA takes place at the polynucleotide level [183], and the reactions are catalysed by specific enzymes, the DNA methylases (methyl-transferases). The source of methyl groups is methionine, in the form of an intermediate with a high free energy of hydrolysis, S-adenosyl-L-methionine [184]. The mechanism is illustrated in Fig. 8.8.

At least one, and possibly two, DNA methylases have been

S-adenosylmethionine (SAM) + base (in DNA or RNA) →[methylase] S-adenosylhomocysteine + methylated base (in DNA or RNA)

cytosine (in RNA or DNA) + SAM → S-adenosylhomocysteine + 5-methyl-cytosine (in RNA or DNA)

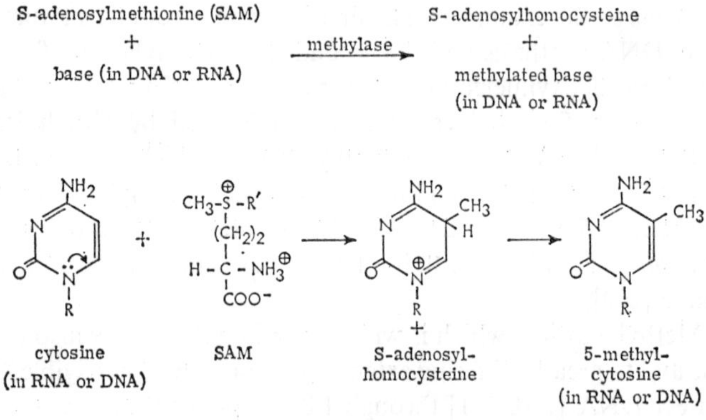

Fig. 8.8 *The nucleic acid methylase reaction*

purified from *Esch. coli W* [185] and four DNA methylases have been purified from *Haemophilus influenza* Rd [186]. These enzymes methylate specific sequences corresponding to those cleaved by the restriction endonucleases (see below and Table 8.4).

Certain bacteriophages induce DNA methylase activities which are distinct from those of their host cells and which may therefore play a role in the infective process. Among these are T1, T2, and T4, while T3 produces an enzyme which cleaves S-adenosyl-methionine and thereby inhibits DNA methylation [187, 188].

8.5.4 *Mammalian DNA methylase activity*

This is normally found in the chromatin fraction of the cell [189–193] and the majority of methyl groups are added shortly after the DNA is made [194, 195]. However, there is a delay before methyla-

TABLE 8·4

DNA restriction and modification enzymes

Bacterial strain	Enzyme	Restriction and modification site	Other enzymes recognizing identical sequences ϕ
Class I			
Esch. coli B	EcoB	contains GA*, A*A, CA* and A*C†	
Esch. coli PI	EcoPI	AGA*TC†	
Class II			
Esch. coli RI	EcoRI	G↓AA*TTC	
Esch. coli RII	EcorRII	↓CC*TGG and ↓CC*AGG	
Bacillus subtilis X5	BsuI	GG↓CC	HaeIII, HhgI, SfaI
Bacillus trevis		GC*TGC and GC*AGC	
Bacillus amyloliquifaciens	BamI	GGATCC	
Haemophilus influenzae Re	HincII	GTPy↓PuA*C	HindII
Haemophilus influenzae Rd	HindI	CA*CNGTG	
Haemophilus influenzae Rd	HindIII	A*↓AGCTT	HinbIII, HsuIII, BbrI
Haemophilus influenzae Rd	HindIV	contains GA*T	
Haemophilus parainfluenzae	HpaI	GTT↓AAC	ApoI (HincII)
Haemophilus parainfluenzae	HpaII	C↓CGG	HapII, MnoI
Haemophilus haemolyticus	HhaI	GCG↓C	
Anabaena variabilis	AvaI	CGPu↓PyCG	
Anabaena variabilis	AvaII	GTC↓GAC and GAC↓GTC	(HincII)
Arthiobacter luteus	AluI	AGCT	

When known, modified bases are all starred and the restricted site is arrowed; † indicates that the restriction site of class I enzymes is indefinite ϕ. The bacterial strains are *Haemophilus aegyptius* (Hae), *Haemophilus haemoglobinophilus* (Hhg), *Streptococcus faecalis* (Sfa), *Haemophilus aphrophilus* (Hap), *Maraxella nonliquifaciens* (Mno), *Arthrobacter polychromogenes* (Apo), *Haemophilus suis* (Hsu), *Bordatella bronchioseptica* (Bbr.)

tion is complete and homologous DNA will therefore act as substrate for mammalian DNA methylase [192, 194, 196].

One interesting feature of methylation of DNA concerns the doublet CpG. This doublet occurs with low frequency in mammalian DNA's (see Chapter 5) and is present as 5-methyl CpG primarily; this accounts for most, of the 5-methylcytosine in mammalian DNA [197]. A similar situation applies in sea urchin DNA [198].

8.5.5 *Function of DNA methylation*

Methylation of specific sites in DNA by a host-specific methylase now seems to form the modification role in modification–restriction phenomena observed in bacteria [176] (see below). In *Esch. coli*, continuing DNA synthesis *in vivo* appears to require the formation of a normally methylated DNA template [199]. However, mutants of *Esch. coli* K12 with highly undermethylated DNA grow normally [200] and removal of 6-methyladenine from T2 and T4 DNA has no effect on several biological properties [201].

8.6 DNA restriction and modification

8.6.1 *The phenomenon*

When a bacteriophage is transferred from growth on one host to a different host, its efficiency is frequently impaired several thousand-fold. However, when those bacteriophages which do survive and multiply are used to reinfect the second host they now grow normally.

The initial poor growth is caused by the action on the bacteriophage DNA of highly specific bacterial endonucleases (known as restriction endonucleases). Following restriction, the invading bacteriophage DNA is rapidly degraded to nucleotides by exonuclease action, although certain regions may be rescued by recombination [202].

Host (i.e. bacterial) DNA is not degraded because the nucleotide sequence which is recognized by the restriction endonucleases has been modified by methylation (see above). The few molecules of bacteriophage DNA which survive the initial infection do so because they are themselves modified by the host methylase before the restriction enzyme has time to act. Similarly, methylation of progeny DNA renders it resistant in the second infection [203].

A related phenomenon is the degradation of cytosine containing

NUCLEASES AND RELATED ENZYMES

DNA in *Esch. coli* infected with T-even bacteriophage whose own DNA contains hydroxymethylcytosine (see p. 145).

8.6.2 Restriction endonucleases

There are two classes of restriction enzymes. Those responsible for restriction in *E. coli* strains B and K are representative of class I and most of the remainder so far investigated are in Class II [204–206].

(a) *Class I* (e.g. EcoB, EcoK, EcoPI) are complex multifunctional proteins which cleave unmodified DNA in the presence of S-adenosyl-L-methionine (SAM), ATP, and Mg^{2+} [29]. These enzymes, which are also methylases and ATPases, have two or more non-identical subunits and molecular weights around 300 000 [206]. The enzymes bind to DNA at a specific site in the presence of SAM [208]. If this site is methylated in both strands of the DNA the enzyme does not recognize it. If the site is methylated on one strand only (as would be the case with DNA immediately following synthesis) the enzyme binds to this site and methylates the second strand. If the site is unmodified in both strands the enzyme is triggered into its restriction mode. However, restriction does not occur at the binding site but at other sites situated apparently at random. Cleavage occurs in two stages (first one strand, followed some time later by a break on the opposite strand) and at this point the enzyme ceases to be a nuclease (i.e. it performs only one restriction event) and becomes a vigorous ATPase. About 10^5 ATP molecules are hydrolysed for each restriction event [202]. (EcoPI differs from EcoK in that it does not become an ATPase.) The requirement for SAM in the initial binding reaction is perhaps a safety device to prevent the breakdown of host DNA formed when conditions of methionine deprivation have limited methylation [202].

(b) *Class II* (e.g. EcoRI, HpaII) are simpler enzymes which require only Mg^{2+} for activity. They have molecular weights ranging from 20 000 to 100 000 and EcoR1 has two identical subunits. These enzymes, which are neither methylases nor ATPases, recognize a specific site on the DNA, and if this is unmodified cleavage occurs at this site. DNA modified on one strand is not a substrate for restriction, but is a substrate for a complementary methylase which recognizes the same specific nucleotide sequence. All sites are 4–6 nucleotide pairs long and have a two-fold rotational symmetry (see Table 8.4) which suggests that the two enzyme subunits may be arranged with two-fold symmetry [206]. Cleavage at the site may be staggered by up to 4 nucleotides (e.g. EcoRI) when identical self complementary, cohesive termini are produced. BsuI and HindII produce even breaks without single-stranded termini (see Table 8.4).

EcoRII is unusual in that the cleavage site is a sequence of five nucleotide pairs and hence not all the ends are mutually cohesive. In this case the complementary methylase modifies two cytosines in different sequences.

8.6.3 *Nomenclature*

The restriction endonucleases are named according to the system described by Smith and Nathans [209]. The first three letters give the name of the bacterium (e.g. Eco for *Esch. coli*) and the fourth indicates the strain (e.g. EcoR, Hind). Where more than one restriction enzyme is found in a particular strain these are indicated by Roman numerals (e.g. HinaI, HinaII).

8.6.4 *Applications*

Since their discovery the class II restriction enzymes have become increasingly used as tools for the biochemist. Because of their ability to make relatively few specific cuts they can be used as a first step in the sequencing of DNA (see Chapter 5) or in the isolation of specific genes. However, their most important contribution so far has been in the field of chromosome mapping. Thus the small chromosome of the tumour virus SV40 (simian virus 40) is cleaved by each restriction enzyme into a small number of discrete pieces which are readily separated by electrophoresis in gels of polyacrylamide or agarose [206, 210, 211]. The size of the fragments can be determined from their speed of migration in the gel or by direct length measurements using the electron microscope. The order of the fragments in the genome can be determined by a variety of techniques such as by analysis of partial digests or by successive cleavage by multiple restriction endonucleases [206, 212]. The knowledge of the chromosome map thus gained may be used to compare closely related organisms and to localize various functions to discrete regions of the chromosome. Thus it has been shown that early viral functions (see Chapter 7) are coded for by the continuous fragments A, H, I, and B on the HindII/HindIII restriction map of

SV40 (see Fig. 8.9) [213]. Similarly the origin and direction of replication of several viral and plasmid chromosomes have been discovered in particular by using EcoRI [214, 215] (see Chapter 11).

As cleavage by many of these restriction enzymes is staggered, this leads to production of fragments with 'sticky' ends (i.e. termini with overlapping self-complementary sequences) which can be rejoined by DNA ligase (see Chapter 11). Moreover, fragments from different genomes can be joined together to form hybrid genomes, and a number of potential uses can be envisaged for such hybrids.

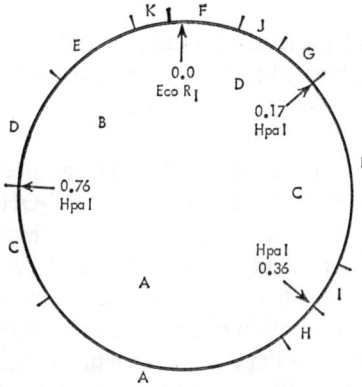

Fig. 8.9 *Map of the SV40 chromosome* [211, 212] *The fragments enumerated outside the circle in order of decreasing size are produced by the joint action of HindII and HindIII. Those enumerated within the circle are produced by the* HpaI *and* EcoRI

Thus mammalian genes could be coupled to bacteriophage or plasmid chromosomes, and the cloned hybrids used to study the control of gene expression and possibly to act as a source of large quantities of specific mammalian proteins. Imagine the benefits which might accrue if *Esch. coli* synthesized large quantities of insulin. However, this type of biochemical engineering also presents potential hazards such as the transfer of drug resistance factors to haemolytic streptococci, or the amplification in *Esch. coli* (an indigenous inhabitant of the human gut) of potentially tumorigenic genes from animal viruses. Such hazards have recently been delineated and the necessary precautions indicated [216, 217].

8.7 Enzymic modification of RNA

It is now clear that most cellular RNA species, with the notable exception of the small 5S ribosomal RNA, contain modified nucleosides to varying extents.

8.7.1 *Transfer RNA*

Since the discovery of tRNA in 1958 the presence of a large variety of minor nucleosides has been considered to be one of its characteristics, e.g. dihydrouridine, pseudouridine, ribothymidine (see Chapter 6 for location in tRNA molecules) (for reviews see [218–222]).

Other modified nucleosides found in tRNA can be classed as follows: (i) those located in the first position of the anticodon, e.g. uridin-5-oxyacetic acid, 5-methylaminomethyl-2-thiouridine, inosine. The first of these can recognize A, G, or U in the wobble position of the codon (see Chapter 13) whereas inosine can recognize C, U, and A; (ii) those located next to the 3'-OH end of the anticodon, e.g. 6-(isopent-2-enyl)adenosine, 2-methylthio-6-(isopent-2-enyl)adenosine, 6-methyladenosine, 2-methyladenosine, 1-methylguanosine. Such modified nucleosides adjacent to the anticodon may facilitate the formation of precise codon–anticodon base pairs by stabilizing the three-dimensional structure of the anticodon loop; (iii) those located elsewhere in the tRNA molecule, e.g. 7-methylguanosine, 4-thiouridine, 2-dimethylguanosine, 1-methyladenosine, and 5-methylcytosine. These are generally present in single-stranded parts of the clover-leaf structure, and although their role is unknown they may serve to stabilize the three-dimensional structure by preventing incorrect base-pairing [221].

The tRNA methylases catalyse the transfer of an intact methyl group from S-adenosyl-L-methionine to a C, N, or O atom of a purine or pyrimidine base or of ribose (Fig. 8.8). The activity was first described by Borek [223] and is present predominantly in the soluble or cytoplasmic fraction of cells, although tRNA synthesis is nuclear [224–228, 189].

Several tRNA methylases are present in any one cell system. For example, Hurwitz and his colleagues [224] separated six activities from *Esch. coli*. Eight activities have been studied in yeast [229] and a growing number have been purified from higher eukaryotic sources [230–234], including mitochondria [235]. tRNA methylases will methylate homologous substrate only if it is methyl-deficient

[223, 236] and can act on heterologous substrates to variable extents.

Several neoplastic tissues, both experimentally produced and spontaneous [225, 237–239], contain elevated levels of tRNA methylase activity with some changes in specificity [240, 241]. However, whether this differential activity is a reflection of new tRNA synthesis or of increased levels of methylation in tRNA remains to be established. Hormone treatment [244] and virus [245] and bacteriophage [242, 243] infection affect tRNA methylase activity. T2 causes changes in base-specific methylation after infection [242] and T4 has a similar affect. A reduced activity is the result of induction in a lysogenic strain of *Esch. coli* K12λ$^+$ [243].

Induction by ultraviolet irradiation of *Esch. coli* K12λ$^+$ leads to the development of a dialysable inhibitor of methylation [243]. The activity seems to be directed against uracil methylase.

Animal tissues have been shown to contain a complex inhibitory activity to tRNA methylases, distinct from the above activity. It is largely absent from certain tumour and embryonic tissues and is influenced by hormones (for reviews see [246, 247]).

Another type of modification studied at the enzymic level is the formation of 4-thiouridine, another constituent of some tRNA's.

A tRNA sulphur transferase has been isolated from *Esch. coli* which catalyses the *in vitro* transfer of sulphur from L-cysteine into tRNA uracil. ATP and Mg^{2+} are required [222]. (Isopent-2-enyl)-adenosine, on the other hand, is formed in tRNA by the action of isopent-2-enyl pyrophosphate tRNA-transferase. No cofactors are required. Pseudouridine formation in tRNA probably occurs by the enzymic modification of uridine residues in the RNA chain. Enzymes capable of carrying out this conversion have been detected in *Esch. coli* [222] and in the cytoplasm of mammalian cells [222]. The exact mechanism is not known. Whether it involves the cleavage of an N-glycoside bond in uridine, rotation of uracil residue, and the formation of a C-glycosidic linkage remains to be seen.

8.7.2 *Ribosomal RNA*

In *Esch. coli*, the lighter species of rRNA has been shown to contain approximately 20 per cent more methyl groups than the heavier species [248, 249]. In bacterial rRNA, methylation of bases is about four times more frequent than methylation of sugar [248], while in plants [250, 100] and mammals [251, 226] many more sugar residues than base residues are methylated. In HeLa cells 70 and 46 per cent of nucleotides are methylated in 28S and 18S rRNA respectively [252]. In general, the formation of N^6-dimethyladenine seems to distinguish 18S from 28S rRNA in mammalian cells [220, 252].

Methylation of 45S ribosomal precursor RNA (p. 330) takes place at the site of its synthesis in the nucleolus [253], and HeLa cells deprived of methionine produce undermethylated rRNA but do not form ribosomes [254]. Resistance to the antibiotic kasugamycin has been correlated with the lack of a methylated base in ribosomal RNA [255].

The ribosomal RNA methylases, which act at the polynucleotide level using S-adenosylmethionine in the same way as the tRNA methylases, were first isolated from *Esch. coli* [249, 256, 257] and since then have been demonstrated in a variety of systems. Both sugar and base moieties can be modified *in vitro* [259]. Recently these methylases have been located as component parts of nascent ribosomal particles [259] in *Esch. coli* cells. In mammalian cells they appear to be associated with nucleoli [260], possibly in some association with nascent ribosomal particles by analogy with the *Esch. coli* situation.

8.7.3 *Messenger RNA*

As will be detailed more extensively in Chapter 12, recent data also show messenger RNA's of viral and eukaryotic origin to contain methylated nucleosides. Very briefly, the 5'-terminus contains 7-methylguanosine linked through its 5'-OH via a tri-(or pyro-)-phosphate group to an O'-methylated nucleoside. Methylases thought to be involved in this complex modification have so far been isolated from reovirus and vaccinia virions [261–263].

8.8 Phosphatases (phosphomonoesterases) and kinases

Phosphomonoesterases remove as inorganic orthophosphate the terminal monoesterified phosphate group from mononucleotides or oligonucleotides. The 5'-nucleotidases which have been prepared from seminal plasma and snake venom remove the phosphate group from nucleoside 5'-phosphates [2]. Rye-grass contains a 3'-nucleotidase [264]. The alkaline phosphatase from *Esch. coli* hydrolyses a wide range of compounds containing monoesterified phosphate. *Esch. coli* contains a second 3'-nucleotidase (in addition to the exonuclease III) which may remove the 2'- or 3'- but not the 5'-terminus of DNA or RNA [265] and acts on all the natural mononucleotides. A similar activity is induced after T-even bacteriophage infection; it attacks 3'-monophosphates or 3'-phosphoryl termini of DNA and may be involved in preparing a priming site for the polymerization step in DNA repair (see Chapter 11).

Another type of enzyme which may be involved in DNA repair is *polynucleotide (5-hydroxyl) kinase* which has been detected in T2 and T4 infected *Esch. coli* [256, 267] and also in rat liver nuclei [268]. This enzyme specifically catalyses the incorporation of ^{32}P from (^{32}P-γ)ATP into 5'-hydroxyl groups of DNA [267]. Such a phosphorylation of 5'ends may be a prelude to DNA ligase action in a repair process (see Chapter 11). RNA can also act as acceptor of phosphate groups, and the enzyme together with ^{32}P-ATP of very high specific activity has been used to label the digestion products of non-radioactive RNA prior to sequence analysis by standard 'fingerprinting' techniques [269, 270] (see also Chapter 6).

REFERENCES

[1] Laskowski, M., Sr. (1967) *Adv. Enzymology,* **29,** 165
[2] Colowick, S. P. and Kaplan, N. O. (1955) *Methods in Enzymology,* Vol. 2, pp. 427–450 and 561–570: (1963) Vol. 6, pp. 40–55. New York: Academic Press
[3] Cantoni, G. L. and Davies, D. R. (Eds.) (1966) *Procedures in Nucleic Acid Research.* New York: Harper and Row
[4] Shugar, D. and Sierakowska, H. (1967) *Progress in Nucleic Acid Research and Molecular Biology,* Vol. 7, p. 369 (J. N. Davidson and W. E. Cohn, Eds.). New York: Academic Press
[5] Lehman, I, R. (1967) *Ann. Rev. Biochem.,* **36,** 645
[6] Barnard, E. A. (1965) *Ann. Rev. Biochem.,* **38,** 677
[7] Hilmoe, R. J., Heppel, L. A., Springhorn, S. S. and Koshland, D. E., Jr. (1961) *Biochim. Biophys. Acta,* **53,** 214
[8] de Meuron-Landolt, M. and Privat de Garilhe, M. (1964) *J. Amer. Chem. Soc.,* **78,** 4642
[9] Grossman, R. and Moldave, K. (Eds.) (1967–68) *Methods in Enzymology,* Vols. 12A and 12B; (1971) Vols. 20 and 21. New York: Academic Press
[10] Khorana, H. G. (1961) *The Enzymes* (2nd Edition), Vol. 5, p. 79 (P. D. Boyer, H. Lardy and K. Myrbäck, Eds.). New York: Academic Press
[11] Anfinsen, C. B. and White, F. H. (1961) *The Enzymes* (2nd Edition), Vol. 5, p. 95 (P. D. Boyer, H. Lardy and K. Myrbäck, Eds.). New York: Academic Press
[12] Witzel, H. (1963) *Progress in Nucleic Acid Research,* Vol. 2, p. 221 (J. N. Davidson and W. E. Cohn, Eds.). New York: Academic Press
[13] Jones, W. (1920) *Amer. J. Physiol.,* **52,** 203
[14] Enzyme Nomenclature (1965), Recommendations of the International Union of Biochemistry. Amsterdam: Elsevier

[15] Dubos, R. J. and Thompson, R. H. S. (1938) *J. Biol. Chem.*, **124**, 501
[16] Kunitz, M. (1940) *J. Gen. Physiol.*, **24**, 15
[17] McDonald, M. R. (1955) *Methods in Enzymology*, Vol. 2, p. 427 (S. P. Colowick and N. O. Kaplan, Eds.). New York: Academic Press
[18] Smyth, D. G., Stein, W. H. and Moore, S. (1963) *J. Biol., Chem.*, **238**, 227
[19] Stein, W. H. (1964) *Fed. Proc.*, **23**, 599
[20] Findlay, D., Herries, D. G., Mathias, A. P., Rabin, B. R. and Ross, C. A. (1962) *Biochem. J.*, **85**, 152
[21] Bernhard, S. (1968) *The Structure and Function of Enzymes*. New York: Benjamin
[22] Gutfreund, H. (1965) *An Introduction to the Study of Enzymes*. Oxford: Blackwell
[23] Scheraga, H. A. and Rupley, J. A. (1962) *Adv. Enzymology*, **24**, 161
[24] Kartha, G., Bello, J. and Harker, D. (1967) *Nature*, **213**, 862
[25] Wyckoff, H. W., Hardman, K. D., Allewell, N. M., Inagami, T., Johnson, L. N. and Richards, F. M. (1967) *J. Biol. Chem.*, **242**, 3984
[26] Gutte, B. and Merrifield, R. B. (1969) *J. Amer. Chem. Soc.*, **91**, 501
[27] Hirschmann, R., Nutt, R. F., Veber, D. F., Vitali, R. A., Varga, S. L., Jacob, T. A., Holly, F. W. and Denkewalter, R. G. (1969) *J. Amer. Chem. Soc.*, **91**, 507
[28] Bernd, G. and Merrifield, R. B. (1971) *J. Biol. Chem.*, **246**, 1922
[29] Markham, R. and Smith, J. D. (1952) *Biochem. J.*, **52**, 552
[30] Cohn, W. E. and Volkin, E. (1953) *J. Biol. Chem.*, **203**, 319
[31] Brown, D. M., Dekker, C. A. and Todd, A. R. (1952) *J. Chem. Soc.*, 2715
[32] Beers, R. F. (1960) *J. Biol. Chem.*, **235**, 2393
[33] MacFayden, D. A. (1934) *J. Biol. Chem.*, **107**, 297
[34] Crook, E. M., Mathias, A. P. and Rabin, B. R. (1960) *Biochem. J.*, **74**, 234
[35] Kunitz, M. (1946) *J. Biol. Chem.*, **164**, 563
[36] Dickman, S. R., Aroskar, J. P. and Kropf, R. B. (1956) *Biochim. Biophys. Acta*, **21**, 539
[37] Zittle, C. A. and Reading, E. H. (1945) *J. Biol. Chem.*, **160**, 519
[38] Roth, J. S. and Milstein, S. W. (1952) *J. Biol. Chem.*, **196**, 489
[39] Egami, F., Takahashi, K. and Uchida, T. (1964) *Progress in Nucleic Acid Research*, Vol. 3, p. 59 (J. N. Davidson and W. E. Cohn, Eds.). New York: Academic Press
[40] Uchida, T. and Egami, F. (1967) *J. Biochem.*, **61**, 44
[41] Glitz, D. G. and Dekker, C. A. (1964) *Biochem.*, **3**, 1391
[42] De Lamirande, G., Allard, C., DaCosta, H. C. and Cantero, A. (1954) *Science*, **119**, 351
[43] Roth, J. S. (1954) *J. Biol. Chem.*, **208**, 180
[44] Reid, E. and Nodes, J. T. (1959) *Ann. N.Y. Acad. Sci.*, **81**, 618
[45] Winicov, I. and Perry, R. P. (1974) *Biochem.*, **13**, 2908
[46] Prestayko, A. W., Lewis, B. C. and Busch, H. (1975) *Biochim. Biophys. Acta*, **319**, 323
[47] Mirault, M. E. and Scherrer, K. (1972) *Eur. J. Biochem.*, **28**, 197
[48] Bothwell, A. L. M. and Altman, S. (1975) *J. Biol. Chem.*, **250**, 1460
[49] Neu, H. C. and Heppel, L. A. (1964) *J. Biol. Chem.*, **239**, 3893
[50] Gesteland, R. (1966) *J. Mol. Biol.*, **16**, 67
[51] Robertson, H. D., Webster, R. E. and Zinder, N. D. (1967) *Virology*, **32**, 718

NUCLEASES AND RELATED ENZYMES

[52] Robertson, H. D., Webster, R. E. and Zinder, N. D. (1968) *J. Biol. Chem.*, **243**, 82
[53] Schweitz, H. and Ebel, J. P. (1971) *Biochemie*, **53**, 582
[54] Robertson, H. D. (1971) *Nature New Biol.*, **229**, 169
[55] Stern, R. (1970) *Biochem. Biophys. Res. Commun.*, **41**, 608
[56] Robertson, H. D. and Matthews, M. B. (1973) *Proc. Nat. Acad. Sci.*, **70** 225
[57] Stern, R. and Wilczeck, J. (1973) *Fed. Proc.*, **32**, 620
[58] Spahr, P. F. and Gesteland, R. F. (1968) *Proc. Nat. Acad. Sci.*, **58**, 876
[59] Gesteland, R. F. and Spahr, P. F. (1969) *Cold Spring Harbor Symp. Quant. Biol.*, **34**, 707
[60] Altman, S. and Smith, J. D. (1971) *Nature New Biol.*, **233**, 35
[61] Robertson, H. D., Altman, S. and Smith, J. D. (1972) *J. Biol. Chem.*, **247**, 5243
[62] Altman, S., Bothwell, A. L. M. and Stark, B. C. (1975) *Brookhaven Symp. Biol.*, **26**, 12
[63] Burdon, R. H. (1975) *Brookhaven Symp. Biol.*, **26**, 138
[64] Stein, H. and Hausen, P. (1969) *Science*, **166**, 393
[65] Hausen, P. and Stein, H. (1970) *Eur. J. Biochem.*, **14**, 278
[66] Roewekamp, W. and Sekeris, C. E. (1974) *Eur. J. Biochem.*, **43**, 405
[67] Keller, W. and Crouch, R. (1972) *Proc. Nat. Acad. Sci.*, **69**, 3360
[68] Banks, G. R. (1974) *Eur. J. Biochem.*, **47**, 499
[69] Leis, J. P., Berkower, I. and Hurwitz, J. (1973) *Proc. Nat. Acad. Sci.*, **70**, 466
[70] Weatherford, S. C., Weisberg, L. S., Achord, D. T. and Apirion, D. (1973) *Biochem. Biophys. Res. Commun.*, **69**, 1307
[71] Stavrinopoulos, J. S. and Chargaff, E. (1973) *Proc. Nat. Acad. Sci.*, **70**, 1959
[72] Haberkern, R. C. and Cantoni, G. L. (1973) *Biochem.*, **12**, 2389
[73] Spahr, P. F. and Schlessinger, D. (1963) *J. Biol. Chem.*, **238**, PC2251
[74] Singer, M. F. and Tolbert, G. (1964) *Science*, **145**, 593
[75] Schedl, P., Primakoff, P. and Roberts, J. (1975) *Brookhaven Symp. Biol.*, **26**, 53
[76] Castles, J. J. and Singer, M. (1969) *J. Mol. Biol.*, **40**, 1
[77] Venkov, P., Schlessinger, D. and Longo, D. (1971) *J. Biol.*, **108**, 601
[78] Lazarus, H. M. and Sporn, M. B. (1967) *Proc. Nat. Acad. Sci.*, **57**, 1386
[79] Perry, R. P. and Kelley, D. E. (1972) *J. Mol. Biol.*, **70**, 265
[80] Niyogi, S. K. and Datta, A. K. (1975) *J. Biol. Chem.*, **250**, 7307
[81] Datta, A. K. and Niyogi, S. K. (1975) *J. Biol. Chem.*, **250**, 7313
[82] Grandgenett, D., Gerard, G. and Green, M. (1972) *J. Virol.*, **10**, 1136
[83] Grandgenett, D., Gerard, G. and Green, M. (1973) *Proc. Nat. Acad. Sci.*, **70**, 230
[84] Deutscher, M. P. (1970) *J. Biol. Chem.*, **245**, 4225
[85] Hotchkiss, R. D. (1948) *J. Biol. Chem.*, **175**, 315
[86] Roth, J. S. (1959) *Ann. N.Y. Acad. Sci.*, **81**, 611
[87] Shortman, K. (1962) *Biochim. Biophys. Acta*, **55**, 88
[88] Singer, B., Fraenkel-Conrat, H. and Tsugita, A. (1961) *Virology*, **14**, 54
[89] Daigneauet, R., Bellemare, G. and Cousineau, G. H. (1971) *Lab. Practice*, **20**, 487
[90] Rosén, C. G. and Fedorscsàk, I. (1966) *Biochim. Biophys. Acta*, **130**, 401
[91] Solymosy, F., Hüvös, P., Gulyas, A., Kapovits, I., Gaal, O., Bagi, G. and Garkas, G. L. (1971) *Biochim. Biophys. Acta*, **238**, 406

[92] Holley, R. W., Apgar, J. and Merrill, S. H. (1961) *J. Biol. Chem.*, **236**, PC42
[93] Laskowski, M. (1971) *The Enzymes* (3rd Edition), Vol. IV, p. 289 (P. D. Boyer, Ed.). New York: Academic Press
[94] Cotton, F. A. and Hazen, E. E. (1971) *The Enzymes* (3rd Edition), Vol. IV, p. 153 (P. D. Boyer, Ed.). New York: Academic Press
[95] Anfinsen, C. B., Cuatrecasas, P. and Taniuchi, H. (1971) *The Enzymes*, Vol. IV, p. 177 (P. D. Boyer, Ed.). New York: Academic Press
[96] Richards, G. M., du Vair, G. and Laskowski, M. (1965) *Biochem.*, **4**, 501
[97] Linn, S. and Lehman, I. R. (1965) *J. Biol. Chem.*, **240**, 1287, 1294
[98] Lehman, I. R. (1960) *J. Biol. Chem.*, **235**, 1474
[99] Shisido, K. and Ando, T. (1972) *Biochim. Biophys. Acta*, **287**, 477
[100] Isaksson, L. A. and Phillips, J. H. (1968) *Biochim. Biophys. Acta*, **155**, 63
[101] Mikulski, A. J. and Laskowski, M. (1970) *J. Biol. Chem.*, **245**, 5026
[102] Laskowski, M. (1971) *The Enzymes* (3rd Edition), Vol. IV, p. 313 (P. D. Boyer, Ed.). New York: Academic Press
[103] Cohn, W. E., Volkin, E. and Khym, J. X. (1957) *Biochem. Prep.*, **5**, 49
[104] Butler, G. C. (1955) *Methods in Enzymology*, Vol. 2, p. 561 (S. P. Colowick and N. O. Kaplan, Eds.). New York: Academic Press
[105] Bernardi, R. and Bernardi, G. (1971) *The Enzymes* (3rd Edition), Vol. IV, p. 329 (P. D. Boyer, Ed.). New York: Academic Press
[106] Heppel, L. A. and Hilmoe, R. J. (1955) *Methods in Enzymology*, Vol. 2, p. 565 (S. P. Colowick and N. O. Kaplan, Eds.). New York: Academic Press
[107] McDonald, M. R. (1955) *Methods in Enzymology*, Vol. 2, p. 437 (S. P. Colowick and N. O. Kaplan, Eds.). New York: Academic Press
[108] McCarty, M. (1946) *J. Gen. Physiol.*, **29**, 123
[109] Kunitz, M. (1948) *Science*, **108**, 19
[110] Polson, A. (1956) *Biochim. Biophys. Acta*, **22**, 61
[111] Bollum, F. J. (1965) *J. Biol. Chem.*, **240**, 2599
[112] Tamm, C. and Chargaff, E. (1951) *Nature*, **168**, 916
[113] Gilbert, L. M., Overend, W. G. and Webb, M. (1951) *Exp. Cell Res.*, **2**, 349
[114] Young, E. T. and Sinsheimer, R. L. (1965) *J. Biol. Chem.*, **240**, 1274
[115] Tamm, C. Shapiro, H. S. and Chargaff, E. (1952) *J. Biol. Chem.*, **199**, 313
[116] Matsuda, M. and Ogoshi, H. (1966) *Biochim. Biophys. Acta*, **119**, 210
[117] Worthington Biochemical Corporation, Catalogue
[118] Sandeen, G. and Zimmerman, S. B. (1966) *Anal. Biochem.*, **14**, 269
[119] Bernardi, G. (1971) *The Enzymes* (3rd Edition), Vol. IV, p. 271 (P. D. Boyer, Ed.). New York: Academic Press
[120] Bernardi, G. (1965) *J. Mol. Biol.*, **13**, 603
[121] Lehman, I. R. (1971) *The Enzymes* (3rd Edition), Vol. IV, p. 251 (P. D. Boyer, Ed.). New York: Academic Press
[122] Potter, J. L. and Laskowski, M. (1959) *J. Biol. Chem.*, **234**, 1263
[123] Yasmineh, W. G. and Gray, E. D. (1968) *Biochem.*, **7**, 105
[124] Lehman, I. R. (1963) *Progress in Nucleic Acid Research*, Vol. 2, p. 84 (J. N. Davidson and W. E. Cohn, Eds.). New York: Academic Press
[125] Lehman, I. R., Roussos, G. G. and Pratt, E. A. (1962) *J. Biol. Chem.*, **237**, 819
[126] Lehman, I. R. (1963) *Methods in Enzymology*, Vol. 6, p. 44 (S. P. Colowick and N. O. Kaplan, Eds.). New York: Academic Press

[127] Radloff, R., Bauer, W. and Vinograd, J. (1967) *Proc. Nat. Acad. Sci.,* **57,** 1514
[128] Studier, F. W. (1965) *J. Mol. Biol.,* **11,** 373
[129] Friedberg, E. C. and Goldthwaite, D. A. (1968) *Cold Spring Harbor Symp. Quant. Biol.,* **33,** 271
[130] Friedberg, E. C. and Goldthwaite, D. A. (1969) *Proc. Nat. Acad. Sci.,* **62,** 934
[131] Yajko, D. M. and Weiss, B. (1975) *Proc. Nat. Acad. Sci.,* **72,** 688
[132] Ljungquist, S., Nyberg, B. and Lindahl, T. (1975) *FEBS Lett.,* **57,** 169
[133] Buttin, G. and Wright, M. R. (1968) *Cold Spring Harbor Symp. Quant. Biol.,* **33,** 259
[134] Anai, M., Takakata, H. and Takagi, Y. (1970) *J. Biol. Chem.,* **245,** 767
[135] Anai, M., Hirahashi, T., Yamauaka, M. and Takagi, Y. (1970) *J. Biol. Chem.,* **245,** 775
[136] Goldmark, P. J. and Linn, S. (1970) *Proc. Nat. Acad. Sci.,* **67,** 434
[137] Hurwitz, J., Becker, A., Gefter, M. and Gold, M. (1967) *J. Cell. Comp. Physiol.,* Suppl. 1, **70,** 181
[138] Sadowski, P., Ginsberg, B., Yudelevitch, A., Fesnier, L. and Hurwitz, J. (1968) *Cold Spring Harbor Symp. Quant. Biol.,* **33,** 165
[139] Sadowski, P. D. and Hurwitz, J. (1969) *J. Biol. Chem.,* **244,** 6182, 6192
[140] Henry, J. J. and Knippers, R. (1974) *Proc. Nat. Acad. Sci.,* **71,** 1549
[141] Burlingham, B. T., Doerfler, W., Pettersson, U. and Philipson, L. (1971) *J. Mol. Biol.,* **60,** 45
[142] Pogo, B. G. T. and Dales, S. (1969) *Proc. Nat. Acad. Sci.,* **63,** 820
[143] Lehman, I. R. (1960) *J. Biol. Chem.,* **235,** 1479
[144] Lehman, I. R. (1963) *Methods in Enzymology,* Vol. 6, p. 40 (S. P. Colowick and N. O. Kaplan, Eds.). New York: Academic Press
[145] Lehman, I. R. and Richardson, C. C. (1964) *J. Biol. Chem.,* **239,** 233
[146] Richardson, C. C. and Kornberg, A. (1964) *J. Biol. Chem.,* **239,** 242
[147] Richardson, C. C., Lehman, I. R. and Kornberg, A. (1964) *J. Biol. Chem.,* **239,** 251
[148] Lehman, I. R. and Nussbaum, A. L. (1964) *J. Biol. Chem.,* **239,** 2628
[149] Kellock, M. G., Smellie, R. M. S. and Davidson, J. N. (1962) *Biochem. J.,* **84,** 112P
[150] Burdon, M. G., Smellie, R. M. S. and Davidson, J. N. (1964) *Biochim. Biophys. Acta,* **91,** 46
[151] Healy, J. W., Stollar, D. and Levine, L. (1963) *Methods in Enzymology,* Vol. 6, p. 49 (S. P. Colowick and N. O. Kaplan, Eds.). New York: Academic Press
[152] Lacks, S. and Greenberg, B. (1967) *J. Biol. Chem.,* **242,** 3108
[153] Okazaki, T. and Kornberg, A. (1964) *J. Biol. Chem.,* **239,** 259
[154] Radding, C. M., Szpirer, J. and Thomas, R. (1967) *Proc. Nat. Acad. Sci.,* **57,** 277
[155] Jorgensen, S. E. and Koerner, J. F. (1966) *J. Biol. Chem.,* **241,** 3090
[156] Lieberman, R. P. and Oishi, M. (1974) *Proc. Nat. Acad. Sci.,* **71,** 4816
[157] Kornberg, A. (1974) *DNA Synthesis.* San Francisco: Freeman
[158] Cozzarelli, N. R., Kelly, R. B. and Kornberg, A. (1969) *J. Mol. Biol.,* **45,** 513
[159] Kelly, R. B., Atkinson, M. R., Huberman, J. A. and Kornberg, A. (1969) *Nature,* **224,** 495
[160] Klenow, H. and Overgaard-Hansen, K. (1970) *FEBS Lett.,* **6,** 25
[161] Oleson, A. E. and Koerner, J. F. (1964) *J. Biol. Chem.,* **239,** 2935
[162] Short, E. C., Jr. and Koerner, J. F. (1969) *J. Biol. Chem.,* **244,** 1487

[163] Trilling, D. M. and Aposhian, H. V. (1968) *Proc. Nat. Acad. Sci.*, **60**, 214
[164] McAuslan, B. R. (1971) in: *Strategy of the Viral Genome*, p. 25. Ciba Symposium Volume. Edinburgh: Livingstone; London: Churchill
[165] Morrison, J. M. and Keir, H. M. (1968) *J. Gen. Virol.*, **3**, 337
[166] Lindahl, T., Gally, J. A. and Edelman, G. M. (1969) *J. Biol. Chem.*, **244**, 5014
[167] Lindahl, T., Gally, J. A. and Edelman, G. M. (1969) *Fed. Proc.*, **28**, 348
[168] Lindberg, U. (1964) *Biochim. Biophys. Acta*, **72**, 237
[169] Dunn, D. B. and Smith, J. D. (1958) *Biochem. J.*, **68**, 627
[170] Wyatt, G. R. (1951) *Biochem. J.*, **48**, 584
[171] Wyatt, G. K. and Linzen, B. (1965) *Biochim. Biophys. Acta*, **103**, 588
[172] Tewari, K. K. and Wildman, S. G. (1966) *Science*, **153**, 1269
[173] Ray, D. S. and Hanawalt, P. C. (1964) *J. Mol. Biol.*, **9**, 812
[174] Salomon, R., Kaye, A. M. and Herzberg, M. (1969) *J. Mol. Biol.*, **43**, 581
[175] Dawid, I. B., Brown, D. D. and Reader, R. H. (1970) *J. Mol. Biol.*, **51**, 341
[176] Arber, W. and Linn, S. (1969) *Ann. Rev. Biochem.*, **38**, 467
[177] Haussman, R. and Gold, M. (1966) *J. Biol. Chem.*, **241**, 1985
[178] Hudnik-Plevnik, T. A. and Melechen, N. E. (1967) *J. Biol. Chem.*, **242**, 4118
[179] Gefter, M., Haussman, R., Gold, M. and Hurwitz, J. (1966) *J. Biol. Chem.*, **241**, 1995
[180] Kaye, A. M. and Winocour, E. (1967) *J. Mol. Biol.*, **24**, 475
[181] Low, M., Hay, J. and Keir, H. M. (1969) *J. Mol. Biol.*, **46**, 205
[182] Low, M., Mechie, M. and Hay, J. (1971) *Biochem. J.*, **124**, 63P
[183] Borek, E. and Srinivasan, P. R. (1966) *Ann. Rev. Biochem.*, **35**, 275
[184] Gold, M., Hurwitz, J. and Anders, M. (1964) *Proc. Nat. Acad. Sci.*, **50**, 164
[185] Gold, M. and Hurwitz, J. (1964) *J. Biol. Chem.*, **239**, 3858
[186] Roy, P. H. and Smith, H. O. (1973) *J. Mol. Biol.*, **81**, 427
[187] Gold, M., Haussman, R., Maitra, U. and Hurwitz, J. (1964) *Proc. Nat. Acad. Sci.*, **52**, 292
[188] Krueger, D. H., Presber, W., Hansen, S. and Rosenthal, H. A. (1975) *J. Virol.*, **16**, 453
[189] Burdon, R. H., Martin, B. T. and Lal, B. (1967) *J. Mol. Biol.*, **38**, 357
[190] Sheid, B., Srinivasan, P. R. and Borek, E. (1968) *Biochem.*, **7**, 280
[191] Kalousek, F. and Morris, N. R. (1969) *J. Biol. Chem.*, **244**, 1157
[192] Adams, R. L. P., Turnbull, J., Smillie, E. J. and Burdon, R. H. (1974) 9th F.E.B.S. Meeting, Symposium on Post Synthetic Modification of Macromolecules, p. 39
[193] Turnbull, J. F. and Adams, R. L. P. (1975) Abstracts of 10th F.E.B.S. Meeting, No. 178
[194] Burdon, R. H. and Adams, R. L. P. (1969) *Biochim. Biophys. Acta*, **174**, 322
[195] Kappler, J. W. (1970) *J. Cell Physiol.*, **75**, 21
[196] Adams, R. L. P. (1971) *Biochim. Biophys. Acta*, **254**, 205
[197] Doskocil, J. and Sormova, Z. (1965) *Coll. Czech. Chem. Commun.*, **30**, 38
[198] Grippo, P., Iaccarino, M., Parisi, E. and Scarano, E. (1968) *J. Mol. Biol.*, **36**, 195
[199] Lark, C. (1968) *J. Mol. Biol.*, **31**, 401
[200] Marinus, M. G. and Morris, N. R. (1973) *J. Bact.*, **114**, 143

NUCLEASES AND RELATED ENZYMES

[201] Novogrodsky, A., Gefter, M., Maitra, U., Gold, M. and Hurwitz, J. (1966) *J. Biol. Chem.*, **241**, 1977
[202] Arber, W. (1974) *Progr. Nucleic Acid Res. Mol. Biol.* (W. E. Cohn, Ed.), **14**, 1
[203] Murray, K. and Old, R. W. (1974) *Progr. Nucleic Acid Res. Mol. Biol.* (W. E. Cohn, Ed.), **14**, 117
[204] Boyer, H. (1971) *Ann. Rev. Microbiol.*, **25**, 153
[205] Meselson, M., Yuan, R. and Heywood, J. (1972) *Ann. Rev. Biochem.*, **41**, 447
[206] Nathans, D. and Smith, H. O. (1975) *Ann. Rev. Biochem.*, **44**, 273
[207] Meselson, M. and Yuan, R. (1968) *Nature*, **217**, 1110
[208] Yuan, R., Bickle, T. A., Ebbers, W. and Brack, C. (1975) *Nature*, **256**, 556
[209] Smith, H. O. and Nathans, D. (1973) *J. Mol. Biol.*, **81**, 419
[210] Sugisaki, H. and Takanami, M. (1973) *Nature New Biol.*, **246**, 138
[211] Sharp, P. A., Sugden, B. and Sambrook, J. (1973) *Biochem.*, **12**, 3055
[212] Danna, K. J., Sack, G. H., Jr. and Nathans, D. (1973) *J. Mol. Biol.*, **78**, 363
[213] Khoury, G., Martin, M. A., Lee, T. N. H., Danna, K. J. and Nathans, D. (1973) *J. Mol. Biol.*, **78**, 377
[214] Fareed, G. C., Garon, C. F. and Salzman, N. P. (1972) *J. Virol.*, **10**, 484
[215] Lovett, M. A., Katz, L. and Helinski, D. R. (1974) *Nature*, **251**, 337
[216] Berg, P., Baltimore, D., Boyer, H. W., Cohen, S. N., Davis, R. W., Hogness, D. S., Nathans, D., Roblin, R., Watson, J. D., Weissman, S. and Zinder, N. D. (1974) *Science*, **185**, 303
[217] Report of the Working Party on the Experimental Manipulation of the Genetic Composition of Microorganisms, Chairman Lord Ashley (1974) London: H.M. Stationery Office
[218] Hall, R. H. (1971) *The Modified Nucleosides in Nucleic Acid*. New York: Columbia Univ. Press
[219] Goldwasser, E. and Heinrikson, R. L. (1966) *Progr. Nucleic Acid Res. Mol. Biol.*, **5**, 399
[220] Zachau, H. G. (1969) *Angew. Chem. Internat. Ed. Enzymol.*, **8**, 711
[221] Nishimura, S. (1972) *Progr. Nucleic Acid. Res. Mol. Biol.*, **12**, 49
[222] Soll, D. (1971) *Science*, **173**, 293
[223] Mandel, L. R. and Borek, E. (1963) *Biochem.*, **2**, 555
[224] Hurwitz, J., Gold, M. and Anders, M. (1964) *J. Biol. Chem.*, **239**, 3462
[225] Tsutsui, E., Srinivasan, P. R. and Borek, E. (1966) *Proc. Nat. Acad. Sci.*, **56**, 1003
[226] Lane, B. G. and Tamaoki, T. (1969) *Biochim. Biophys. Acta*, **179**, 332
[227] Baguley, B. C. and Staehelin, M. (1968) *Biochem.*, **7**, 45
[228] Rodeh, R., Geldman, M. and Littauer, U. Z. (1967) *Biochem.*, **6**, 451
[229] Svensson, I., Björk, G. K. and Lundahl, P. (1969) *Eur. J. Biochem.*, **9**, 216
[230] Agris, P. F., Sprermulli, L. M. and Brown, G. M. (1974) *Arch. Biochem. Biophys.*, **162**, 38
[231] Kraus, J. and Staehlin, M. (1974) *Nucleic Acids Res.*, **1**, 1455
[232] Kraus, J. and Staehlin, M. (1974) *Nucleic Acids Res.*, **1**, 1479
[233] Wierzbicka, H., Jakubowski, J. and Pawelkiewicz, J. (1975) *Nucleic Acids Res.*, **2**, 101
[234] Smolnar, N., Hellman, U. and Svensson, I. (1975) *Nucleic Acids Res.*, **2**, 993
[235] Smolnar, N. and Svensson, I. (1974) *Nucleic Acids Res.*, **1**, 707

[236] Gold, M., Hurwitz, J. and Anders, M. (1963) *Biochem. Biophys. Res. Commun.*, **11**, 107
[237] Hacker, B. and Mandel, L. R. (1969) *Biochim. Biophys. Acta*, **190**, 38
[238] Silber, R., Goldstein, B., Berman, E., Decter, J. and Friend, C. (1967) *Cancer Res.*, **27**, 1264
[239] Borek, E. (1971) *Cancer Res.*, **31**, 596
[240] Mittleman, A., Hall, R. H., Yohn, D. S. and Grace, J. T. (1967) *Cancer Res.*, **27**, 1409
[241] Burdon, R. H. (1971) *Progr. Nucleic Acid Res. Mol. Biol.*, **11**, 33
[242] Wainfan, E., Srinivasan, P. R. and Borek, E. (1965) *Biochem.*, **4**, 2845
[243] Wainfan, E., Srinivasan, P. R. and Borek, E. (1966) *J. Mol. Biol.*, **22**, 349
[244] Turkington, R. W. (1969) *J. Biol. Chem.*, **244**, 5140
[245] Van de Woude, G. F., Arlinghaus, R. B. and Polatnick, J. (1967) *Biochem. Biophys. Res. Commun.*, **29**, 483
[246] Kerr, S. J., Sharma, O. K. and Borek, E. (1971) *Cancer Res.*, **31**, 633
[247] Kerr, S. J. and Borek, E. (1973) *Adv. Enzyme Regulation*, **11**, 63
[248] Starr, J. L. and Fefferman, R. (1964) *J. Biol. Chem.*, **239**, 3457
[249] Srinivasan, P. R., Nofal, S. and Sussman, C. (1964) *Biochem. Biophys. Res. Commun.*, **16**, 82
[250] Lane, B. G. (1965) *Biochem.*, **4**, 212
[251] Wagner, E. K., Penman, S. and Ingram, V. (1967) *J. Mol. Biol.*, **29**, 271
[252] Maden, B. E. H. and Salim, M. (1974) *J. Mol. Biol.*, **88**, 133
[253] Greenberg, H. and Penman, S. (1966) *J. Mol. Biol.*, **21**, 527
[254] Vaughan, M. H., Soiero, R., Warner, J. and Danell, J. E. (1967) *Proc. Nat. Acad. Sci.*, **58**, 1527
[255] Hesler, T. L., Davies, J. E. and Dahlberg, J. E. (1971) *Nature New Biol.*, **233**, 12
[256] Gordon, J. and Borman, H. G. (1964) *J. Mol. Biol.*, **9**, 638
[257] Hurwitz, J., Anders, M., Gold, M. and Smith, J. (1965) *J. Biol. Chem.*, **240**, 1256
[258] Nichols, J. L. and Lane, B. G. (1968) *Can. J. Biochem.*, **46**, 108
[259] Thammana, P. and Held, W. A. (1974) *Nature*, **251**, 682
[260] Culp, L. A. and Brown, G. M. (1970) *Arch. Biochem. Biophys.*, **137**, 222
[261] Faust, M. and Millward, S. (1974) *Nucleic Acids Res.*, **1**, 1739
[262] Shatkin, A. (1974) *Proc. Nat. Acad. Sci.*, **71**, 3204
[263] Wei, C.-M. and Moss, B. (1974) *Proc. Nat. Acad. Sci.*, **71**, 3014
[264] Schuster, L. and Kaplan, N. O. (1955) *Methods in Enzymology*, Vol. 2, p. 551 (S. P. Colowick and N. O. Kaplan, Eds.). New York: Academic Press
[265] Becker, A. and Hurwitz, J. (1967) *J. Biol. Chem.*, **242**, 936
[266] Novagrodsky, A. and Hurwitz, J. (1965) *Fed. Proc.*, **24**, 602
[267] Richardson, C. C. (1965) *Proc. Nat. Acad. Sci.*, **54**, 158
[268] Novagrodsky, A., Tal, M., Traub, A. and Hurwitz, J. (1966) *J. Biol. Chem.*, **241**, 2933
[269] Labrie, F. and Sanger, F. (1969) *Biochem. J.*, **114**, 29P
[270] Szekely, M. and Sanger, F. (1969) *J. Mol. Biol.*, **43**, 607
[271] Chase, J. W. and Richardson, C. C. (1974) *J. Biol. Chem.*, **248**, 4545
[272] Heppel, L. A. (1966) *Procedures in Nucleic Acid Res.*, **1**, 31

CHAPTER 9

The Metabolism of Nucleotides

9.1 Anabolic pathways
A study of the biosynthesis of the nucleic acids involves consideration of several different aspects – the biosynthesis of the purine and pyrimidine ring systems, the origin of the sugar components, and the biosynthesis of the polynucleotides themselves.

9.2 The biosynthesis of the purines
Purines are synthesized in the cell in the form of their nucleoside monophosphates. The subject has been so extensively reviewed [1–8] that only brief outline need be given here.

It is known from experiments with isotopes that the sources of the atoms in the purine ring are as shown in Fig. 9.1.

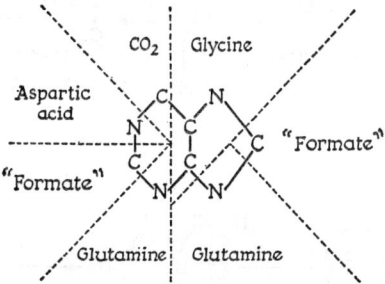

Fig. 9.1 *Sources of the atoms in the purine ring*

The starting material in purine biosynthesis is 5'-phosphoribosyl-1-pyrophosphate (PRPP) (Fig. 9.2) which accepts the γ-amino group of glutamine to give 5-phosphoribosylamine (PRA) (Fig. 9.3) under the influence of the enzyme *phosphoribosyl pyrophosphate amidotransferase* (*amidophosphoribosyl transferase*, E.C.2.4.2.14). Glycine then reacts with PRA to give glycinamide ribonucleotide (GAR) a nucleotide-like compound in which the amide of glycine takes the place of the usual purine or pyrimidine base. The reaction sequence continues by formylation from N^5,N^{10}-methenyl-tetrahydrofolic

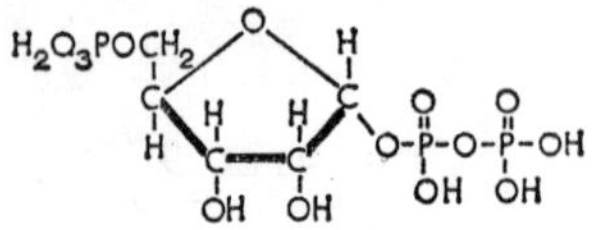

Fig. 9.2 5′-Phosphoribosyl-1-pyrophosphate (PRPP)

acid to give formylglycinamide ribonucleotide (formyl-GAR), then amination from glutamine to give formylglycinamidine ribonucleotide (formyl-GAM). Ring closure ensues, producing the imidazole ring compound 5-aminoimidazole ribonucleotide (AIR). Carboxylation of this compound gives 5-aminoimidazole-4-carboxylic acid ribonucleotide (carboxy-AIR). The corresponding amide, 5-aminoimidazole-4-carboxamide ribonucleotide (AICAR) is produced in the subsequent two reactions via an intermediate compound, 5-aminoimidazole-4-succinocarboxamide ribonucleotide (succino-AICAR). The purine ring system is completed when N^{10}-formyl-tetrahydrofolic acid donates its formyl group to the 5-amino group of the imidazole carboxamide ribonucleotide. The complete parent ribonucleotide is inosinic acid (inosine 5′-monophosphate, IMP).

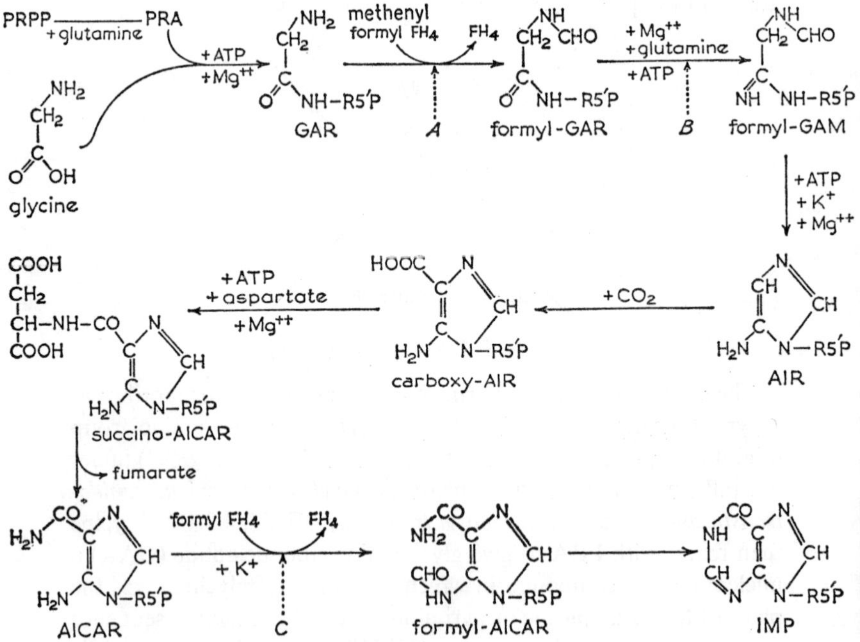

Fig. 9.3 Biosynthetic pathway for purines

THE METABOLISM OF NUCLEOTIDES

Amination of IMP to AMP proceeds in two stages with the intermediate formation of adenylosuccinic acid (Fig. 9.4). This reaction, in which the amino group of aspartate is transferred to C-6 of IMP to give AMP, resembles the reaction above in which 5-aminoimidazole-4-carboxamide ribonucleotide is formed from 5-aminoimidazole carboxylic acid ribonucleotide (Fig. 9.3). One difference, however, is the requirement for GTP as coenzyme in the reaction forming adenylosuccinic acid from IMP.

The formation of GMP from IMP is also a two-stage reaction in

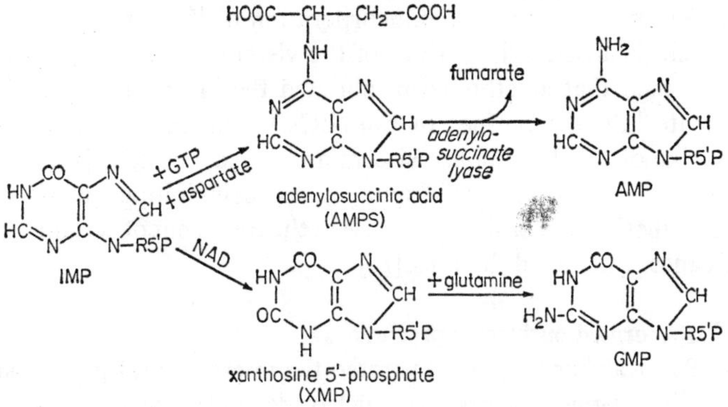

Fig. 9.4 *Formation of AMP and GMP from IMP*

which xanthosine 5′-monophosphate (XMP) is initially formed and then animated to give GMP (Fig. 9.4).

The two purine mononucleotides AMP and GMP are phosphorylated by kinases through the diphosphate stage to give ATP and GTP.

The complete pathway for purine biosynthesis is summarized in Fig. 9.3.

The importance of glutamine:

$$H_2N-CO-CH_2-CH_2-CH(NH_2)-COOH$$

in this scheme is obvious and the powerful antimetabolites azaserine:

$$\bar{N}{=}\overset{+}{N}{=}CH-CO-O-CH_2-CH(NH_2)-COOH$$

and 6-diazo-5-oxonorleucine (DON):

$$\bar{N}{=}\overset{+}{N}{=}CH-CO-CH_2-CH_2-CH(NH_2)-COOH$$

act by blocking glutamine utilization in purine synthesis at position B. Both compounds specifically inhibit the conversion of formyl-GAR into formyl-GAM [9].

Moreover, derivatives of tetrahydrofolic acid are involved in the introduction of carbons 2 and 8 in the purine ring at positions A and C respectively in Fig. 9.3 and, as will appear later, in the biosynthesis of the methyl group of thymine (p. 208). It is therefore not surprising to find that folic acid antagonists may affect the incorporation of precursors into the nucleic acid molecule. In 1950 Skipper *et al.* [10] observed that *aminopterin* and *amethopterin* (*methotrexate*) inhibited the incorporation of ^{14}C-formate into the combined nucleic acid purines of the viscera. In later experiments they noted that amethopterin inhibited the incorporation of ^{14}C-formate into the purines of both RNA and DNA and into the thymine of DNA in mouse viscera and leukaemic cells [11].

This action of folic acid antagonists in inhibiting nucleic acid biosynthesis is the basis of their use in the treatment of certain forms of cancer and allied diseases [12].

9.3 Preformed purines as precursors

In 1947 Kalckar [13] demonstrated the interaction of purine bases and ribose 1-phosphate to yield nucleosides and inorganic phosphate. Such reactions, which are forms of transglycosidation, are reversible and are catalysed by enzymes termed *nucleoside phosphorylases* (E.C.2.4.2.1):

$$\text{Hypoxanthine} + \text{ribose 1-phosphate} \rightleftharpoons \text{Inosine} + \text{Pi}$$

5-Aminoimidazole-4-carboxamide can take the place of a purine base in this type of reaction.

If the nucleosides so produced could be phosphorylated by ATP under the influence of appropriate phosphokinases, a route would be open for the biosynthesis of ribonucleotides from preformed purines.

A much more important mechanism for the conversion of bases into nucleotides involves 5-phosphoribosylpyrophosphate (PRPP). Under the influence of enzymes termed *nucleotide pyrophosphorylases* by Kornberg [14], bases can react with PRPP to form nucleotides and pyrophosphate [15].

The reactions catalysed by *nucleotide pyrophosphorylases* are illustrated in Fig. 9.5. One such enzyme, *adenine phosphoribosyl-*

THE METABOLISM OF NUCLEOTIDES

transferase (E.C.2.4.2.7), converts adenine into AMP in the presence of PRPP. A second enzyme, *hypoxanthine-guanine phosphoribosyltransferase* (E.C.2.4.2.8) converts hypoxanthine and guanine into IMP and GMP respectively in the presence of PRPP and also converts the drugs 6-mercaptopurine and azathioprine (imuran) into

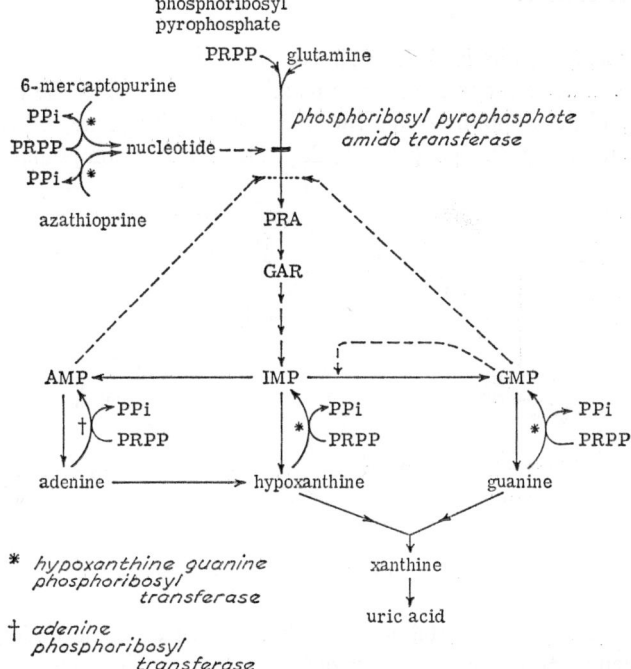

Fig. 9.5 *Reactions catalysed by nucleotide pyrophosphorylases. Feed-back controls are shown by the broken lines*

the corresponding nucleotides which in turn inhibit phosphoribosyl pyrophosphate amidotransferase and so prevent biosynthesis of purines [16, 17]. Since such drugs inhibit purine (and therefore nucleic acid) biosynthesis they are sometimes used as immunosuppressive or cancerostatic agents [18]. Azathioprine is also of value in the treatment of gout by inhibiting purine formation.

In the rare condition in children known as the *Lesch–Nyhan syndrome* there is a deficiency of the enzyme hypoxanthine-guanine phosphoribosyltransferase [19–22]. Consequently the condition is associated with excessive uric acid synthesis and is resistant to the action of azathioprine [23].

THE BIOCHEMISTRY OF THE NUCLEIC ACIDS

An important mechanism for the phosphorylation of nucleosides is to be found in the 'nucleoside transferase' systems. These enzymes, which are found in bacteria, plants, and animals, catalyse the transfer of phosphoric acid from low-energy organic phosphates to a wide variety of nucleosides with the formation of the 5'-nucleotides.

9.4 The biosynthesis of the pyrimidines

The complete series of enzymic reactions giving rise to the parent pyrimidine mononucleotide (uridine 5'-monophosphate, UMP) is shown in Fig. 9.6 [3, 6–8, 24, 25]. The starting compounds are

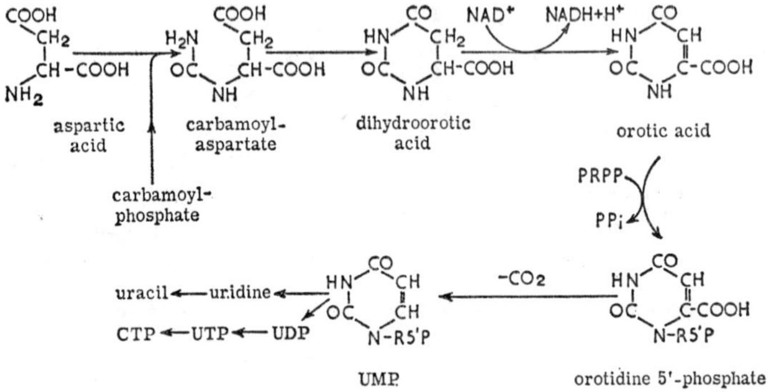

Fig. 9.6 *Biosynthetic pathway for pyrimidines*

aspartic acid and carbamoylphosphate which combine under the influence of *aspartate carbamoyltransferase* to form carbamoyl-aspartate. Formation of the pyrimidine ring is then effected by the action of *dihydro-orotase* giving dihydro-orotic acid, dehydrogenation of which produces the important pyrimidine intermediate orotic acid. A pyrophosphorylase reaction then follows in which orotic acid accepts a ribose 5-phosphate group from PRPP. The resulting product is orotidine 5'-monophosphate (OMP); inorganic pyrophosphate is eliminated. Decarboxylation of orotidine 5'-monophosphate gives uridine 5'-monophosphate (UMP) the parent pyrimidine nucleotide.

UMP is then converted by kinases through uridine 5'-diphosphate (UDP) into uridine 5'-triphosphate (UTP).

THE METABOLISM OF NUCLEOTIDES

9.5 The biosynthesis of cytosine derivatives

Conversion of uracil into cytosine takes place at nucleoside triphosphate level [26–29] under the influence of the enzyme *CTP synthetase* (E.C.6.3.4.2):

$$UTP + NH_3 + ATP \longrightarrow CTP + ADP + Pi$$

9.6 The biosynthesis of deoxyribonucleotides

The conversion of ribose into deoxyribose takes place at nucleotide level without breakage of the glycosidic linkage [29–31] since such compounds as uniformly labelled ^{14}C-cytidine are incorporated into the dCMP residues of DNA without change in the relative specific activities of sugar and base. This conversion appears to take place at nucleoside diphosphate level [29, 63–65].

Two related but readily distinguishable systems have been purified for the reduction of the ribosyl moiety of ribonucleotides to the corresponding deoxyribosyl derivative [31, 32]. The system from *Esch. coli* uses ribonucleoside *di*phosphates as substrates; that from *Lactobacillus leichmannii* uses ribonucleoside *tri*phosphates and requires a cobamide coenzyme. The mechanism of reduction in animal cells is less well established but appears not to require cobamide [33] and to resemble that in *Esch. coli*.

The mechanism for the reduction of CDP to dCDP in *Esch. coli* is known as the *thioredoxin system*. Thioredoxin is a sulphur-containing protein with some 108 amino acid residues [34]. It is reduced by NADPH under the influence of *thioredoxin reductase*, a flavoprotein containing FAD. The reduced thioredoxin then reduces CDP to the deoxy derivative, becoming itself reoxidized to thioredoxin. The complete system requires the participation of four different proteins, thioredoxin, thioredoxin reductase, and two proteins known as B1 and B2 which are believed to be non-identical subunits of the ribonucleotide reductase and to be involved in the allosteric regulation of ribonucleotide reduction [35].

Ribonucleotide reduction is inhibited by hydroxyurea [36, 37].

dUMP is produced by deamination of dCMP under the influence of *dCMP deaminase* (E.C.3.5.4.6) which is discussed further on page 218. The complete route for the conversion of UMP into dUMP is probably as follows:

$$UMP \rightarrow UTP \rightarrow CTP \rightarrow CDP \rightarrow dCDP \rightarrow dCMP \rightarrow dUMP$$

In *Esch. coli*, which lacks dCMP deaminase, dUMP is produced from UMP by a tortuous route thus:

$$UMP \to UDP \to dUDP \to dUTP \to dUMP$$

[29] (Fig. 9.8).

9.7 The biosynthesis of thymine derivatives

The essential step in the formation of thymine nucleotides is the methylation of deoxyuridine monophosphate (dUMP) to produce thymidine monophosphate (dTMP) (dUMP → dTMP) under the influence of an enzyme system which is frequently referred to as *thymidylate synthetase*. The process is elaborate and takes place in several stages. The source of the additional carbon atom at C-5 is N^5,N^{10}-methylene tetrahydrofolic acid [38, 39]. The reaction is as follows:

N^5,N^{10}-methylene tetrahydrofolate + dUMP →
$$\text{dihydrofolate} + \text{dTMP}$$

The dihydrofolate is reduced again to tetrahydrofolate under the influence of *dihydrofolate reductase*:

$$\text{dihydrofolate} + NADPH + H^+ \to \text{tetrahydrofolate} + NADP^+$$

This reaction is powerfully inhibited by the folic acid analogues *aminopterin* and *amethopterin* (*methotrexate*) which therefore inhibit the formation of thymine derivatives.

Extracts of *Esch. coli* infected with a T-even phage contain the enzyme *deoxycytidylate hydroxymethylase* which brings about the formation of 5-hydroxymethyl deoxycytidylic acid from formaldehyde and deoxycytidylic acid in the presence of N^5,N^{10}-methylene tetrahydrofolic acid [40] (pp. 144, 293).

9.8 The formation of nucleoside triphosphates

In the biosynthesis of RNA and DNA the substrates for the appropriate polymerases which are discussed in Chapters 11 and 12 are the ribonucleoside 5′-triphosphates and the deoxyribonucleoside 5′-triphosphates. These are formed from the corresponding nucleoside monophosphates by the appropriate *kinases* in the presence of ATP [8, 29]. Some of these kinases are discussed on p. 212 but preliminary mention might be made at this stage of the important kinases which convert thymidine into its triphosphate (dTTP).

Thymidine is readily incorporated into cells which are synthesizing DNA, and the incorporation of labelled thymidine has been very extensively used in studies on the biosynthesis of DNA (for which it is a specific precursor). Thymidine labelled with ^{15}N or ^{14}C has been shown to be incorporated readily into DNA in the rat [41], the chick embryo [42], bone marrow cells [43], onion root rips [44], and tissue cultures [45], while thymidine labelled with tritium has been used in the autoradiographic study of chromosome reproduction (for which it has special advantages) [46–49]. It has also been used extensively to study DNA biosynthesis in cell-free preparations from mammalian tissues (see Chapter 11).

The incorporation of thymidine into DNA involves several stages in which it is converted by a series of kinases (p. 212) into dTTP by stepwise phosphorylation, thus:

$$\text{thymidine} \rightarrow \text{dTMP} \rightarrow \text{dTDP} \rightarrow \text{dTTP}$$

Each step is catalysed by a separate kinase [50, 51].

9.9 The control of nucleotide biosynthesis

It is clear from the work described earlier in this chapter that the formation of nucleotides for nucleic acid biosynthesis is the culmination of a long and complex chain of enzyme reactions which are under an elaborate system of control [52, 53].

The biosynthesis of DNA can be considered as taking place in four main stages: (1) the biosynthesis of purine and pyrimidine ribonucleoside monophosphates; (2) the conversion of these ribonucleotides into the corresponding deoxyribonucleotides; (3) the phosphorylation of the deoxyribonucleoside monophosphates to the corresponding triphosphates; and (4) the polymerization of the deoxyribonucleoside triphosphates to yield polydeoxyribonucleotides in the presence of an appropriate DNA template (Fig. 9.7).

Much of this investigation of control mechanisms has been carried out on rapidly growing cells and tissues such as strains of cells growing in tissue culture, tumours, and liver tissue regenerating after partial hepatectomy.

The formation of purine and pyrimidine ribonucleotides is controlled by well-known feed-back mechanisms [54–58]. Positive feed-back mechanisms are illustrated by the production of the appropriate enzyme in the presence of the corresponding substrate. The mechanism of positive feed-back is, of course, the phenomenon of enzyme adaptation which is well known in micro-organisms but which is also widely recognized in mammalian systems [59].

In the operation of negative feed-back mechanisms the products of a series of enzyme reactions can affect either the activity, or the amount produced, of one or more of the enzymes of the chain.

For example both AMP and GMP inhibit the first specific enzyme in the purine biosynthetic pathway, phosphoribosyl pyrophosphate amindotransferase (E.C.2.4.2.14) (p. 202) [60], and are involved in other control mechanisms [61]. CTP exercises an allosteric regulation of pyrimidine biosynthesis by directly inhibiting the action of

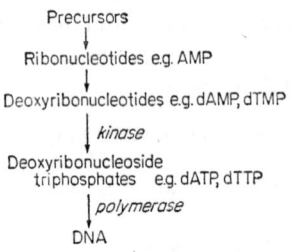

Fig. 9.7 *Stages in the biosynthesis of DNA*

the first specific enzyme in the biosynthetic pathway, aspartate carbamoyltransferase (E.C.2.1.3.2) (p. 206), which catalyses the condensation of carbamoyl phosphate and aspartate to yield carbamoyl aspartate [54, 55] (Fig. 9.8).

UMP exercises a feed-back inhibition on orotidine monophosphate decarboxylase [62].

The *amount* of enzyme produced by an enzyme-forming system may also be controlled by negative feed-back mechanisms. For example pyrimidine derivatives repress the formation in *Esch. coli* of the three enzymes asparate carbamoyltransferase (E.C.2.1.3.2), dihydro-orotase, and dihydro-orotic dehydrogenase [54].

Another control mechanism operates in the conversion of ribose derivatives into deoxyribose derivatives [32]. The conversion of cytidine diphosphate (CDP) into deoxycytidine diphosphate (dCDP), in extracts of chick embryos, and hence, ultimately, the biosynthesis of DNA, is inhibited by low (5×10^{-5} M) concentrations of dATP [63].

In Ehrlich ascites cells, a similar inhibitory effect of dATP on the reduction of guanosine nucleotide to deoxyguanosine nucleotide has been recorded [64], and this is almost certainly the explanation of the observation that DNA synthesis in such cells [65] and also in chick embryos [65] is inhibited by deoxyadenosine which is readily converted into its triphosphate.

According to Reichard, the reduction of pyrimidine ribonucleoside diphosphates in *Esch. coli* is initiated by ATP. Of the two products dCDP and dUDP, the latter is transformed into dTTP in a series of steps which have already been discussed. As the concentration of dTTP increases, it initiates the reduction of purine ribonucleotides to dADP and dGDP which are phosphorylated to dATP and dGTP. The latter causes a further stimulation of purine ribonucleotide reduction, especially the reduction of ADP.

Finally dATP, as it accumulates, acts as a general feed-back-inhibitor, depressing the reduction of all four ribonucleoside diphosphates and so preventing an accumulation of deoxyribonucleotides in the cell [31, 63, 66].

A similar mechanism probably operates in animal cells.

It would appear, therefore, that DNA synthesis takes place most readily at critical concentrations of the deoxyribonucleoside triphosphates, and that these compounds are components of a homeostatic mechanism [63] (Fig. 9.8).

A third regulating mechanism operates in relation to the formation of deoxyuridine monophosphate (dUMP) which is the immediate precursor of the thymidine series of nucleotides. There is some evidence that uridine monophosphate (UMP) may be a direct precursor of dUMP but in most tissues this conversion also takes place by a more indirect route involving the cytidine nucleotides (Fig. 9.8) [67, 68], the amination of the uridine series to the cytidine series probably taking place at ribonucleoside triphosphate level [69, 70]. The enzyme dCMP deaminase (dCMP aminohydrolyase, E.C. 3.5.4.5) then brings about the conversion of dCMP into dUMP. This enzyme is very low in most adult mammalian tissues except thymus and bone marrow [68] which synthesize DNA readily, but is active in such rapidly growing tissues as rat and chick embryos and regenerating rat liver [71–74] and may be involved in a rate-limiting step in DNA biosynthesis [71, 75, 76]. It is abundant in certain tumours such as the Novikoff hepatoma, but not in other hepatomas, i.e. Morris 5123 and Dunning L-C18. The significance of these observations has been discussed by Potter and his colleagues in relation to the deletion hypothesis [58, 77].

The dCMP deaminase is present in small amounts in normal rat liver [78] but increases in regenerating rat liver about 12 hours after partial hepatectomy, reaches a maximum at about 48 hours, and then declines. Its elevation before that of the other enzymes involved in thymidine monophosphate (dTMP) formation and phosphorylation is consistent with a sequential induction of the enzymes concerned in DNA biosynthesis [72]. The first of these other enzymes is thymidylate synthetase, which converts dUMP into dTMP and which also rises in regenerating rat liver shortly after the deaminase. It has been suggested that dUMP may activate or induce the formation of thymidylate synthetase which in turn, through its product of

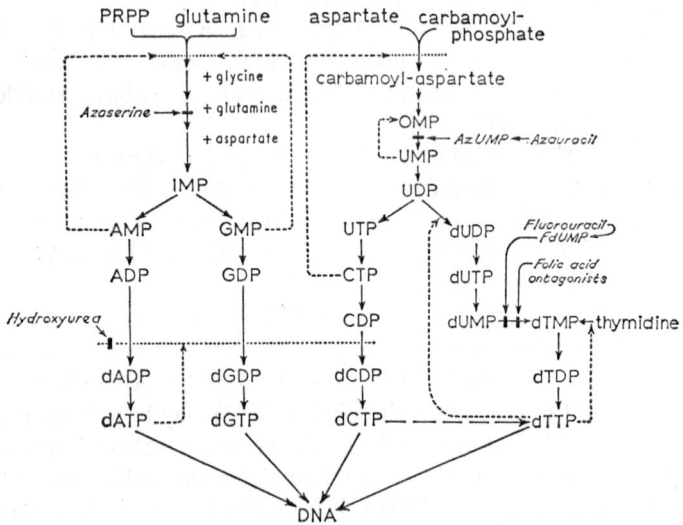

Fig. 9.8 *Pathways involved in the biosynthesis of the deoxyribonucleoside triphosphates. Feed-back mechanisms are indicated by broken lines. The blocking action of some antimetabolites is indicated*

dTMP, may induce the formation of dTMP kinase [72]. This kinase and related enzymes must now be considered.

The kinases phosphorylating the thymidine nucleotides to dTTP are active in such tissues as thymus and bone marrow in which DNA synthesis is pronounced, and low in non-proliferating tissues such as liver. In liver regenerating after partial hepatectomy, however, kinase activity is high, especially at about 48 hours after operation [79–82]. In this respect the dTMP kinase system differs from the kinases involved in the phosphorylation of dAMP, dGMP, and dCMP, which are of comparable activity in normal and regenerating liver [79, 80].

One of the most important kinases is the thymidine kinase which phosphorylates thymidine to dTMP and has many of the attributes of rate-determining enzyme for DNA biosynthesis. Its activity is greatly enhanced in regenerating rat liver [79, 80, 82], in livers of rats on a high-protein diet [83], and in rapidly proliferating tissues in general [79, 83]. It increases in cultured mammalian cells after infection with DNA viruses [84–88] which may also cause its appearance in mutant strains of fibroblasts in which it is absent [89]. Thymidine kinase is inhibited by dTTP [90–96] and by dCTP [94, 95]. dTTP therefore exercises a feed-back inhibition on thymidine kinase without influencing dTMP kinase [92] and on the deamination of dCMP [75, 91, 97, 98] (Fig. 9.8). This mechanism consequently prevents large accumulations of dTTP, either from thymidine or from CDP by way of dUMP.

A study of the kinases in regenerating rat liver at various times after partial hepatectomy shows that the appearance of the enzymes tends to be sequential [79]. By 24 hours the thymidine kinase has begun to increase sharply, reaching maximum activity in about 30 hours. It is followed closely by the dTMP kinase, while the peak of the overall reaction for the incorporation of ^{3}H-thymidine is at about 48 hours. Subsequently all kinases decline in activity. A similar sequential pattern of kinase appearance is also found in cultures of the L strain of mouse subcutaneous fibroblasts [79]. Such cells in the resting phase after exhaustion of the medium show either no kinase activity or only thymidine kinase activity. On inoculation into fresh medium, a period of rapid growth occurs after a lag phase. The thymidine kinase appears early, rises sharply, and remains elevated. The dTMP kinase activity rises later, reaches its peak of activity during the early part of the growth phase, and declines before growth (increase in cell number) has stopped. This pattern can be greatly modified by adding thymidine to the cultures, without affecting the growth rates. While there is little effect on the enzymes during the early period of growth, the thymidine kinase activity remains elevated for a much longer period in the test cultures, and dTMP kinase rises to a second peak during the period when its activity in the control culture declines. These elevations in the kinase activities can be produced only during the growth phase, and are apparently due to true enzyme induction, although it must be kept in mind that dTMP kinase is a very labile enzyme which is stabilized in the presence of thymidine or thymidylate [8].

While normal liver is lacking in the kinases for phosphorylating the thymine nucleotides, it contains the enzymes required for the reductive catabolism of thymine. In hepatomas such as the Novikoff and the Dunning, the catabolic enzymes are lacking and thymine derivatives can therefore be diverted along the anabolic pathway to dTTP. The importance of the absence of catabolic enzymes in rapidly proliferating tissues which are active in DNA biosynthesis has been fully discussed by Potter [58] in the exposition of the *deletion hypothesis*.

9.10 General aspects of catabolism

The precise mechanisms of degradation of nucleic acids are still far from clear but many cells and tissues contain enzymes such as ribonucleases, deoxyribonucleases, phosphodiesterases, and nucleotidases (Chapter 8), all of which are concerned in breaking down nucleic acids. It seems clear that RNA and DNA are hydrolysed by nucleases and diesterases first to oligonucleotides and eventually to mononucleotides and nucleosides, and that the glycosidic linkages between the purine or pyrimidine bases and the sugar moieties are cleaved either hydrolytically or phosphorolytically to yield the free

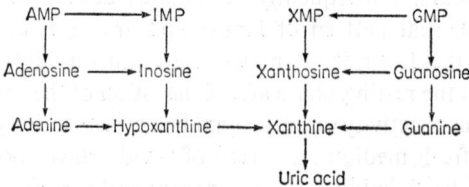

Fig. 9.9 *The degradation of purines at the levels of nucleotides, nucleosides, and bases*

purine and pyrimidine bases [99–101]. The nature and function of nucleases and phosphodiesterases has been considered in Chapter 8, and the pyrophosphatases and phosphatases that attack nucleotides have been reviewed extensively [102–104].

It is recognized that cellular DNA tends to be strongly conserved and that the amount of degradation of DNA in normal circumstances is small. Some species of RNA have a relatively short life span and are therefore degraded quite rapidly. There is also evidence that nucleic acids may not undergo total degradation but that some of the intermediate products such as nucleotides and nucleosides may be reutilized by so-called 'salvage' pathways.

9.11 Purine catabolism

The breakdown of purine nucleotides has been studied extensively. Adenine and its nucleosides and nucleotides for instance can be deaminated hydrolytically under the influence of the enzymes *adenine deaminase, adenosine deaminase,* and *adenylate deaminase* to yield hypoxanthine, inosine, or inosine monophosphate respectively (Fig. 9.9) and guanine nucleotides are similarly attacked by guanine, guanosine or guanylate deaminases to yield xanthine or its

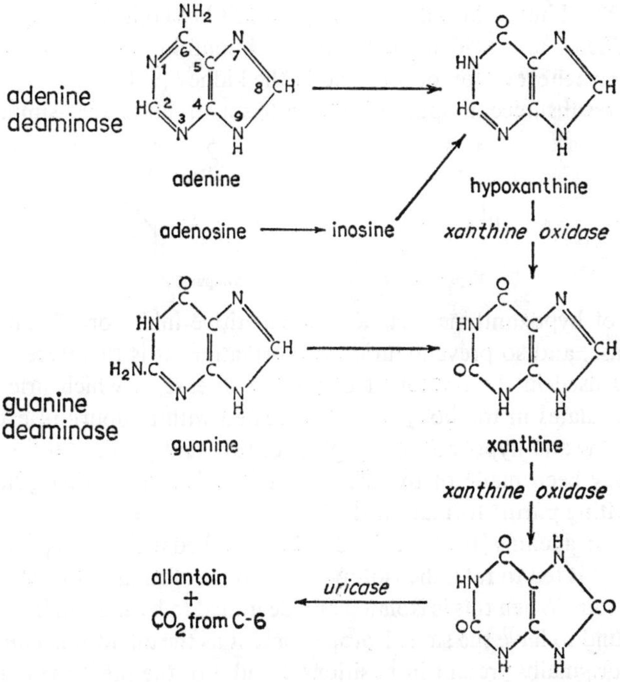

Fig. 9.10 *The catabolism of purines*

ribose derivatives (Fig. 9.9). Hypoxanthine and xanthine are then oxidized under the influence of *xanthine oxidase* to yield uric acid (Fig. 9.10). Although the distribution of the enzymes involved is far from uniform in different species, this scheme of purine degradation appears to be of fairly general application, and recent experiments with ^{15}N have shown that, as might be expected, the administration of labelled purines to animals is followed by the appearance of the isotope in the excreted uric acid or in its further degradation products. Uric acid itself is excreted by only a few mammals, since most non-uricotelic animals are provided with the enzyme *uricase*, which oxidizes uric acid to the much more soluble *allantoin*, and under certain conditions to other end-products as well [105]. Man and certain higher apes, however, are unable to bring about this step owing to absence of uricase from their tissues, and in them the end-product of purine metabolism is uric acid itself which is excreted in the urine along with very much smaller amounts of xanthine and hypoxanthine [106]. The Dalmatian coach-hound is peculiar in that it excretes uric acid in preference to allontoin, owing to lack of tubular reabsorption of uric acid in the kidney [99].

The substance allopurinol, which has a structure very similar to

Hypoxanthine   Allopurinol

that of hypoxanthine, acts as a competitive inhibitor of xanthine oxidase, and so prevents uric acid formation. It is therefore sometimes used in the treatment of gout, a disease in which uric acid accumulates in the body. Patients treated with allopurinol excrete xanthine and hypoxanthine in place of uric acid [106]. Mention has already been made of the use of azathioprine in treating gout by inhibiting purine formation [107].

When adenine [108] or uric acid [109] labelled with ^{15}N in positions 1 and 3 is fed to rats, the isotope is, of course, found in the excreted allantoin. When this is isolated and degraded to hydantoin the latter is found to have the same isotope content as the allantoin. Thus the ^{15}N originally present in positions 1 and 3 of the purine ring must have become uniformly distributed between the iminazole and urea moieties of the allantoin. This suggests that conversion of uric acid into allantoin involves the formation of a symmetrical intermediate such as hydroxyacetylene-diureinecarboxylic acid [108–110].

allantoin   hydantoin   hydroxyacetylene diureinecarboxylic acid

In fishes, in amphibia, and in more primitive organisms allantoin is broken down by *allantoinase* to allantoic acid, and this in turn may be degraded by *allantoicase* to urea and glyoxylic acid. The main nitrogenous excretory product in the spider is not uric acid but guanine. These aspects of comparative biochemistry are discussed in detail in the books by Baldwin [111] and by Florkin [112].

$$\underset{\text{allantoic acid}}{\overset{NH_2}{\underset{CO}{|}}\underset{\underset{H}{|}}{N}\overset{COOH}{\underset{CH}{|}}\underset{\underset{H}{|}}{N}\overset{NH_2}{\underset{CO}{\diagdown}}}$$

It is in the birds and the uricotelic reptiles that uric acid formation is most pronounced since in them uric acid rather than urea is the main nitrogenous excretory product. In most birds, uric acid production can be shown to take place in the liver, since hepatectomy is followed by cessation of uric acid synthesis and a rise in the blood ammonia level. The obvious inference that in birds and reptiles uric acid is derived ultimately from ammonia is supported by isotopic experiments. Urea does not act as a precursor of uric acid except insofar as it may give rise to ammonia. While the liver of the fowl or goose contains all the enzymes required for uric acid formation, that of the pigeon is lacking in xanthine oxidase. In the pigeon, therefore, hypoxanthine is produced in the liver, and is oxidized to uric acid in the kidney where xanthine oxidase is present.

Intravenous administration to normal human subjects of uric acid labelled with ^{15}N and examination of the excretion of isotope in the urine have shown that the injected uric acid is promptly diluted by a miscible pool of uric acid amounting to about 1 gram [113–115]. Since the rate of formation of uric acid calculated from the rate of fall in isotope concentration exceeds the rate of excretion of uric acid by 20 per cent or more, it would appear that some uric acid undergoes catabolic breakdown in man.

In the pathological condition known as gout [116], uric acid is deposited in the joints, particularly in the great toe, and under the skin as nodules called *tophi*. In this disease the miscible pool of uric acid in the human body is increased to as much as 15 times the normal value [113, 117]. Administration of ^{15}N-glycine to a gouty human subject has revealed a more rapid incorporation of isotope into the excreted uric acid as compared with the normal, although the

THE BIOCHEMISTRY OF THE NUCLEIC ACIDS

excretory patterns for total nitrogen, urea, and ammonia are unchanged [118]. It has therefore been suggested that in gout the mechanism of transformation of dietary glycine to uric acid is more rapid than normal, so that overproduction occurs with consequent increase in the size of the miscible pool of uric acid.

Excessive amounts of uric acid are excreted by children suffering from Lesch–Nyhan syndrome mentioned on p. 205.

9.12 Pyrimidine catabolism

The catabolism of pyrimidine nucleotides, like that of purine nucleotides, involves dephosphorylation, deamination, and cleavage of glycosidic bonds, and many of the phosphatases that act upon purine nucleotides act also on the corresponding pyrimidine derivatives. As with purine nucleosides, so pyrimidine nucleosides may be

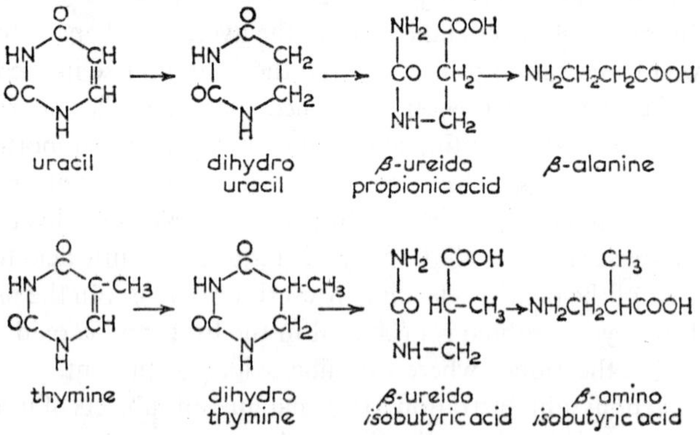

Fig. 9.11 *Catabolic pathways for uracil and thymine*

hydrolysed to form pyrimidine bases and sugar, or they may be involved in phosphorolytic cleavage [99].

Cytosine can be deaminated by cytosine deaminase; this has been demonstrated in yeasts and other micro-organisms [119], and cytosine nucleosides are broken down to uridine nucleosides by cytidine deaminase which is widespread in animal tissues [120] as well as in bacteria.

The catabolic pathways for uracil [121] and for thymine [122, 123] in mammalian tissues involve reduction of the pyrimidines to the dihydro derivatives, ring opening to give the appropriate ureido-acid, and the removal of ammonia and CO^2 to give β-alanine or its methylated derivative (Fig. 9.11).

REFERENCES

[1] Hartman, S. C. and Buchanan, J. M. (1959) *Ann. Rev. Biochem.*, **28**, 365
[2] Buchanan, J. M. and Hartman, S. C. (1959) *Adv. Enzymology*, **21**, 199
[3] Reichard, P. (1955) *The Nucleic Acids*, Vol. 2, p. 277 (E. Chargaff and J. N. Davidson, Eds.). New York: Academic Press
[4] Schlenk, F. (1955) *The Nucleic Acids*, Vol. 2, p. 309 (E. Chargaff and J. N. Davidson, Eds.). New York: Academic Press
[5] Flaks, J. G. and Lukens, L. N. (1963) *Methods in Enzymology*, Vol. 6, p. 52 (S. P. Colowick and N. O. Kaplan, Eds.). New York: Academic Press
[6] Warren, L. (1961) *Metabolic Pathways*, Vol. 2, p. 459 (D. M. Greenberg, Ed.). New York: Academic Press
[7] Schulman, M. P. (1961) *Metabolic Pathways*, Vol. 2, p. 389 (D. M. Greenberg, Ed.). New York: Academic Press
[8] Grav, H. J. (1967) *Methods in Cancer Research*, **3**, 243. New York: Academic Press
[9] Levenberg, B., Melnick, I. and Buchanan, J. M. (1957) *J. Biol. Chem.*, **225**, 163
[10] Skipper, H. E., Mitchell, J. H. and Bennett, L. L. (1950) *Cancer Res.*, **10**, 510
[11] Skipper, H. E., Bennett, L. L. and Law, L. W. (1952) *Cancer Res.*, **12**, 677
[12] Rhoads, C. P. (Ed.) (1955) *Anti-metabolites and Cancer* (American Association for the Advancement of Science)
[13] Kalckar, H. (1947) *Symp. Soc. Exp. Biol.*, **1**, 38
[14] Kornberg, A., Lieberman, I. and Simms, E. S. (1954) *J. Amer. Chem. Soc.*, **76**, 2027
[15] Murray, A. W. (1971) *Ann. Rev. Biochem.*, **40**, 811
[16] Kelley, W. N., Rosenbloom, F. M., Henderson, J. F. and Seegmiller, J. E. (1967) *Proc. Nat. Acad. Sci.*, **57**(6), 1735
[17] Kelley, W. N., Rosenbloom, F. M. and Seegmiller, J. E. (1967) *J. Clin. Invest.*, **46**(9), 1518
[18] Hitchings, G. (1967) *Fed. Proc.*, **26**, 958
[19] Fujimoto, W. Y., Subak-Sharpe, J. H. and Seegmiller, J. E. (1971) *Proc. Nat. Acad. Sci.*, **68**, 1516
[20] Rubin, C. S., Dancis, J., Yip, L. C., Bowinski, R. C. and Balis, M. E. (1971) *Proc. Nat. Acad. Sci.*, **68**, 1461
[21] Boyle, J. A. (1970) *Science*, **169**, 688
[22] Kelley, W. N. (1968) *Fed. Proc.*, **27**, 1047
[23] Seegmiller, J. E., Rosenbloom, F. M. and Kelley, W. N. (1967) *Science*, **155**, 1682
[24] Reichard, P. (1959) *Adv. Enzymology*, **21**, 263
[25] Crosbie, G. W. (1960) *The Nucleic Acids*, Vol. 3, p. 323 (E. Chargaff and J. N. Davidson, Eds.). New York: Academic Press
[26] Lieberman, I. (1955) *J. Amer. Chem. Soc.*, **77**, 2661
[27] Lieberman, I. (1956) *J. Biol. Chem.*, **222**, 765
[28] Kammen, H. O. and Hurlbert, R. B. (1959) *Cancer Res.*, **19**, 654
[29] Rose, I. A. and Schweigert, B. S. (1953) *J. Biol. Chem.*, **202**, 635
[30] Brown, G. B. (1954) *Ann. N.Y. Acad. Sci.*, **60**, 185
[31] Larsson, A. and Reichard, P. (1967) *Progress in Nucleic Acid Research and Molecular Biology*, Vol. 7, p. 303 (J. N. Davidson and W. E. Cohn, Eds.). New York: Academic Press

[32] Reichard, P. (1967) *The Biosynthesis of Deoxyribose*, New York: Wiley
[33] Gershman, H., Simon, M. and Abeles, R. H. (1971) *Biochim. Biophys. Acta*, **246**, 169
[34] Stryer, L., Holmgren, A. and Reichard, P. (1967) *Biochem.*, **6**, 1016
[35] Henderson, J. F. and Paterson, A. R. P. (1973) *Nucleotide Metabolism*. New York: Academic Press
[36] Adams, R. L. P. and Lindsay, J. G. (1966) *J. Biol. Chem.*, **242**, 1314
[37] Bono, V. H., Jr. and Wells, J. H. (1968) *Proc. Am. Assoc. Cancer Res.* **9**(26), 7
[38] Kornberg, A. (1957) *The Chemical Basis of Heredity* (W. D. McElroy and B. Glass, Eds.). Baltimore: Johns Hopkins
[39] Friedkin, M. (1963) *Ann. Rev. Biochem.*, **32**, 185
[40] Flaks, J. G. and Cohen, S. S. (1957) *Biochim. Biophys. Acta*, **25**, 667
[41] Reichard, P. and Estborn, B. (1951) *J. Biol. Chem.*, **188**, 839
[42] Friedkin, M., Tilson, D. and Roberts, De W. (1956) *J. Biol. Chem.*, **220**, 627
[43] Friedkin, M. and Wood, H. (1956) *J. Biol. Chem.*, **220**, 639
[44] McQuade, H. A., Friedkin, M. and Atchison, A. A. (1956) *Exp. Cell Res.*, **11**, 249
[45] Lu, K. H. and Winnick, T. (1954) *Exp. Cell Res.*, **7**, 238
[46] Taylor, J. H., Woods, P. S. and Hughes, W. L. (1957) *Proc. Nat. Acad. Sci.*, **43**, 122
[47] Painter, R. B., Forro, F., Jr. and Hughes, W. L. (1958) *Nature*, **181**, 328
[48] Firket, H. and Verly, W. G. (1958) *Nature*, **181**, 274
[49] Gall, J. G. (1957) *Biol. Bull.*, **113**, 322
[50] Grav, H. J. and Smellie, R. M. S. (1963) *Biochem. J.*, **89**, 486
[51] Grav, H. J. and Smellie, R. M. S. (1965) *Biochem. J.*, **94**, 518
[52] Davidson, J. N. (1962) *The Molecular Basis of Neoplasia*, p. 420. University of Texas Press
[53] Blakley, R. L. and Vitols, E. (1968) *Ann. Rev. Biochem.*, **37**, 201
[54] Changeux, J. P. (1965) *Sci. Amer.*, **212**(4), 36
[55] Gerhart, J. C. and Pardee, A. B. (1964) *Fed. Proc.*, **23**, 727
[56] Potter, V. R. (1958) *Fed. Proc.*, **17**, 691
[57] Potter, V. R. (1957) *Univ. Michigan Med. Bull.*, **23**, 401
[58] Potter, V. R. (1962) *The Molecular Basis of Neoplasia*, p. 367. University of Texas Press
[59] Knox, W. E., Auerbach, V. H. and Lin, E. C. C. (1956) *Physiol. Rev.*, **36**, 164
[60] Wyngaarden, J. B. and Ashton, D. M. (1959) *Nature*, **183**, 747
[61] Bojarski, T. B. and Hiatt, H. H. (1960) *Nature*, **188**, 1112
[62] Creasey, W. A. and Handschumacher, R. E. (1961) *J. Biol. Chem.*, **236**, 2050
[63] Reichard, P., Canellakis, Z. N. and Canellakis, E. S. (1960) *Biochim. Biophys. Acta*, **41**, 558
[64] Munch-Petersen, A. (1960) *Biochem. Biophys. Res. Commun.*, **3**, 392
[65] Maley, G. F. and Maley, F. (1960) *J. Biol. Chem.*, **235**, 2964
[66] Moore, E. C. and Hurlbert, R. B. (1966) *J. Biol. Chem.*, **241**, 4802
[67] Hecht, L. I. and Potter, V. R. (1956) *Cancer Res.*, **16**, 999
[68] Potter, V. R., Pitot, H. C., McElya, A. B. and Morse, P. A. (1960) *Fed. Proc.*, **19**, 312
[69] Lieberman, I. (1956) *J. Biol. Chem.*, **222**, 765
[70] Hurlbert, R. B. and Kammen, H. O. (1960) *J. Biol. Chem.*, **235**, 443

[71] Maley, G. F. and Maley, F. (1959) *J. Biol. Chem.*, **234**, 2975
[72] Maley, F. and Maley, G. F. (1960) *J. Biol. Chem.*, **235**, 2968
[73] Scarano, E., Talarico, M., Bonaduce, L. and de Petrocellis, B. (1960) *Nature*, **186**, 237
[74] Maley, G. F. and Maley, F. (1964) *J. Biol. Chem.*, **239**, 1168
[75] Maley, F. and Maley, G. F. (1964) *Metabolic Control Mechanisms in Animal Cells*, National Cancer Institute, Monograph No. 13, p. 117
[76] Geraci, G., Rossi, M. and Scarano, E. (1967) *Biochem.*, **6**, 183
[77] Potter, V. R. (1964) *Cancer Res.*, **24**, 1085
[78] Maley, G. F. and Maley, F. (1961) *Biochim. Biophys. Acta*, **47**, 181
[79] Weissman, S. M., Smellie, R. M. S. and Paul, J. (1960) *Biochim. Biophys. Acta*, **45**, 101
[80] Bollum, F. J. and Potter, V. R. (1959) *Cancer Res.*, **19**, 561
[81] Canellakis, E. S., Jaffe, J. J., Mantsavinos, R. and Krakow, J. S. (1959) *J. Biol. Chem.*, **234**, 2096
[82] Weissman, S. M., Paul, J., Thomson, R. Y., Smellie, R. M. S. and Davidson, J. N. (1960) *Biochem. J.*, **76**, 1P
[83] Gebert, R. A. and Potter, V. R. (1964) *Fed. Proc.*, **23**, 268
[84] Kit, S. (1963) *Viruses, Nucleic Acids and Cancer*, p. 296. Baltimore: Williams and Wilkins
[85] Kit, S., Dubbs, D. R. and Piekarski, L. J. (1962) *Biochem. Biophys. Res. Commun.*, **8**, 72
[86] McAuslan, B. R. and Joklik, W. K. (1962) *Biochem. Biophys. Res. Commun.*, **8**, 486
[87] Keir, H. M. (1968) *Soc. Gen. Microbiol. Symp.*, **18**, 67
[88] Klemperer, H. G., Haynes, G. R., Shedden, W. I. H. and Watson, D. H. (1967) *Virology*, **21**, 120
[89] Dubbs, D. R. and Kit, S. (1964) *Virology*, **22**, 493
[90] Potter, V. R. (1964) *Metabolic Control Mechanisms in Animal Cells*, National Cancer Institute, Monograph No. 13, p. 111
[91] Maley, F. and Maley, G. F. (1962) *Biochem.*, **1**, 847
[92] Ives, D. H., Morse, P. A., Jr. and Potter, V. R. (1963) *J. Biol. Chem.*, **238**, 1467
[93] Breitman, T. R. (1963) *Biochim. Biophys. Acta*, **67**, 153
[94] Bresnick, E. and Karjala, R. J. (1964) *Cancer Res.*, **24**, 841
[95] Bresnick, E., Thompson, U. B., Morris, H. P. and Liebelt, A. G. (1964) *Biochem. Biophys. Res. Commun.*, **16**, 278
[96] Okazaki, R. and Kornberg, A. (1964) *J. Biol. Chem.*, **239**, 275
[97] Maley, G. F. and Maley, F. (1962) *J. Biol. Chem.*, **237**, PC3311
[98] Scarano, E., Geraci, G. and Rossi, M. (1964) *Biochem. Biophys. Res. Commun.*, **16**, 239
[99] Henderson, J. F. and Paterson, A. R. P. (1973) *Nucleotide Metabolism*, p. 152. New York: Academic Press
[100] Smellie, R. M. S. (1955) *The Nucleic Acids* (E. Chargaff and J. N. Davidson, Eds.), Vol. II, p. 393
[101] Potter, V. R. (1960) *Nucleic Acid Outlines*, p. 217. Minneapolis, Burgess
[102] Kielley, W. W. (1961) *The Enzymes* (2nd Edition) (P. D. Boyer, H. Hardy and K. Myreback, Eds.), Vol. 5, p. 149. New York: Academic Press
[103] Morton, R. K. (1965) *Comprehensive Biochemistry*, 16, 55
[104] Bodansky, O. and Schwartz, M. K. (1968) *Adv. Clin. Chem.*, **11**, 277
[105] Canellakis, E. S. and Cohen, P. P. (1955) *J. Biol. Chem.*, **213**, 385
[106] Balis, E. W. (1968) *Fed. Proc.*, **27**, 1067
[107] Sorensen, L. B. (1966) *Proc. Nat. Acad. Sci.*, **55**, 571

[108] Brown, G. B., Roll, P. M., Plent, A. A. and Cavalieri, L. F. (1948) *J. Biol. Chem.*, **172**, 469
[109] Brown, G. B., Roll, P. M. and Cavalieri, L. F. (1947) *J. Biol. Chem.*, **171**, 835
[110] Dalgliesh, C. E. and Neuberger, A. (1954) *J. Chem. Soc.,* 3407
[111] Baldwin, E. (1949) *An Introduction to Comparative Biochemistry*. London: Cambridge University Press
[112] Florkin, M. (1949) *Biochemical Evolution*. New York: Academic Press
[113] Benedict, J. D., Forsham, P. H. and Stetten, D. (1949) *J. Biol. Chem.*, **181**, 183
[114] Buzard, J., Bishop, C. and Talbott, J. H. (1952) *J. Biol. Chem.*, **196**, 179
[115] Green, W., Bendich, A., Bodansky, O. and Brown, G. B. (1950) *J. Biol. Chem.*, **183**, 21
[116] Wyngaarden, J. B. (1966) *Adv. Metabolic Disorders*, **2**, 1
[117] Bishop, C., Garner, W. and Talbott, J. H. (1951) *J. Clin. Invest.*, **30**, 879
[118] Benedict, J. D., Roche, M., Yu, T. F., Bien, E. J., Gutman, A. B. and Stetten, D. (1952) *Metabolism*, **1**, 3
[119] O'Donovan, G. A. and Neuhard, J. (1970) *Bact. Rev.*, **34**, 278
[120] Wisdom, G. B. and Orsi, B. A. (1969) *Eur. J. Biochem.*, **7**, 223
[121] Schulman, M. P. (1954) *Chemical Pathways of Metabolism* (D. M. Greenberg, Ed.), Vol. II, p. 223. New York: Academic Press
[122] Canellakis, E. S. (1957) *J. Biol. Chem.*, **227**, 701
[123] Fink, K., Cline, R. E., Henderson, R. B. and Fink, R. M. (1956) *J. Biol. Chem.*, **221**, 425

CHAPTER 10

The Genetic Function of DNA

10.1 Introduction

The function of a DNA molecule is to carry the genetic information of a cell in such a way that this information can be passed uncorrupted from one generation to the next. For this to occur the DNA must be a stable molecule which can be *exactly* duplicated so that the two daughter cells arising at mitosis may each receive identical copies. Mechanisms must exist to correct for any errors which may occur in the duplicating process, or any damage arising through the action of chemicals or radiation.

The information present in the DNA must be made available to the cell by an ordered mechanism which does not affect the integrity of the DNA; conversely, the act of duplication of the DNA must not drastically interfere with the expression of the information contained therein.

These are the attributes of the carrier of genetic information but, as described in Chapter 1, there was in the 1930's and 1940's some controversy as to the nature of the genetic material. This chapter deals with some of the experiments which convinced people that DNA was the sought after carrier of genetic information, and leads on to delineate the nature of the genetic code and the possibilities provided by DNA in fulfilling its role.

10.2 Bacterial transforming factors

One of the most striking achievements of the science of immunology has been the classification of pneumococci (*Diplococcus pneumoniae*) into a number of different types each characterized by the ability to synthesize a specific serologically distinct and chemically distinct capsular polysaccharide. In 1928 Griffith observed that a particular strain of pneumococci cultivated *in vitro* under specific conditions lost the ability to form the appropriate polysaccharide and consequently grew on solid media in so-called 'rough' colonies in contrast to the 'smooth' glistening colonies formed by encapsulated cells. If a living culture of such unencapsulated cells was injected into mice together with killed encapsulated pneumococci of type III, the organisms subsequently recovered from the animals were live virulent pneumococci of the encapsulated type III. It appeared therefore that some material present in the dead type III organisms had endowed the unencapsulated pneumococci with the capacity to synthesize the characteristic type III polysaccharide.

During the next five years it was shown that such pneumococcal transformation could be produced *in vitro*, that a cell-free extract could replace killed cells as the transforming agent, and that organisms which had undergone transformation did not spontaneously revert to their original type.

The chemical nature of the active principle remained obscure until 1944 when Avery, McLeod, and McCarty [1], at the Rockefeller Institute in New York, showed that DNA extracted from encapsulated smooth strains of pneumococcus type III could, on addition to the culture medium, transform unencapsulated 'rough' cells into the fully encapsulated smooth type III. The smooth cells so developed could propagate indefinitely in the same form, producing more DNA with the same capabilities. The pneumococcal DNA had therefore initiated its own reduplication as well as inducing the specific inheritable property of capsule synthesis. In other words, it had executed two functions usually associated with the gene.

It is clearly important to establish whether or not the chemical substance responsible for such a transformation is DNA alone, and to eliminate the possibility that some other factor, such as associated protein or carbohydrate, might be involved. It is known that the DNA of purified transforming principle contains no chemically detectable protein and no serologically detectable protein and that, although it is not inactivated by proteolytic enzymes, it is inactivated by deoxyribonuclease. Moreover, the only amino acid present in hydrolysates of the transforming principle is glycine known to be derived from the breakdown of adenine. It therefore seems certain that the transforming principle is DNA alone.

These observations stimulated further research into bacterial transformation, from which it emerged that the reaction was not limited to pneumococci but could be produced in a wide range of bacteria, e.g. *Haemophilus influenzae* [2, 3], *Esch. coli* [4], and the meningococcus. Nor were these transformations limited to changes in serological type since they could also be used to endow bacteria with resistance to specific drugs or antibiotics or the ability to utilize particular nutrients. More than 20 different capsular transformations have been recognized, involving at least 15 different polysaccharide antigens, and many hundreds of biochemically distinct characters have been introduced *in vitro* by the application of bacterial DNA to the cells of the homologous species. A few of these transformations are shown in Table 10.1.

TABLE 10.1
Some characters transferred in bacterial transformation brought about by DNA preparations in vitro

Capsular antigens	
	D. pneumoniae
	H. influenzae
	N. meningtides
	Esch. coli
Drug resistance	
Penicillin	D. pneumoniae
Streptomycin	D. pneumoniae
Streptomycin	H. influenzae
Sulphanilamide	D. pneumoniae

While the proportion of treated cells which may develop a new characteristic after exposure to appropriate DNA is usually small, figures as high as 17 per cent have been recorded by Hotchkiss [5–7]. Among the factors influencing the yield is the capacity of the recipient strain to be transformed, since some strains are much more susceptible than others. A second factor is the concentration of DNA to which the cells are exposed and the length of exposure. The optimum concentration appears to be about 0.5×10^{-6} grams per ml and the most appropriate time for transformation to occur is just after cell division. When pneumococci are cooled to a temperature at which growth is arrested and then rewarmed so that they start to divide synchronously, transformations are exceptionally numerous.

Just how the transforming DNA enters the cell to be transformed is not fully understood, but it is known that the acquisition of the new characteristic induced by the DNA requires a period up to 1 hour and that after acquiring the new DNA a cell multiplies more slowly for some time than do its unchanged neighbours. The establishment of the mechanism for duplicating the new DNA requires still longer. The mechanism of the transformation reaction has been recently reviewed [107].

It is possible to transfer more than one inheritable characteristic to susceptible bacteria in a single DNA preparation, e.g. one specimen of DNA may carry the three characteristics of resistance to penicillin, resistance to streptomycin, and the ability to form a capsule in pneumococci. Such a specimen of DNA might bring about transformation in 5 per cent of the recipient cells. Of the cells transformed, 98 per cent would acquire only one of the three characteristics, 2 per cent would acquire two of the characteristics, and only 0·01 per cent would acquire all three. Clearly, therefore, the DNA preparation cannot convey a complete set of the donor's characteristics to the recipient, although certain characteristics appear to be linked. For example, the DNA factors responsible for streptomycin resistance tend to be coupled with those responsible for the ability of pneumococci to use mannitol as a source of energy.

An allied phenomenon of great medical importance is the transfer of multiple drug resistance, to three or four different antibiotics, from, say, an infecting *Shigella* strain of bacteria in the gut to the normal *Esch. coli* of the gut flora and so to the many different species of intestinal bacteria in general [8–11]. Such a spread of antibiotic resistance is mediated by elements resembling the bacterial sex factors and referred to as *resistance transfer factors*. Many strains of *Staphylococcus* [12] carry genes controlling drug resistance on extra-chromosomal factors (*plasmids*) [13] which may be transferred rapidly from cell to cell in a process akin to transduction (see below).

Presumably all 'transformations' are essentially processes by which bacteria are endowed with enzyme-synthesizing capacities which they did not previously possess. The first direct proof of the presence of such a new enzyme in an organism after treatment with a DNA transforming factor was made by Marmur and Hotchkiss [14] who demonstrated that the ability to oxidize mannitol could be transferred to a non-utilizing strain of pneumococcus by culturing it in the presence of DNA prepared from a strain which possessed the power of utilizing this sugar. The organisms so transformed differ from the parents in possessing the new enzyme mannitol phosphate dehydrogenase.

The subject of bacterial transformations has been surveyed in several reviews [4, 6, 8, 15–17].

An analogous somatic transformation has been reported by Szybalski for mammalian cells [18]. He isolated DNA from cultures of human cells of the strain D98S which contain the enzyme IMP-pyrophosphorylase (E.C.2.4.2.8) responsible for the reaction:

$$\text{hypoxanthine} + \text{PRPP} \rightleftharpoons \text{IMP} + \text{PPi}$$

which is discussed further in Chapter 9. The addition of this DNA to cultures of cells of the strain D98/AH-2 which are deficient in the enzyme resulted in the appearance of IMP-pyrophosphorylase-positive genetically transformed cells detected under highly selective conditions. The transforming activity was abolished by deoxyribonuclease but not by ribonuclease.

One problem with eukaryotic DNA for use in transformation is that the required gene represents only a minute fraction of the total DNA and so its concentration is very low. One way of increasing the concentration of a particular gene within a DNA sample is by *transduction* using a bacteriophage.

10.3 Transduction phenomena

Evidence of a process analogous to bacterial transformation has been obtained from the study of the bacteriophage systems discussed in Chapter 7. When a thymine-less mutant of *Esch. coli* is infected with T2 bacteriophage the organisms can be induced to synthesize thymine. Since the lack of ability to bring about this process is presumably due to the absence of an appropriate enzyme, and since infection of the bacterial cell by bacteriophage involves the transference of phage DNA and not protein to the host cell, it would appear that the phage DNA has brought about this process of transduction by carrying the necessary mechanism for the establishment of the appropriate enzyme for thymine synthesis. Before the bacteriophage can induce the synthesis of a new substance in the bacterial host it must have been cultivated initially in a host with the ability to carry out this synthesis. The process therefore is analogous to that operating in the case of transforming principles [6, 19]. This phenomenon of transduction is discussed further in Chapter 7.

A much more dramatic example of transduction has been described by Merril [108] who used human fibroblasts from a patient suffering from congenital galactosaemia. Such cells are characterized by absence of the enzyme galactose 1-phosphate uridyl transferase (GPU transferase) which is responsible for one step in the conversion of galactose into glucose. When such cells were exposed to DNA from bacteriophage λ grown in *Esch. coli* which carried the gene for the synthesis of the transferase, they developed the ability to synthesize the missing enzyme. The appropriate piece of DNA had apparently been transferred from the *Esch. coli* cell by the bacteriophage and had been incorporated into the genome of the recipient fibroblasts. This was confirmed by labelling the RNA of the recipient cells and showing that as much as 0·2 per cent could hybridize (p. 319) with λ DNA. Furthermore the ability to synthesize the transferase is passed on to succeeding generations of fibroblasts.

A similar type of result has been reported by Harris and his colleagues [20] who used a strain of mouse fibroblast cells lacking the enzyme IMP-pyrophosphorylase (Chapter 9). By the technique of cell fusion with chick erythrocyte cells they were able to show that the interpolated genetic material was expressed and replicated so that the missing enzyme could be produced in the mouse fibroblasts.

Such results have profound implications since they open a door to the prospect of what has been termed genetic engineering or gene therapy [21, 22] (see Chapter 8).

10.4 The Hershey–Chase experiment

The classic experiment of Hershey and Chase [23–25] showed that on infection of *Esch. coli* with T2 bacteriophage only the DNA enters the bacterium. Thus only the DNA replicates and carries the information to specify new virus. To demonstrate this they grew T2 phage in the presence of ^{32}P-labelled phosphate and ^{35}S-labelled amino acids. This resulted in phage containing DNA labelled with ^{32}P (there is no sulphur in DNA) and protein labelled with ^{35}S. The labelled virus was then allowed to infect unlabelled bacteria and the progeny isolated. It was found that much of the ^{32}P (i.e. the DNA) was present in the progeny virus, but none of the ^{35}S-labelled protein. Vigorous shaking of the culture within a few minutes of infection was found to dislodge the empty ^{35}S-labelled protein coat of the virus from the bacteria. On centrifugation this coat remained in the supernatant fraction when the bacteria sedimented. The bacteria carried the ^{32}P-labelled viral DNA which replicated and produced new phage particles. This shows that the viral DNA carries the genetic material and that the coat protein is not essential once the DNA is inside the bacterium.

10.5 The metabolic stability of DNA

It has also been argued that the metabolism of DNA affords evidence of its genetic function. The early work of the Hevesy group [26–28], of Brown and his collaborators [29, 30], and of others revealed negligible incorporation of ^{32}P or labelled adenine into the DNA of the nuclei of non-dividing cells. Subsequent work with ^{14}C-formate as a precursor has made it clear that the resting rat or rabbit liver *in vivo* or *in vitro* incorporates only negligible amounts of the isotope into the DNA purines [31, 32]. In this work the most rigorous precautions are necessary to ensure that the purines isolated for counting are completely free from radioactive contaminants such as traces of amino acids.

Evidence of the same kind has also been obtained from a human patient suffering from leukaemia who was treated with therapeutic doses of ^{32}P for one year and failed to show any measurable uptake of the isotope into the DNA of brain, cartilage, and skeletal muscle [33]. Similar results have been obtained by autoradiography.

A more direct method of approaching the question of metabolic stability is to allow growing cells to incorporate a labelled precursor into their DNA and then to observe the degree to which the isotope is retained after its administration is discontinued. In experiments of this type with *Esch. coli* [34, 35] it was found that when DNA had been labelled no replacement of phosphorus atoms or of purine and pyrimidine carbon in the DNA took place during the course of further growth.

The same type of result has been obtained with rat liver regenerating after partial hepatectomy. During this phase of regeneration, cell growth is rapid and mitotic activity is intense with a peak at about 24 hours after operation. When the synthesis of DNA as calculated from ^{32}P incorporation is compared with the net increase in DNA, good agreement is obtained; this suggests that the synthesis of DNA is irreversible [36, 37]. It has been found that ^{32}P [38] and ^{14}C-adenine [30] incorporated into DNA during the regeneration period are retained over long periods of time in a manner which suggests pronounced biochemical stability [39].

Similar results have been found with tissue cultures of Earl's 'L' cells which were allowed to incorporate ^{32}P or ^{14}C-formate over a period of several days before being transferred to an inactive medium in which they were maintained for several generations. Although further cell division occurred during this period of 'decorporation' none of the isotope already incorporated into the DNA was lost [40, 41]. Similar results have been obtained with ascites tumour cells [42].

The balance of evidence thus at present suggests that although DNA may not be metabolically completely inert it is much more stable than other components of the cell. This is precisely what one would expect if DNA were the genetic material of the cell [43].

These results, however, do not exclude the possibility that in certain types of cell a portion of the DNA may be metabolically labile [44, 45]. Such DNA has been described in several sources including the oocytes of certain insects and amphibia [46, 47], the growing tissues of higher plants [48], and bacterial populations [49].

Of particular interest is the DNA of the nucleolar organizer of amphibia such as *Xenopus* and *Rana*. This DNA contains many copies of the genes coding for ribosomal RNA. During oogenesis this DNA undergoes multiple replication to produce thousands of additional copies. The rest of the DNA is not replicated at this time. Each of the copies of ribosomal DNA organizes an independent nucleolus which engages in the rapid synthesis of ribosomal RNA. At the first meiosis this extra DNA is discarded.

When DNA is damaged as a result of the action of mutagenic chemicals or irradiation, the cell has a mechanism whereby the damage can be repaired (see Chapter 11). This mechanism involves the removal of damaged regions and their replacement by the synthesis of new DNA.

DNA can thus be labelled during three periods: premitotic synthesis, formation of metabolic DNA, and renewal or repair.

10.6 The DNA of the cell nucleus

The most characteristic constituent of the cell nucleus is DNA, which forms about 30 per cent of the dry weight.

It was first pointed out by Boivin *et al*. [50, 51] and Mirsky and Ris [52, 53] that, while the mean amount of DNA in the nucleus varies quite widely from species to species, it is apparently constant for the nuclei of the different somatic tissues of a given species. On the other hand, the amount of DNA in sperm nuclei, which contain the haploid number of chromosomes, is approximately half that found in the somatic cell nuclei of the same species. These observations have been confirmed by other workers [54–57] who have discussed their evolutionary significance [51], and are illustrated in Table 10.2. The

THE GENETIC FUNCTION OF DNA

TABLE 10.2

Mean DNA content of cell nuclei in picograms (10^{-12} g) per nucleus
(Figures from various authors)

	Rat	Fowl	Ox	Frog	Toad	Carp
Liver	9·4	2·6	6·4	15·7	—	3·3
Kidney	6·7	2·3	6·3	—	—	—
Spleen	6·5	2·6	—	—	—	—
Lung	6·7	—	—	—	—	—
Leucocytes	6·6	—	—	—	—	—
Erythrocytes	—	2·6	—	15·0	7·3	3·5
Heart	6·5	2·5	—	—	—	—
Pancreas	7·3	2·7	—	—	—	—
Brain	—	2·3*	—	—	—	—
Muscle	—	2·5*	—	—	—	—
Sperm	—	1·3	2·8	—	3·7	1·6

* Chick embryo

figures for rat tissues given in this table show a considerably higher value (9·4 pg) for nuclei from liver than for those from other tissues (6·7 pg). The reason for this is discussed below. The range of values found in a variety of biological material is shown in Table 10.3.

The DNA in the different tissues of any one species is not only the same in amount per cell but also shows qualitative similarities. Thus the DNA from different tissues from the mouse cannot readily be distinguished by measurements of average nucleotide composition and chromatographic behaviour or density [59, 60].

An exception to this general rule is 'satellite' DNA. In addition

TABLE 10.3

DNA complement in picograms per particle or per cell (haploid values)[58]

Material	pg(10^{-12} g)
T2 bacteriophage	0·0002
Esch. coli	0·01
Sponge	0·06
Coelenterate	0·3
Echinoderm	0·9
Teleosts	0·5–1·5
Birds	1·0–2·0
Turtle	2·5
Mammals	2·9–3·2
Frog	7·5
Dipnoi	50
Amphiuma	84

to the bulk of the DNA in a cell, a small amount of DNA with a different buoyant density can be separated out by centrifugation to equilibrium in gradients of caesium chloride (see Chapter 4). These satellite DNA's are known as simple sequence DNA's since they are made of short lengths of DNA repeated many thousands of times [103]. The function of these satellite DNA's is unknown but it is reported [104] that the germ line cells of several species contain satellites which are not present in the somatic cells.

The data given in Table 10.2 are mean values for large numbers of nuclei in bulk, and consequently they give no indication of the variations in DNA content from one nucleus to another within such a population. To overcome this difficulty attempts have been made to determine the DNA content of individual cell nuclei by the quantitative microspectrophotometric method (p. 27), using Feulgen staining methyl green staining, or ultraviolet absorption [61–65].

The results of a spectrophotometric estimation are shown in Fig. 10.1. The upper diagram is derived from 50 rat kidney nuclei which apparently fall into one compact group. The lower diagram is derived from 50 liver nuclei which clearly fall into two main groups – a

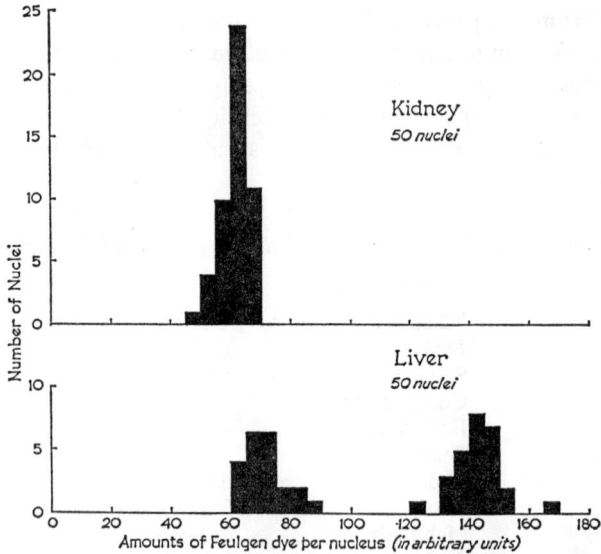

Fig. 10.1 *The DNA content of isolated rat kidney and liver nuclei as determined by quantitative microspectrophotometry after Feulgen staining (from data obtained by R. Y. Thomson)*

smaller group with a mean value similar to that for kidney nuclei and a larger and more scattered group with a mean value double that of the nuclei in the lower peak and in the kidney. This peak represents tetraploid nuclei (nuclei with a double chromosome complement), while the lower peak represents ordinary diploid nuclei (with the normal chromosome content) such as are found in most other somatic tissues. Since rodent liver is known to exhibit the phenomenon of polyploidy (the occurrence of nuclei with multiples of the diploid number of chromosomes), it is not surprising to find these two classes of nuclei by microspectrophotometric analysis, and the presence of so many tetraploid nuclei is responsible for the high figure found by chemical analysis of rat liver nuclei in bulk [61, 66].

This high figure, undergoes no statistically significant change, in spite of great alterations in the nutritional condition of the animal [55].

Such polyploidy is also found in the human liver. Schwartz [67] has found that up to the age of 6 the human liver contains only diploid nuclei but that a definite tetraploid class is developed between the years 11 and 14 and is well established at the age of 20. At this age also octoploid nuclei make their appearance and all three classes coexist in the liver for the rest of the life span.

The concept of a fixed amount of DNA per nucleus holds for bacterial, as well as for animal, cells. Caldwell and Hinshelwood [68] have shown that the DNA content of individual cells of *Bact. lactis aerogenes* grown under a variety of conditions remains unchanged although the RNA content and the cell size undergo wide variations.

The stability in the DNA content of the cell nucleus has been utilized in two ways. In the first place it enables us to calculate the number of cells in a piece of tissue by determining the total DNA content [54]. Secondly, DNA can act as a reference substance in terms of which the chemical composition of a tissue may be expressed. This usage enables us to avoid some of the fallacies inherent in the customary procedure of expressing the composition of a tissue in terms of wet or dry weight [54, 55]. In general, however, the DNA content of a cell nucleus correlates very well with the chromosome number, as has been very beautifully demonstrated in ascites tumour cells [69].

That the chromosomes are the carriers of genetic information was indicated by their behaviour at cell division when identical chromosome sets pass to each daughter cell. This relationship has been reinforced by the analysis of chromosome abnormalities and their relation to genetic disease [106]. For example, a deficiency in part of the human chromosome 22 is often found in patients with leukaemia (*Philadelphia chromosome*) and a deficiency in part of chromosome 5 is associated with the *cri du chat syndrome*. More dramatically an extra chromosome 21 produces *Down's syndrome* (mongolism) a condition found in one in every 600 newborn children. An extra female sex chromosome (XXY) results in sterile males exhibiting *Klinefelter syndrome*.

10.7 Changes during mitotic division

Since the DNA content of the cell nucleus is characteristic for the species, and since it varies within only very narrow limits, it is clear that in a rapidly growing tissue there must come a time in the course of the mitotic cycle when the DNA content of a dividing cell doubles itself, and the stage at which this doubling occurs appears to be interphase or just before prophase [70–72]. Evidence strongly supporting this view has been provided by Richards [73] using a photoelectric scanning device to enable much more accurate measurements to be made on cells in mitosis than had hitherto been possible. Some of his results are shown in Fig. 10.2 for the Ehrlich ascites tumour. In this material the non-tumour inflammatory cells (shaded in Fig. 10.2) which are present in small numbers among the tumour cells form a useful marker for the diploid value of DNA. The tumour cells themselves are tetraploid.

The results of a considerable number of incorporation experiments have shown that, in general, tissues in which cell division is infrequent, e.g. adult liver and kidney, incorporate labelled nucleic acid precursors into their DNA to only a very slight extent. Incorporation is, however, quite considerable in tissues such as intestinal mucosa, thymus, spleen, bone marrow, and appendix, in which mitoses are numerous [74] (Fig. 10.3). In mouse liver considerable diurnal variations in the incorporation of ^{32}P into both DNA and RNA have been noted [75] and the DNA exhibits metabolic heterogeneity [76].

In eukaryotic cells DNA synthesis is restricted to a period of several hours midway through interphase. In 1953 Howard and Pelc [77], as a result of autoradiographic experiments following the incorporation of ^{32}P-phosphate into the DNA of bean root tip cells

(*Vicia faba*), concluded that the period between mitotic divisions could be divided into three parts. Immediately following and immediately prior to mitosis are times when DNA synthesis is not occurring. These are the first and second gaps (G_1 and G_2). They are separated by the DNA-synthetic period (S) and the whole cell cycle is represented in Fig. 10.4. Generalizing from a large number of observations from higher animals and plants, one can say that the S-period lasts from 6 to 12 hours, the G_2-period from 3 to 8 hours,

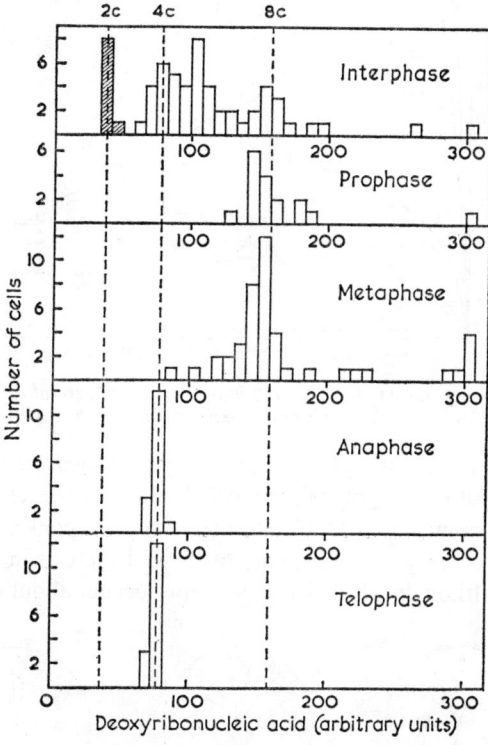

Fig. 10.2 *The DNA content, expressed in arbitrary units, of a tetraploid strain of Ehrlich ascites tumour cells. The tumour cells are unshaded. The non-tumour inflammatory cells which accompany them and which serve as a useful marker for the diploid value of DNA are shaded* [73]

but that the G_1 period is highly variable, being absent in some cells and lasting for several days in other cells [81]. By using microspectrophotometric methods [61–65] it has been shown that during the S-period the DNA content of the cell doubles and at mitosis the

DNA is shared equally between the two daughter cells. Thus, in eukaryotes DNA duplication and cell division occur strictly alternately. Should cell division fail to occur, binucleate or tetraphoid cells result.

In bacteria under slow growth conditions (i.e. doubling times

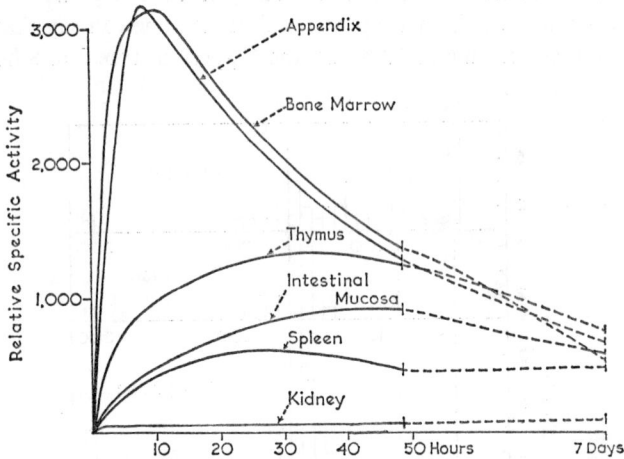

Fig. 10.3 *Incorporation of ^{32}P in vivo into the DNA of several rabbit tissues at different time intervals after administration of the isotope* [74]

greater than 60 minutes) there may be a gap between the completion of a round of DNA synthesis and cell division. However, as the rate of growth is increased DNA synthesis becomes continuous, i.e. it does not occupy a restricted part of the cell cycle as in eukaryotes [82, 83]. Although cell division generally occurs about 60 minutes

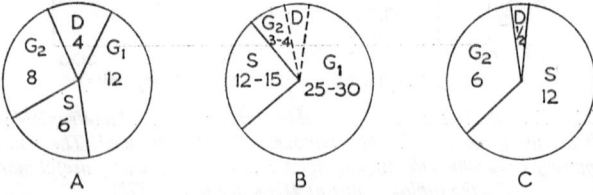

Fig. 10.4 *Mitotic cycles deduced from autoradiograph studies* [78]

A. *Bean root meristem* [77]. B. *Human bone marrow* [79]. C. *Mouse Ehrlich ascites tumour* [80]. *The figures denote times in hours. The total lengths of the mitotic cycles are* A, *30 hours;* B, *40–45 hours;* C, *18 hours.* D = *mitotic divisions;* S = *period of uptake of isotope in DNA;* G_1 *and* G_2 = *periods in early and late interphase during which DNA does not become labelled.*

THE GENETIC FUNCTION OF DNA

after the onset of a round of DNA synthesis [105], several rounds may be in process simultaneously each spanning up to three division cycles. This results in the presence in some cells immediately prior to division of two copies of part of the DNA (the last to be duplicated) yet eight copies of other regions (those duplicated first) (see Fig. 10.5).

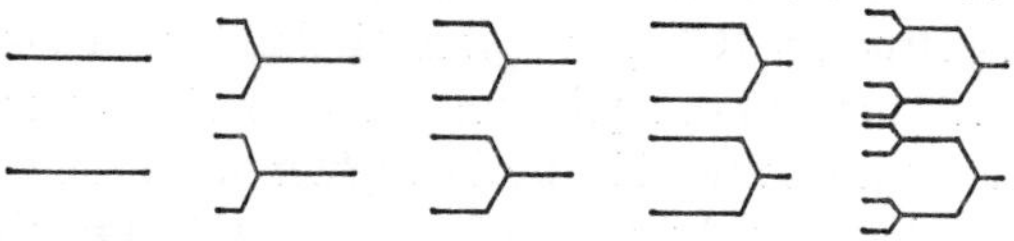

60	50	40	30	25

CELL DOUBLING TIME (MIN.)

Fig. 10.5 *Diagrammatic representation of chromosome structure at cell division in* Esch. coli *growing at different rates* [83]

10.8 Mutations

In the DNA molecule the genetic information is carried encoded in the sequence in which the four bases are arranged along the chain. Recent experiments on the nature of the genetic code have shown how different nucleotide sequences in the DNA correspond to different amino acid sequences in proteins (see Chapter 13).

A mutation is a change in the sequence of nucleotides in the DNA. The change may occur spontaneously (a spontaneous mutation will occur once for approximately every 10^9 base pairs replicated during DNA synthesis [84, 85]) or it may be induced by the action of a mutagen (see Chapter 11). Some mutations produce lethal effects, others create minor changes, whilst others are silent (i.e. have no effect; see Chapter 13). Any change is, however, transmitted to future generations by the mechanisms described in Chapter 11.

Treatment of cells with mutagens such as proflavin leads to addition (insertion) or removal (deletion) of bases from DNA [86] The study of such mutants confirmed the nature of the triplet code (see Chapter 13).

Base substitutions are readily brought about by the action of nitrous acid [87] which causes deamination of cytosine to uracil and of adenine to hypoxanthine. With the analysis of the corresponding changes in the amino acid sequence of the *Esch. coli* protein tryptophan synthetase [88] came the final proof that DNA is the genetic material, that each enzyme is coded for by a particular region or gene (the one gene one enzyme theory put forward by Beadle in 1941

[89]), and that the gene and protein are colinear [88] (see Chapter 13).

10.9 Summary and possibilities

We have then an impressive body of evidence in favour of the view that DNA is the carrier of genetic information. In terms of its complementary structure (Chapter 5) it is admirably suited to play such a part, and its method of replication (Chapter 11) whereby a parent molecule gives rise to two identical daughter molecules ensures that each cell produced as the result of mitotic division receives exactly the same complement of DNA both qualitatively and quantitatively as was contained in the parent. The constancy of the amount in all resting somatic cells of a given species, the presence of double this amount in cells about to divide, and of half this amount in sperm cells with half the normal chromosome complement is confirmatory evidence, although it is by no means conclusive. This holds also for the known relationship between the DNA content of a cell and the chromosome number, and for the well-known fact that DNA is metabolically stable and located in the chromosomes. The observation that DNA, in the absence of protein, can act as an infective agent (p. 137) which can convey biological information is further confirmatory evidence as is also the evidence derived from the study of bacterial transformation.

One of the difficulties which had to be overcome before DNA was accepted as the genetic material was that its structure was believed to be an orderly repeat of a tetranucleotide (see Chapter 5). How could such a simple molecule carry the information to specify the sequence of amino acids in the far more complicated proteins?

However, we now know that in DNA the genetic message is carried in the *sequence* in which the four bases or nucleotides are arranged along the polynucleotide chain (Chapter 5). The number of possible ways in which the four nucleotides can be arranged along the DNA chain is astronomically great (p. 365). For example, a short chain of only 1500 nucleotides (i.e. the length of an average gene) could be arranged in 4^{1500} different ways, which is far in excess of the number of proteins known. It has been stated by Beadle [90] that the information in a human egg cell is contained in approximately 100 000 genes consisting of 5×10^9 DNA bases. This would make 1 700 000 000 three-letter words which would, in turn, make up 1000 books of 600 pages with 500 words per page. There is therefore

THE GENETIC FUNCTION OF DNA

ample scope within the structure of DNA to account for species variation and for variations within a species.

Experiments on the hybridization of radioactive fragments of DNA from various species with DNA's from the same or other species on columns of DNA in agar (p. 62) have shown the existence of homologies among polynucleotide sequences in the DNA's of such widely diverse forms of life as fish and man. As might be expected, human DNA shows a more obvious relationship to that of the rhesus monkey than it does to that of the normal rat or rabbit, and virtually no relationship to that of *Esch. coli*. Similarly, mouse DNA shows a relationship to that of the rat and hamster which is more pronounced than its relationship to man or the rabbit [91]. The significance of the difference between DNA's of related animals has been discussed by Walker [92]. Indeed the DNA pattern of a species may ultimately come to be of great taxonomic value [93].

As a carrier of genetic information DNA serves two main functions: (1) to make exact copies of itself in the process of *duplication* or *replication* (Chapter 11); (2) to pass on the information coded in it to messenger RNA in the process of *transcription* (Chapter 12) so that the messenger RNA in its turn may translate the information in the four-letter language of the nucleic acids into the twenty-letter language of the amino acids and proteins (Chapter 13).

This concept is illustrated in Fig. 10.6 and is the basis of the

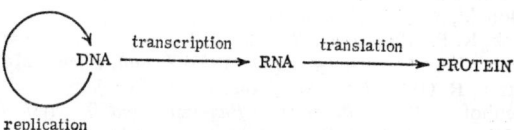

Fig. 10.6 *The Central Dogma*

'Central Dogma' put forward by Crick initially in 1958 [94] and later modified [95] to the form shown in Fig. 10.7. The modification was brought about following the discovery of the RNA-dependent DNA polymerase (p. 296).

The role of DNA in heredity has been discussed in many reviews [5, 6, 8, 17, 44, 90, 96–100]. A critical note has been introduced by Commoner [101, 102].

THE BIOCHEMISTRY OF THE NUCLEIC ACIDS

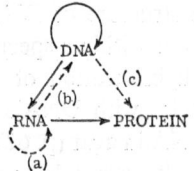

Fig. 10.7 *The present-day concept of the Central Dogma* [95]. *Solid arrows indicate general transfers while broken arrows refer to special cases:* (a) *the synthesis of viral RNA on a viral RNA template* (p. 345); (b) *the action of the RNA-dependent DNA polymerase* (p. 296); (c) *a reaction for which the evidence is obscure*

REFERENCES

[1] Avery, O. T., McLeod, C. M. and McCarty, M. (1944) *J. Exp. Med.*, **79**, 137
[2] Alexander, H. E. and Leidy, G. (1951) *J. Exp. Med.*, **93**, 345
[3] Zamenhof, S., Leidy, G., Alexander, H. E., Fitzgerald, P. L. and Chargaff, E. (1952) *Arch. Biochem. Biophys.*, **40**, 50
[4] Boivin, A. (1947) *Cold Spring Harbor Symp. Quant. Biol.*, **12**, 7 (1948) *C. R. Soc. Biol.*, Paris, **142**, 1258
[5] Hotchkiss, R. D. (1955) *The Nucleic Acids* (E. Chargaff and J. N. Davidson, Eds.), Vol. II, p. 435. New York: Academic Press
[6] Hotchkiss, R. D. (1956) *Enzymes: Units of Biological Structure and Function* (O. H. Gaebler, Ed.), p. 119. New York: Academic Press
[7] Hotchkiss, R. D. and Weiss, E. (1956) *Sci. Amer.*, **195**, 48
[8] Hayes, W. (1968) *The Genetics of Bacteria and their Viruses* (2nd Ed.). Oxford: Blackwell
[9] Anderson, E. S. (1967) *Ann. Inst. Pasteur*, **112**, 547
[10] Datta, N. (1962) *J. Hyg. Camb.*, **60**, 301
[11] Meynell, E. and Datta, N. (1967) *Nature*, **214**, 885
[12] McDonald, S. (1966) *Lancet*, 1107
[13] Novick, R. P. (1969) *Bact. Rev.*, **33**, 210
[14] Marmur, J. and Hotchkiss, R. D. (1953) *J. Biol. Chem.*, **214**, 383
[15] Austrian, R. (1952) *Bact. Rev.*, **16**, 31
[16] Zamenhof, S. (1956) *Progress in Biophysics and Biophysical Chemistry*, Vol. VI, p. 86 (J. A. V. Butler, Ed.). London: Pergamon
[17] Peacocke, A. R. and Drysdale, R. B. (1965) *The Molecular Basis of Heredity*. London: Butterworth
[18] Szybalska, E. H. and Szybalski, W. (1962) *Proc. Nat. Acad. Sci.*, **48**, 2026
[19] Zinder, N. D. (1958) *Sci. Amer.*, **199**(5), 38
[20] Schwartz, A. G., Cook, P. R. and Harris, H. (1971) *Nature New Biol.*, **230**, 5
[21] Aposhian, H. V. (1970) *Persp. Biol. Med.*, **14**, 98
[22] Qasba, P. K. and Aposhian, H. V. (1971) *Proc. Nat. Acad. Sci.*, **68**, 2345
[23] Hershey, A. D. and Chase, M. (1952) *Cold Spring Harbor Symp. Quant. Biol.*, **16**, 471
[24] Hershey, A. D. and Chase, M. (1952) *J. Gen. Physiol.*, **26**, 36
[25] Hershey, A. D. (1953) *Cold Spring Harbor Symp. Quant. Biol.*, **18**, 135
[26] Hevesy, G. and Ottesen, J. (1943) *Acta Physiol Scand.*, **5**, 237
[27] Andreasen, E. and Ottesen, J. (1945) *Acta Physiol. Scand.*, **10**, 257
[28] Hevesy, G. C. (1951) *J. Chem. Soc.*, 1618

[29] Furst, S. S., Roll, P. M. and Brown, G. B. (1950) *J. Biol. Chem.*, **183**, 251
[30] Furst, S. S. and Brown, G. B. (1951) *J. Biol. Chem.*, **181**, 239
[31] Sibatani, A. (1957) *Biochim. Biophys. Acta*, **25**, 592
[32] Fresco, J. R. and Bendich, A. (1960) *J. Biol. Chem.*, **235**, 1124
[33] Osgood, E. E., Li, J. G., Tivey, H., Duerst, M. L. and Seaman, A. J. (1951) *Science*, **114**, 95
[34] Hershey, A. D. (1954) *J. Gen. Physiol.*, **38**, 145
[35] Fujisawa, Y. and Sibatani, A. (1954) *Experientia*, **10**, 178
[36] Nygaard, O. and Rusch, H. P. (1955) *Cancer Res.*, **15**, 240
[37] Ives, D. H. and Barnum, C. P. (1962) *J. Biol. Chem.*, **237**, 2604
[38] Brues, A. M., Tracy, M. M. and Cohn, W. E. (1944) *J. Biol. Chem.*, **155**, 519
[39] Fresco, J. R., Bendich, A. and Russell, P. J. (1955) *Fed. Proc.*, **14**, 214
[40] Healy, G. M., Simonovitch, L., Parker, R. C. and Graham, A. F. (1956) *Biochim. Biophys. Acta*, **20**, 425
[41] Thomson, R. Y., Paul, J. and Davidson, J. N. (1958) *Biochem. J.*, **69**, 553
[42] Laszlo, R., Forssberg, A. and Klein, G. (1956) *J. Nat. Cancer Inst.*, **17**, 37
[43] Hughes, W. L. (1959) *Kinetics of Cellular Proliferation* (F. Stohlman, Ed.). New York: Grune and Stratton
[44] Muller, H. J. (1960) *Persp. Biol. Med.*, **5**, 1
[45] Pelc, S. R. (1964) *J. Cell Biol.*, **22**, 21
[46] Lima-de-Faria, A. (1962) *Chromosoma*, **13**, 47
[47] Gall, J. (1969) *Genetics*, (Suppl.) **61**, 121
[48] Sampson, M., Katoh, A., Hotta, Y. and Stern, H. (1963) *Proc. Nat. Acad. Sci.*, **50**, 459
[49] Contois, D. E. and Seymour, W. F. K. (1964) *Biochem. Biophys. Res. Commun.*, **16**, 124
[50] Vendrely, R. and Vendrely, C. (1948) *Experientia*, **4**, 434
(1949) *ibid.*, **5**, 327
[51] Boivin, A., Vendrely, R. and Vendrely, C. (1948) *C. R. Acad. Sci.*, **226**, 1061
[52] Mirsky, A. E. and Ris, H. (1949) *Nature*, **163**, 666
[53] Mirsky, A. E. and Ris, H. (1951) *J. Gen. Physiol.*, **34**, 451
[54] Davidson, J. N. and Leslie, I. (1950) *Nature*, **165**, 49
(1950) *Cancer Res.*, **10**, 587
[55] Thomson, R. Y., Heagy, F. C., Hutchison, W. C. and Davidson, J. N. (1953) *Biochem. J.*, **53**, 460
[56] Vendrely, R. (1955) *The Nucleic Acids*, Vol. 2, p. 155 (E. Chargaff and J. N. Davidson, Eds.). New York: Academic Press
[57] Sober, H. A. (1968) *Handbook of Biochemistry*. Cleveland: Chemical Rubber Co.
[58] Sinsheimer, R. (1957) *Science*, **125**, 1123
[59] Kit, S. (1960) *Arch. Biochem. Biophys.*, **87**, 318, 330
[60] Kit, S. (1960) *J. Biol. Chem.*, **235**, 1756
[61] Pollister, A. W., Swift, H. and Alfert, M. (1951) *J. Cell. Comp. Physiol.*, **38**, Suppl. 1, 101
[62] Kurnick, N. B. (1950) *Exp. Cell. Res.*, **1**, 151
[63] Leuchtenberger, C., Leuchtenberger, R., Vendrely, C. and Vendrely, R. (1952) *Exp. Cell. Res.*, **3**, 240
[64] Schrader, R. and Leuchtenberger, C. (1950) *Exp. Cell. Res.*, **1**, 421
[65] Killander, D. and Zetterburg, A. (1965) *Exp. Cell Res.*, **38**, 272
[66] Frazer, S. C. and Davidson, J. N. (1953) *Exp. Cell. Res.*, **4**, 316

[67] Schwartz, F. J. (1956) *Chromosoma*, **8**, 53
[68] Caldwell, P. C. and Hinshelwood, C. (1950) *J. Chem. Soc.*, 1415
[69] Richards, B. M., Walker, P. B. M. and Deeley, E. M. (1956) *Ann. N.Y. Acad. Sci.*, **63**, 931
[70] Swift, H. (1950) *Proc. Nat. Acad. Sci.*, **36**, 643
[71] Alfert, M. (1950) *J. Cell. Comp. Physiol.*, **36**, 281
[72] Walker, P. M. B. and Yates, H. (1952) *Proc. Roy. Soc.*, B, **140**, 274
[73] Richards, B. M. (1955) *Nature*, **175**, 259
[74] Smellie, R. M. S., Humphrey, G. H., Kay, E. R. M. and Davidson, J. N. (1955) *Biochem. J.*, **60**, 177
[75] Barnum, C. P., Jardetsky, C. D. and Halberg, F. (1958) *Amer. J. Physiol.*, **195**, 301
[76] Morin, G. A., Zajdela, F. and Costerousse, O. (1957) *Exp. Cell. Res.*, **13**, 204
[77] Howard, A. and Pelc, S. E. (1953) *Heredity*, Suppl. 6, 261
[78] Howard, A. (1956) *Ciba Symposium on Ionizing Radiations and Cell. Metabolism*, p. 196. London: Churchill
[79] Lajtha, L. G., Oliver, R. and Ellis, F. (1954) *Brit. J. Cancer*, **8**, 367
[80] Hornsey, S. and Howard, A. (1956) *Ann. N.Y. Acad. Sci.*, **63**, 915
[81] Mitchison, J. M. (1971) *The Biology of the Cell Cycle*. Cambridge University Press
[82] Abbo, F. E. and Pardee, A. B. (1960) *Biochim. Biophys. Acta*, **39**, 478
[83] Cooper, S. and Helmstetter, C. E. (1968) *J. Mol. Biol.*, **31**, 519
[84] Freeze, E. (1963) *Molecular Genetics*, Part I, p. 207 (J. H. Taylor, Ed.). New York: Academic Press
[85] Pauling, L. (1964) *Bull. N.Y. Acad. Med.*, **40**, 334
[86] Goldberg, I. H., Reich, E. and Rabinowitz, M. (1963) *Nature*, **199**, 44
[87] Wittman, H. G. and Wittman-Liebold, B. (1963) *Cold Spring Harbor Symp. Quant. Biol.*, **28**, 589
[88] Yanofsky, C. (1967) *Sci. Amer.*, **216**(5), 80
[89] Beadle, G. W. and Tatum, E. L. (1941) *Proc. Nat. Acad. Sci.*, **27**, 499
[90] Tatum, E. L. (1964) *Bull. N.Y. Acad. Med.*, **40**, 361
[91] Hoyer, H. H., McCarthy, B. J. and Bolton, E. T. (1964) *Science*, **144**, 959
[92] Walker, P. M. B. (1968) *Nature*, **219**, 228
[93] Simpson, G. G. (1962) *Proc. Nat. Acad., Sci.*, **102**, 497
[94] Crick, F. (1958) *Symp. Soc. Exp. Biol.*, **12**, 138
[95] Crick, F. (1970) *Nature*, **227**, 561
[96] Kendrew, J. (1966) *The Thread of Life*. London: Bell
[97] Perutz, M. F. (1962) *Proteins and Nucleic Acids*. Amsterdam: Elsevier
[98] Dunn, L. C. (1964) *Bull. N.Y. Acad. Med.*, **40**, 325
[99] Beadle, G. S. (1963) *Genetics and Modern Biology*, Jayne Lectures for 1962
[100] Crick, F. H. C. (1964) Proc. 6th Internat. Cong. Biochem., 33, 109
[101] Commoner, B. (1964) *Nature*, **202**, 960; **203**, 486
[102] Commoner, B. (1968) *Nature*, **220**, 334
[103] Southern, E. M. (1970) *Nature*, **227**, 294
[104] Hennig, W. (1975) *Trans. Biochem. Soc.* (Edinburgh Meeting)
[105] Pardee, A. B. and Rosengurt, E. (1975) MTP International Review of Science: Biochemistry Series One, Vol. 2 (C. F. Fox, Ed.), p. 155
[106] de Grouchy, J. (1974) *The Cell Nucleus*, Vol. 2 (H. Busch, Ed.), p. 371. New York: Academic Press
[107] Notani, N. K. and Setlow, J. K. (1974) *Progr. Nucleic Acid Res. Mol. Biol.*, **14**, 39
[108] Merril, C. R., Geier, M. R. and Petricciani, J. C. (1971) *Nature*, **233**, 398

CHAPTER 11

Replication of DNA

11.1 Introduction

Each daughter cell produced at cell division contains an identical copy of the genetic material. Since DNA is now known to carry in its sequence of nucleotides the genetic information or 'code', the question of the way in which it is reproduced in the cell has attracted a great deal of attention. Like most biological phenomena, DNA replication can be studied at various levels of cellular disorganization, Section 11.2 deals with studies on intact prokaryotic and eukaryotic cells. Although results from such studies obviously reflect the true *in vivo* processes they are often difficult to analyse; this is partly because the immediate precursors of DNA are nucleotides to which the living cell is impermeable. In order to circumvent this problem and to have a system of complexity intermediate between the whole cell and the purified enzyme, permealized (e.g. toluenized) bacteria, cell lysates, and isolated nuclei have been investigated in an attempt to mimic the *in vivo* situation (see Section 11.3).

Studies with purified enzymes have indicated the detailed mechanism of the polymerization reaction and illustrated certain unexpected limitations on the ability of DNA polymerases to replicate DNA. These are considered in Section 11.4.

Our increased understanding of DNA replication has come about over the last few years as a result of the intensive use of bacterial mutants, especially temperature-sensitive mutants, i.e. bacteria which grow normally at low temperatures but cease to grow at higher temperatures.

The study of *Esch. coli* mutants defective in DNA synthesis has demonstrated that a number of proteins are essential for DNA replication. These, together with the genes which specify them, are listed in Table 11.1. Many of these proteins have been purified and used with some success in attempts to reconstruct a DNA synthesizing system.

THE BIOCHEMISTRY OF THE NUCLEIC ACIDS

TABLE 11.1

Esch. coli *proteins essential for normal DNA synthesis; where no enzymic function is known for a protein it is known simply as, for example the dna B gene product*

Gene	Enzyme	Effect on DNA synthesis of changing to non-permissive temperature
dna A	Membrane protein? Anti-repressor?	Current round of replication completed but no further initiation
dna B	DNA-dependent ATPase	Synthesis stops immediately; very short pieces produced
dna C/D		Current round completed but no further initiation
dna E (pol C)	DNA polymerase III	Stops immediately
dna F (nrd)	Ribonucleotide reductase	Stops immediately
dna G	New RNA polymerase	Stops immediately; cannot initiate short pieces
pol A	DNA polymerase I	Slow joining of short pieces; susceptible to u.v.
pol B	DNA polymerase II	Normal
lig	DNA ligase	Slow joining of short pieces
dna S	——	Accumulates very short pieces
——	DNA binding protein	——

11.2 DNA replication – *in vivo* studies

11.2.1 *Semi-conservative replication*

At the end of their paper [5] suggesting the double-stranded structure for DNA with its complementary base pairs, Watson and Crick wrote: 'It has not escaped our notice that the specific pairing we have postulated suggests a possible copying mechanism for the genetic material.'

Let us suppose that a short length of a DNA double helix has the nucleotide sequence shown in Fig. 11.1 (top), and further that the two strands can be untwisted and separated from one another to form two single chains, as in Fig. 11.1 (middle), and that each base in the single strands can attach to itself the complementary deoxyribonucleotide by the same hydrogen bonding which exists in the intact DNA double helix. Finally, if these attached mononucleotides

REPLICATION OF DNA

are polymerized to form a polynucleotide chain as in Fig. 11.1 (bottom), the end result will be the formation of two complete DNA double helices identical with each other and with the original molecule. One strand of each daughter molecule will be derived from the original DNA molecule; the other will be the product of the new synthesis.

This mechanism is described as *semi-conservative* to distinguish it from the other possible mechanisms. In the *conservative* mechanism the two strands do not come apart but act together as a template to form a completely new double helical molecule; in this case one daughter molecule would be wholly new and the other totally derived from the parent. In the *dispersive* mechanism the parental

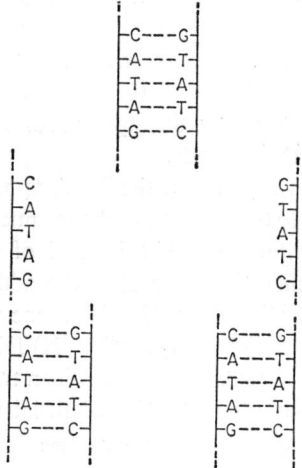

Fig. 11.1 *Schematic illustration of the separation of the two strands of a portion of DNA with the formation of a new strand on each*

molecule is partly degraded and the fragments are incorporated into two new daughter double helices. These possibilities are illustrated in Fig. 11.2.

Convincing evidence for semi-conservative replication of DNA was obtained by Meselson and Stahl [6] who grew *Esch. coli* in a medium containing $^{15}NH_4Cl$ of 96·5 per cent isotopic purity for fourteen generations so as to label the DNA very heavily with ^{15}N. The cells were then transferred to a medium containing $^{14}NH_4Cl$ and samples of bacteria were withdrawn at intervals for several generations. Each sample was lysed by means of sodium dodecyl sulphate

THE BIOCHEMISTRY OF THE NUCLEIC ACIDS

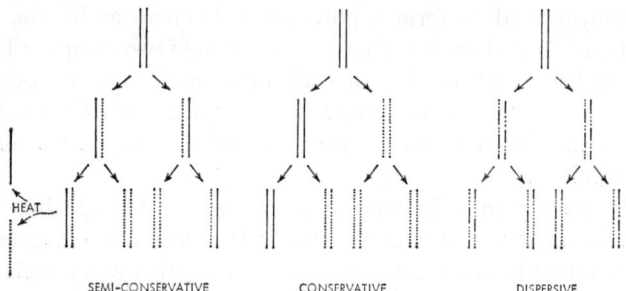

Fig. 11.2 *Possible mechanisms of replication. The dotted lines represent filial strands. For details see text*

and was centrifuged in a concentrated solution of caesium chloride at 140 000 g for 20 hours to enable the DNA to attain sedimentation equilibrium. The bands of DNA were found in the CsCl gradient in the region of density 1·71 g/cm^3 and were well isolated from all other macromolecular components of the bacterial lysate. Ultraviolet absorption photographs taken during the course of the run revealed the positions of the DNA bands.

At the start of the experiment the DNA appeared as one single band corresponding to the heavy ^{15}N-labelled nucleic acid (Fig. 11.3). Macromolecules containing half this level of ^{15}N then began to appear, and one generation time after the addition of ^{14}N these hybrid (light-heavy, ^{14}N-^{15}N) molecules alone were present. Subsequently a mixture of light-heavy (^{14}N-^{15}N) DNA and unlabelled (light-light, ^{14}N only) DNA was found. When two generation times had elapsed after the addition of ^{14}N, the half-labelled and unlabelled

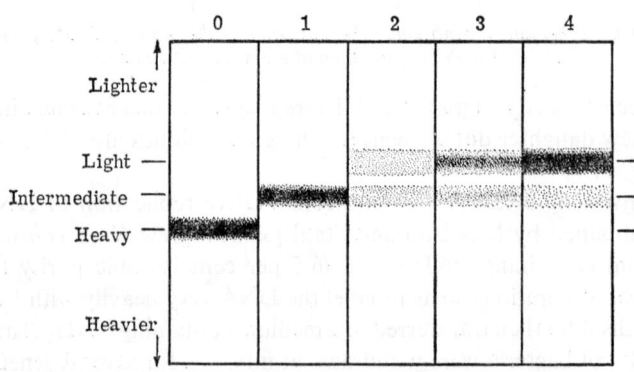

Fig. 11.3 *The pattern of results from the Meselson and Stahl experiment. For explanation see text*

DNA molecules were present in equal amounts (Fig. 11.3). During subsequent generations the unlabelled DNA accumulated. Moreover, when the hybrid ^{14}N-^{15}N molecules were heated they separated to give a ^{14}N strand and a ^{15}N strand.

Such experiments indicate that in DNA synthesis each existing DNA molecule is split into two subunits (Fig. 11.2), each subunit going to a different daughter molecule. The other subunit of each daughter molecule is the product of new synthesis. These subunits do not undergo any fragmentation but remain intact for many generations.

Experiments at the chromosomal level also support the view that DNA replication is semi-conservative. When the replication of the *Esch. coli* chromosome is followed by autoradiography after labelling of the DNA with tritium, the amount of tritium per unit length of new synthesized DNA is consistent with the presence of only one newly synthesized strand in a daughter chromosome [7, 8].

In plants to which tritiated thymidine has been given as a specific precursor of DNA during the period of DNA synthesis, both the daughter chromatids are found to be labelled at the time of cell division. At the next round of duplication, however, after withdrawal of the ^{3}H-thymidine, these two chromosomes each produce one labelled and one unlabelled chromatid as would be expected [9].

11.2.2 *Initiation and direction of replication*

(a) *Site of initiation.* The results of the Meselson–Stahl experiment suggested that DNA replication would be found to be a sequential process and that successive rounds of replication would not begin at random positions on the chromosome. It is now clear that chromosomes, from most viruses and prokaryotes, whether containing linear or cyclic DNA, initiate their replication at one specific site. For small DNA molecules, or for fragments of larger molecules, these sites can be visualized using the electron microscope when they appear as double-stranded 'bubbles' (Plate III and Fig. 11.9) [10–13, 17]. On closer examination it is apparent that on one side of each fork the DNA is only single-stranded (see p. 253).

Cairns was able to visualize the whole of the replicating *Esch. coli* chromosome autoradiographically by growing the bacteria for several generations in the presence of tritiated thymidine (Fig. 11.4), and he showed that the chromosome exists as a continuous piece of double-stranded cyclic DNA.

That the initiation of the replication bubble occurs at a unique site has been shown by a number of different techniques for viral, plasmid, bacterial, and mitochondrial DNA. Thus initiation of replication of the linear T7 bacteriophage chromosome occurs 17 per cent from one end [12] and Schnös and Inman [15] were able to show that initiation of replication occurred at a particular site with reference to the partial denaturation map of bacteriophage λ

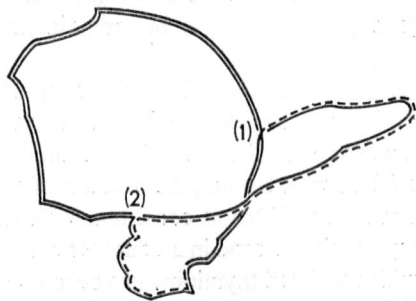

Fig. 11.4 *Diagrammatic representation of an autoradiograph of the chromosome of* Esch. coli *labelled with tritiated thymidine for two generations of replication* [14]. *The chromosome has been two-thirds duplicated and the two replication forks are indicated as (1) and (2). By courtesy of Dr John Cairns*

[16]. In the plasmid Col E1 [10] and in the simian virus 40 [17] chromosome replication is always initiated at a fixed distance from the site of action of the EcoRI restriction endonuclease (Plate III, see Chapter 8). In exponentially growing *Esch. coli* and *B. subtilis* genetic analysis has shown that, relative to non-growing cells, certain genes are present in greater than single copies [18, 19]. The interpretation of this finding is that genes replicated early will be present at twice the number of copies (or even four or eight times in rapidly growing cells) relative to late replicated genes (Fig. 10.5).

(b) *Direction of replication.* Further studies of replicating bubbles (both by electron microscopy and by genetic analysis) show that in most cases the bubbles are expanding in both directions, i.e. the bubble is made up of two replicating forks moving away from a central point of initiation. There are a few exceptions to this, e.g. the plasmid Col E1 where replication is unidirectional [10] and the rate and extent of movement of the two forks are not necessarily the same. Thus in *B. subtilis* one fork moves only approximately one-fifth of the way round the chromosome and then stops and waits for the other fork to traverse four-fifths of the chromosome [19, 416].

REPLICATION OF DNA

Replicating DNA from eukaryotes shows a large number of replication bubbles on electron micrographs [13], and fibre autoradiographic studies suggest that DNA from mammalian cells is composed of many tandemly joined sections each about 15–60 μm long [20, 21]. These sections or 'replicons' are separately replicated bidirectionally from their origin.

(c) *The 'swivel'*. Thus replication is sequential, moving bidirectionally from a fixed starting point. It involves the movement of a replication fork along the chromosome, leaving behind two separated daughter helices (Fig. 11.5). It is not easy, however, to envisage a means by which the two strands of the long DNA helix could

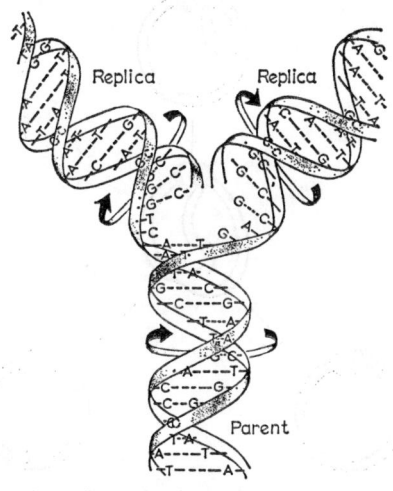

Fig. 11.5 *A portion of native, double-helical DNA shown in the process of replication* [22]. *The arrows indicate the directions of rotation that are necessary to allow unwinding of the parent helix with concomitant formation of the daughter helices. The two newly synthesized strands are growing in the downwards direction*

become untwisted, although several ingenious solutions have been suggested [23–27]. For complete strand separation, the body of the molecule must revolve once about its axis for every ten base pairs to come apart, and it has been calculated that such a rotation at 10 000 revolutions per minute is energetically not impossible. It is, however, difficult to imagine such a rotation taking place with molecules a millimetre long, and even more difficult to envisage the mechanism operating with circular DNA molecules, although a solution has been offered by Cairns (Fig. 11.6) [7, 8, 14] who proposed a swivel mechanism. The swivel would act at the starting point, allowing free rotation of the unduplicated part of the molecule to occur. More recent work would support the action of the swivel at the point of replication [28].

The swivel, whose nature was never explicit, may result from the

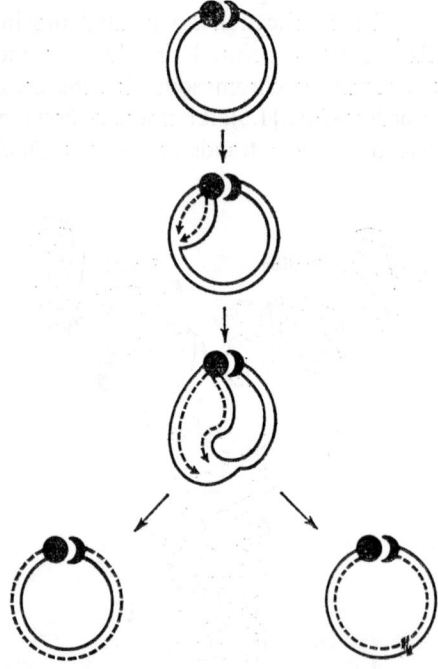

Fig. 11.6 *Replication of the cyclic molecule of DNA in the* Esch. coli *cell according to the mechanism proposed by Cairns* [8]. *The two DNA chains are represented by concentric circles joined at a ball-and-socket joint 'swivel'. Duplication starts at the swivel and proceeds counterclockwise. The newly formed DNA chains are indicated by broken lines and the arrowheads mark the replicating fork which is shown in greater detail in Fig. 11.7*

alternating action of endonuclease and ligase which could respectively break and rejoin the cyclic DNA molecules. Alternatively a specific 'DNA untwisting enzyme' may perform the function of the swivel, and such an enzyme has been reported in *Esch. coli* and mouse embryo cells [29, 30].

(d) *The signal for initiation.* Implicit in the fact that initiation of DNA synthesis occurs at a unique site on the chromosome is the suggestion that this site has a distinctive structure which allows the binding of the replicative machinery and the unwinding of the double helix. Beyond this very little is known of the requirements for initiation or the mechanism by which it is brought about. One suggestion is that one or both of the DNA strands is broken by the action of a specific endonuclease and the resulting 3'-OH ends act as primers for DNA polymerase action (see p. 261). This would imply a covalent attachment of parental and daughter strands at the point of initiation, and this, despite earlier reports, appears not to occur [31, 32]. If the attachment is transient it may however escape detection.

That specific proteins are required to initiate replication is indicated by the use of inhibitors and mutants. Thus, in *Esch. coli*, inhibition of protein synthesis by chloramphenicol, or starvation of an auxotroph for an essential amino acid, allows completion of the current round of replication but prevents reinitiation (33). Mutations on the dna A or dna C-D gene of *Esch. coli* (see Table 11.1) have similar effects, and changes in membrane proteins are implicated in at least the dna A gene mutants. Initiation of phage λ replication involves the synthesis of a specific RNA primer [417].

11.2.3 *Rate of replication*

The rate of replication of DNA is dependent on the temperature and the supply of nutrients, particularly deoxyribonucleotides. The minimum time required to replicate the *Esch. coli* chromosome is about 40 minutes, implying a rate of synthesis of about 1700 base pairs per second [7]. As replication is bidirectional the rate at each fork is about 850 base pairs per second or 14 μm per minute. (Compare this with the rate of transcription which proceeds at 35–40 nucleotides per second in *Esch. coli*.) In mammalian cells the rate of fork movement is only 0·5–1·2 μm per minute or about 60 base pairs per second [35]. However, this slower rate is compensated for by the smaller size of the replication unit in eukaryotes (on average 50 μm as compared with 1100 μm in *Esch. coli*) and the fact that about 5000 replicating forks (out of a total of 35 000 per cell) are simultaneously active [36]. In certain early embryos such as that of the sea urchin the S-period lasts for only 20 minutes during which time the whole genome is replicated. This rapid synthesis of DNA is perhaps brought about both by increasing the rate of fork movement and by increasing the number of replicating sections simultaneously active (deoxyribonucleotide pools are much greater than in mammalian cells [37]).

11.2.4 The replication fork

(a) *Discontinuous synthesis.* As described above, replication of DNA occurs by means of a 'fork' which grows along the DNA molecule resulting in semi-conservative replication. However, as the two DNA chains are antiparallel (i.e. one is running $5' \to 3'$ and the other $3' \to 5'$) this poses the problem as to whether the two daughter strands are synthesized by two different mechanisms: one by adding nucleotides to a growing 3'-end and the other to a growing 5'-end. Addition to the 5'-end may involve deoxyribonucleoside 3'-triphosphates or, alternatively, may result in DNA molecules having terminal 5'-triphosphates (see Fig. 11.7). Such termini have

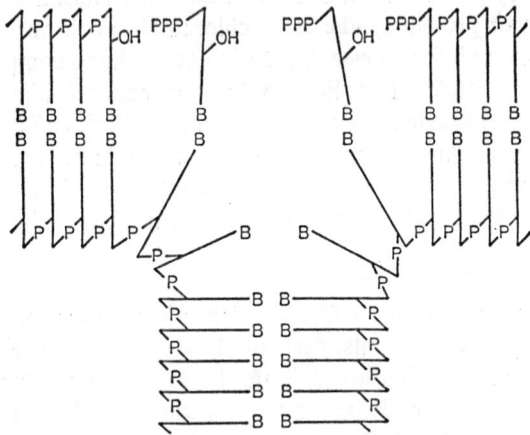

Fig. 11.7 *A possible mechanism of DNA chain growth*

not been found and, moreover, addition of nucleotides to a 5'-triphosphate terminated primer is incompatible with the proof-reading function of DNA polymerase (see p. 277). This function requires the removal of a mismatched terminal residue. If this residue held the energy for the polymerization, i.e. if it were a triphosphate, then this energy would be dissipated on proof-reading and polymerization would halt.

Other evidence argues against the $5' \to 3'$ extension of DNA chains.

(1) Many DNA polymerases have been investigated and they all add deoxyribonucleoside 5'-phosphates on to a 3'-OH group on the growing DNA chain, i.e. the primer (see p. 261).

(2) No enzymes have been found which will generate or poly-

merize deoxyribonucleoside 3′-triphosphates [38, 39], and neither have these putative substrates been found in cells.

(3) Both daughter strands have been shown to grow in the 5′→3′ direction (see p. 255).

The alternative explanation put forward by Okazaki [40, 41–42] is that both chains are synthesized by the same mechanism but that one is made 'backwards' in short pieces which are subsequently joined together by DNA ligase (see Fig. 11.8). Although evidence for

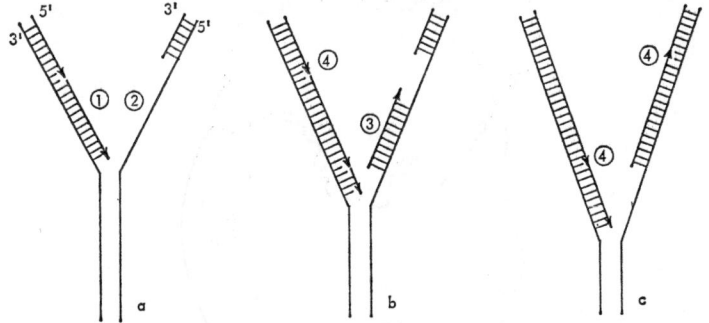

Fig. 11.8 *A working model of DNA synthesis. Synthesis occurs in the 5′→3′ direction down the left-hand strand of the parental DNA molecule (1). This synthesis may be continuous or discontinuous and it exposes the right-hand strand of the parental DNA molecule. This single-stranded region (2) is probably stabilized by association with DNA-binding proteins. At some point DNA synthesis starts on the right-hand strand (3) which is copied backwards (i.e. 5′→3′) until the growing strand has filled the gap. DNA ligase then joins the newly synthesized pieces (4)*

this mechanism is not apparent from the Cairns autoradiograph, careful investigation of electron micrographs (which give a resolution of about 100 nucleotides) shows that on one side of a replication fork the DNA is single-stranded (Fig. 11.9) and this supports the 'discontinuous' mechanism. It is this model which forms the basis of discussion in this section, but other models of DNA replication have not been excluded and are discussed in Section 11.5.

(b) *Okazaki pieces*. The major support for discontinuous synthesis stems from the work of Okazaki with *Esch. coli* and T4 infected *Esch. coli*. He showed that the most recently synthesized DNA i.e. that labelled with a brief pulse of tritiated thymidine, can be isolated *after denaturation* as short pieces now known as Okazaki pieces [40]. In this system the short pieces sediment at 8–10S on a gradient of alkaline sucrose, representing chain lengths of 1000–2000 nucleotides. In animal cells the fragments are much shorter, being only about 100–200 residues long, i.e. 4–5S [43, 44], and recent evidence suggests that the 10S pieces of *Esch. coli* may be formed by the joining of several 3–5S pieces [45–48].

In order to detect Okazaki pieces it is essential to lower the temperature to slow down the rate of reaction and to give a *very brief pulse* of tritiated thymidine. Under these conditions considerably more than half the radioactivity is found in small pieces, showing that

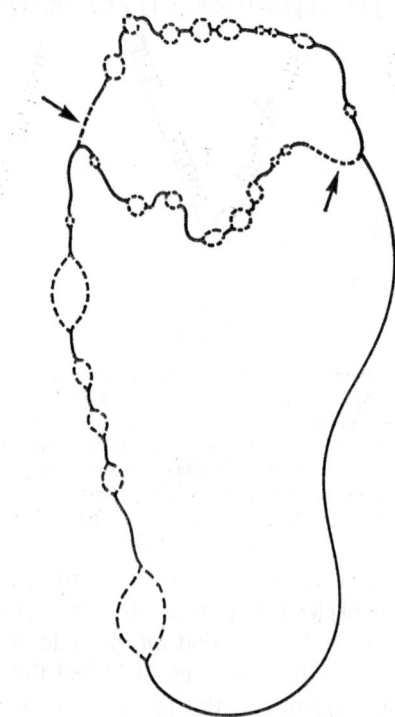

Fig. 11.9 *A partial denaturation map of the replicating λ chromosome. Broken lines represent single-stranded regions of DNA, and the arrows indicate such regions occurring at the replication fork. Drawn from the electron micrographs of Schnös and Inman* [15]

Okazaki pieces are made on both sides of the replication fork. (There is, however, some controversy over this; see p. 258.) Moreover, the Okazaki pieces accumulate in mutant *Esch. coli* cells deficient in DNA ligase or DNA polymerase I, implying that their subsequent incorporation into high molecular weight DNA is dependent on the action of these two enzymes [49–51].

(c) *Direction of chain growth.* When bacteriophage T4, growing at 8°C to reduce the rate of DNA synthesis, is incubated with ^{14}C-thymidine for 150 seconds, and for the final 6 seconds with ^{3}H-thymidine, Okazaki pieces can be isolated which apparently contain tritium at only the 3'-end [52, 53]. This was shown by degrading the isolated Okazaki pieces with exonuclease 1 of *Esch. coli* (which degrades single-stranded DNA from the 3'-end; see Chapter 8) when the ^{3}H label is released before the ^{14}C label. The complementary experiment using a nuclease from *B. subtilis* which acts from the 5'-end causes release of much of the ^{14}C before the ^{3}H is rendered acid-soluble.

These experiments demonstrate that Okazaki pieces are made in the $5' \to 3'$ direction and support the discontinuous mechanism as outlined in Fig. 11.8. However, Diaz and Werner [54] have recently thrown doubt on the interpretation of these results and suggested that the 10S Okazaki pieces arise through the joining of many smaller chains.

(d) *Initiation of Okazaki pieces.* In wild type *Esch. coli* rifampicin (an inhibitor of *Esch. coli* RNA polymerase) inhibits replication of phage M13 and certain plasmids. Replication is normal in mutants with a rifampicin-resistant RNA polymerase [55]. Moreover, some RNA synthesis has been shown to be essential for the *in vitro* conversion of M13 single strands into the duplex form [56]. Such evidence, considered with the fact that all DNA polymerases require a 3'-hydroxyl priming end (see p. 261), suggested that this end may be provided by an oligoribonucleotide. That rifampicin does not inhibit replication of *Esch. coli* DNA or the DNA of other phages such as ØX174 was interpreted to mean that in these cases a second RNA polymerase (resistant to the drug) was involved in the synthesis of the priming oligoribonucleotides. Indeed, the product of the dna G gene (see Table 11.1) is believed to be such an enzyme and has been shown *in vitro* under specific conditions to synthesize short lengths of RNA on single-stranded DNA of phage G4 [57] (see p. 260). Although not required for the initial stages of M13 replication, the dna G gene product is essential for duplication of the replicative form [1].

Okazaki fragments containing oligoribonucleotides at their 5'-end have now been isolated from both prokaryotic and eukaryotic systems. However, the experimental proof of their existence has been subject to much controversy because non-covalent RNA–DNA interactions may have given spurious results in early experiments [3, 58]. However, new methods (Fig. 11.10) to detect RNA covalently linked to the 5'-end of DNA molecules [2–4] have shown that, although not all Okazaki pieces have ribonucleotides at their 5'-end, the shortest ones do. Moreover, RNA-linked Okazaki pieces accumulate in mutant *Esch. coli* cells deficient in either the

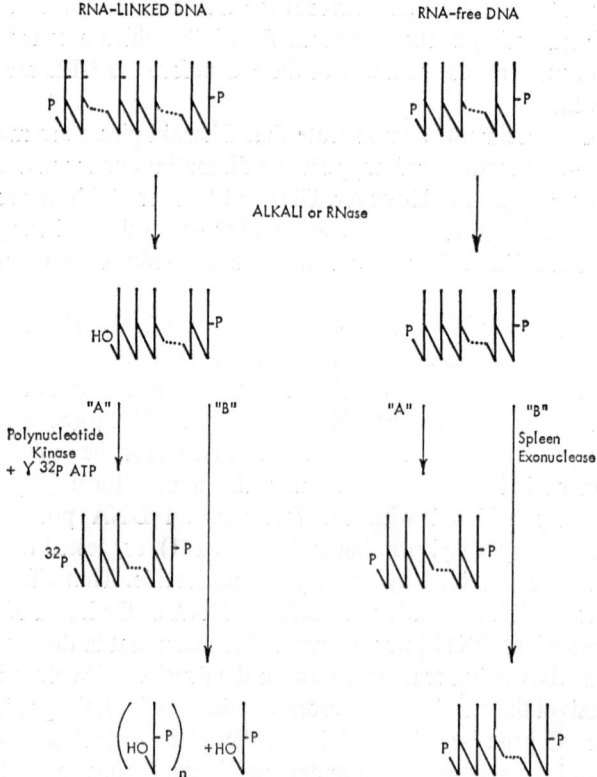

Fig. 11.10 *Methods to detect RNA-linked Okazaki pieces produced [2–4] in vivo. The isolated Okazaki pieces are first treated with polynucleotide kinase and non-radioactive ATP to ensure that all 5'-ends are phosphorylated. Subsequent treatment with alkali or ribonuclease removes any RNA from the 5'-end and leaves a 5'-hydroxyl group. This group can be detected using polynucleotide kinase and γ-^{32}P-ATP (Method A). However, this method is not selective for nascent DNA. Method B overcomes this disadvantage by starting with tritiated Okazaki pieces resulting from a pulse labelling experiment. Spleen exonuclease is used under conditions where only 5'-OH terminated DNA (i.e. that initially linked to RNA) is degraded*

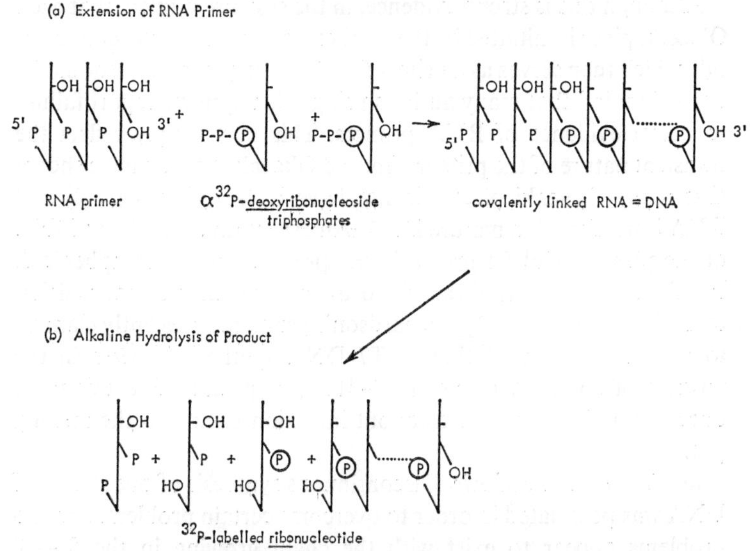

REPLICATION OF DNA

(a) Extension of RNA Primer

Fig. 11.11 *The formation of the ^{32}P-labelled ribonucleotide is indicative of the occurrence of a covalently linked RNA–DNA molecule where the radioactive phosphate forms the bridge between the two nucleic acids. (P) = radioactive phosphate*

polymerase or the $5' \rightarrow 3'$ exonuclease of DNA polymerase I (see p. 272).

The use of toluenized *Esch. coli* enables α-^{32}P-phosphate labelled *deoxyribo*nucleoside triphosphates to penetrate the cell membrane when transfer of radioactivity to a *ribo*nucleotide occurs [59]. This is definite evidence for a covalent attachment of RNA to DNA (see Fig. 11.11). When DNA synthesis is studied in nuclei isolated from polyoma infected cells (see p. 260) the nascent DNA chains are found with a length of about ten ribonucleotides at their 5'-ends [44, 60]. When α-^{32}P-labelled deoxyribonucleoside triphosphates are injected into the slime mould *Physarum polycephalum* [61] there is transfer of the ^{32}P to ribonucleotides, which is indicative, as described above, of a covalent attachment of RNA and DNA.

Although the early experiments with prokaryotes suggested that the RNA primer was of the order of 50–100 nucleotides long, and that the sequence of the RNA–DNA junction was, at least partially, specific [59], this has not been found in later experiments [4]. In animal viruses the RNA primers appear to be only about ten nucleotides long and do not have a specific base sequence [44, 60].

Although this is strong evidence, in the systems studied, that each Okazaki piece is initiated by the synthesis of a short oligoribonucleotide which then serves as a primer for DNA polymerase, it should be borne in mind that many authors using other systems report failure to detect evidence of RNA primers. This may simply reflect the transient nature of the primers. Indeed Okazaki's group have shown that many Okazaki pieces do not have RNA primers [3, 4] and RNA is not found in mature DNA. An exception to this is the DNA of the plasmid Col E1 grown in the presence of chloramphenicol. In this case it is inferred that the drug prevents the normal excision of the RNA primers [62]. Richardson's group have recently claimed to have obtained initiation of T7 DNA synthesis *in vitro* in the absence of an RNA primer [409–411] (see p. 261), and this may, once again, throw our ideas about initiation back into the melting pot.

(e) *Continuous synthesis.* Discontinuous synthesis of one strand of DNA was postulated in order to overcome certain problems, but no problems appear to exist with the chain growing in the $5' \rightarrow 3'$ direction, and so there is no *a priori* reason why this chain should not be made continuously. Indeed, this may well be the case when DNA synthesis is proceeding rapidly. Under these conditions nascent DNA is found in both large and small pieces. However, when deoxyribonucleotides are limiting and the rate of chain elongation slowed, both daughter strands appear to be made discontinuously. A possible explanation is that there is competition between the propagation of the growing chain and the initiation of new chains [63, 64].

Although the single-stranded regions on the 'discontinuous' side of the replication fork may be up to the length of an Okazaki piece, those on the 'continuous' side will tend to be small if they occur at all. This will have two consequences. Firstly, on electron micrographs single-stranded gaps will be apparent on only one side of the fork. Secondly, any Okazaki pieces formed on the 'continuous' side will be very rapidly ligated (thereby making them even more difficult to detect).

11.3 DNA replication – *in vitro* studies

To help understand a process as complicated as DNA replication it is often an advantage to work with a system simpler than the intact cell; some reference has already been made to results obtained in this way. The geneticist indicates the involvement of a number of gene products in DNA replication, but in order that the biochemist may identify these proteins and delineate their role it is essential to circumvent the barrier of the cell membrane, and if possible to remove extraneous material. This has been done in a number of ways [65, 66].

11.3.1 *Permeable cells*

Treatment of *Esch. coli* with lipid solvents (ether [67] or toluene [68]) whilst inhibiting cell division appears to leave the cells and constituent enzymes intact yet renders the membrane permeable to small molecules including nucleotide precursors. Such cells when incubated with the four deoxyribonucleoside triphosphates, ATP, and Mg^{2+} continue semiconservative replication. The products of such replication are the 10S Okazaki pieces which are only joined into high molecular weight DNA in the presence of NAD, the cofactor for *Esch. coli* DNA ligase. It was with toluenized cells and $\alpha\text{-}^{32}P$-labelled deoxyribonucleotides that Okazaki was able to show the covalent linkage of RNA and DNA (see p. 257).

11.3.2 *Cell lysates*

If the cell membrane is disintegrated a lysate results which, under certain conditions, is capable of maintaining DNA synthesis [69, 70]. It is very important not to dilute the macromolecular constituents of the cell but to maintain them near their *in vivo* concentration. One way in which this has been achieved is by lysis of cells on a cellophane disc held on an agar surface. Small molecules are then allowed to diffuse through the disc into the concentrated lysate. Semiconservative DNA synthesis occurs in this and similar systems if the lysate is prepared from wild type *Esch. coli*, but when it is prepared from *Esch. coli* cells defective in DNA synthesis then the lysate reflects the *in vivo* capabilities of the cells. Thus lysates prepared from mutants temperature-sensitive in the genes dna B, dna E, or dna G (fail to extend growing DNA chains at the non-permissive temperature; Table 11.1) only synthesize DNA at low temperatures. In lysates prepared from mutants temperature-sensitive in the genes dna A or dna C/D (fail to initiate DNA synthesis at non-permissive temperatures) DNA synthesis fails to occur only if the cells are maintained at the non-permissive temperature for one generation time prior to lysis [71].

This system has proved very useful in the purification of the products of these dna genes [65, 66]. Thus fractions containing the dna G gene product from wild type cells, when added to a lysate from a mutant temperature-sensitive in dna G, were able to restore DNA synthesis at the non-permissive temperature. Although the products of the dna genes have all been at least partially purified, functions for all these proteins have not yet been established. However, they have been used with some success in reconstruction experiments using fully defined components (see below). Permealysed cells [72] and lysates from eukaryotic cells [73–75] also show a limited ability to continue DNA synthesis when provided with the four deoxyribonucleoside triphosphates, ATP, and Mg^{2+}. In this case, however, isolation of the nuclei results in a marked loss of activity which can be restored by addition of cytoplasmic components [76–78]. So far little progress has been made in the elucidation of the nature or function of these cytoplasmic components although the system using cells infected with the small tumour viruses (polyoma, SV40) appears promising. Here the product DNA is well defined and deficiencies in the system are immediately obvious.

11.3.3 *Reconstruction experiments*

The ultimate aim of these experiments is to achieve the ordered replication of a complex DNA molecule such as the *Esch. coli* chromosome using totally defined components. This is far from being achieved. However, the requirements for the conversion of single-stranded bacteriophage DNA into the double-stranded replicative form have been delineated and the reaction has been successfully carried out using purified components. Thus, according to Kornberg's group [79], conversion of single-stranded M13 DNA into the duplex form requires RNA polymerase and *Esch. coli* unwinding protein for initiation, DNA polymerase III* (the dna E gene product), and copolymerase III* (see p. 280) for extension, DNA polymerase I to remove the RNA primer and fill the gap, and DNA ligase to seal the new strand and form the covalently closed duplex [80]. The same reaction with bacteriophage ØX174 is more complicated in that the products of the dna A, dna B, and dna C–D genes are also required [70, 81, 82] and the traditional *Esch. coli* RNA polymerase has to be replaced with the dna G gene product (probably a rifampicin-resistant RNA polymerase) [57]. The *in vivo* conversion of single-stranded G4 DNA into the replicative form has been achieved using simply the dna G protein, the DNA unwinding protein (see p. 288), and DNA polymerase III* [83]. DNA polymerase I and ligase can then be used to seal the final gap. In this system the dna G protein has been shown to synthesize the RNA primer [57].

The *in vitro* replication of the linear double-stranded T7 chromosome is also being studied [409–412]. The products of the host dna genes are not required but two other host proteins are, together with six viral specified proteins. The product of viral genes 5 (the DNA polymerase) and 4 together with *Esch. coli* thioredoxin form a complex which is able to initiate the synthesis of Okazaki fragments.

11.4 Enzymes of DNA synthesis

11.4.1 *Mechanism of action of DNA polymerase*

DNA polymerase is the name given to an enzyme which catalyses the synthesis of DNA from its deoxyribonucleotide precursors. Such enzymes have been purified from various sources (bacterial, plant, and animal) but, apart from some details discussed below, the mechanism of the polymerization reaction is common to all preparations.

DNA polymerase catalyses the formation of a phosphodiester bond between the 3'-OH group at the growing end of a DNA chain (the primer) and the 5'-phosphate group of the incoming deoxyribonucleotide triphosphate. Growth is in the $5' \rightarrow 3'$ direction, and the order in which the deoxyribonucleotides are added is dictated by base pairing to a template DNA chain. Thus, as well as the four triphosphates and Mg^{2+} ions, the enzyme requires both primer and template DNA. No DNA polymerase has been found which is able to initiate DNA chains.

The simplest case to consider is that in which a single-stranded template has bound to it a growing strand of primer terminating at the growing point in a 3'-hydroxyl group (Fig. 11.12(1)). Such a situation is also illustrated in Fig. 8.7 for a piece of double-stranded DNA partially degraded by exonuclease III.

The polymerase binds to the single-stranded template in the region of the 3'-hydroxyl end of the primer (Figs. 11.12 and 11.13). An incoming deoxyribonucleoside triphosphate containing a base which can pair with the corresponding base on the template becomes attached to the triphosphate binding site (Fig. 11.13). The poly-

THE BIOCHEMISTRY OF THE NUCLEIC ACIDS

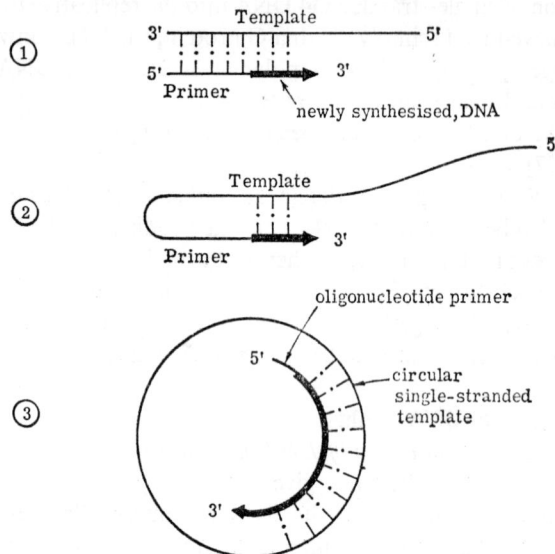

Fig. 11.12 *Mechanism for the replication of single-stranded DNA*

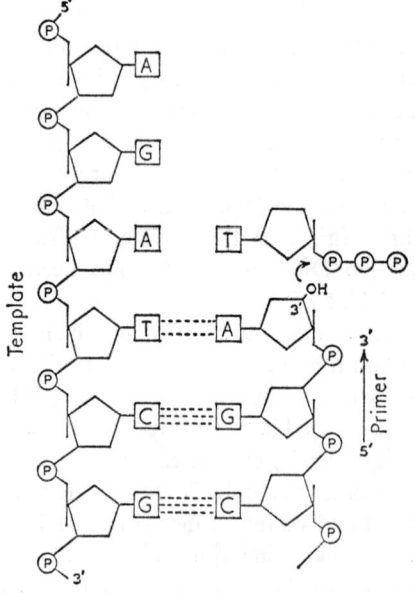

Fig. 11.13 *Greater detail of the mechanism shown in* Fig. 11.12(1)

merase, perhaps after a conformational change, then catalyses a nucleophilic attack by the 3'-hydroxyl group of the primer on the α-phosphate of the activated deoxyribonucleoside 5'-phosphate (Fig. 11.14). Inorganic pyrophosphate is released, a phosphodiester bond is formed, and the chain is lengthened by one unit. The enzyme moves along the template by the distance of one unit and the newly added nucleotide with its 3'-hydroxyl group now occupies the primer terminus site. The process is then repeated until the enzyme reaches the end of the template strand.

In the replication of single-stranded DNA the 3'-hydroxyl terminus may loop back on itself and serve as a priming strand as

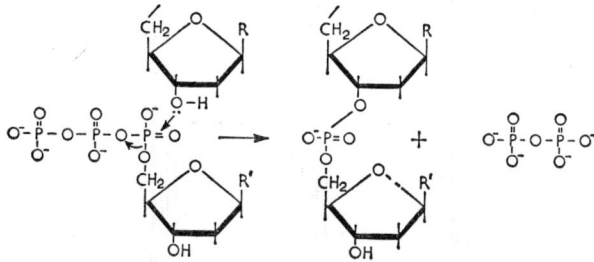

Fig. 11.14 *The mechanism of the action of DNA polymerase*

shown in Fig. 11.12(2), or short lengths of oligonucleotide may act as primers by becoming hydrogen bonded to the template.

When a cyclic single-stranded DNA is used as template replication can occur only in the presence of short oligonucleotides which become attached by base-pairing to the template and serve as primers (Fig. 11.12(3)).

The replication of double-stranded DNA is more complex and can only be initiated at nicks or gaps. The exposed 3'-OH group acts as a primer and the new growing strand will fill the gap and then may continue displacing the complementary strand from the 5'-end (Fig. 11.15). This can lead to complications since at some point the enzyme may leave the original template strand and begin to copy the complementary strand so that a branched structure is formed (Fig. 11.15). Branched structures have been seen under the electron microscope [84–86]. Such strand switching is also postulated to occur *in vivo* (see p. 289).

When the growing chain reaches the free 5'-end of the displaced strand it doubles back on itself, once more displacing the parental strand and making a strand complementary to itself (Fig. 11.15). Thus sections of the product are covalently linked and self-complementary, and this gives the product the property of rapid renaturability.

When native DNA is treated with exonuclease III, a partially single-stranded molecule is produced (Fig. 8.7). It can be repaired and restored to the native double-stranded form by the polymerase.

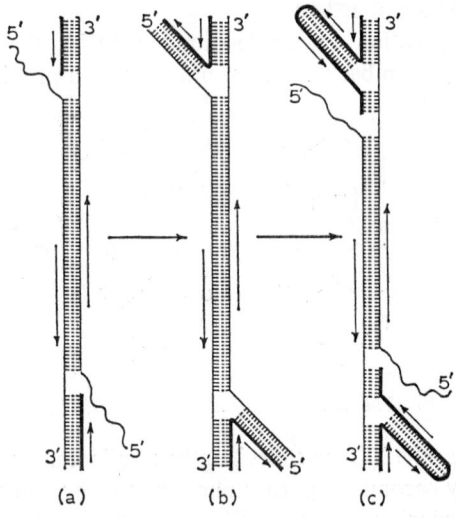

Fig. 11.15 *Kornberg's model to explain the non-denaturability and branching of DNA synthesized in vitro. The start of a new chain is from the 3'-hydroxyl terminus at each end of the helix. The arrows show the direction of synthesis and the polarity of the strand. The template is shown by thin lines and the product by heavy lines. Broken lines denote hydrogen bonds between strands. At certain points the newly synthesized strand may serve as a template and produce a hairpin-like loop* [84, 87]

The newly synthesized DNA is covalently attached to the 3'-hydroxyl end of the degraded strand of the template. The fully repaired DNA resembles the original native DNA in appearance in electron micrographs, in density gradient analysis, in denaturability, and in genetic activity [87, 88]. The synthesis of DNA which follows the repair phase results in the production of a structure that is not covalently linked to the template and resembles the product obtained with a native DNA template in non-denaturability, in branched appearance, and in lack of genetic activity.

11.4.2 Esch. coli *DNA polymerase* I (E.C.2.7.7.7)

Since the greatest amount of effort has been put into the study of DNA polymerase I from *Esch. coli* (the Kornberg enzyme) this will be considered in some detail and then comparisons will be made with other enzymes. Kornberg himself has recently written a book on the subject [89].

(a) *Historical*. In 1956 Kornberg and his collaborators used cell-free extracts from exponentially growing cultures of *Esch. coli* to demonstrate that, in the presence of ATP and Mg^{2+}, ^{14}C-thymidine, a specific precursor for DNA, was incorporated into an acid-precipitable material which was judged to be DNA from the fact that it was no longer acid-precipitable after treatment with deoxyribonuclease. Using this incorporation method for assay purposes, the bacterial extract was fractionated and it soon became evident that crude extracts contained a kinase system which converted thymidine into thymidine triphosphate (dTTP). In subsequent work labelled dTTP was used as substrate and the enzyme responsible for its incorporation into DNA has been purified to homogeneity as judged by sedimentation, chromatography, and electrophoresis [89–92].

With a partially purified enzyme preparation it was shown that, for good incorporation of dTTP (labelled either with ^{14}C in C-2 of the thymine or with ^{32}P in the innermost phosphate group), the 5'-triphosphates of all four deoxyribonucleosides which normally occur in DNA had to be present together with Mg^{2+} ions and a denatured DNA template (calf thymus DNA was used in early experiments). Under these conditions excellent incorporation was observed as is illustrated in Table 11.2. Omission of any one triphosphate or Mg^{2+} or of DNA template, or pretreatment of the template with deoxyribonuclease, reduced the incorporation to a very low level. Deoxyribonucleoside diphosphates could not replace the triphosphates. With the purified enzyme it was also possible by chemical analysis to demonstrate net synthesis of DNA, and increases of DNA by a factor of 10–20 could be obtained indicating that 90–95 per cent of the isolated DNA was derived from the added deoxyribonucleoside triphosphates.

As well as DNA polymerase activity, the purified enzyme shows several other activities each of which is associated with the same enzyme molecule. Thus exonuclease (p. 272) activity is purified along with the polymerase activity in a ratio which is not altered by fractionation. The significance of this association is discussed later but it is important to note that 5′→3′ exonuclease activity is absent from highly purified preparations of DNA polymerase from *B. subtilis* [94].

(b) *Evidence for the copying of the template.* The DNA polymerase is an enzyme with the unusual property of taking directions from a template and faithfully reproducing the sequence of nucleotides in

TABLE 11.2

Incorporation of ^{32}P-dCTP into DNA (Bessman et al. [93])

Additions	μμmoles ^{32}P-DNA
dCTP (DNA omitted)	0·0
dCTP	2·5
dCTP + dGTP	5·1
dCTP + dGTP + dTTP	15·7
dCTP + dGTP + dTTP + dATP	3300

The incubation mixture (0·3 ml) contained 5 mμmoles of dCTP ($7·2 \times 10^7$ c.p.m. per mole) and 5 mμmoles of each of the other deoxynucleoside triphosphates as indicated, together with 1·0 μmole of $MgCl_2$, 20 μmoles of glycine buffer (pH 9·2), 10 μg of DNA, and 3 μg of purified 'polymerase' from *Esch. coli*. The incubation was carried out at 37° for 30 minutes.

the product. The evidence for this rests on several experimental observations.

(1) The most significant fact is that the enzyme can faithfully copy the nucleotide sequence of the single-stranded cyclic DNA bacteriophage ØX174. The product can be converted into biologically active material by subsequent cyclization by the enzyme polynucleotide ligase. This experiment was done in 1967 by Kornberg and his colleagues [95–97]. They incubated ØX174 DNA in the form of (+) circles (i.e. those present in the virus particles) with *Esch. coli* DNA polymerase and ligase, four deoxyribonucleoside triphosphates, NAD, Mg^{2+} ions, and boiled *Esch. coli* extract (containing oligonucleotides which provide initiation sites [98]); the product was a duplex circle with all the properties of the natural RF as shown by electron microscopy and contour length. The polymerase had synthesized a linear (−) strand using the (+) circle as template, and the ligase had effected the joining of the two ends of this new strand.

By carrying out the same procedure with 5-bromodeoxyuridine triphosphate in place of dTTP in the triphosphate mixture, so as to replace the methyl group in thymine with the much heavier bromine atom, and by using ^{32}P-labelled dATP, a type of RF was produced in which the (−) circle was much denser than the template (+) circle and was labelled with ^{32}P. The duplex circles were exposed to pancreatic DNase to an extent just sufficient to produce a single scission in one of the strands in about half the molecules (Fig. 11.16). The mixture of intact and 'nicked' molecules was denatured by heating to yield cyclic and linear (+) strands and cyclic and linear (−) strands, the (−) strands being labelled with ^{32}P and bromouracil. On fractionation by equilibrium density gradient centrifugation in caesium chloride, three peaks were obtained corresponding to single-stranded (−) (circular and linear) DNA containing bromouracil, a duplex hybrid (RF), and single-stranded (+) (cyclic and linear) DNA containing neither ^{32}P nor bromouracil.

The completely synthetic (−) circles, free of natural DNA, were found to be capable of infecting spheroplasts and of giving rise to a new generation of normal virus particles in spite of the fact that they contained an unnatural base (Fig. 11.16).

When the completely synthetic heavy (−) circles were used as templates in a system analogous to that just described but with dTTP in place of dBrUTP, the product was a completely synthetic cyclic RF which could be separated into its component (+) and (−) circles (Fig. 11.16). The (+) circles were infective and were identical with natural ØX174 DNA. The complete synthesis *in vitro* of infective DNA identical with the natural material had thus been achieved [97]. These findings show that DNA polymerase I is an enzyme which has all the requirements necessary for the replication of DNA.

(2) When *Esch. coli* polymerase is used to prepare DNA under conditions such that only 5 per cent of the sample produced comes from the template, the product has many of the same physical and chemical properties as DNA isolated from natural sources. The product appears to have a hydrogen-bonded structure similar to that of natural DNA and undergoes molecular melting (p. 100) in the same way.

It shows the same equivalence of adenine to thymine and guanine to cytosine that characterizes natural DNA. Moreover the characteristic ratio of A · T pairs to G · C pairs of a given DNA primer is imposed on the product whether the net DNA increase is 1 per cent or 1000 per cent. The base ratios in the product are not distorted when widely differing molar proportions of substrate are used. This is best illustrated by the use of the synthetic template poly(dA-dT) (see p. 274) when, from a mixture of all four deoxyribonucleoside triphosphates, only dAMP and dTMP are incorporated into the product.

(3) The nucleotide sequence in the template DNA is reproduced

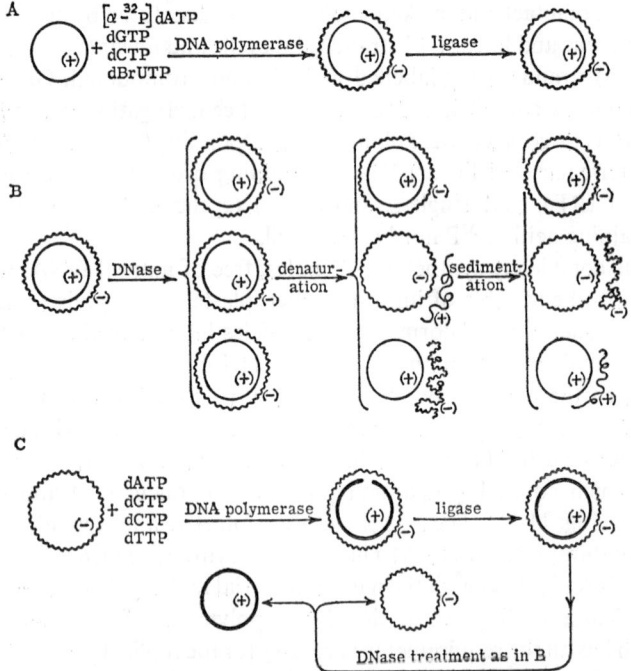

Fig. 11.16 *The schematic representation of the preparation of synthetic cyclic molecules of ØX174 DNA* [99]
A. *The preparation of partially synthetic RF with the aid of the DNA polymerase and the ligase. The synthetic (−) circle is labelled with bromodeoxyuridine and with ^{32}P and is indicated by the wavy line*
B. *The isolation of synthetic (−) circles labelled with ^{32}P and bromodeoxyuridine*
C. *The use of synthetic (−) circles to prepare synthetic (+) circles. These synthetic circles have been shown to be infective. For further details, see text*

in the product. This has been established by the Kornberg group by using the technique of *nearest-neighbour sequence* analysis [100–102] (see also p. 319). The partially purified *Esch. coli* enzyme is incubated with a particular template DNA and all four deoxyribonucleoside triphosphates, one of which, say dATP, is labelled with ^{32}P in the innermost phosphate. During the synthetic reaction this ^{32}P be-

REPLICATION OF DNA

comes the bridge between the nucleoside of that labelled triphosphate (A) and the nearest-neighbour nucleotide containing the base Z at the growing end of the polynucleotide chain (Fig. 11.17A). After the synthetic reaction is complete the DNA is isolated and degraded with micrococcal DNase (p. 170) and spleen phosphodiesterase (p. 173) to yield the deoxyribonucleoside 3'-monophosphates. The ^{32}P is thus transferred to the 3'-carbon of the neighbouring nucleotide in the chain (Z in Fig. 11.17), i.e. the one with which the labelled

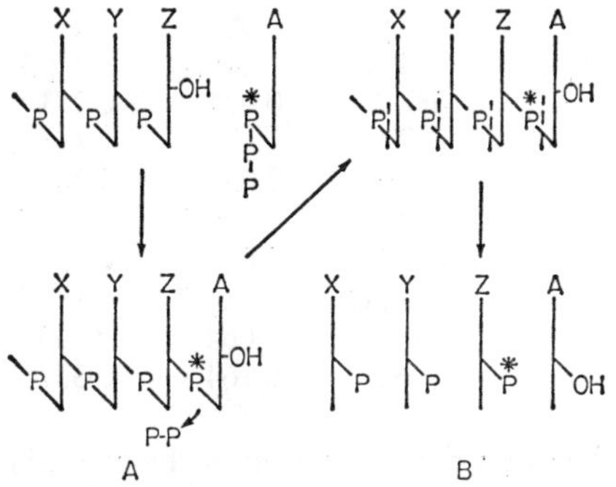

Fig. 11.17 *Illustration of the method of nearest-neighbour sequence analysis*

triphosphate has reacted (Z might be any one of the four bases). The four deoxyribonucleoside 3'-monophosphates are isolated by paper electrophoresis, and their radioactivities measured to give the relative frequency with which the nucleotide originally labelled in position 5' locates itself next to another nucleotide in the new chain. This procedure is carried out four times with a different labelled triphosphate each time so as to determine the relative frequencies of all sixteen possible nearest-neighbour (or dinucleotide) sequences [102]. The results of such an experiment with DNA from *M. phlei* as primer are shown in Table 11.3.

They illustrate several points:

(i) All sixteen possible nearest-neighbour sequences are present and they occur with widely varying frequencies.

(ii) The results show a very striking deviation from the nearest-neighbour frequencies predicted if the arrangement of mononucleotides were completely random. Thus the frequency of TpA in the first row is quite different from that of ApT in the second row whereas these two frequencies would have to be identical in a random

THE BIOCHEMISTRY OF THE NUCLEIC ACIDS

assembly. The nucleotides have therefore been assembled in accordance with a definite pattern.

(iii) The sums of the four columns show the equivalence of A to T and of G to C in the product and indicate both the validity of the

TABLE 11.3

Nearest-neighbour frequencies of Mycobacterium phlei *DNA* [103]

Labelled triphosphate	Deoxyribonucleoside 3'-phosphate isolated			
	Tp	Ap	Cp	Gp
dATP	TpA	ApA	CpA	GpA
	0·012	0·024	0·063	0·065
dTTP	TpT	ApT	CpT	GpT
	0·026	0·031	0·045	0·060
dGTP	TpG	ApG	CpG	GpG
	0·063	0·045	0·139	0·090
dCTP	TpC	ApC	CpC	GpC
	0·061	0·064	0·090	0·122
Sums	0·162	0·164	0·337	0·337

analytical method and the replication of the overall composition of the primer DNA.

(iv) The results indicate that base pairing occurs in the newly synthesized DNA and that its two strands are of opposite polarity [104]. According to the Watson–Crick model the two strands of the double helix are of opposite polarity (p. 90), and it is presumed that each can act as a template for the formation of a new chain so as to

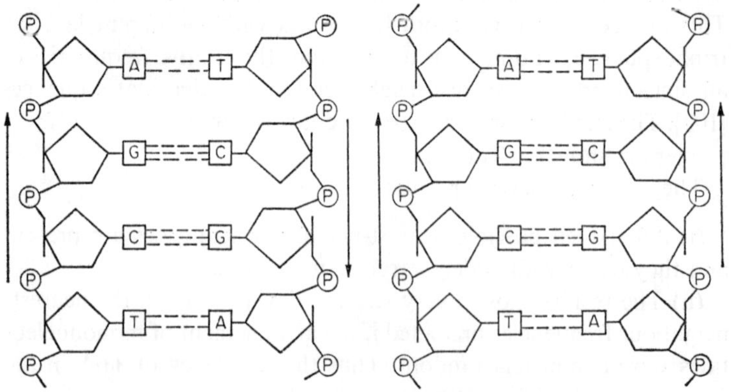

Fig. 11.18 *Possible structures of the DNA molecule showing opposite polarity of strands (left) and the same polarity of strands (right)*

give precise replication with the formation of two daughter helices identical with each other and with the parent helix. The results of nearest-neighbour sequence analysis support this mechanism. For example, the frequencies of ApA and TpT sequences are equivalent, and so are the frequencies of GpG and CpC. The matching of the other sequences depend upon whether the strands of the double helix are of similar or opposite polarity (Fig. 11.18).

If the strands are of opposite polarity the following matching sequences can be predicted:

CpA	and	TpG
GpA	and	TpC
CpT	and	ApG
GpT	and	ApC

whereas if the strands are of the same polarity the matching sequences would be:

TpA	and	ApT
CpA	and	GpT
GpA	and	CpT
TpG	and	ApC
ApG	and	TpC
CpG	and	GpC

The results in Table 11.3 favour the helix with strands of opposite polarity. The ApA and TpT, and the CpC and GpG sequences match similarly in both models.

(v) The nearest-neighbour frequencies measured by the method described above are those of the newly synthesized DNA. To verify that they are an accurate reflection of those in the original DNA template, an enzymically synthesized sample of calf thymus DNA in which only 5 per cent of the total DNA consisted of the original template was itself used as template in a sequence analysis. The results showed good agreement between the sequence frequencies of the products primed by native DNA and by enzymically produced DNA (Table 11.4) whereas DNA's from other sources gave quite different results.

It can be concluded therefore that the polymerase yields a DNA product with strands of opposite polarity and that the sequence of bases is faithfully reproduced.

(c) *The chemical nature of DNA polymerase I.* The purified

Kornberg enzyme is a protein of mol. wt. 109 000 in the form of a single polypeptide chain [91]. This chain can be unfolded in guanidine-HCl–mercaptoethanol so as to denature the protein. When the reagent is diluted out renaturation occurs with restoration of activity.

The protein migrates as a single band on SDS-acrylamide gel electrophoresis; it contains only one sulphydryl group and one disulphide group; the residue at the amino terminal end is methionine. One *Esch. coli* cell contains about 400 molecules of enzyme, each

TABLE 11.4

Nearest-neighbour frequencies of native and enzymically synthesized calf thymus DNA [102]

Nearest-neighbour sequence		Native calf thymus DNA as template		Enzymically synthesized calf thymus DNA as template	
ApA	TpA	0·089	0·053	0·088	0·059
ApG	TpG	0·072	0·076	0·074	0·076
ApC	TpC	0·052	0·067	0·051	0·064
ApT	TpT	0·073	6·087	0·075	0·083
GpA	CpA	0·064	0·064	0·063	0·078
GpG	CpG	0·050	0·016	0·057	0·011
GpG	CpC	0·044	0·054	0·042	0·055
GpT	CpT	0·056	0·067	0·056	0·068

spherical and of diameter 6·5 nm. It can form dimeric forms which can be visualized in the electron microscope [105].

Treatment of the enzyme with the protease *subtilisin* from *B. subtilis* breaks it into two fragments, a larger fragment of mol. wt. 76 000 which retains polymerase activity and 3' → 5' nuclease activity (see below) but not the 5' → 3' nuclease activity, and a smaller fragment of mol. wt. 34 000 which retains nuclease 5' → 3' activity in the presence of DNA [106–108].

(d) *The active centre of the enzyme.* On the basis of binding experiments Kornberg has concluded that the multiple functions of the enzyme include the following operations [109].

(i) Extension of a DNA chain in the 5' → 3' direction by the addition to the 3'-hydroxyl terminus of mononucleotides from deoxyribonucleoside triphosphates at the rate of 1000 nucleotides per minute.

(ii) Hydrolysis of a DNA chain from the 3'-hydroxyl end in the 3'→5' direction to yield 5'-monophosphates (the exonuclease II action referred to on p. 181).

(iii) Hydrolysis of a DNA chain from the 5'-phosphate (or 5'-hydroxyl) terminus in the 5'→3' direction to yield mainly 5'-monophosphates (see p. 182).

(iv) Pyrophosphorolysis of a DNA chain from the 3'-end; this is essentially the reversal of the polymerization reaction.

(v) Exchange of inorganic pyrophosphate with the terminal pyrophosphate group of a deoxyribonucleoside triphosphate.

These last two reactions are simply the result of reversal of the polymerization reaction and are of doubtful *in vivo* significance since they require a high concentration of inorganic pyrophosphate.

Kornberg envisages the active centre of the enzyme as a specially adapted polypeptide surface comprising at least five major sites as shown in Fig. 11.19 [109].

(1) A site for the binding of the template chain in the region where base pairs are formed and for a few nucleotides on each side of this.

(2) A site for the growing primer chain which is, of course, base-paired to the template.

(3) A site for the special recognition of the terminal 3'-hydroxyl group of the primer. This point is the start of the 3'→5' hydrolytic cleavage.

(4) A triphosphate binding site for which all four triphosphates compete.

(5) A site which allows for the 5'→3' cleavage of a 5-phosphoryl terminated chain. It is presumably this area that is broken off by subtilisin.

These sites determine the nature of the DNA which can bind to the enzyme. For example, linear single-stranded DNA binds readily on site (1) whereas an intact linear duplex such as the DNA bacteriophage T7 does not bind if it has been prepared with great care so as to avoid internal breaks. An intact circular duplex such as plasmid DNA or ØX174 replicative form DNA (p. 295) does not bind to the enzyme until a 'nick' has been introduced in one of the strands by an appropriate nuclease yielding 3'-hydroxyl and 5'-phosphate termini. Such nicks are active points for replication whereas nicks introduced by micrococcal nuclease with 5'-hydroxyl and 3'-phosphoryl termini are not replication points although they bind the enzyme. One molecule of enzyme is bound at each nick in either case.

(e) *Synthetic DNA's*. When the DNA polymerase I is incubated without primer in the presence of dATP, dTTP, and Mg^{2+} an interesting polymer is formed containing adenine and thymine nucleotides [100, 110]. Polymer formation occurs only after a lag period of several hours and then takes place rapidly until 60–80 per cent of the triphosphates have been utilized. When the polymer is

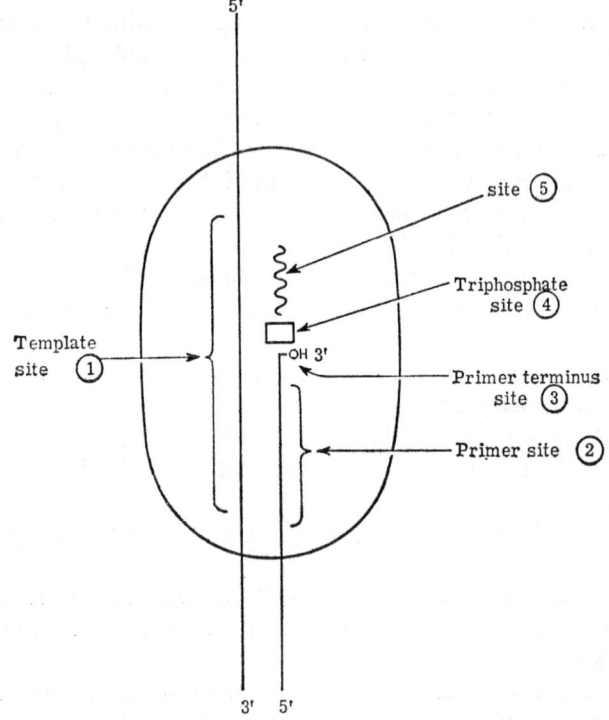

Fig. 11.19 *Binding sites on the active centre of the Kornberg enzyme*

isolated and used as primer in a DNA polymerase reaction containing two of the four deoxyribonucleoside triphosphates, there is prompt and extensive synthesis of an identical polymer.

The product of the reaction whether unprimed or primed contains equal amounts of adenine and thymine. Nearest-neighbour frequency analysis shows that the frequencies of ApT and TpA are each 0·500 whereas the sequences ApA and TpT are undetectable. The polymer therefore contains alternating residues of A and T.

The molecular weight calculated from the sedimentation value and reduced viscosity is between 2 and 8×10^6. The polymer melts sharply at 71° with an increase of 37 per cent in the absorbance at 260 nm. The process is completely reversible on cooling. Such physical data, including the X-ray diffraction pattern, suggest that

```
A  T  A  T  A  T  A  T  A
•  •  •  •  •  •  •  •  •
T  A  T  A  T  A  T  A  T
```

```
G  G  G  G  G  G  G  G
•  •  •  •  •  •  •  •
C  C  C  C  C  C  C  C
```

Fig. 11.20 *Poly d(A-T) · poly d(T-A) (upper) and poly(dG) · poly(dC) (lower)*

the molecule is a long fibrous double-stranded structure with the strands joined by hydrogen bonds between adenine residues in one chain and thymine residues in the other (Fig. 11.20). The occurrence of such a polymer in nature has already been mentioned (p. 85).

Chemically synthesized oligonucleotides [111–113] containing sequences of 6–14 (d(A-T)$_3$ to d(A-T)$_7$) alternating deoxyadenylate and deoxythymidylate residues prime the synthesis of poly d(A-T)

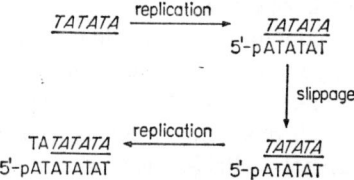

Fig. 11.21 *The possible mechanism of d(A-T)$_3$ as a template (shown underlined in italics). The newly formed strand (roman type) moves by slippage across the template, exposing a new pair of bases on which further replication may occur*

[114]. The relative priming capacity of these oligonucleotides is directly related to their size.

Kornberg and his colleagues [115, 116] have suggested a mechanism whereby an oligonucleotide such as d(A-T)$_4$ might prime the reaction (Fig. 11.21). The first step involves replication of the template with a new strand starting at the 3'-hydroxyl end. This newly formed helix then melts and reanneals so as to expose a segment of template for further replication. This slippage process results in the shift of the new strand along the template by one AT notch at a time so that the correct base pairing is maintained. A repetition of such successive steps of replication and slippage leads eventually to the formation of a long d(A-T) polymer. The process is critically temperature dependent; d(A-T)$_4$ for example primes optimally at 10° whereas d(A-T) linked by a greater number of hydrogen bonds to the replica to be slipped is more effective at 37°.

A somewhat similar polymer [100] is formed when *Esch. coli* polymerase is incubated with high concentrations of dGTP and dCTP in the presence of Mg^{2+}. Again, a lag period of several hours is found which can be abolished by using the product of the reaction as primer. This product contains guanine and cytosine, not necessarily in equal amounts, and nearest-neighbour sequence analysis shows that the frequencies of GpG and CpC are each 0·500. Mild acid hydrolysis releases all the dGMP but none of the dCMP. Sedimentation and viscosity measurements yield values similar to those found with the poly d(A-T) but the T_m value is much higher (83°). These observations are consistent with the view that the molecule consists of two homopolymers, one containing only guanine and the other only cytosine, hydrogen-bonded throughout their lengths (Fig. 11.20). That the two chains need not necessarily be of the same length is indicated by the lack of equivalence of guanine and cytosine and by the observation that dGTP and dCTP are not necessarily incorporated to the same extent.

Similar polymers containing base analogues may also be formed. For example, bromouracil may replace thymine in both strands of poly d(A-T)·poly d(T-A) with the formation of poly d(A-BrU)·poly d(BrU-A). When such replacement occurs in one strand only the hybrid poly d(A-T)·poly d(BrU-A) is produced [100].

Khorana and his colleagues have used chemically synthesized short-chain polydeoxyribonucleotides with repeating sequences to serve as templates for the synthesis of long-chain double-stranded polydeoxyribonucleotides [117].

For example, d(T-T-C)$_4$ and d(A-A-G)$_3$ have been synthesized chemically. When they are incubated with DNA polymerase in the presence of dTTP, dATP, dCTP, and dGTP the product is poly d(T-T-C)·poly d(G-A-A). For such reactions to proceed, chemically synthesized segments corresponding to both strands are required and the minimal size of the two complementary segments used as templates varies between 8 and 12 units. Template strands are continually exposed by the slippage of the two oligonucleotides with respect to each other. Synthesis is extensive and the products are of high molecular weight. They are double-stranded with sharp melting transitions, and nearest-neighbour analysis shows that the individual strands contain appropriate repeating nucleotide sequences. They may be used as templates for further synthesis.

(f) *The $3' \rightarrow 5'$ exonuclease and proof-reading.* Before it was realized that this was an integral function of the polymerase the $3' \rightarrow 5'$ exonuclease activity was ascribed to an exonuclease II (see p. 179). It appears that this exonuclease functions to recognize and eliminate a non base-paired terminus on the primer DNA [108]. When *Esch. coli* DNA polymerase I is provided with 4 triphosphates and a primer/template with a mismatched end, the non-matching terminus is removed by the $3' \rightarrow 5'$ exonuclease before polymerization begins. (In the absence of the triphosphates the exonuclease continues to remove nucleotides from the frayed ends of the DNA molecules.) Thus the nuclease is able to correct errors which may occur in the polymerization process. This proof-reading function dramatically increases the fidelity of the base pairing mechanism [79].

In addition to the two mechanisms already mentioned for ensuring that the correct base is inserted into the new DNA molecule (selection by polymerase and checking by $3' \rightarrow 5'$ exonuclease), a third proof-reading mechanism is thought to exist. The dna B gene product, which acts as a DNA-dependent ATPase, is essential for elongation [407], and Hopfield [408] has suggested that such an ATPase would be required for the kinetic proof-reading mechanism he proposes. Alberts has shown that a similar protein blocks polymerization by bacteriophage T4 polymerase. Selection of the correct triphosphate activates an ATPase and allows elongation to occur.

(g) *The $5' \rightarrow 3'$ exonuclease.* This activity, which is present in the smaller fragment released from DNA polymerase I by *subtilisin* treatment (see p. 272), cleaves base-paired regions of DNA, releasing oligonucleotides from 5'-ends. Because of its ability to jump several bases at a time, this nuclease can act on DNA molecules containing mismatched bases or distortions which render them unsuitable as substrates for polymerase [118]. It may thus serve a function, for instance, in the elimination of thymine dimers from DNA exposed to ultraviolet radiation (see p. 303).

When *Esch. coli* polymerase I binds to a nick on double-stranded DNA two reactions occur simultaneously. Polymerization extends the 3'-OH end and $5' \rightarrow 3'$ exonuclease degrades the 5'-phosphate terminus. This results in *nick translation*, a process which may only end when the enzyme reaches the end of the DNA molecule [119]. Alternatively nick translation may end if a long 5'-terminus is released from the nick thereby rendering it no longer susceptible to the exonuclease. The branched product typical of reaction with native DNA primer/templates would then be formed (see p. 263).

(h) *The role of the Kornberg enzyme* in vivo. Although the DNA polymerase I is exceedingly effective in the copying of a single-stranded DNA template when provided with a primer, it is much less effective with double-stranded DNA. This observation and other considerations have led to doubts as to the role of the Kornberg enzyme (DNA polymerase I) in the replication of DNA *in vivo* and to the suggestion that it is concerned merely in maintenance and repair of DNA (p. 303). These considerations may be summarized as follows:

(1) The purified enzyme cannot replicate double-stranded DNA semi-conservatively to yield a biologically active product.

(2) The purified enzyme catalyses the incorporation of 1000 nucleotides per minute per mole of enzyme whereas the estimated *in vivo* rate of incorporation is 100 times faster.

(3) Mutants of *Esch. coli* have been isolated which contain apparently normal Kornberg enzyme but are defective in DNA duplication. This demonstrates that other enzymes (which *may* include other polymerases) are required for *in vivo* DNA synthesis, and this may help to explain the deficiencies enumerated under (1) and (2) above where only the purified Kornberg polymerase was present. It is unfortunate that so many attempts have been made to play down its biological role, for it is in many respects a unique enzyme the study of which has yielded much valuable information about enzyme action.

(4) Nevertheless the evidence against the Kornberg enzyme being *essential* for replication *in vivo* is strong and is based on the properties of a mutant of *Esch. coli* (*pol A1* or *pol A$^-$*) isolated by Cairns and de Lucia [120, 121]. This mutant and several others discovered later multiply normally but contain 1 per cent or less of the Kornberg enzyme activity present in wild-type cells. Such mutants, however, show a reduced ability to join Okazaki fragments [122] (p. 253) and an increased sensitivity to ultraviolet [120, 121] and ionizing radiation [123] and to alkylating reagents [120, 121, 124]. This implies a role for the enzyme in 'gap' filling, and perhaps also in excision of RNA primers and mismatched base pairs. However, these functions

are not completely lacking in mutants lacking DNA polymerase I, and it is suggested that they may also be performed to a limited extent by other enzymes, e.g. polymerases II and III and the rec B and C nuclease (see p. 305). Double mutants of pol A rec B are non-viable [125–127], suggesting an essential function which can be carried out by more than one enzyme.

11.4.3 Esch. coli *DNA polymerase II*

The discovery of the *pol A*⁻ mutant of *Esch. coli* which grows well and replicates its DNA in the usual manner in spite of the absence of the Kornberg enzyme gave an impetus to the search for another enzyme apparatus which can synthesize DNA. Two further polymerases, designated polymerase II and III, were found in extracts of pol A⁻ cells. These enzymes had not been detected previously because, in extracts of wild-type cells, they show little activity relative to DNA polymerase I when single-stranded or nicked DNA is used as template. DNA polymerase II and III show significant activity only with a 'gapped' DNA template (Table 11.5). The enzymes can be separated from one another by chromatography on phosphocellulose [128], DEAE-cellulose [129], or DNA-agarose [130].

Purified DNA polymerase II has a mol. wt. of 90 000–120 000 [128, 131] and is homogeneous as judged by SDS–polyacrylamide gel electrophoresis [128]. It synthesizes DNA in the $5' \to 3'$ direction and for maximal activity it requires all four triphosphates, Mg^{2+}, NH_4^+, and a native DNA template containing single-stranded gaps 50–200 bases long [129, 130]. The rate of reaction falls off with longer gaps, but may be restored by addition of *Esch. coli* unwinding protein (see p. 288). The enzyme also requires a 3'-OH primer. It is sensitive to sulphydryl reagents and is not affected by antiserum to DNA polymerase I [128]. The purified enzyme, however, like polymerase I, only synthesizes DNA at rates a fraction of those found *in vivo*. It is inhibited by the powerful antileukaemic agent Ara-CTP, the triphosphate of Ara-C (1-β-D-arabinofuranosylcytosine). The enzyme also possesses $3' \to 5'$ exonuclease activity, but no $5' \to 3'$ exonuclease activity [134].

The *in vivo* function of *Esch. coli* DNA polymerase II is unknown and mutants lacking the enzyme appear normal in all respects. However, double mutants lacking both polymerase I and polymerase II join Okazaki fragments even more slowly than do mutants lacking polymerase I alone [135].

11.4.4 Esch. coli *DNA polymerase III*

In contrast to mutants with defective DNA polymerase I or II, *Esch. coli* with a temperature sensitive mutation in the gene for DNA polymerase III (dnaE) are not viable at the restrictive temperature [136] and lysates prepared from them are defective in DNA synthesis [134, 137]. Complementation of such lysates with DNA polymerase III purified from normal cells restores their DNA synthetic ability. This is strong evidence that, unlike DNA polymerases I and II, polymerase III is *essential* for DNA synthesis.

DNA polymerase III has been purified about 20 000 fold and shown to be highly sensitive to salt in dilute solutions (i.e. normal assay conditions) but not when assayed by complementation (i.e. at high protein concentration) [129, 130, 134, 138]. Although there appear to be only about 10 molecules per bacterial cell, its high rate of polymerization of nucleotides (Table 11.5) shows it to be capable of its proposed role in DNA synthesis. The best template for DNA polymerase III is double-stranded DNA with many small gaps containing 3'-OH primary ends.

According to Kornberg's group [139] DNA polymerase III is a dimer of two identical subunits each of molecular weight 90 000. However, Richardson's group [398, 399] claim that the two subunits have molecular weights of 140 000 and 40 000 [397].

Like the other two *Esch. coli* polymerases it has a $3' \rightarrow 5'$ exonuclease activity, and like polymerase I it also shows $5' \rightarrow 3'$ exonuclease activity [398, 399].

DNA polymerase III cannot use as template long single-stranded DNA molecules even when provided with a primer. However, a more complex form of the enzyme known as polymerase III* (which may be a dimeric form of polymerase III) in the presence of other proteins (copolymerase III* or factor I) and lipids and ATP is able to extend an RNA primer hydrogen bonded to circular, single-stranded M13, G4, or ØX174 DNA [74, 79, 83, 139–142, 398, 400]. The reaction ceases when the growing DNA chain has passed around the circle and reached the 5'-end of the RNA primer. At this stage addition of DNA polymerase I will cause removal of the RNA and its replacement by DNA, and the final nick can be sealed by DNA ligase to yield a covalently closed, double-stranded, replicative form of the molecule (see Chapter 5).

B. subtilis, a bacterium widely divergent from *Esch. coli*, also has three DNA polymerases which closely resemble those of *Esch. coli*.

TABLE 11.5
DNA polymerases of Esch. coli

Polymerase (Gene)	Molecular weight	Molecules per cell	Nucleotides polymerized/sec (a) per enzyme molecule (b) per bacterial cell	Direction of (a) Polymerization (b) exonuclease action	Template (all require 3'-OH Primer)
I (pol A)	109 000	400	(a) 16–20 (b) 8000	(a) $5' \rightarrow 3'$ (b) $3' \rightarrow 5'$ and $5' \rightarrow 3'$	Denatured Nicked Gapped
II (pol B)	120 000	17–100	(a) 2–5 (b) 500	(a) $5' \rightarrow 3'$ (b) $3' \rightarrow 5'$	Gapped
III (dna E)	180 000	10	(a) 250–1000 (b) 10 000	(a) $5' \rightarrow 3'$ (b) $3' \rightarrow 5'$ and $5' \rightarrow 3'$	Gapped

They differ in that polymerase I and II and possibly also III are devoid of nuclease activity and, in addition, DNA polymerase III is sensitive to 6-(*p*-hydroxyphenylazo)uracil (HPUra), an antibiotic active against gram-positive bacteria. DNA synthesis in lysates of *B. subtilis* is also inhibited by HPUra [143].

11.4.5 *DNA polymerases in eukaryotes*
While the DNA polymerases of micro-organisms have been studied more intensively, similar enzymes are present in eukaryotic cells. In general they resemble the bacterial enzymes in their requirement

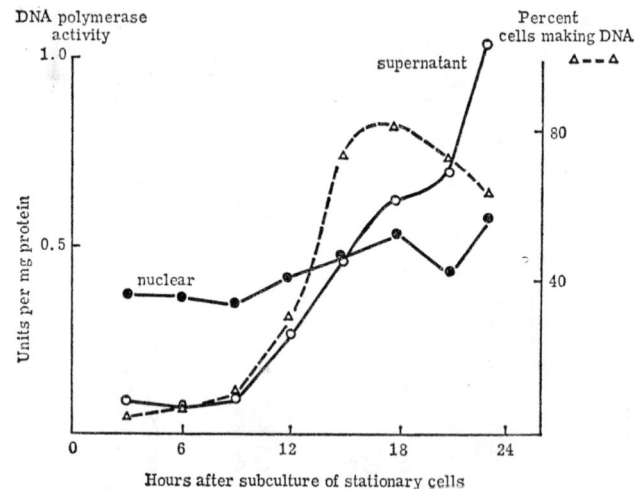

Fig. 11.22 *Variation in the activities of nuclear (β) and superanatant (α) DNA polymerases with growth phase of cultured mouse L cells*

for a template, a 3'-OH primer, and four deoxyribonucleoside triphosphates [403–405]. On fractionation of animal cells DNA polymerase activity is found in several fractions, but until recently it was the enzyme recovered in the supernatant fraction which alone was investigated [144–146].

This soluble enzyme, DNA polymerase α, is most active in extracts of rapidly proliferating cells (Table 11.6) and has been extensively investigated in ascites tumour cells [147, 148] in normal, developing, and regenerating rat liver [149–153] and several hepatomas [153–155], in calf thymus [151, 156–159], in phytohaemagglutinin stimulated lymphocytes [160–162], in brain [163, 164] and in cultured cells [165–170] as well as in chick embryo [169], sea urchins [170], and Euglena [171]. It is this enzyme (DNA polymerase α) which varies in activity with the growth rate of the cells (Fig. 11.22) [149, 152, 153, 160, 162, 163, 170, 174].

Because of its lability and apparent heterogeneity, DNA polymerase α has not been purified to the same extent as *Esch. coli* DNA polymerase I despite the fact that a 200-fold purification of the calf thymus enzyme was achieved in 1965 [157] when it was separated from a terminal transferase (see below).

DNA polymerase α sediments in sucrose gradients as a broad peak in the 6–8S region and on gel filtration activity is found in a number of regions [151]. It has been suggested that the molecular

TABLE 11.6

The incorporation of ^{3}H-thymidine into DNA by soluble extracts prepared from sonically disrupted rabbit and chicken tissues [147]

Tissue	DNA specific activity (counts/min/μmole DNAP)
Rabbit bone marrow	24750
Rabbit thymus	9275
Rabbit appendix	1458
Rabbit liver	131
Rabbit kidney	71
Chick embryo	12025

The reaction mixture contained 0·1M-phosphate buffer (pH 8·1), 50 μmoles glucose/ml, 2·5 μmoles ATP/ml, 2·5 μmoles NAD/ml, 4 μmoles $MgCl_2$/ml, 500 μg DNA/ml, 20 μmoles NaCl/ml, and 2 μCi ^{3}H-thymidine/ml. Incubation time 2 hours.

weight of the enzyme is between 165 000 and 175 000 but that (a) it is associated with other proteins, (b) it is susceptible to terminal proteolysis, and (c) it readily aggregates. This enzyme has a pH optimum in the range 6·5–8·0 and is highly susceptible to inhibition by thiol-active reagents (i.e. N-ethylmaleimide and *p*-chloromercuribenzoate). DNA polymerase α shows optimal activity with a gapped DNA template but shows considerable activity with denatured DNA and with synthetic single-stranded templates provided with either oligoribo- or deoxyribo-nucleotide primers, e.g. poly(dT) · oligo(dA) or poly(dT) · oligo(A) [146, 182]. In contrast to the enzymes from *Esch. coli* no nuclease activity is associated with the eukaryotic enzymes. Although polymerase α is normally recovered in supernatant fractions following cell homogenization, there is evidence to suggest that it may be located within the nucleus *in vivo* [413].

Unlike DNA polymerase α, the DNA polymerase recovered in the nuclear fraction (DNA polymerase β) shows little correlation between activity and the growth rate of the cells (Fig. 11.22). This enzyme, which has been purified to homogeneity, has a molecular weight of 43 000 [175] and a pH optimum of around 8·6–9·0 [176]. It shows optimal activity with native DNA activated by limited treatment with DNase I (to produce single-stranded nicks and short gaps bearing 3′-OH priming termini) [177, 178] and shows negligible activity with denatured DNA. DNA polymerase β is relatively insensitive to thiol-active reagents but resembles the DNA polymerase α in being devoid of nuclease activities. Thus neither enzyme has the 'proof-reading' capacity of the bacterial enzymes, and in fact both will add on to a mismatched primer terminus [175, 179].

DNA polymerase α cannot use an oligoribonucleotide primer but, unlike polymerase α, given an oligodeoxyribonucleotide primer it is able to copy a ribonucleotide template (e.g. poly A oligo dT) [146, 180–182]. This is the preferred template for a third enzyme (DNA polymerase γ, mol. wt. 119 000) which Weissbach's group has studied. The activity of DNA polymerase γ is very low but it does vary throughout the cell cycle [182–184].

Both mitochondria [185–188] and chloroplasts [189, 190] possess the enzymic mechanisms for synthesizing their own DNA. The DNA polymerase from rat liver mitochondria has been partially purified and completely separated from the nuclear enzyme which it resembles in most of its requirements [191, 192].

In addition to replicative DNA polymerase activity, which requires the presence of all four deoxyribonucleoside triphosphates, extracts contain a separable enzyme responsible for the addition of nucleotidyl units to the *ends* of polynucleotide chains. This second activity, which was originally described by Krakow in 1962 [193], does not require a template strand but catalyses the incorporation of nucleotide units from single triphosphates into terminal positions in the DNA primer molecule. It is not further stimulated by the addition of the other three triphosphates but it is stimulated by cysteine. It has been called the 'terminal enzyme' [194, 195] and may be used in the biosynthesis of homopolymers of deoxyribonucleotides [157, 196].

In measuring the activities of these enzymes the assay mixture for the 'replicative' enzyme contains all four deoxyribonucleoside triphosphates, one of them, dTTP, being radioactively labelled. The assay mixture for the 'terminal' enzyme contains only one triphosphate; radioactive dTTP has been used routinely, but dATP is more effective with the purified enzyme.

It is possible to distinguish between the two enzymes by the use of actinomycin D which is well known as an inhibitor of both the DNA polymerase and the DNA-dependent RNA polymerase (p. 300). Its action is to block the surface of the priming strand of DNA by binding to guanine residues, and it is for this reason a powerful inhibitor of the replicative enzyme. On the other hand, it would be reasonable to suppose that actinomycin D would exercise a much less pronounced effect on 'terminal' incorporation since direct interference would arise only in primer molecules bearing deoxyguanylyl residues at or near the 3'-hydroxy-terminal residues. The results clearly indicate the sharp distinction between the 'terminal' and 'replicative' enzymes which can be revealed by actinomycin D [103].

Unlike the replicative enzyme, which is stimulated by low levels of EDTA, the terminal enzyme is completely inhibited by micromolar concentrations of EDTA which is believed to exert its effect by binding Zn^{2+}.

The terminal transferase has been purified to homogeneity from calf thymus. It can be dissociated by sodium dodecyl sulphate into two subunits of molecular weight 8000 and 26 500 daltons [197].

The activity of the terminal transferase is very low in all tissues except thymus [199] and acute leukaemic lymphoblasts (thought generally to be a tumour of thymic origin) [200, 201]. To what extent this reflects the peculiar physiological functions of the thymus is still unknown.

11.4.6 *DNA ligases*

The ligase or joining enzymes catalyse the repair of a single-stranded phosphodiester bond cleavage of the type introduced by endonuclease. They were first described in *Esch. coli* [202–206] and have since been described in both animal [207] and plant [208] cells.

DNA ligases catalyse the formation of a phosphodiester bond between the free 5'-phosphate end of an oligo- or poly-nucleotide and the 3'-OH group of a second oligo- or poly-nucleotide positioned next to it (Fig. 11.23). A ligase–AMP complex seems to be an obligatory intermediate and is formed by reaction with NAD in the case of *Esch. coli* and *B. subtilis* [204, 209, 210] and with ATP in mammalian and phage-infected cells [203, 205, 211, 212] (Fig. 11.23). The adenyl group is then transferred from the enzyme to the 5'-phosphoryl terminus of the DNA. The activated phosphoryl group is then attacked by the 3'-hydroxyl terminus of the DNA to form a phosphodiester bond.

DNA ligase will close single-strand breaks in double-stranded

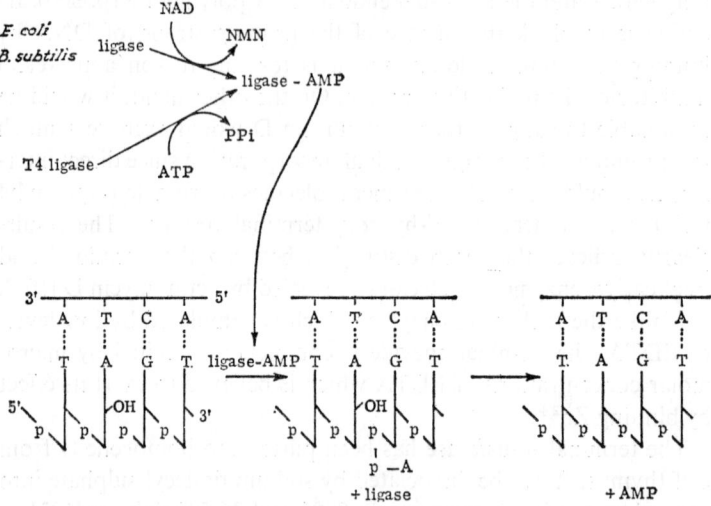

Fig. 11.23 *Postulated action of DNA ligases*

DNA or in either strand of a polyribonucleotide–polydeoxyribonucleotide hybrid polymer; double-stranded ribopolymers are not substrates [213–216]. Breakage of a single phosphodiester bond without removal of a nucleotide to give 5'-phosphoryl and 3'-OH termini is essential for repair by ligase activity (see Fig. 11.24). Reaction is independent of the base composition around the cleavage point [217].

The enzymes from *Esch. coli* or bacteriophage T4 will join short oligodeoxynucleotides in the presence of a long complementary strand; e.g. $d(T-G)_3$, $d(T-G)_4$, or $d(T-G)_5$ can be joined in the presence of poly(dC-dA) [218].

(a) *Assay of DNA ligase*. DNA ligase has been assayed in a variety of systems, including the formation of covalently closed circles of double-stranded DNA, restoration of transforming activity of nicked DNA (e.g. [210]), and the formation of phosphatase-resistant radioactive phosphate [219] (see Fig. 11.24).

(b) *Role of DNA ligase.* The nature of the reaction catalysed by DNA ligase has made it an important feature of several models for DNA synthesis, for DNA repair, and for genetic recombination [220]. In all of these cases, its postulated role is in re-establishing

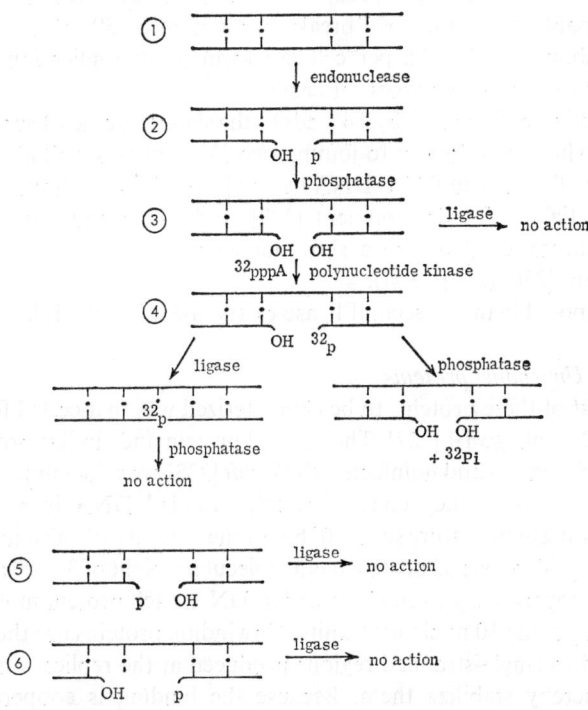

Fig. 11.24 *An assay for DNA ligase. The substrate (4) is formed by treating double-stranded DNA (1) with DNase I to give single-stranded breaks (2), removing the 5'-phosphate residues with phosphatase (3) and replacing them with ^{32}P-phosphate groups with the aid of polynucleotide kinase and γ-labelled ^{32}P-ATP. Forms (3), (5), and (6) are inactive as substrates for the ligase*

continuity by joining a stretch of newly synthesized DNA to pre-existing DNA by the formation of a phosphodiester bond (see p. 254).

It has, however, proved difficult to define a situation in which DNA ligase could be shown to be absolutely necessary for any one of these functions. Nevertheless, since *Esch. coli* ligase-deficient mutants selected in different ways exhibit abnormal u.v. sensitivity [221, 222], it appears that ligase has some function in the repair process. These mutants also accumulate Okazaki pieces at the restrictive temperature. That these pieces are eventually joined is probably a reflection of the presence of a few molecules of ligase still remaining in the mutant cells [223–225]. There are normally about 300 molecules of ligase per cell, and it has been calculated that these are capable of sealing 7500 breaks per minute at 30° [389]. Since only about 200 breaks per cell are formed on replication each minute there is a vast excess of enzyme.

Infection of *Esch. coli* by T4 leads to the synthesis of a T4 specified ligase which, in addition to joining two DNA chains, will also join RNA to DNA in an RNA–DNA hybrid [389, 401, 402]. Some ligase is essential for T4 development [222], and bacteriophages with a temperature-sensitive ligase show increased susceptibility to u.v. radiation [226] (see p. 304).

It is possible that a second ligase exists in *Esch. coli* [47].

11.4.7 *Unwinding proteins*

The first of these proteins to be characterized was that coded for by gene 32 of phage T4 [227]. The gene 32 protein and similar proteins isolated from T7 and uninfected *Esch. coli* [228] are called unwinding proteins because they convert double-stranded DNA into single strands at a temperature some 40° below the normal melting temperature (T_m). The gene 32 protein has a molecular weight of 35 000 and it binds cooperatively to single-stranded DNA, each protein molecule covering some 10 nucleotide units. Unwinding proteins are thought to bind to single-stranded regions produced at the replication fork and thereby stabilize them. Because the binding is cooperative, protein molecules line up side-by-side and cause the unwinding of the DNA. There are from 300 to 800 molecules of unwinding protein per uninfected *Esch. coli* cell, which is enough to cover about 1600 nucleotides at each replication fork [132, 228]. A similar length of T4 DNA can be covered by the 10 000 molecules of gene 32 protein present in an infected cell. (There are about 60 T4 replication forks per infected cell [227].)

Purified polymerases are stimulated in a very specific manner by the presence of their complementary unwinding proteins. Thus the T4 unwinding protein will stimulate only the T4 polymerase and the T7 unwinding protein will stimulate the T7 polymerase [132, 229,

230]. The *Esch. coli* unwinding protein stimulates *Esch. coli* polymerase II and III* but does not stimulate polymerases I or III [231] (see p. 280). It is suggested that the polymerase and unwinding protein form a specific complex that interacts with DNA and other proteins during replication [83, 232].

Unwinding proteins have also been isolated from eukaryotic cells [233–235].

Another type of protein, an untwistase, present in prokaryotes and eukaryotes, is one which reduces the number of negative superhelical turns in closed circular viral DNA [30, 236, 237]. This protein could act at or near the replication fork and thereby eliminate the need for continuous rapid rotation of the DNA molecule, i.e. it could act as the swivel (see p. 249).

11.5 Models of DNA synthesis

11.5.1 *Discontinuous synthesis*

Although the previous discussion has been based on the so-called 'discontinuous model' of DNA replication (Fig. 11.8), other models have been proposed and not effectively excluded. One must bear in mind that none of these models may be appropriate or alternatively that each may apply in different situations. The evidence in favour of the discontinuous model is that it is consistent with electron microscope data and the finding that nascent DNA can be recovered in short Okazaki pieces with RNA primers at their 5'-ends.

11.5.2 *Knife and fork model* (Fig. 11.25)

This is a variant of the discontinuous model. Synthesis of DNA occurs in the $5' \rightarrow 3'$ direction down the left-hand strand of the parental DNA molecule (Fig. 11.25a). This synthesis occurs by continuous extension of the daughter DNA chain, and exposes a single-stranded region on the right-hand strand. At some point the daughter strand switches from the left-hand to the right-hand strand where synthesis now occurs backwards (i.e. $5' \rightarrow 3'$) to produce the fork. It is the same daughter strand which is being elongated so there is no requirement for a primer. A nuclease (knife) then nicks the new DNA at the fork thus generating a 3'-OH priming end on the left-hand daughter strand and an Okazaki piece (Fig. 11.25c). This continues to be extended and is then joined by DNA ligase to the right-hand daughter strand.

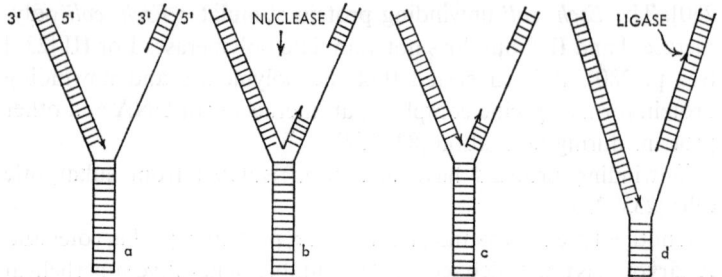

Fig. 11.25 *The knife and fork model; for an explanation see the text*

According to this model RNA primers are not required and only 50 per cent of nascent DNA should be found in Okazaki pieces. Moreover, the form shown in Fig. 11.26c has not been seen in the electron microscope [12]. This may, however, simply reflect its transient nature which may also explain the inability to detect significant amounts of nascent DNA which is of the hairpin structure shown in Fig. 11.25b [238, 239].

11.5.3 *Pre-fork replication* (Fig. 11.26)

In 1967 Haskell and Davern proposed a model whereby single-stranded 'nicks' are introduced ahead of the replication fork at alternating sites on the parental DNA [240] (N_1 in Fig. 11.26a). The 3'-OH ends thus generated serve as primers for DNA polymerase action (Fig. 11.26b). The daughter strands are quickly released from the parental primers by a second nuclease event (N_2 in Fig. 11.26c), and the original nick is resealed using DNA ligase (L in Fig. 11.26d). The daughter strands continue to elongate in the $5' \rightarrow 3'$ direction until they meet the 5'-end of the preceding daughter section when DNA ligase acts to seal the two sections together (L in Fig. 11.26f).

This model postulates a transient covalent attachment of parental and daughter DNA which has been reported to occur in an *in vivo* replicating system from *Esch. coli* [241, 242]. Moreover, after this transient association, all nascent DNA should be present in Okazaki pieces. Evidence from electron microscope studies only supports this model in part in that single-stranded regions are found on one side of the fork (Fig. 11.26 b and d) but triple and four-stranded regions ahead of the fork have not been found. (Possible exceptions to this are the D-loops found in replicating mitochondrial and λ DNA [243, 244].).

REPLICATION OF DNA

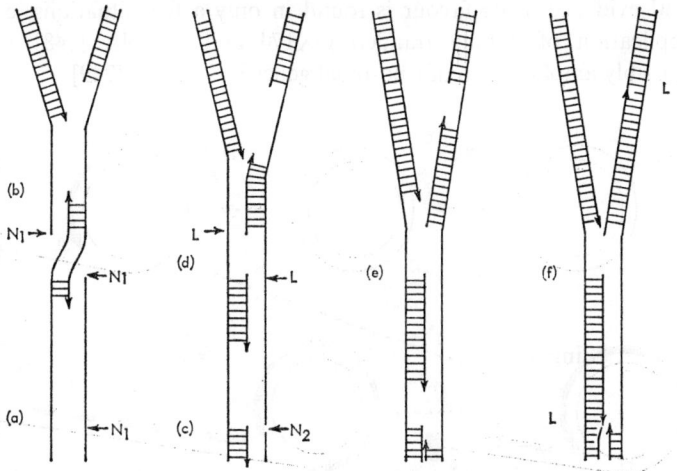

Fig. 11.26 *Pre-fork replication; for an explanation see the text. For clarity only newly synthesized duplex is shown hydrogen-bonded*

11.5.4 *The rolling circle model*

The rolling circle model [245] was developed in an attempt to explain the structure of the replicating DNA molecule in bacteriophages (ØX174, T_4, and λ) and in *Esch. coli*. Replication is envisaged as starting with a covalently linked cyclic (circular) DNA duplex. In the case of ØX174 this would be the double-stranded RF. The (+) strand is opened at a specific point by an endonuclease to expose a 3'-OH and a 5'-phosphate group. The latter is anchored to a site on the cell membrane and the chain is elongated from the 3'-OH end, using the (−) strand, which remains closed, as an endless template (Fig. 11.27). The process is continuous, the new (+) strand being continuously copied from the 3'-OH end. As the long thread of covalently-linked (+) strands peels off, it is used as a template for the synthesis of progeny (−) strands, perhaps as fragments as described by Okazaki [41]. Finally the tail is cut into appropriate lengths and ring-closure effected by DNA ligase.

The rolling circle model is based on the discovery of (+) DNA strands longer than the length of a mature viral genome, (−) strands that are covalently closed single-stranded circles, and 3'-termini of the long (+) strands lying upon the template rings while the 5'-ends are free in solution [246, 247].

The rolling circle model has not received universal acceptance and evidence in its favour is found in only a few situations, e.g. replication of double-stranded ØX174 and λ DNA [248] and possibly amplification of ribosomal genes in *Xenopus* [249].

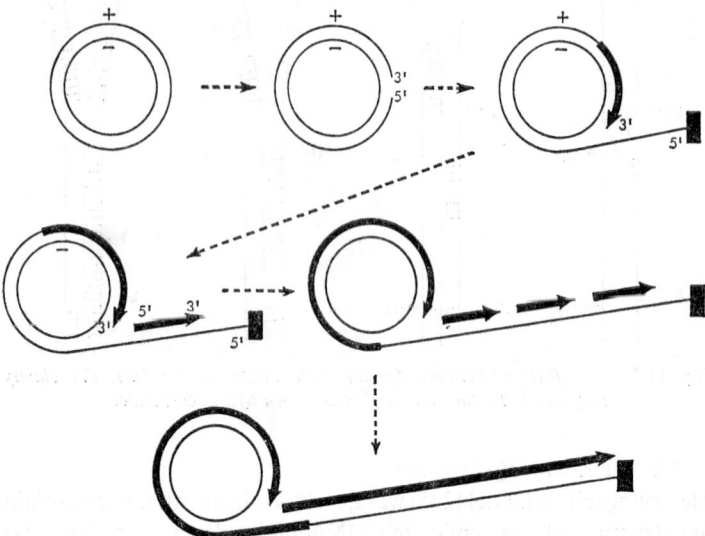

Fig. 11.27 *The rolling circle model for DNA replication. For explanation see text*

11.5.5 Termination

It is assumed that synthesis of Okazaki pieces ends when the 3'-end of the growing chain reaches the 5'-end of the adjacent chain. If this 5'-end consists of ribonucleotides these must be removed and the gap filled before DNA ligase can join the Okazaki pieces together. With cyclic DNA molecules a continuation of this process will produce two complete daughter molecules. However, not all chromosomes are cyclic, though it has been suggested that linear DNA molecules, such as that of the bacteriophage λ, circularize as a condition for replication [248, 250].

Watson [251] has put forward a mechanism whereby the complete replication of phage T7 DNA can be accomplished without circularization. With a linear molecule single-stranded regions remain at one end of each daughter molecule following replication. This is because 3'→5' growth cannot be initiated at the extreme 5'-end of the template. However, T7 DNA has an identical sequence of about

260 base pairs at each end of the molecule, and when these ends are left single-stranded following replication they are able to form intermolecular hydrogen bonds, giving rise to double-length concatameric molecules. A bacteriophage T7 mutant with a defective gene 2 protein is unable to form concatamers and fails to form complete T7 DNA molecules. By the successive action of ligase, nuclease, and polymerase the missing regions can be added and the two complete daughter molecules produced.

11.6 Synthesis of DNA in virus infected cells
11.6.1 *Double-stranded DNA viruses*

When cells of *Esch. coli* are infected with the T-even bacteriophages, the economy of the cells is completely altered so as to lead to the production of new phage DNA which differs from the host DNA in containing hydroxymethylcytosine in place of cytosine (p. 145). These changes result in the production in the infected cell of a series of new and interesting enzymes [100, 177, 252–261] (Fig. 11.28).

(*a*) Within a few minutes of infection a hydroxymethylase (the product of phage T4 gene 42) appears which brings about the conversion of dCMP into hydroxymethyldeoxycytidine monophosphate (CH_2OH-dCMP or dHMCMP).

(*b*) At about the same time a *kinase* is produced which phosphorylates dHMCMP to the corresponding triphosphate dHMCTP.

Neither of these new enzymes is found in cells infected with bacteriophage T5 which does not contain HMC.

The kinases for dTMP and dGMP are also greatly increased but not that for dAMP. This increase is due to the production of new enzymes which can be distinguished from the kinases of the host cell prior to infection.

(*c*) The formation of host DNA is prevented by the appearance of a pyrophosphatase (the product of phage T4 gene 56) which converts dCTP into pyrophosphate and dCMP which then acts as a substrate for the hydroxymethylase.

(*d*) Five distinct glucosyltransferases are known to be induced after infection with T-even phages for the purpose of transferring glucose residues from uridine diphosphate glucose to the HMC of phage DNA in the proportions shown in Table 7.3. The glucosylase found in T2-phage-infected cells transfers a glucose residue to HMC in the α-configuration. Two glucosylating enzymes are produced after T4 infection; one adds a glucosyl group in α-linkage to HMC while the second also adds a glucose group but in β-configuration. After T6 infection two glucosyltransferases are also produced. One adds a monoglucosyl residue to HMC in α-linkage while the other reacts with the monoglucosylated groups on HMC to add a second glucose residue, the linkage between the residues being of the β-configuration.

(e) The products of T4 genes 46 and 47 are thought to be a nuclease which degrades DNA containing cytosine. T4 DNA which contains hydroxymethylcytosine is resistant.

(f) A new polymerase is also formed after phage infection [260]. In T4 it is the product of gene 43 and is essential for replication of bacteriophage DNA. In many ways it resembles *Esch. coli* DNA

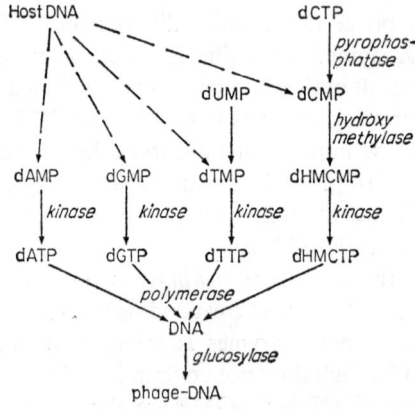

Fig. 11.28 *Key enzymes involved in the formation of the DNA of T-even phages*

polymerase I from which it is separated by phosphocellulose chromatography. However, it cannot use nicked DNA as template/primer but requires single-stranded DNA with short primers. It has a powerful $3' \rightarrow 5'$ exonuclease but lacks a $5' \rightarrow 3'$ exonuclease [262]. Bacteriophage T7 induces a DNA polymerase which is the product of the phage gene 5 (mol. wt. 84 000) and host thioredoxin (mol. wt. 11 500) [410].

The enzymic effects of phage infection are summarized in Fig. 11.28. The general principle would appear to hold for most cells, mammalian or bacterial, that viral infection results in the formation of new polymerases (DNA polymerase and RNA polymerase, p. 402) rather than in an increased amount of the polymerase characteristic of the host cell [263–265].

11.6.2 Single-stranded DNA viruses

One problem of obvious interest is how the single-stranded DNA of phages ØX174, G4, and M13 replicates [79, 83]. ØX174 DNA, which contains 5500 nucleotide residues, can be an excellent template in the DNA polymerase reaction *in vitro*, and the product is a duplex each strand of which can act as a template in further replications. The existence of such a double-stranded 'replicative form' (RF) *in vivo* has also been proved by infecting bacterial cells in a $^{31}P,^{14}N$-medium with phage labelled with both ^{32}P and ^{15}N. The DNA replicating in such infected cells was isolated as two bands on density-gradient centrifugation, both containing some of the radioactivity of the parental phage. The heavier band was of the density expected of ^{15}N-labelled ØX174 in the usual single-stranded form. The lighter band had the density expected of a hybrid consisting of one strand of ^{15}N-DNA and one of ^{14}N-DNA and behaved on heating in the manner expected of double-stranded DNA. Its composition indicated matching base pairs. This RF, like single-stranded ØX174 DNA, can be used to infect bacterial spheroplasts and consists of intact double-stranded circles [266–269].

To convert the single strand into the RF, these phages make use of the host enzymes, but some, like M13, appear to use the system present in *Esch. coli* for replicating the extrachromosomal plasmids (see p. 255) rather than the bacterial chromosomal DNA [79]. Phage-induced enzymes are essential for RF multiplication which may take place by a rolling circle mechanism (see Section 11.5.4).

It is obvious that at some stage in the development of the phage there must be a reversion to the single-stranded stage but since there is no pool of single-stranded molecules in the infected cell it would appear that the single strands must be incorporated into progeny particles as fast as they are formed. Phage-induced proteins are essential to stabilize the progeny single strands and prevent synthesis of (−) strands [270, 271].

It is customary to refer to the original infecting circle as the (+) strand and to the complementary circle formed by the action of DNA polymerase as the (−) strand. Together they constitute the RF which is a covalent duplex circle from which the (+) and (−) strands can be separated physically provided that one (or both) is 'nicked' with DNase. Both (+) and (−) may be infective provided that they are in the circular form. The double helical RF replicates in the conventional manner within the cell from which ultimately infective particles are released containing only (+) strands encased in coat protein.

11.6.3 *RNA viruses and reverse transcriptase*

The enzymes so far discussed copy DNA strands in the synthesis of DNA, but as long ago as 1963 Cavalieri [272] showed that an RNA template, poly(A-U), could serve as a template for the synthesis of a strand of DNA, poly(dT-dA), under the influence of DNA polymerase I from *Esch. coli*. Nevertheless excitement amounting almost to hysteria was created in 1970 by the announcement of the existence in certain RNA viruses of RNA-dependent DNA polymerases which use RNA as a template for the synthesis of DNA. These enzymes were discovered simultaneously by Temin [273] in the virus particles (virions) of Rous sarcoma virus (RSV) and by Baltimore [274] in Rauscher mouse leukaemia (R-MLV) virus. The observation was confirmed for more than half a dozen RNA viruses by Spiegelman [275–277], and the phenomenon of Teminism as it came to be called was hailed both as an example of the reversal of the Central Dogma (p. 239), a suggestion which was vigorously contested by Crick [278], and as an important break-through in cancer research since the RNA viruses involved were oncogenic, i.e. capable of bringing about malignant change (see Chapter 7).

On treating the viral 70S RNA genome with agents which disrupt hydrogen bonds it dissociates to 35S pieces and to several molecules of tRNA [279]. The unique RNA directed DNA polymerase (molecular weight 70 000) is also present in the virion. The enzyme appears to use one of the tRNA molecules (tRNA^{Try}) [280] as a primer to synthesize DNA on the template 70S single-stranded RNA, and this distinguishes it from any cellular enzyme. With synthetic primer/templates the enzyme shows a marked preference for poly(rA) · oligo(dT) or poly(rC) · oligo(dG) [281]. Ribonucleoside triphosphates are without effect as substrates and the process is not sensitive to actinomycin D.

The immediate product of the reaction is a double-stranded RNA–DNA hybrid which is the result of the synthesis of a complementary strand of DNA on the single-stranded viral RNA as template or at least on part of it. Hybridization experiments have confirmed that the DNA strand is complementary to the viral RNA.

The enzyme also shows ribonuclease H and DNA-dependent DNA polymerase activities which convert the hybrid into a duplex

DNA. A particularly good template for this enzyme is the synthetic hybrid poly(dC) · poly(rG) [277]. The polymerase then proceeds to replicate the duplex DNA so as to provide more copies which may be integrated into the genome of the host cell. The enzyme replicates gapped double-stranded DNA templates better than single-stranded DNA templates but cannot use nicked double-stranded DNA.

The presence of RNA-dependent DNA polymerase has now been reported in many viruses, oncogenic and otherwise [282–286], in spontaneous mammary carcinoma [287], and in virus-like particles isolated from human milk [287]. Enzyme activity can also be readily detected in normal mouse and human cells, adult and foetal, as well as in human tumour cells not associated with any RNA-containing virus [288] (see p. 152). The pattern which is emerging is of a group of enzymes all of which can produce DNA from a variety of RNA templates [289–291]. However, cellular RNA directed enzymes differ from the viral enzymes in preferring Mn^{2+} to Mg^{2+} and in their failure to use natural RNA's as templates.

A number of reviews on the subject have appeared recently [292–295].

11.7 Mutations and mutagens

Alterations in the base pattern of DNA may arise in various ways. For example, existing bases may be replaced by others, or they may be deleted, or new bases may be inserted in the DNA chain. Occasional mistakes in the normal duplication of DNA give rise to *spontaneous mutations* but such mistakes are surprisingly rare [296, 297]. The frequency of such mutations depends on conditions of temperature, pH, composition of growth medium, and the like, but it can be greatly increased by exposure of cells to ultraviolet and ionizing radiations (p. 302) or to certain types of chemical which are known collectively as *mutagens*. Such substances include base analogues, some dyes of the acridine series, alkylating agents, certain antibiotics, urethane, hydroxylamine, and nitrous acid. This last substance has been used very effectively in studying mutations in certain viruses such as TMV (p. 371).

Mutagenic substances are the subject of an extensive literature [99, 296, 298–307] which reflects the considerable effort now being devoted to attempts to inhibit cell division, especially in neoplastic tissues, by the use of compounds which might be expected to inhibit nucleic acid biosynthesis. Research in this field has been stimulated by the hope of finding a basis for an improved therapy for cancer. Reference has already been made (Chapter 9) to the use of such compounds as azaserine and the folic acid antagonists in preventing the synthesis of the purine and pyrimidine nucleotides. Some of the other substances which have been used to prevent nucleic acid biosynthesis and to bring about mutations artificially are discussed below.

11.7.1 Base and nucleoside analogues

Some of the artificially produced base analogues are incorporated into RNA and DNA and may have powerful mutagenic effects [296, 301, 302, 304, 308, 309]. Among the most important analogues are the halogenated pyrimidines, and those bases where nitrogen has been substituted for a —CH= group (see Fig. 11.29).

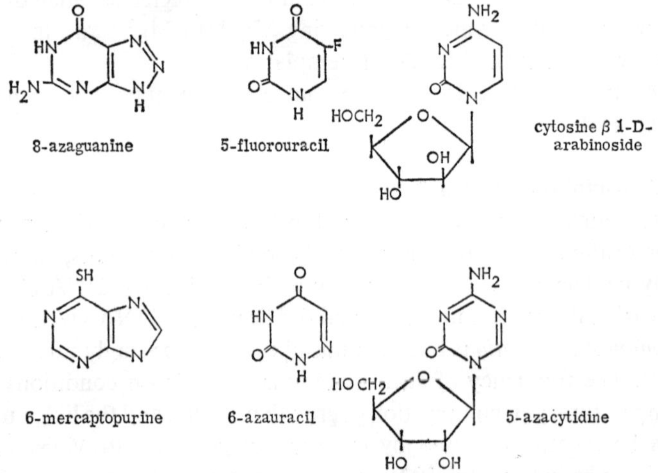

Fig. 11.29 *Structures of some purine and pyrimidine analogues*

The action of these unnatural bases (Fig. 11.29) seems, at least in some cases, to be twofold:

(1) They generally block some stage in the biosynthesis of the normal purine and pyrimidine nucleotides. Thus 8-azaguanine inhibits the biosynthesis of GMP and 6-mercaptopurine blocks the conversion of IMP into AMP [310]. In general, these inhibitions are brought about only after the inhibitor itself has been converted into its nucleotide. Thus 6-azauracil is converted first into its nucleoside (Aza-U) then into its nucleotide (Aza-UMP) which inhibits the

action of orotidine 5'-phosphate decarboxylase (Fig. 9.6) and so prevents pyrimidine biosynthesis [304, 311, 312]. 5-Fluorouracil, which has proved to be a potent inhibitor of the growth of certain tumours, is converted first into its ribonucleotide (F-UMP) and then into its deoxyribonucleotide (F-dUMP) which exerts its main effect by inhibiting conversion of dUMP into dTMP [304, 313, 314] (p. 208) and hence inhibiting DNA synthesis (Fig. 9.8).

When 5-fluorodeoxyuridine is added to cells in culture it is converted into 5F-dUMP which blocks DNA synthesis. As no other processes are effected, cells progress around the cycle and accumulate at the beginning of the S-phase (p. 236). The inhibition can be overcome by the addition of thymidine with the result that a population of cells in synchronized growth result [315, 316].

(2) They are themselves, after conversion into nucleotides, incorporated to varying degrees into RNA and/or DNA although the incorporation may take an abnormal form. Thus 8-azaguanine can be incorporated at the expense of guanine into the RNA of TMV [317] and, to a much larger extent, into the RNA of *B. cereus* [318]. Only very small amounts are incorporated into the DNA.

5-Azacytidine is incorporated into RNA but this rapidly interferes with protein synthesis [162, 304]. 5-Azadeoxycytidine is incorporated into DNA but this renders the cells non-viable [319]. 5-Bromouracil can replace thymine in DNA where it normally base-pairs with adenine. However, in its rare enol-state (which it assumes more readily than does thymine) it may pair with guanine instead of adenine so bringing about the base-pair transition A-T into G-C. DNA containing 5-bromouracil instead of thymine is very susceptible to breakage at light-induced bromouracil dimers [320] (see below).

The D-arabinosyl nucleosides are effectively analogues of deoxyribonucleosides (e.g. cytosine β-D-arabinoside is incorporated in place of deoxycytidine into DNA where it causes chain termination or a marked reduction in the rate of further chain extension [320–322]).

11.7.2 *Alkylating agents*
The alkylating agents exert a variety of biological effects including mutagenesis, carcinogenesis, and tumour growth inhibition [296, 298, 300, 304, 306, 323, 375]. They all carry one, two, or more alkyl groups in reactive form and include the well-known compounds

sulphur mustard or di-(2-chloroethyl) sulphide and nitrogen mustard or methyl-di-(2-chloroethyl)amine.

The action of the alkylating agents on DNA is complex. They are known to react with purine bases, particularly with guanine at the N-7 atom, and the bifunctional alkylating agents (i.e. those with two reactive alkyl groups) may thus bring about cross-linking between the opposing strands in the DNA molecule. Alkylation of purines in position 7 also gives rise to unstable quaternary nitrogens so that the alkylated purine may separate from the deoxyribose leaving a gap which might interfere with DNA replication or cause the incorporation of the wrong base [296, 323]. The phosphate groups may also be alkylated. The phosphate triester so formed is unstable and may hydrolyse between the sugar and the phosphate so that the DNA chain is broken.

11.7.3 *Antibiotics and allied agents*
Some of the carcinostatic antibiotics have been of great value in the study of the nucleic acid biosynthesis [324]. One of the most useful is

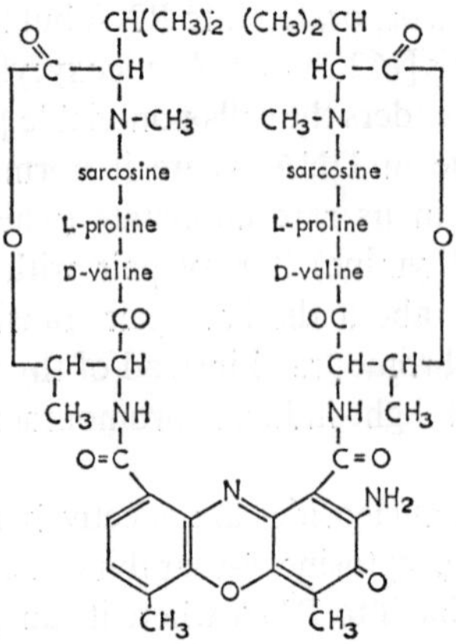

Fig. 11.30 *Structure of actinomycin D*

actinomycin D (Fig. 11.30) which forms complexes with the deoxyguanosine residues in DNA and so blocks it as a template. Actinomycin D therefore inhibits both the DNA polymerase and the DNA-dependent RNA polymerase (p. 322), the former being much less sensitive to its action than the latter [325–330]. At a concentration of actinomycin of 1·0 μM, for example, the DNA-dependent

RNA polymerase is almost completely inhibited whereas the DNA polymerase is only slightly affected [326, 331]. At lower concentrations the synthesis of the different sorts of RNA is affected to different extents, ribosomal RNA being particularly sensitive [332]. In the complex formed between actinomycin and DNA the drug is pictured as being intercalated between alternating dG-dC nucleotide pairs with the cyclic peptide in the groove hydrogen-bonded to guanine [307].

Mitomycin C inhibits bacterial DNA synthesis by causing covalent cross-linking of the complementary DNA strands [333–335]. The drug is reduced in the cell to produce an active bifunctional alkylating agent which cross-links guanine residues [336].

Phleomycin attaches covalently to thymine residues in DNA [337] to inhibit DNA polymerase action [338]. As with bleomycin, which binds in a similar way, this results in single-strand breaks in the DNA [339].

The antibiotic streptonigrin inhibits both DNA and RNA synthesis by inhibiting the respective polymerases [340]. Edeine and nalidixic acid inhibit bacterial DNA synthesis by unknown mechanisms. Antibiotic action is discussed further on p. 386 and is reviewed in [414].

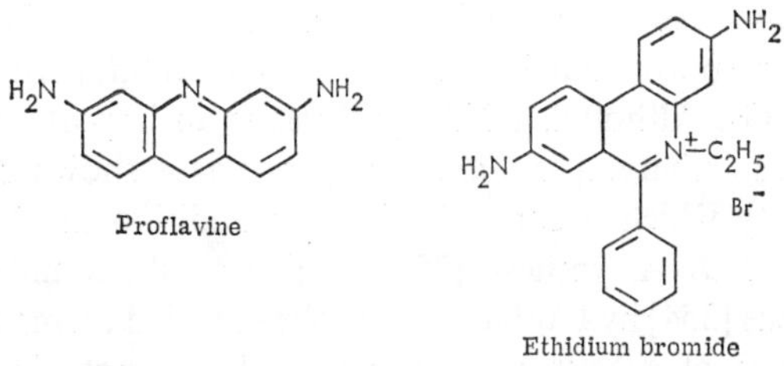

Fig. 11.31 Structures of proflavine and ethidium bromide

11.7.4 Dyes

Proflavine (Fig. 11.31), one of the acridine series of dyes, inhibits DNA-dependent RNA biosynthesis by its molecules becoming intercalated between adjacent nucleotide-pair pairs in the DNA molecule [327, 341, 342] (p. 91). Its mutagenic action on pages has been applied effectively in the study of coding triplets (p. 360). The acriflavines (also acridine dyes) ethidium bromide (Fig. 11.31) and propidium di-iodide also intercalate between DNA nucleotide pairs and in so doing bring about some unwinding of the double helix. This unwinding is resisted in closed circular supercoiled molecules (see p. 98) which therefore bind less dye and are therefore considerably denser than open circle (nicked) molecules from which they can be separated in gradients of caesium chloride containing the dye [343].

11.7.5 *The effects of ionizing radiations*

It has long been known that rapidly growing tissues are much more sensitive to the action of X-rays than are adult tissues, and it is generally recognized that irradiation exerts a pronounced inhibitory influence on the process of cell division. It might therefore be expected that irradiation would exert an appreciable effect on the metabolism and biosynthesis of the nucleic acids.

The effect of ionizing radiations on living cells is the subject of a vast literature which cannot be discussed here (for summaries see [344, 345]) but those aspects of the problem which concern nucleic acid metabolism have been reviewed by several authors [346–350].

Pelc and Howard [351] first showed that incorporation of ^{32}P into DNA of bean root (*Vicia faba*) cells was inhibited if the root tips were irradiated during the G_1 period but that much higher doses were required if the exposure was during the S period (see p. 236). In growing cells there is an association between the onset of DNA synthesis and resistance to irradiation [352]. Thereafter survival rates remain approximately constant throughout this period and then decline in G_2. Although DNA synthesis is an important factor in radiation sensitivity, it is not the only one as is shown by changes in sensitivity which occur following the addition in G_1 phase of inhibitors of DNA synthesis [353]. Similar results found with bone marrow cells [354] by Lajtha and his colleagues led them to postulate the existence of a 'system connected with but not identical with DNA synthesis which is more radiosensitive than the process of DNA synthesis' [355], and studies with regenerating rat liver have given strong support of this opinion [356]. The most sensitive period in the cell cycle is the period immediately prior to mitosis. Cells irradiated at this time with as little as 9 rads show a delay in entering mitosis [357].

What may be of importance is the distinction between rapidly growing cells which appear to possess throughout the cell cycle all the enzymes required for DNA synthesis, and non-growing (G_0) cells which must synthesize these enzymes when stimulated to grow.

REPLICATION OF DNA

Regenerating liver and primary cell cultures fall into this second category and the initiation of DNA synthesis is very sensitive to X-radiation in such systems.

11.8 Radiation injury and repair

Large doses of ultraviolet radiation can damage living cells by causing the formation of chemical bonds between adjacent pyrimidine nucleotides in the DNA. Two pyrimidine bases joined in this way in one strand form what is known as a *dimer*, and of the three possible types of pyrimidine dimer, the thymine dimer, is formed most readily (Fig. 11.32). The presence of such dimers blocks the

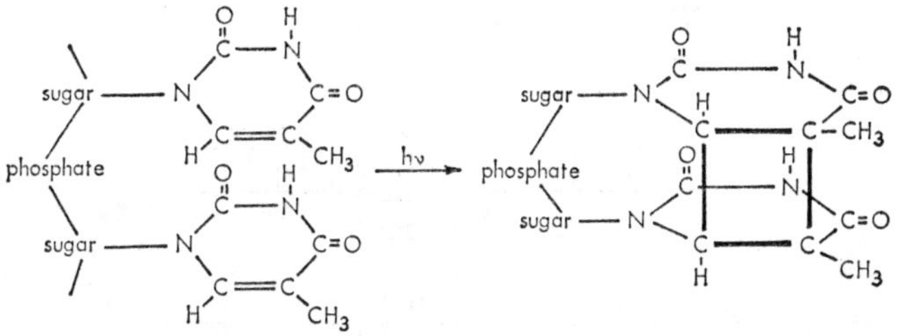

Fig. 11.32 *The formation of a thymine dimer under the influence of ultraviolet light*

action of the DNA polymerase and so prevents replication [350, 358–363]. Damage resulting from exposure to ultraviolet light or to any other chemical or physical agent (e.g. alkylating agents or nitrous acid) must be repaired if the DNA is to retain its genetic integrity.

When bacteria damaged by ultraviolet light are exposed to an intense source of visible light (wavelengths between 320 and 370 nm) a large proportion of the damaged cells recover. This process is known as *photoreactivation* and is due to the activation by visible light of an enzyme which cleaves the pyrimidine dimers and restores the two bases to their original form. This photo-reactivating enzyme has been obtained in pure form [364], and in *Esch. coli* it is the product of the *phr* gene [365].

A second kind of repair mechanism known as *dark reactivation* involves a series of enzymes which act in a more complex way by excising the dimers and then repairing the resulting gaps. It is probable that this mechanism takes place in four separate stages (Fig. 11.33). (i) An endonuclease recognizes the local distortion and breaks the adjoining phosphodiester bond so as to introduce a nick on the 5'-side of the dimer with a 3'-hydroxyl terminus at the nick. (ii) A second enzyme excises a short stretch of the DNA strand including the dimer. (iii) DNA polymerase uses the intact complementary strand as template to synthesize a piece of DNA to fill the gap. (iv) The repair is completed by ligase action [362, 366–368].

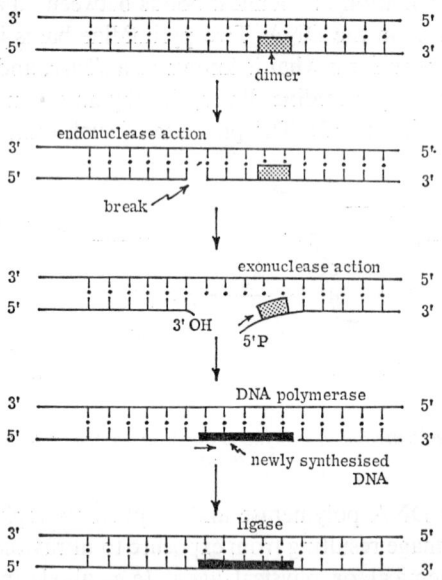

Fig. 11.33 *The repair of DNA damaged by ultraviolet light. The action of u.v. has been to produce a dimer which is excised by the sequential action of an endonuclease and an exonuclease leaving a short gap which is filled in by the action of DNA polymerase. The final join is effected by polynucleotide ligase*

In strains of *Esch. coli* which are uvr^+ the dimers are excised in the form of short oligonucleotides and the products of genes *uvr* A and *uvr* B are implicated in the endonucleolytic step [369]. DNA treated with monofunctional alkylating agents, such as methyl methanesulphonate, loses purine residues, and such damaged sites are recognized by a different endonuclease (endonuclease II in *Esch. coli*) and an incision is made on the 5'-side of the damaged area [396]. DNA polymerase I may be the enzyme which first excises the portion of the affected strand (including, on average, 30 nucleotides [370]) by virtue of its $5' \rightarrow 3'$ nuclease action (p. 277) and then fills the gap [371–373]. *pol* A^- mutants show increased u.v. sensitivity [372, 374] and, although they do manage to repair lesions, the patch size is considerably bigger than in the wild type [376, 377]. This may be because DNA polymerases II and III take over the role of polymerase I or that repair is occurring by a recombination mechanism (see below) [415].

The mechanism of repair synthesis is defective in the skin fibroblasts of patients suffering from the condition known as *xeroderma pigmentosum* [378–382]. The repair endonuclease is missing and such people are therefore abnormally sensitive to exposure to sunlight and tend readily to develop skin cancer.

A third mechanism for the repair of damaged DNA which is of importance in the absence of excision repair is repair by *recombination*. (This process, whose main function is probably in the assortment of genes, also plays a role in the insertion of DNA into the bacterial chromosome during transformation and transduction; p. 223.) In this type of repair a preformed patch supplied by another copy of the DNA is used to replace the defective segment [383]. In *Esch. coli* recombination requires the products of the genes *rec* A, *rec* B, and *rec* C and leads to large repair patches. The *rec* B and *rec* C genes specify a nuclease (exonuclease V) which can act as an endo- or an exo-nuclease [384, 385] and the *rec* A gene may have a control function [386]. The product of *uvr* A is also required as is an unwinding protein, DNA polymerase II or III and DNA ligase [396].

The non-viability of *pol* A *rec* A double mutants and some *pol* A *rec* B double mutants points to the existence of two alternative mechanisms of dark repair of damaged DNA. In the absence of both mechanisms (i.e. in the double mutant) damaged DNA accumulates resulting in cell death.

REFERENCES

[1] Ray, D. S., Duebner, J. and Suggs, S. (1975) *J. Virol.,* **16,** 348
[2] Okazaki, R., Hirose, S., Okazaki, T., Ogawa, T. and Kurosawa, Y. (1975) *Biochem. Biophys. Res. Commun.,* **62,** 1018
[3] Kurosawa, Y., Ogawa, T., Hirose, S., Okazaki, T. and Okazaki, R. (1975) *J. Mol. Biol.,* **96,** 653
[4] Okazaki, R., Okazaki, T., Hirose, S., Sugino, A., Ogawa, T., Kurosawa, Y., Shinozaki, K., Tamanoi, F., Seki, T., Machida, Y., Fujiyama, A. and Kohara, Y. (1975) *DNA Synthesis and its Regulation* (M. Goulian and P. Hanwalt, Eds.), Vol. III (F. Fox, Series Ed.), ICN–UCLA Symposium on Molecular and Cellular Biology. California: Benjamin
[5] Watson, J. D. and Crick, F. H. C. (1953) *Nature,* **171,** 737
[6] Meselson, M. and Stahl, F. W. (1958) *Proc. Nat. Acad. Sci.,* **66,** 671
[7] Cairns, J. (1963) *J. Mol. Biol.,* **6,** 208
[8] Cairns, J. (1966) *Sci. Amer.,* **214**(1), 36

[9] Taylor, J. H. (1963) *Molecular Genetics*, Part I, p. 65 (J. H. Taylor, Ed.). New York: Academic Press
[10] Lovett, M. A., Katz, L. and Helinski, D. R. (1974) *Nature*, **251**, 337
[11] Hirt, B. (1969) *J. Mol. Biol.*, **40**, 141
[12] Wolfson, J. and Dressler, D. (1972) *Proc. Nat. Acad. Sci.*, **69**, 2682
[13] Kriegstein, H. J. and Hogness, D. S. (1974) *Proc. Nat. Acad. Sci.*, **71**, 135
[14] Cairns, J. (1963) *Cold Spring Harbor Symp. Quant. Biol.*, **28**, 43
[15] Schnös, M. and Inman, R. B. (1970) *J. Mol. Biol.*, **51**, 61
[16] Chattoraj, D. K. and Inman, R. B. (1974) *Methods in Molecular Biology*, Vol. 7 (R. B. Wickner, Ed.). New York: Dekker
[17] Fareed, G. C., Garon, C. F. and Satzman, N. P. (1972) *J. Virol.*, **10**, 484
[18] Masters, M. and Broda, P. (1971) *Nature New Biol.*, **232**, 137
[19] Hara, H. and Yoshikawa, H. (1973) *Nature New Biol.*, **244**, 200
[20] Hand, R. and Tamm, I. (1973) *J. Cell Biol.*, **58**, 410
[21] Huberman, J. A. and Tsai, A. (1973) *J. Mol. Biol.*, **75**, 5
[22] Stent, G. S. (1963) *Molecular Biology of Bacterial Viruses*. San Francisco: Freeman
[23] Levinthal, C. and Crane, H. R. (1956) *Proc. Nat. Acad. Sci.*, **42**, 436
[24] Longuet-Higgins, H. C. and Zimm, B. H. (1960) *J. Mol. Biol.*, **2**, 1
[25] Fong, P. (1964) *Proc. Nat. Acad. Sci.*, **52**, 239, 641
[26] Freese, E. B. and Freese, E. (1963) *Biochem.*, **2**, 707
[27] Butler, J. A. V. (1963) *Nature*, **199**, 68
[28] Jaenisch, R., Mayer, A. and Levine, A. (1971) *Nature New Biol.*, **233**, 72
[29] Wang, J. C. (1971) *J. Mol. Biol.*, **55**, 523
[30] Champoux, J. J. and Dulbecco, R. (1972) *Proc. Nat. Acad. Sci.*, **69**, 143
[31] Yoshikawa, H. (1970) *J. Mol. Biol.*, **47**, 403
[32] Stein, G. H. and Hanawalt, P. C. (1972) *J. Mol. Biol.*, **64**, 393
[33] Lark, K. G. (1972) *J. Mol. Biol.*, **64**, 47
[34] Huberman, J. A. and Riggs, A. D. (1968) *J. Mol. Biol.*, **32**, 327
[35] Housman, D. and Huberman, J. A. (1975) *J. Mol. Biol.*, **94**, 173
[36] Lewin, B. (1974) *Gene Expression*, Vol. 2. John Wiley and Son
[37] Woodland, H. R. and Pestell, R. Q. W. (1972) *Biochem. J.*, **127**, 597
[38] Mitra, S. and Kornberg, A. (1966) American Heart Assoc. Symposium, *Macromolecular Metabolism*, p. 59
[39] Price, T. D., Darmstadt, R. A., Hinds, H. A. and Zamenhof, S. (1967) *J. Biol. Chem.*, **242**, 140
[40] Okazaki, R., Okazaki, T., Sakabe, K., Sugimoto, K., Kainuma, R., Sugino, A. and Iwatsuki, N. (1968) *Cold Spring Harbor Symp. Quant. Biol.*, **33**, 129
[41] Okazaki, R., Okazaki, T., Sakabe, K., Sugimoto, K. and Sugino, A. (1968) *Proc. Nat. Acad. Sci.*, **59**, 598
[42] Sugimoto, K., Okazaki, T., Imae, Y. and Okazaki, R. (1969) *Proc. Nat. Acad. Sci.*, **63**, 1343
[43] Gautschi, J. R. and Clarkson, J. M. (1975) *Eur. J. Biochem.*, **50**, 403
[44] Magnussen, G., Pigiet, V., Winnacker, E. L., Abrams, R. and Reichard, P. (1973) *Proc. Nat. Acad. Sci.*, **70**, 412
[45] Hess, U., Dürwald, H. and Hoffmann-Berling, H. (1973) *J. Mol. Biol.*, **73**, 407
[46] Diaz, A. T., Weiner, D. and Werner, R. (1975) *J. Mol. Biol.*, **95**, 45
[47] Konrad, E. B. and Lehman, I. R. (1975) *Proc. Nat. Acad. Sci.*, **72**, 2150
[48] Lark, K. G. and Wechster, J. A. (1975) *J. Mol. Biol.*, **92**, 145
[49] Newman, J. and Hanawalt, P. (1968) *Cold Spring Harbor Symp. Quant. Biol.*, **33**, 145

[50] Sugimoto, K., Okazaki, T. and Okazaki, R. (1968) *Proc. Nat. Acad. Sci.*, **60,** 1356
[51] Okazaki, R., Sugimoto, K., Okazaki, T., Imac, Y. and Sugino, A. (1970) *Nature*, **228,** 223
[52] Okazaki, T. and Okazaki, R. (1969) *Proc. Nat. Acad. Sci.*, **64,** 1242
[53] Sugino, A. and Okazaki, R. (1972) *J. Mol. Biol.*, **64,** 61
[54] Diaz, A. T. and Werner, R. (1975) *J. Mol. Biol.*, **95,** 63
[55] Brutlag, D., Schekman, R. and Kornberg, A. (1971) *Proc. Nat. Acad. Sci.*, **68,** 2826
[56] Wickner, W., Brutlag, D., Schekman, R. and Kornberg, A. (1972) *Proc. Nat. Acad. Sci.*, **69,** 965
[57] Bouché, J. P., Zeckel, K. and Kornberg, A. (1975) *J. Biol. Chem.*, **250,** 5995
[58] Sugino, A., Hirose, S. and Okazaki, R. (1972) *Proc. Nat. Acad. Sci.*, **69,** 1863
[59] Sugino, A. and Okazaki, R. (1973) *Proc. Nat. Acad. Sci.*, **70,** 88
[60] Reichard, P., Eliasson, R. and Söderman, G. (1974) *Proc. Nat. Acad. Sci.*, **71,** 4901
[61] Huberman, J. A. and Horwitz, H. (1973) *Cold Spring Harbor Symp. Quant. Biol.*, **38,** 233
[62] Blair, D. G., Sherratt, D. J., Clewel, D. B. and Helinski, D. R. (1972) *Proc. Nat. Acad. Sci.*, **69,** 2518
[63] Olivera, B. M. and Bonhoeffer, F. (1972) *Nature New Biol.*, **240,** 233
[64] Olivera, B. M., Lark, K. G., Herrmann, R. and Bonhoeffer, F. (1973) *DNA Synthesis* in vitro (R. D. Wells and R. B. Inman, Eds.), p. 214. University Park Press
[65] Wells, R. D. and Inman, R. B. (1973) *DNA Synthesis* in vitro. University Park Press
[66] Wickner, R. B. (1974) *Methods in Molecular Biology*, Vol. 7, *DNA Replication*. New York: Dekker
[67] Vosberg, H. P. and Hoffmann-Berling, H. (1971) *J. Mol. Biol.*, **58,** 739
[68] Moses, R. E. and Richardson, C. C. (1970) *Proc. Nat. Acad. Sci.*, **67,** 674
[69] Schaller, H., Otto, B., Nüsslein, V., Huf, J., Herrmann, R. and Bonhoeffer, F. (1972) *J. Mol. Biol.*, **63,** 183
[70] Schekman, R., Wickner, W., Westergaard, O., Brutlag, D., Geider, K., Bertsch, L. L. and Kornberg, A. (1972) *Proc. Nat. Acad. Sci.*, **69,** 2691
[71] Nusslein, V. and Klein, A. (1974) *Methods in Molecular Biology*, Vol. 7 (R. B. Wickner, Ed.). New York: Dekker
[72] Burgoyne, L. A. (1972) *Biochem. J.*, **130,** 959
[73] Hunter, T. and Francke, B. (1974) *J. Virol.*, **13,** 125
[74] Winnacker, E. L., Magnussen, G. and Reichard, P. (1972) *J. Mol. Biol.*, **72,** 523
[75] De Pamphilis, M. L., Beard, P. and Berg, P. (1975) *J. Biol. Chem.*, **250,** 4340
[76] Hershey, H. V., Stieber, J. F. and Mueller, G. C. (1973) *Eur. J. Biochem.*, **34,** 383
[77] Lynch, W. E., Brown, R. F., Umeda, T., Langreth, S. G. and Lieberman, I. (1970) *J. Biol. Chem.*, **245,** 3911
[78] De Pamphilis, M. L. and Berg, P. (1974) *J. Biol. Chem.*, **250,** 4348
[79] Kornberg, A. (1974) *DNA Synthesis*. San Francisco: Freeman
[80] Geider, K. and Kornberg, A. (1974) *J. Biol. Chem.*, **249,** 3999
[81] Schekman, R., Weiner, A. and Kornberg, A. (1974) *Science*, **186,** 987
[82] Wickner, R. B., Wright, M., Wickner, S. and Hurwitz, J. (1972) *Proc. Nat. Acad. Sci.*, **69,** 3233

THE BIOCHEMISTRY OF THE NUCLEIC ACIDS

[83] Zechel, K., Bouché, J. P. and Kornberg, A. (1975) *J. Biol. Chem.*, **250**, 4684
[84] Inman, R. B., Schildkraut, C. L. and Kornberg, A. (1965) *J. Mol. Biol.*, **11**, 285
[85] Sibatani, A. (1964) *J. Theor. Biol.*, **7**, 588
[86] Crick, F. H. C. (1974) *Proc. Plenary Sessions 6th Internat. Cong. Biochem.*, New York, p. 109
[87] Schildkraut, C. L., Richardson, C. C. and Kornberg, A. (1964) *J. Mol. Biol.*, **9**, 24
[88] Richardson, C. C., Inman, R. B. and Kornberg, A. (1964) *J. Mol. Biol.*, **9**, 46
[89] Lehman, I. R., Bessman, M. J., Simms, E. S. and Kornberg, A. (1958) *J. Biol. Chem.*, **233**, 163
[90] Richardson, C. C., Schildkraut, C. L., Aposhian, H. V. and Kornberg, A. (1964) *J. Biol. Chem.*, **239**, 222
[91] Jovin, T. M., Englund, P. T. and Bertsch, L. L. (1969) *J. Biol. Chem.*, **244**, 2996
[92] Lehman, I. R. (1963) *Methods in Enzymology*, Vol. 6, p. 34 (S. P. Colowick and N. O. Kaplan, Eds.). New York: Academic Press
[93] Bessman, M. J., Lehman, I. R., Simms, E. S. and Kornberg, A. (1958) *J. Biol. Chem.*, **233**, 171
[94] Okazaki, T. and Kornberg, A. (1964) *J. Biol. Chem.*, **239**, 259
[95] Kornberg, A. (1968) *Sci. Amer.*, **219**(4), 64
[96] Goulian, M. and Kornberg, A. (1967) *Proc. Nat. Acad. Sci.*, **58**, 1723
[97] Goulian M., Kornberg, A. and Sinsheimer, R. L. (1967) *Proc. Nat. Acad. Sci.*, **58**, 2321
[98] Goulian, M. (1968) *Cold Spring Harbor Symp. Quant. Biol.*, **33**, 11
[99] Wagner, R. P. and Mitchell, H. K. (1964) *Genetics and Metabolism* (2nd Edition). London: Wiley
[100] Kornberg, A. (1961) *The Enzymatic Synthesis of DNA*. London: Wiley
[101] Josse, J. (1962) *The Molecular Basis of Neoplasia*, p. 91. University of Texas Press
[102] Josse, J., Kaiser, A. D. and Kornberg, A. (1961) *J. Biol. Chem.*, **236**, 864
[103] Becker, A. and Hurwitz, J. (1971) *Progr. Nucleic Acid Res. Mol. Biol.* (J. N. Davidson and W. E. Cohn, Eds.), **11**, 423
[104] Chargaff, E., Buchowicz, J., Türler, H. and Shapiro, H. S. (1965) *Nature*, **206**, 145
[105] Griffith, J., Huberman, J. A. and Kornberg, A. (1971) *Proc. Nat. Acad. Sci.*, **55**, 209
[106] Setlow, O., Brutlag, D. and Kornberg, A. (1972) *J. Biol. Chem.*, **247**, 224
[107] Setlow, P. and Kornberg, A. (1972) *J. Biol. Chem.*, **247**, 232
[108] Brutlag, D. and Kornberg, A. (1972) *J. Biol. Chem.*, **247**, 241
[109] Kornberg, A. (1969) *Science*, **163**, 1410
[110] Schachman, H. K., Adler, J., Radding, C. M., Lehman, I. R. and Kornberg, A. (1960) *J. Biol. Chem.*, **235**, 3242
[111] Byrd, C., Ohtsuka, E., Moon, M. W. and Khorana, H. G. (1965) *Proc. Nat. Acad. Sci.*, **53**, 79
[112] Khorana, H. G. (1965) *Fed. Proc.*, **24**, 1473
[113] Khorana, H. G., Jacob, H., Moon, T. M., Kossel, H., Narang, S. A. and Ohtsuka, E. (1967) *J. Amer. Chem. Soc.*, **89**, 2154
[114] Schwartz, M. N., Trantner, A. and Kornberg, A. (1962) *J. Biol. Chem.*, **237**, 1961

[115] Kornberg, A., Bertsch, L., Jackson, J. F. and Khorana, H. G. (1964) *Proc. Nat. Acad. Sci.,* **51,** 315
[116] Kornberg, A. (1965) *Evolving Genes and Proteins,* p. 403 (V. Bryson and H. J. Vogel, Eds.). New York: Academic Press
[117] Gupta, N. K., Ohtsuka, E., Sgaramella, V., Büchi, H., Kumar, A., Weber, H. and Khorana, H. G. (1968) *Proc. Nat. Acad. Sci.,* **60,** 1338
[118] Cozzarelli, N. R., Kelly, R. B. and Kornberg, A. (1969) *J. Mol. Biol.,* **45,** 513
[119] Kelly, R. B., Cozzarelli, N. R., Deutscher, M. P., Lehman, I. R. and Kornberg, A. (1970) *J. Biol. Chem.,* **245,** 39
[120] de Lucia, P. and Cairns, J. (1969) *Nature,* **224,** 1164
[121] Gross, J. and Gross, M. (1969) *Nature,* **224,** 1166
[122] Okazaki, R., Arisawa, M. and Sugino, A. (1971) *Proc. Nat. Acad. Sci.,* **68,** 2954
[123] Kato, T. and Konda, S. (1970) *J. Bact.,* **104,** 871
[124] Smirnov, G. B., Favorskaya, Y. N. and Skavronskaya, A. G. (1971) *Mol. Gen. Genetics,* **111,** 357
[125] Monk, M. and Kinross, J. (1972) *J. Bact.,* **109,** 971
[126] Strike, P. and Emmerson, P. T. (1972) *Mol. Gen. Genetics,* **116,** 177
[127] Emmerson, P. T. and Strike, P. (1973) *DNA Replication and the Cell Membrane* (A. Kolber and M. Kohiyama, Eds.)
[128] Kornberg, T. and Gefter, M. L. (1971) *Proc. Nat. Acad. Sci.,* **68,** 761
[129] Kornberg, T. and Gefter, M. L. (1972) *J. Biol. Chem.,* **247,** 5369
[130] Otto, B., Bonhoeffer, F. and Schaller, H. (1973) *Eur. J. Biochem.,* **34,** 440
[131] Wickner, R. B., Ginsburg, B., Berkower, I. and Hurwitz, J. (1972) *J. Biol. Chem.,* **247,** 489
[132] Sigal, N., Delius, H., Kornberg, T., Gefter, M. L. and Alberts, B. (1972) *Proc. Nat. Acad. Sci.,* **69,** 3537
[133] Rama Reddy, G. V., Goulian, M. and Hendler, S. S. (1971) *Nature New Biol.,* **234,** 286
[134] Gefter, M. L. (1974) *Progr. Nucleic Acid Res. Mol. Biol.,* **14,** 101
[135] Tait, R. C. and Smith, D. W. (1974) *Nature,* **249,** 116
[136] Gefter, M. L., Hirota, Y., Kornberg, T., Wechsler, S. A. and Barnoux, C. (1971) *Proc. Nat. Acad. Sci.,* **68,** 3150
[137] Klein, A., Nusslein, V., Otto, B. and Powling, A. (1973) *DNA Synthesis in vitro,* p. 185 (R. D. Wells and R. B. Inman, Eds.). University Park Press
[138] Otto, B. (1973) *Biochem. Soc. Trans.,* **1,** 629
[139] Wickner, W., Schekman, R., Geider, K. and Kornberg, A. (1973) *Proc. Nat. Acad. Sci.,* **70,** 1764
[140] Wickner, W. and Kornberg, A. (1974) *J. Biol. Chem.,* **249,** 6244
[141] Hurwitz, J., Wickner, S. and Wright, M. (1973) *Biochem. Biophys. Res. Commun.,* **51,** 257
[142] Hurwitz, J. and Wickner, S. (1974) *Proc. Nat. Acad. Sci.,* **71,** 6
[143] Gass, K. B. and Cozzarelli, N. R. (1974) *Methods in Enzymology,* Vol. XXIX, p. 17
[144] Fansler, B. S. (1974) *Internat. Rev. Cytology,* Suppl. 4, 363
[145] Craig, R. K. and Keir, H. M. (1974) *The Cell Nucleus,* Vol. III (H. Busch, Ed.), p. 35. Academic Press
[146] Bollum, F. J. (1975) *Progr. Nucleic Acid Res. Mol. Biol.,* **15,** 109
[147] Smellie, R. M. S., Keir, H. M. and Davidson, J. N. (1959) *Biochim. Biophys. Acta,* **35,** 389
[148] Shepherd, J. B. and Keir, H. M. (1966) *Biochem. J.,* **99,** 443

[149] Chang, L. M. S. and Bollum, F. J. (1972) *J. Biol. Chem.*, **247**, 7948
[150] deRecondo, A. M., Lepesant, J. A., Fichot, O., Grasset, L., Rossignol, J. M. and Cazillis, M. (1973) *J. Biol. Chem.*, **248**, 131
[151] Holmes, A. M., Hesslewood, I. P. and Johnston, I. R. (1974) *Eur. J. Biochem.*, **43**, 487
[152] Ove, P., Jenkins, M. D. and Laszlo, J. (1970) *Cancer Res.*, **30**, 535
[153] Baril, E. F., Jenkins, M. D., Brown, O. E., Laszlo, J. and Morris, H. P. (1973) *Cancer Res.*, **33**, 1187
[154] Ove, P., Coetzee, M. L. and Morris, H. P. (1973) *Cancer Res.*, **33**, 1272
[155] Iwamura, Y., Ono, T. and Morris, H. P. (1968) *Cancer Res.*, **28**, 2466
[156] Bollum, F. J. (1963) *Progress in Nucleic Acid Research*, Vol. 1, p. 1 (J. N. Davidson and W. E. Cohn, Eds.). New York: Academic Press
[157] Yoneda, M. and Bollum, F. J. (1965) *J. Biol. Chem.*, **240**, 3385
[158] Chang, L. M. S. and Bollum, F. J. (1973) *J. Biol. Chem.*, **248**, 3398
[159] Bollum, F. J., Chang, L. M. S., Tsiapalis, C. M. and Dorson, J. W. (1974) *Methods in Enzymology*, Vol. XXIX, p. 70 (L. Grossman and K. Moldave, Eds.)
[160] Loeb, L. A. and Agarwal, S. S. (1971) *Exp. Cell Res.*, **66**, 299
[161] Smith, R. G. and Gallo, R. C. (1972) *Proc. Nat. Acad. Sci.*, **69**, 2879
[162] Zain, B. S., Adams, R. L. P. and Imrie, R. C. (1973) *Cancer Res.*, **33**, 40
[163] Chiu, J. F. and Sung, S. C. (1972) *Biochim. Biophys. Acta*, **262**, 397
[164] Stambolova, M. A., Cox, D. and Mathias, A. P. (1974) *Biochem. J.*, **140**, 65
[165] Gold, M. and Helleiner, C. W. (1964) *Biochim. Biophys. Acta*, **80**, 193
[166] Adams, R. L. P., Henderson, M. A. L., Wood, W. and Lindsay, J. G. (1973) *Biochem. J.*, **131**, 237
[167] Craig, R. K. and Keir, H. M. (1975) *Biochem. J.*, **145**, 215
[168] Sedwick, W. D., Wang, T. S. F. and Korn, D. (1972) *J. Biol. Chem.*, **247**, 5026
[169] Weissbach, A., Schlabach, A., Fridlander, B. and Bolden, A. (1971) *Nature New Biol.*, **231**, 167
[170] Chang, L. M. S., Brown, M. and Bollum, F. J. (1973) *J. Mol. Biol.*, **74**, 1
[171] Brun, G., Rougeon, F., Lauber, M. and Chapeville, F. (1974) *Eur. J. Biochem.*, **41**, 241
[172] Loeb, L. A. and Fansler, B. (1970) *Biochim. Biophys. Acta*, **217**, 50
[173] McLennan, A. G. and Keir, H. M. (1975) *Nucleic Acid Research*, **2**, 223
[174] Lindsay, J. G., Berryman, S. and Adams, R. L. P. (1970) *Biochem. J.*, **119**, 839
[175] Wang, T. S. F., Sedwick, W. D. and Korn, D. (1974) *J. Biol. Chem.*, **249**, 841
[176] Chang, L. M. S. (1975) *Methods in Enzymology*, Vol. XXIX, p. 81 (L. Grossman and K. Moldave, Eds.)
[177] Aposhian, H. V. and Kornberg, A. (1962) *J. Biol. Chem.*, **237**, 519
[178] Loeb, L. A. (1969) *J. Biol. Chem.*, **244**, 1672
[179] Chang, L. M. S. (1973) *J. Biol. Chem.*, **248**, 6983
[180] Wickremasinghe, R. G. and Johnston, I. R. (1974) *Biochim. Biophys. Acta*, **361**, 37
[181] Chang, L. M. S. (1974) *J. Biol. Chem.*, **249**, 7441
[182] Spadari, S. and Weissbach, A. (1975) *Proc. Nat. Acad. Sci.*, **72**, 503
[183] Spadari, S. and Weissbach, A. (1974) *J. Mol. Biol.*, **86**, 11
[184] Spadari, S. and Weissbach, A. (1974) *J. Biol. Chem.*, **249**, 5809
[185] Ter Schegget, J. and Borst, P. (1971) *Biochim. Biophys. Acta*, **246**, 239
[186] Ter Schegget, J. and Borst, P. (1971) *Biochim. Biophys. Acta*, **246**, 249

[187] Ter Schegget, J., Flavell, E. A. and Borst, P. (1971) *Biochim. Biophys. Acta,* **254,** 1
[188] Kasamatsu, H., Robberson, D. L. and Vinograd, J. (1971) *Proc. Nat. Acad. Sci.,* **68,** 2252
[189] Tewari, K. K. and Wildman, S. G. (1967) *Proc. Nat. Acad. Sci.,* **58,** 689
[190] Spencer, D. and Whitfield, P. R. (1967) *Biochem. Biophys. Res. Commun.,* **28,** 538
[191] Meyer, R. R. and Simpson, M. V. (1970) *J. Biol. Chem.,* **245,** 3426
[192] Fry, M. and Weissbach, A. (1973) *Biochem.,* **12,** 3602
[193] Krakow, J. S., Coutsogeorgopoulos, C. and Canellakis, E. S. (1962) *Biochim. Biophys. Acta,* **55,** 639
[194] Keir, H. M. and Smith, M. J. (1963) *Biochim. Biophys. Acta,* **68,** 589
[195] Keir, H. M., Shepherd, J. B. and Hay, J. (1963) *Biochem. J.,* **89,** 9P
[196] Bollum, F. (1965) *J. Biol. Chem.,* **240,** 2599
[197] Chang, L. M. S. and Bollum, F. J. (1971) *J. Biol. Chem.,* **246,** 909
[198] Chang, L. M. S. and Bollum, F. J. (1970) *Proc. Nat. Acad. Sci.,* **65,** 1041
[199] Chang, L. M. S. (1971) *Biochem. Biophys. Res. Commun.,* **44,** 124
[200] Srivastava, B. I. S. (1974) *Cancer Res.,* **34,** 1015
[201] Srivastava, B. I. S. (1975) *Res. Commun. Chem. Path. Pharm.,* **10,** 715
[202] Gellert, M. (1967) *Proc. Nat. Acad. Sci.,* **57,** 48
[203] Weiss, B. and Richardson, C. C. (1967) *Proc. Nat. Acad. Sci.,* **57,** 1021
[204] Olivera, B. M. and Lehman, I. R. (1967) *Proc. Nat. Acad. Sci.,* **57,** 1426
[205] Becker, A., Lyn, G., Gefter, M. and Hurwitz, J. (1967) *Proc. Nat. Acad. Sci.,* **58,** 1966
[206] Little, J. W., Zimmerman, S. B., Oshinsky, C. K. and Gellert, M. (1967) *Proc. Nat. Acad. Sci.,* **58,** 2004
[207] Spadari, S., Ciarrocchi, G. and Falaschi, A. (1971) *Eur. J. Biochem.,* **22,** 75
[208] Howell, S. H. and Hecht, N. B. (1971) *Biochim. Biophys. Acta,* **240,** 343
[209] Zimmerman, S. B., Little, J. W., Oshinsky, C. K. and Gellert, M. (1967) *Proc. Nat. Acad. Sci.,* **57,** 1841
[210] Laipis, P. J., Oliver, B. M. and Ganesan, A. T. (1969) *Proc. Nat. Acad. Sci.,* **62,** 289
[211] Lindahl, T. and Edelman, G. M. (1968) *Proc. Nat. Acad. Sci.,* **61,** 680
[212] Tsukada, K. and Ichimura, M. (1971) *Biochem. Biophys. Res. Commun.,* **42,** 1156
[213] Richardson, C. C., Masamura, Y., Live, T. R., Jacquemin-Sablon, A., Weiss, B. and Fareed, G. C. (1968) *Cold Spring Harbor Symp. Quant. Biol.,* **33,** 151
[214] Olivera, B. M., Hall, Z. W., Anraku, Y., Chien, J. R. and Lehman, I. R. (1968) *Cold Spring Harbor Symp. Quant. Biol.,* **33,** 27
[215] Gellert, M., Little, J. W., Oshinsky, C. K. and Zimmerman, S. B. (1968) *Cold Spring Harbor Symp. Quant. Biol.,* **33,** 21
[216] Kleppe, K., Van de Sande, J. H. and Khorana, H. G. (1970) *Proc. Nat. Acad. Sci.,* **67,** 68
[217] Weiss, B., Jacquemin-Sablon, A., Live, T. R., Fareed, G. C. and Richardson, C. C. (1968) *J. Biol. Chem.,* **243,** 4543
[218] Gupta, N. K., Ohtsuka, E., Weber, H., Chang, S. H. and Khorana, H. G. (1968) *Proc. Nat. Acad. Sci.,* **60,** 285
[219] Gefter, M. L., Becker, A. and Hurwitz, J. (1967) *Proc. Nat. Acad. Sci.,* **58,** 240
[220] Goulian, M. (1971) *Ann. Rev. Biochem.,* **40,** 855
[221] Pauling, C. and Hamm, L. (1968) *Proc. Nat. Acad. Sci.,* **60,** 1495

[222] Gellert, M. and Bullock, M. L. (1970) *Proc. Nat. Acad. Sci.*, **67**, 1580
[223] Pauling, C. and Hamm, L. (1969) *Proc. Nat. Acad. Sci.*, **64**, 1195
[224] Konrad, E. B., Modrich, P. and Lehman, I. R. (1973) *J. Mol. Biol.*, **77**, 519
[225] Gottesman, M. M., Hicks, M. L. and Gellert, M. (1973) *J. Mol. Biol.*, **77**, 531
[226] Baldy, M. W. (1968) *Cold Spring Harbor Symp. Quant. Biol.*, **33**, 333
[227] Alberts, B. M. and Frey, L. (1970) *Nature*, **227**, 1313
[228] Weiner, J. H., Bertsch, L. L. and Kornberg, A. (1975) *J. Biol. Chem.*, **250**, 1972
[229] Huberman, J. A., Kornberg, A. and Alberts, B. M. (1971) *J. Mol. Biol.*, **62**, 39
[230] Reuben, R. C. and Gefter, M. L. (1973) *Proc. Nat. Acad. Sci.*, **70**, 1846
[231] Geider, K. and Kornberg, A. (1974) *J. Biol. Chem.*, **249**, 3999
[232] Barry, J. and Alberts, B. M. (1972) *Proc. Nat. Acad. Sci.*, **69**, 2717
[233] Hotta, Y. and Stern, H. (1971) *Nature New Biol.*, **234**, 83
[234] Salas, J. and Green, H. (1971) *Nature New Biol.*, **229**, 165
[235] Banks, G. R. and Spanos, A. (1975) *J. Mol. Biol.*, **93**, 63
[236] Wang, J. C. (1971) *J. Mol. Biol.*, **55**, 523
[237] Bauer, W. and Vinograd, J. (1970) *J. Mol. Biol.*, **47**, 419
[238] Barzilai, R. and Thomas, C. A. (1970) *J. Mol. Biol.*, **51**, 145
[239] Burger, R. M. (1971) *J. Mol. Biol.*, **56**, 199
[240] Haskell, E. H. and Davern, C. I. (1967) *Proc. Nat. Acad. Sci.*, **64**, 1065
[241] Schekman, R. and Kornberg, A. (1971) *J. Mol. Biol.*, **61**, 471
[242] Geider, K. and Hoffmann-Berling, H. (1971) *Eur. J. Biochem.*, **21**, 374
[243] Kasamatsu, H., Robberson, D. L. and Vinograd, J. (1971) *Proc. Nat. Acad. Sci.*, **68**, 2252
[244] Inman, R. B. and Schnös, M. (1973) *DNA Synthesis* in vitro, p. 437 (R. D. Wells and R. B. Inman, Eds.)
[245] Gilbert, W. (1968) *Cold Spring Harbor Symp. Quant. Biol.*, **33**, 473
[246] Dressler, D. (1971) *Proc. Nat. Acad. Sci.*, **67**, 1934
[247] Dressler, D. and Wolfson, J. (1970) *Proc. Nat. Acad. Sci.*, **67**, 456
[248] Takahashi, S. (1975) *J. Mol. Biol.*, **94**, 385
[249] Hourcade, D., Dressler, D. and Wolfson, J. (1973) *Proc. Nat. Acad. Sci.*, **70**, 2926
[250] Carter, B. J., Shaw, B. D. and Smith, M. G. (1969) *Biochim. Biophys. Acta*, **195**, 494
[251] Watson, J. D. (1972) *Nature New Biol.*, **239**, 197
[252] Kornberg, A., Zimmerman, S. B., Kornberg, S. R. and Josse, J. (1959) *Proc. Nat. Acad. Sci.*, **45**, 772
[253] Somerville, R., Ebisuyaki, K. and Greenberg, G. R. (1959) *Proc. Nat. Acad. Sci.*, **45**, 1240
[254] Mathews, C. K., Brown, F. and Cohen, S. S. (1964) *J. Biol. Chem.*, **239**, 2957
[255] Zimmerman, S. B. and Kornberg, A. (1961) *J. Biol. Chem.*, **236**, 1480
[256] Kornberg, S. R., Zimmerman, S. B. and Kornberg, A. (1961) *J. Biol. Chem.*, **236**, 1487
[257] Zimmerman, S. B., Kornberg, S. R. and Kornberg, A. (1962) *J. Biol. Chem.*, **237**, 512
[258] Josse, J. and Kornberg, A. (1962) *J. Biol. Chem.*, **237**, 1968
[259] Orr, C. W. M., Herriott, S. T. and Bessman, M. J. (1965) *J. Biol. Chem.*, **240**, 4652

[260] Goulian, M., Lucas, Z. J. and Kornberg, A. (1968) *J. Biol. Chem.*, **243**, 627
[261] Prashad, N. and Hosoda, J. (1972) *J. Mol. Biol.*, **70**, 617
[262] Lehman, I. R. (1975) *Methods in Enzymology*, Vol. XXIX, p. 46 (L. Grossman and K. Moldave, Eds.)
[263] Keir, H. M. (1968) *The Molecular Biology of Viruses*, Eighteenth Symposium of The Society for General Microbiology, p. 67 (L. V. Crawford and M. G. P. Stoker, Eds.)
[264] Subak-Sharpe, H. (1968) *The Molecular Biology of Viruses*, Eighteenth Symposium of The Society for General Microbiology, p. 47 (L. V. Crawford and M. G. P. Stoker, Eds.)
[265] Weissbach, A., Hong, S. C. L., Aucker, J. and Muller, R. (1973) *J. Biol. Chem.*, **248**, 6270
[266] Hayashi, M., Hayashi, M. N. and Spiegelman, S. (1963) *Science*, **140**, 1313
[267] Hayashi, M., Hayashi, M. N. and Spiegelman, S. (1964) *Proc. Nat. Acad. Sci.*, **51**, 351
[268] Chandler, B., Hayashi, M., Hayashi, M. N. and Spiegelman, S. (1964) *Science*, **143**, 47
[269] Kleinschmidt, A. K., Burton, A. and Sinsheimer, R. L. (1963) *Science*, **142**, 961
[270] Pratt, D., Laws, P. and Griffith, J. (1974) *J. Mol. Biol.*, **82**, 425
[271] Fidanian, H. M. and Ray, D. S. (1974) *J. Mol. Biol.*, **83**, 63
[272] Cavalieri, L. F. and Carroll, E. (1970) *Biochem. Biophys. Res. Commun.*, **41**, 1055
[273] Temin, H. M. and Mizutani, S. (1970) *Nature*, **226**, 1211
[274] Baltimore, D. (1970) *Nature*, **226**, 1209
[275] Spiegelman, S., Burny, A., Das, M. R., Kaydar, J., Schlom, J., Travnicek, M. and Watson, K. (1970) *Nature*, **227**, 563
[276] Spiegelman, S., Burny, A., Das, M. R., Keydar, J., Schlom, J., Travnicek, M. and Watson, K. (1970) *Nature*, **227**, 1029
[277] Spiegelman, S., Burny, A., Das, M., Keydar, J., Schlom, J., Travnicek, M. and Watson, K. (1970) *Nature*, **228**, 430
[278] Crick, F. (1970) *Nature*, **227**, 561
[279] Erikson, E. and Erikson, R. L. (1971) *J. Virol.*, **8**, 254
[280] Harada, F., Sawyer, R. C. and Dahlberg, J. E. (1975) *J. Biol. Chem.*, **250**, 3487
[281] Baltimore, D. and Smoler, D. (1971) *Proc. Nat. Acad. Sci.*, **68**, 1507
[282] Parks, W. P., Todaro, G. J., Scolnick, E. M. and Aaronson, S. A. (1971) *Nature*, **229**, 258
[283] Lee-Huang, S. and Cavalieri, L. F. (1963) *Proc. Nat. Acad. Sci.*, **50**, 1116
[284] Stone, L. B., Scolnick, E., Takemoto, K. K. and Aaronson, S. A. (1971) *Nature*, **229**, 257
[285] Gallo, R. C. and Sarin, P. S. (1971) *Nature*, **232**, 140
[286] Schlom, J., Harter, D. H., Burny, A. and Spiegelman, S. (1971) *Proc. Nat. Acad. Sci.*, **68**, 182
[287] Schlom, J. and Spiegelman, S. (1971) *Proc. Nat. Acad. Sci.*, **68**, 1613
[288] Scolnick, E. M., Aaronson, S. A., Todaro, G. J. and Parks, W. P. (1971) *Nature*, **229**, 318
[289] Spiegelman, S., Watson, K. B. and Kacian, D. L. (1971) *Proc. Nat. Acad. Sci.*, **68**, 2843
[290] Wolstenholm, G. E. W. and O'Connor, M. (Eds.) (1971) Ciba Foundation Symposium: *The Strategy of the Viral Genome*. London: Churchill, Livingstone

[291] Spiegelman, S. (1971) *Proc. Roy. Soc.,* B, **177,** 87
[292] Scolnick, E. M. (1972) *Current Topics in Biochemistry* (C. B. Anfinsen, R. F. Goldberger and A. N. Schechter, Eds.), NIH Lectures on Biomedical Sciences. Academic Press
[293] Sarin, P. S. and Gallo, R. C. (1973) *Internat. Rev. of Science Series in Biochemistry,* Vol. 6, p. 219 (K. Burton, Ed.). London: Butterworth; Oxford: Medical and Technical Publications
[294] Green, M. and Gerard, G. F. (1974) *Progr. Nucleic Acid Res. Mol. Biol.,* **14,** 187 (W. E. Cohn, Ed.)
[295] Gillespie, D., Saxinger, W. C. and Gallo, R. C. (1975) *Progr. Nucleic Acid Res. Mol. Biol.,* **15,** 1 (W. E. Cohn, Ed.)
[296] Freeze, E. (1963) *Molecular Genetics,* Part 1, p. 207 (J. H. Taylor, Ed.). New York: Academic Press
[297] Pauling, L. (1964) *Bull. N.Y. Acad. Med.,* **40,** 334
[298] Krieg, D. R. (1963) *Progress in Nucleic Acid Research,* Vol. 2, p. 125 (J. N. Davidson and W. E. Cohn, Eds.). New York: Academic Press
[299] Neuss, N., Gorman, M. and Johnson, I. S. (1967) *Methods in Cancer Res.,* Vol. III, p. 633 (H. Busch, Ed.)
[300] Singer, B. (1975) *Progr. Nucleic Acid. Res. Mol. Biol.,* **15,** 219 (W. E. Cohn, Ed.). New York: Academic Press
[301] Bennett, L. L. and Montgomery, J. A. (1967) *Methods in Cancer Res.,* Vol. III, p. 549 (H. Busch, Ed.)
[302] Fox, J. J., Watanabe, K. A. and Bloch, A. (1966) *Progr. Nucleic Acid. Res. Mol. Biol.,* **5,** 251 (J. N. Davidson and W. E. Cohn, Eds.)
[303] Stock, J. A. (1975) *Biology of Cancer* (2nd Ed.), p. 279 (E. J. Ambrose and F. S. C. Roe, Eds.). Ellis Horwood Ltd.
[304] Balis, M. E. (1968) *Antagonists and Nucleic Acids,* Frontiers of Biology, Vol. 10. North Holland Publ. Co.
[305] Berenblum, I. (1974) *Carcinogenesis as a Biological Problem,* Frontiers of Biology, Vol. 34, North Holland Publ. Co.
[306] Lawley, P. D. (1966) *Progress in Nucleic Acid Research,* Vol. 5, p. 89 (J. N. Davidson and W. E. Cohn, Eds.). New York: Academic Press
[307] Sobell, H. M. (1973) *Progr. Nucleic Acid Res. Mol. Biol.,* **13,** 153 (J. N. Davidson and W. E. Cohn, Eds.)
[308] Hitchings, G. H. and Elson, G. B. (1963) *Metabolic Inhibitors,* p. 215 (R. M. Hochster and J. H. Quaster, Eds.). New York: Academic Press
[309] Brockman, R. W. and Anderson, E. P. (1963) *Metabolic Inhibitors,* p. 229 (R. M. Hochster and J. H. Quastel, Eds.). New York: Academic Press
[310] Lasnitski, J., Matthews, R. E. F. and Smith, J. D. (1954) *Nature,* **173,** 346
[311] Hurwitz, J. and Furth, J. J. (1962) *Sci. Amer.,* **206,** 3
[312] Lipmann, F. (1963) *Progress in Nucleic Acid Research,* Vol. 1, p. 135 (J. N. Davidson and W. E. Cohn, Eds.). New York: Academic Press
[313] Loeb, L. A., Fansler, B., Williams, R. and Mazia, D. (1969) *Exp. Cell Res.,* **57,** 298
[314] Heidelberger, C. (1965) *Progr. Nucleic Acid Res. Mol. Biol.,* **4,** 1 (J. N. Davidson and W. E. Cohn, Eds.)
[315] Priest, J. H., Heady, J. E. and Priest, R. E. (1967) *J. Cell Biol.,* **35,** 483
[316] Priest, J. H., Heady, J. E. and Priest, R. E. (1967) *J. Nat. Cancer Inst.,* **38,** 61
[317] Matthews, R. E. F. (1953) *Nature,* **171,** 1065
[318] Hilmoe, R. J. and Heppel, L. A. (1957) *J. Amer. Chem. Soc.,* **79,** 4810
[319] Adams, R. L. P., unpublished results

[320] Roy-Bowman, P. (1970) *Analogues of Nucleic Acid Components.* New York: Springer Verlag
[321] Cohen, S. S. (1966) *Progr. Nucleic Acid Res. Mol. Biol.*, **5**, 1 (J. N. Davidson and W. E. Cohn, Eds.)
[322] Hunter, T. and Francke, B. (1975) *J. Virol.*, **15**, 759.
[323] Lawley, P. D. and Brookes, P. (1961) *Nature*, **192**, 1081
[324] Goldberg, I. H. and Friedman, P. A. (1971) *Ann. Rev. Biochem.*, **40**, 775
[325] Reich, E. and Goldberg, I. H. (1964) *Progress in Nucleic Acid Research*, Vol. 3, p. 184 (J. N. Davidson and W. E. Cohn, Eds.). New York: Academic Press
[326] Keir, H. M., Omura, H. and Shepherd, J. B. (1963) *Biochem. J.*, **89**, 425
[327] Goldberg, I. H., Reich, E. and Rabinowitz, M. (1963) *Nature*, **199**, 44
[328] Reich, E. (1963) *Cancer Res.*, **23**, 1428
[329] Kahan, E., Kahan, F. M. and Hurwitz, J. (1963) *J. Biol. Chem.*, **238**, 2491
[330] Kit, S., Pickarski, L. J. and Dubbs, D. R. (1963) *J. Mol. Biol.*, **7**, 497
[331] Hurwitz, J. and August, J. T. (1963) *Progress in Nucleic Acid Research*, Vol. 1, p. 59 (J. N. Davidson and W. E. Cohn, Eds.). New York: Academic Press
[332] Peniman, S., Vesco, C. and Penman, M. (1968) *J. Mol. Biol.*, **34**, 49
[333] Iyer, V. N. and Szybalski, W. (1963) *Proc. Nat. Acad. Sci.*, **50**, 355
[334] Matsumoto, I. and Lark, K. G. (1963) *Exp. Cell Res.*, **32**, 192
[335] Sekiguchi, M. and Takagi, Y. (1960) *Biochim. Biophys. Acta*, **41**, 434
[336] Goldberg, I. H. and Friedman, P. A. (1971) *Ann. Rev. Biochem.*, **40**, 775
[337] Pietsch, P. and Garrett, H. (1968) *Nature*, **219**, 488
[338] Falaschi, A. and Kornberg, A. (1964) *Fed. Proc.*, **23**, 940
[339] Stern, R., Rose, J. A. and Friedman, R. M. (1974) *Biochem.*, **13**, 307
[340] Mizuno, N. S. (1965) *Biochim. Biophys. Acta*, **108**, 394
[341] Hurwitz, J., Furth, J. J., Malamy, M. and Alexander, M. (1962) *Proc. Nat. Acad. Sci.*, **48**, 1222
[342] Lerman, L. S. (1964) *J. Cell. Comp. Physiol.*, **64**, Suppl. 1, 1
[343] Radloff, R., Bawer, W. and Vinograd, J. (1967) *Proc. Nat. Acad. Sci.*, **57**, 1514
[344] Spear, F. G. (1958) *Internat. Rev. Cytol.*, **7**, 1
[345] Errera, M. (1959) *The Cell*, Vol. 1, p. 695 (J. Brachet and A. E. Mirsky, Eds.). New York: Academic Press
[346] Kanazir, D. (1969) *Progr. Nucleic Acid Res. Mol. Biol.*, **9**, 117 (J. N. Davidson and W. E. Cohn, Eds.). New York: Academic Press
[347] Ord, M. G. and Stocken, L. A. (1968) *Proc. Roy. Soc. Edin.*, B, **70**, 117
[348] Shooter, K. V. (1967) *Progr. Biophys. Mol. Biol.*, **17**, 291
[349] Van Lancker, J. L. (1962) *Fed. Proc.*, **21**, 1118
[350] Setlow, R. B. (1967) *Regulation of Nucleic Acid and Protein Biosynthesis* (V. V. Koningsberger and L. Bosch, Eds.), **10**, 51
[351] Pelc, S. R. and Howard, A. (1955) *Radiation Res.*, **3**, 135
[352] Terasima, T. and Tolmach, L. J. (1963) *Science*, **140**, 490
[353] Sinclair, W. K. (1967) *Proc. Nat. Acad. Sci.*, **58**, 115
[354] Lajtha, L. G., Oliver, R. and Ellis, F. (1954) *Brit. J. Cancer*, **8**, 367
[355] Lajtha, L. G. (1960) *The Nucleic Acids*, Vol. 3, p. 527 (E. Chargaff and J. N. Davidson, Eds.). New York: Academic Press
[356] Sibatani, A. (1957) *Biochim. Biophys. Acta*, **25**, 592
[357] Puck, T. T. and Steffen, J. (1963) *Biophys. J.*, **3**, 379
[358] Rasmussen, R. E. and Painter, R. B. (1964) *Nature*, **203**, 1360

[359] Setlow, R. B. (1968) *Progress in Nucleic Acid Research and Molecular Biology*, Vol. 9, p. 257 (J. N. Davidson and W. E. Cohn, Eds.). New York: Academic Press
[360] Howard-Flanders, P. (1968) *Ann. Rev. Biochem.*, **37**, 175
[361] Pettijohn, D. and Hanawalt, P. (1964) *J. Mol. Biol.*, **9**, 395
[362] Hanawalt, P. C. (1972) *Endeavour*, **31**, 83
[363] Sueoka, N. (1967) *Molecular Genetics*, Part II, p. 1 (J. H. Taylor, Ed.). New York: Academic Press
[364] Sutherland, B. M., Chamberlin, M. J. and Sutherland, J. C. (1973) *J. Biol. Chem.*, **248**, 4200
[365] Setlow, J. K. and Setlow, R. B. (1963) *Nature*, **197**, 560
[366] Hornsey, S. and Howard, A. (1956) *Ann. N.Y. Acad. Sci.*, **63**, 915
[367] Kushner, S. R., Kaplan, J. C., Ono, H. and Grossman, L. (1971) *Biochem.*, **10**, 3325
[368] Kaplan, J. C., Kushner, S. R. and Grossman, L. (1971) *Biochem.*, **10**, 3315
[369] Braun, A. and Grossman, L. (1974) *Proc. Nat. Acad. Sci.*, **71**, 1838
[370] Setlow, R. B. and Carrier, W. L. (1964) *Proc. Nat. Acad. Sci.*, **51**, 226
[371] Kelly, R. B., Atkinson, M. R., Huberman, J. A. and Kornberg, A. (1969) *Nature*, **224**, 495
[372] Monk, M., Peacey, M. and Gross, J. D. (1971) *J. Mol. Biol.*, **58**, 623
[373] Kato, T. and Kondo, S. (1970) *J. Bact.*, **104**, 871
[374] Boyle, J. M., Paterson, M. C. and Setlow, R. B. (1970) *Nature*, **226**, 708
[375] Roberts, J. J. (1975) *Biology of Cancer*, p. 180 (E. J. Ambrose and F. J. C. Roe, Eds.). Ellis Horwood Ltd.
[376] Cooper, P. K. and Hanawalt, P. C. (1972) *J. Mol. Biol.*, **67**, 1
[377] Glickman, B. W. (1974) *Biochim. Biophys. Acta*, **335**, 115
[378] Regan, J. D. (1971) *Science*, **174**, 147
[379] Cleaver, J. E. (1970) *J. Invest. Dermatol.*, **54**, 181
[380] Bootsma, D., Mulder, M. P., Pot, F. and Cohen, J. A. (1970) *Mutation Res.*, **9**, 507
[381] Müller, W. E. G., Yamazaki, Z. I., Zahn, R. K., Brehm, G. and Korting, G. (1971) *Biochem. Biophys. Res. Commun.*, **44**, 433
[382] Cleaver, J. E. (1969) *Proc. Nat. Acad. Sci.*, **63**, 428
[383] Cole, R. S. (1973) *Proc. Nat. Acad. Sci.*, **70**, 1064
[384] Lieberman, R. P. and Oishi, M. (1974) *Proc. Nat. Acad. Sci.*, **74**, 4816
[385] Tomizawa, S. I. and Ogawa, H. (1972) *Nature New Biol.*, **239**, 14
[386] Barbour, S. D. and Clark, A. J. (1970) *Proc. Nat. Acad. Sci.*, **65**, 955
[387] Kornberg, T. and Kornberg, A. (1974) *The Enzymes*, Vol. X, p. 119 (P. D. Boyer, Ed.). New York: Academic Press
[388] Temin, H. M. and Mizutani, S. (1974) *The Enzymes*, Vol. X, p. 211 (P. D. Boyer, Ed.)
[389] Lehman, I. R. (1974) *The Enzymes*, Vol. X, p. 237 (P. D. Boyer, Ed.)
[390] Thomas, C. A. (1966) *Progr. Nucleic Acid Res. Mol. Biol.*, **5**, 315 (J. N. Davidson and W. E. Cohn, Eds.)
[391] Mandel, H. G. (1969) *Progress in Molecular and Subcellular Biology*, **1**, 82 (F. E. Hahn, Ed.)
[392] Wells, R. D. (1971) *Progress in Molecular and Subcellular Biology*, **2**, 21
[393] Bauer, W. and Vinograd, J. (1971) *Progress in Molecular and Subcellular Biology*, **2**, 181
[394] Waring, M. (1971) *Progress in Molecular and Subcellular Biology*, **2**, 216
[395] Hahn, F. E. (1973) *Progress in Molecular and Subcellular Biology*, 3, 1
[396] Grossman, L., Braun, A., Feldberg, R. and Mahler, I. (1975) *Ann. Rev. Biochem.*, **44**, 19

[397] Gefter, M. L. (1975) *Ann. Rev. Biochem.*, **44**, 45
[398] Livingston, D. M., Hinckle, D. C. and Richardson, C. C. (1975) *J. Biol. Chem.*, **250**, 461
[399] Livingston, D. M. and Richardson, C. C. (1975) *J. Biol. Chem.*, **250**, 470
[400] Hurwitz, J. and Wickner, S. (1974) *Proc. Nat. Acad. Sci.*, **71**, 6
[401] Mate, K. and Hurwitz, J. (1974) *J. Biol. Chem.*, **249**, 3680
[402] Sano, H. and Feix, G. (1974) *Biochem.*, **13**, 5110
[403] Bollum, F. J. (1974) *The Enzymes*, Vol. X, p. 145 (P. D. Boyer, Ed.). New York: Academic Press
[404] Loeb, L. A. (1974) *The Enzymes*, Vol. X, p. 174 (P. D. Boyer, Ed.)
[405] Weissbach, A. (1975) *Cell*, **5**, 101
[406] Center, M. S. (1975) *J. Virol.*, **16**, 94
[407] Wickner, S., Wright, M. and Hurwitz, J. (1974) *Proc. Nat. Acad. Sci.*, **71**, 783
[408] Hopfield, J. J. (1974) *Proc. Nat. Acad. Sci.*, **71**, 4135
[409] Modrich, P. and Richardson, C. C. (1975) *J. Biol. Chem.*, **250**, 5508
[410] Modrich, P. and Richardson, C. C. (1975) *J. Biol. Chem.*, **250**, 5515
[411] Hinkle, D. C. and Richardson, C. C. (1975) *J. Biol. Chem.*, **250**, 5523
[412] Strätling, W., Ferdinand, F. J., Krause, E. and Knippers, R. (1973) *Eur. J. Biochem.*, **38**, 160
[413] Foster, D. N. and Gurney, T. (1973) *J. Cell Biol.*, **59**, 103a
[414] Franklin, T. J. and Snow, G. A. (1975) *Biochemistry of Antimicrobial Action* (2nd Ed.). London: Chapman and Hall
[415] Grossman, L. (1974) *Adv. Radiation Biol.*, **4**, 77
[416] O'Sullivan, M. A., Howard, K. and Sueoka, N. (1975) *J. Mol. Biol.*, **99**, 347
[417] Hayes, S. and Sybalski, W. (1975) *DNA Synthesis and its regulation*, p. 486. ICN-UCLA Symposium of Molecular Biology (M. Goulian, P. Hanawalt and C. F. Fox, Eds.) Benjamin

CHAPTER 12

The Biosynthesis of RNA : Transcription

In the process of DNA transcription the positioning of nucleotide units in the RNA molecules that are being made is under the control of the DNA which acts as template. The means by which this template dictates such a sequence involves both base-pairing interactions and specific interactions between proteins and nucleic acids. Additionally each RNA chain is initiated at a specific site on the DNA template and subject to termination at another unique type of site on the template. In other words there are defined units of transcription. It is a selective process. Specific signals in the DNA template are recognized by the transcription apparatus. Initiation is governed by *promoter* regions in the DNA, and a region governing termination is designated a *terminator*.

Transcription is mediated by DNA-dependent RNA polymerases, which have now been isolated from a wide variety of sources, eukaryotic and prokaryotic. The properties of the enzyme from *Esch. coli* have however been the most widely explored. This purified enzyme can carry out the selective transcription of certain DNA's *in vitro*.

The products of DNA-dependent RNA polymerases have ribonucleotide sequences complementary to *one* of the strands of the DNA which was used as 'template'. A guanine residue in the DNA template strand dictates the insertion of a cytosine nucleotide in the RNA strand under construction, whilst a cytosine in the DNA causes a guanine nucleotide to appear in the new RNA strand. Similarly, a thymine in the DNA results in an adenine nucleotide in the RNA, and an adenine in the DNA results in a uracil nucleotide in the RNA.

12.1 The mode of action and structure of *Esch. coli* RNA polymerase
(for reviews see [1–23])

The first indication for the existence of a DNA-dependent RNA synthesizing enzyme in bacterial cells using the four ribonucleoside

5'-triphosphates as substrates came to light in the early 1960's [5]. The actual nucleotide composition of the RNA produced was soon shown to be determined by the nature of the DNA added as template. Moreover, the technique of 'nearest-neighbour analysis' provided compelling evidence that the added DNA template also determined the sequence of the nucleotides in the RNA made [1, 5]. This aspect was amply confirmed by the then newly developed technique of molecular hybridization. For example, Geiduschek, Nakamoto, and Weiss [24] prepared biosynthetic RNA labelled with ^{32}P using as template the DNA of bacteriophage T2. The RNA was isolated by gradient centrifugation in caesium chloride, and to it was added some T2-DNA. The mixture was heated to 100° for 10 minutes to cause separation of the two DNA strands (Fig. 12.1). It was then

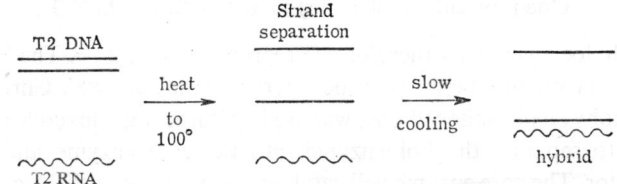

Fig. 12.1 *Formation of RNA–DNA hybrids*

cooled slowly and equilibrated at 90° for 12 hours. The formation of a hybrid during this 'annealing' process was demonstrated both by optical density and by radioactivity. No hybrid formation occurred when DNA's other than that of bacteriophage T2 were employed. This ability of primer T2-DNA to form a specific complex with T2-RNA is proof of the existence of entire nucleotide sequences in T2-DNA and T2-RNA which are capable of binding strongly to each other. The hybrid is relatively resistant to the action of RNase. Indeed, under the conditions employed, complex formation occurs to the exclusion of DNA renaturation, since no material banding at the density of double-stranded DNA is found in the test. This means that every strand of the DNA sample is capable of complex formation, and indicates that RNA molecules have been formed which are complementary to both strands of T2-DNA. Both strands can therefore act as a template either separately or together (see below).

A convenient method of detecting hybrids is to use a nitrocellulose filter [25, 26] which retains RNA hybridized to denatured DNA but allows free RNA to pass through.

The process of specific hybrid formation can be applied in the isolation of complementary RNA by chromatography on columns of DNA attached to cellulose or agar which traps the DNA-specific RNA while allowing other forms of RNA to pass through (p. 61) [26–31]. Different RNA's may also be compared in competition experiments in hybrid formation [32, 33].

The DNA-dependent RNA polymerase has now been extensively purified from *Esch. coli* [20, 23]. The holoenzyme is a complex zinc-containing protein of molecular weight 480 000–500 000 which can be dissociated, for instance in 6M-urea, into a number of polypeptide chain subunits as follows:

> Two α-chains each of mol. wt. 39 000
> One β-chain of mol. wt. 155 000
> One β'-chain of mol. wt. 165 000
> One molecule of σ(sigma)-factor, mol. wt. 95 000

The holoenzyme may therefore be represented as $\alpha_2\beta\beta'\sigma$. The holoenzyme without σ-factor is termed 'core enzyme', or $\alpha_2\beta\beta'$. Chromatography on phosphocellulose was used by Burgess and his colleagues [21] to separate the holoenzyme into the core enzyme and the σ-factor. The core enzyme will catalyse the synthesis of RNA chains from random sites on a DNA template *in vitro* quite well when 'foreign' DNA (for instance that from calf thymus) is added, but ineffectively when the DNA added is that from *Esch. coli* or T4 bacteriophage. Addition of σ-factor however restores the synthetic activity of the core enzyme by activating selective initiation of RNA synthesis [23]. It complexes with the core enzyme to yield holoenzyme which interacts specifically with promoter regions on the DNA template. Using T4 bacteriophage DNA as *in vitro* template the σ-factor addition specifically stimulates the core enzyme into transcribing those bacteriophage genes which correspond precisely to those normally expressed *in vivo* during the early stages of T4 bacteriophage infection of *Esch. coli*.

12.1.1 *Binding and initiation*

The sequence of molecular events for accurate initiation of RNA synthesis can be envisaged as follows [23]. After a series of unproductive random interactions with the DNA template, the holoenzyme 'recognizes' a specific structure, or sequence, within a promoter region. This enables the holoenzyme to bind to the DNA in this region at least an order of magnitude more tightly than in the non-specific interactions mentioned above. The precise structural features of the DNA actually required here are not yet known; however, the sequence [34] of the promoter region for the *Esch. coli* lac operon is shown in Fig. 12.2 (for details of lac operon see Chapter 14).

The next step in the initiation process is a change in the con-

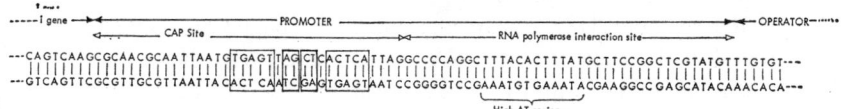

Fig. 12.2 *The lac promoter sequence, showing the RNA polymerase interaction site; for the significance of the CAP binding site see Chapter 14. Boxes indicate regions of symmetry*

formational state of the DNA in this promoter. Once this transition has occurred, the holoenzyme is poised to commence the manufacture of an RNA chain (see Fig. 12.3), but actually using only *one* of the strands as template. The nature of the conformation change

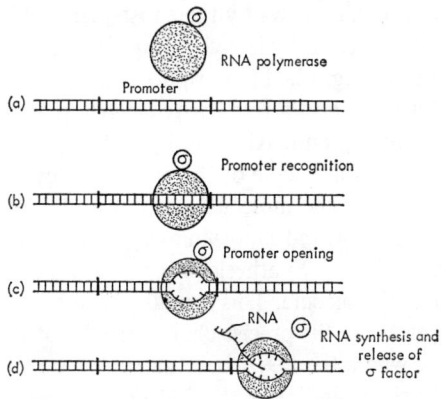

Fig. 12.3 *A diagrammatic representation of the stages in initiation of RNA synthesis*

is thought [35] to involve a local unwinding or strand separation of the DNA double helix over 4–8 base pairs.

Although it is the β'-subunit that is implicated in the binding of the polymerase to the DNA and perhaps has the required specificity, σ appears to bind to the β-subunit [20]. It may be best to consider the σ-factor as an allosteric effector of a multi-subunit enzyme. The β-subunit itself is probably most concerned with at least the initial catalytic reactions, i.e. formation of the first internucleotide linkage of an RNA chain. These initial reactions (as distinct from later chain elongation reactions) can be blocked by the antibiotic rifampicin, and genetic data indicate that rifampicin-resistant mutants carry the resistant phenotype in this β-subunit [37]. The means whereby rifampicin blocks initiation is not yet clear but it does appear to compete out the binding of the polymerase substrates GTP and ATP to the β-subunit [18]. Such nucleoside triphosphates are commonly found at the ends of new RNA chains made by the holoenzyme, e.g. pppGpYp ... and pppApYp ... , etc [37]. Once the initial internucleotide link between either GTP or ATP and the next nucleotide specified by the template has been made, the σ-factor is released, thus possibly reducing the affinity of the polymerase for the promoter site and allowing the remaining core to move along the DNA template, spinning off a strand of RNA as it proceeds. The process can be visualized in the electron microscope [38, 39] (Plate VI).

12.1.2 *Elongation of RNA chains and direction of transcription*
Basically the new chain grows by the subsequent addition of ribonucleotides to the free 3′-hydroxyl group [37]. For instance, the RNA being made in Fig. 12.4 has a triphosphate group at position 5′ on the first nucleotide and a free hydroxyl group at position 3′ on the other, or growing, end. Alkaline hydrolysis of this particular RNA [37] will yield a molecule of the nucleoside cytosine (C) from the 3′-hydroxyl end of the molecule, uridine and adenosine monophosphates (Up and Ap) and a molecule of guanosine tetraphosphate (pppGp) from the 5′-end. Synthesis proceeds from the 5′-end to the 3′-end of the RNA molecule. This can also be shown to be the case *in vivo*. Still unclear are the precise characteristics of DNA structure at the promoter where the DNA is actually used as template rather than just for the tight binding of the RNA polymerase. In any event the nucleotide residue at the 5′-end of the nascent chain always contains a purine base; thus any DNA sequence involved in the actual commencement of template-directed RNA synthesis must contain a thymine or cytosine nucleotide.

Whilst the antibiotic rifampicin only blocks initiation of new chain synthesis, the drug actinomycin D selectively prevents elongation without affecting polymerase–DNA binding, or initiation, by complexing to deoxyguanosine residues on the DNA template and so preventing the movement of the core along the template [40].

THE BIOSYNTHESIS OF RNA: TRANSCRIPTION

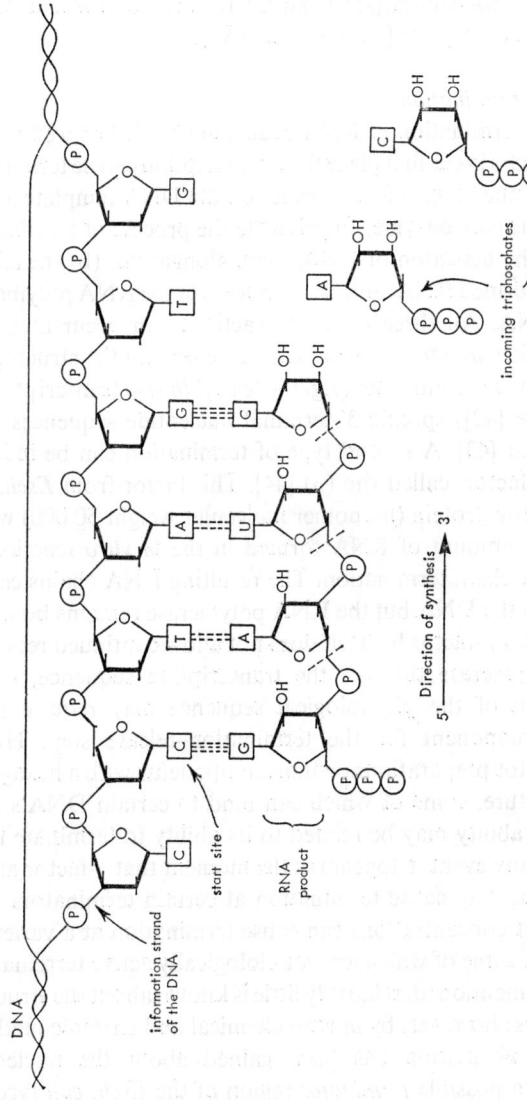

Fig. 12.4 *A diagrammatic representation of the biosynthesis of RNA on one strand of DNA acting as template. The broken lines indicate sites of hydrolysis with alkali*

As the RNA chain is formed it peels off the DNA template and, in *Esch. coli* at least, immediately becomes associated with the ribosomes so that polysomes (see Chapter 13) may be formed before transcription is complete [41] (see Plate VI).

12.1.3 *Chain termination*

The selective termination of RNA chains *in vitro* is less well understood. It appears in the first place that transcription can be terminated directly by virtue of specific sequences on the DNA template acting directly on the core enzyme. In principle the process of termination involves (a) the cessation of RNA chain elongation, (b) the release of the newly formed RNA, and (c) the release of the RNA polymerase from the DNA. All three of these reactions can occur at 'direct terminator' sites *in vitro*. Little is known regarding the structure of such sites, but in certain cases (e.g. the 'early' *in vivo* transcript of T7 bacteriophage [42]) specific 3'-terminal nucleotide sequences have been identified [43]. A second type of termination can be induced *in vitro* by a factor, called rho (ρ) [44]. This factor from *Esch. coli* is an oligomeric protein (monomer molecular weight 50 000) which depresses the amount of RNA formed in the *in vitro* reaction by causing RNA chain termination. The resulting RNA chains can be released from the DNA but the RNA polymerase remains bound to the DNA. Thus ρ-factor by itself does not allow continued recycling of RNA polymerase through the transcription sequence, and a reconstruction of the physiological sequence may require some additional component for the termination–release step. Highly purified ρ-factor preparations contain components with a hexagonal subunit structure, some of which can bind to certain DNA's [45]. This binding ability may be related to its ability to terminate transcription. In any event it appears at the moment that ρ-factor at low concentrations may cause termination at certain terminators used *in vivo*; higher concentrations can cause termination at a variety of sites on DNA some of which are not biologically active terminators.

As already mentioned, relatively little is known about the structure of terminators; however, by *in vitro* chemical and enzymic methods [46], some information has been gained about the nucleotide sequences of a possible *terminator* region of the *Esch. coli* tyrosine suppressor tRNA gene (see Fig. 12.5).

THE BIOSYNTHESIS OF RNA: TRANSCRIPTION

12.2 The eukaryotic nuclear DNA-dependent RNA polymerases
(for reviews see [7, 47–50])

In mammalian cells the RNA polymerase activity detected in nuclei was initially difficult to study by virtue of being tightly bound to the nuclear chromatin complex [51, 52]. Somewhat specialized techniques were required for its solubilization in reasonable yields before purification, similar to that achieved for the *Esch. coli* enzyme, could even commence.

Depending on the origin of the cells used, different approaches to solubilization were employed. Some procedures were mild and involved merely incubating, or homogenizing nuclei, in slightly alkaline buffer. A moderately drastic treatment involved brief sonication in low ionic strength medium. In a more drastic approach, sonication was carried out in a medium of high ionic strength [47].

After solubilization from chromatin, chromatography of the

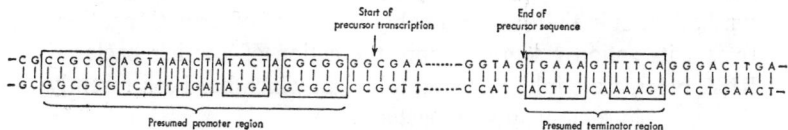

Fig. 12.5 *Nucleotide sequences adjacent to* Esch. coli *tyrosine suppressor tRNA precursor gene* [46]. *Boxes indicate regions of symmetry*

enzyme activity from a variety of animal tissues on DEAE-cellulose, or DEAE-Sephadex, revealed the presence of multiple forms of RNA polymerase. This multiplicity of RNA polymerase has also been observed in lower eukaryotes such as an aquatic fungus, yeast, and maize. The enzyme activities are usually eluted from the columns using a linear gradient of ammonium sulphate or potassium chloride. Depending on the tissue used, two to three discrete peaks of activity are resolvable by this approach [47].

The classes of RNA polymerase separated chromatographically between 0·10M and 0·37M salt concentrations are referred to usually as I, II, and III in order of their elution. To add another level of complexity, classes I, II, and III, which are the major species detected in most eukaryotic cells, have each been further resolved into at least two classes (e.g. III_A and III_B etc.).

Class I RNA polymerases have been established to be of nucleolar origin, whereas classes II and III are of nucleoplasmic origin [47, 53]. Additionally these enzymes operate optimally under somewhat different conditions. The nucleolar enzymes work best at low ionic strength and utilize Mn^{2+} and Mg^{2+} equally well. On the other hand, higher Mn^{2+} concentrations and ionic strengths are required for maximum activity of the nucleoplasmic enzymes. Moreover, the activity of nucleoplasmic class II enzymes is inhibited selectively by α-amanitin (a toxin from the poisonous mushroom *Amanita phalloides*) at concentration as low as 3×10^{-8}M, whereas the nucleolar activity is not affected even at much higher doses [47]. The minor nucleoplasmic class III enzymes also are not affected by α-amanitin [59].

α-Amanitin is specific for the eukaryotic polymerases of class II, as rifampicin is specific for the prokaryotic enzyme (and for the mitochondrial polymerase, as will be seen later). The mushroom toxin appears to bind to the RNA polymerase rather than to the DNA template like actinomycin D. Whereas rifampicin inhibits initiation by the bacterial polymerase, α-amanitin blocks RNA synthesis after initiation, presumably at the level of chain elongation. In fact its action resembles more the action of another bacterial RNA polymerase inhibitor streptolydigin which has been recently shown to block elongation in bacteria.

Since factors such as σ and ρ appear to play an important role in the control of transcription, a search for such factors was made in mammalian systems. A protein factor (mol. wt. 70 000) which can specifically bind to, and stimulate, the activity of nucleoplasmic class II enzymes from rat liver and calf thymus has been tentatively identified in the cytoplasm of rat liver [55]. However, the question of whether or not the factor can promote initiation of RNA synthesis at specific sites on the DNA template cannot yet be answered. With regard to termination in eukaryotes no information is presently available.

Having mentioned a possible factor, another question is how do these enzymes relate structurally to the prokaryotic enzyme described in Section 12.1? The situation is complex but structural analyses [56, 57] have shown that the molecular weights of the large subunits detected in the class III enzymes (138 000 and 155 000) differ from those of the class II enzymes (140 000 and either 170 000 or 205 000, or 240 000) and from those of the class I enzymes (117 000 and 195 000). Some low molecular weight subunits are also unique to each enzyme class. These data clearly distinguish class I, II, and III enzymes on a structural basis. In addition, polypeptides of molecular weight 29 000 and 19 000 were found in all classes, a polypeptide of molecular weight 52 000 was found only in class I and III enzymes, and a polypeptide of molecular weight 41 000 was found only in class II and III enzymes. Thus it appears, from these findings, and the different sensitivities to rifampicin and α-amanitin, that there is no real structural similarity between the eukaryotic polymerases and the core polymerase of prokaryotes. Additionally it appears that the three classes of polymerase are assembled primarily from distinct gene products, and that they are not interconvertible by simple structural alterations.

Despite these advances, the role of these various polymerases in cellular RNA synthesis remains to be clarified. At least in *Esch. coli*, two types of data, (a) that certain temperature sensitive mutants in RNA synthesis can be shown to have extremely heat labile RNA polymerases, and (b) that rifampicin, which blocks all the initiation of cellular RNA synthesis, can be shown to bind to the β-subunit, and that resistant mutants carry the resistant phenotype in the same β-unit, both support the view that the *Esch. coli* enzyme studied *in vitro* is also responsible for all the *Esch. coli* RNA synthesis *in vivo*. In eukaryotes there is no similar direct genetic evidence. It is quite possible that the nucleolar enzymes (I) are involved in the synthesis of the ribosomal precursor RNA whereas the nucleoplasmic enzymes (II and III) make messenger RNA precursors, tRNA, and 5S RNA. Indeed, product analysis indicates enzyme III to be specifically responsible for the biosynthesis of 5S RNA and the precursor to tRNA in mammalian cells [54], thus leaving the possibility that enzyme II is involved in mRNA precursor production.

12.3 Mitochondrial DNA-dependent RNA polymerases

Compared with the nuclear DNA-dependent RNA polymerases, relatively little is known about the properties and functions of the mitochondrial polymerases. Soluble preparations have now been obtained from a number of sources, and the enzymes from *Neurospora crassa* and yeast have been extensively purified [58, 59]. That from rat liver appears closely associated with the mitochondrial membrane, and detergents are required for its extraction [60]. Like the prokaryotic RNA polymerases, the enzymes from rat liver, heart, or *N. crassa* are sensitive to rifampicin. α-Amanitin, on the other hand, has no effect on the mitochondrial enzymes. Despite this similarity to the bacterial polymerase, the mitochondrial enzymes seem to comprise single subunits of relatively low molecular weight (64 000–68 000) [47].

12.4 Post-transcriptional processing

Whilst most of the cellular RNA of eukaryotes is cytoplasmic, the nucleus does contain some RNA (about 5 per cent of the total). It turns out however that its existence is mainly ephemeral, and there is now substantial evidence indicating that it includes several species which are intermediates in biosynthetic pathways leading to the formation of tRNA's, ribosomal RNA's, and messenger RNA's of the cytoplasm [61]. The primary structure of these intermediates is modified in certain instances by cellular enzyme systems subsequent to the completion of their transcription from the DNA template.

Such modifications include (a) 'trimming' or 'tailoring', i.e. alteration to the length of the primary transcription product by scission mechanisms, or (b) the alteration of primary nucleotide sequences as a result of base or sugar modification, or (c) the addition of specific nucleotide sequences. This chemical editorial work carried out by the cell is a prerequisite in the formation of certain nucleic acid species both in eukaryotic and in prokaryotic cells, and the molecular events involved can be collectively described as *post-transcriptional processing events* [61].

12.5 The biosynthesis of rRNA

That the composition and nucleotide sequence of cytoplasmic RNA is under the control of nuclear DNA was supported by molecular hybridization experiments in which it was shown that there were sequences of DNA precisely complementary to the nucleotide sequences of ribosomal RNA, tRNA, and messenger RNA. The technique of molecular hybridization, in which the extent of *in vitro* formation of stable hydrogen bonded complexes between purified RNA molecules and complementary regions on isolated DNA's, after denaturation to single strands, can be determined, has been of great value. The principles and practice involved have been well reviewed [62, 63] (see also Fig. 12.1).

In bacteria, the deoxyribonucleotide sequences complementary to ribosomal RNA are found to amount to approximately 0·1–0·2 per cent of the total DNA genome [64]. Thus the bacterial genome possesses roughly ten copies of the genes for ribosomal RNA. By similar molecular hybridization tests, animal cells possess DNA

THE BIOSYNTHESIS OF RNA: TRANSCRIPTION

complementary to several hundred copies of ribosomal RNA (e.g. 260 in *Drosophila*, 900 in *Xenopus*, 1100 in HeLa cells [65]).

Whilst there is RNA in the nucleus of eukaryotes, most of this is located in the nucleolus (about 3 per cent of the total cellular RNA). In fact the nucleolus is very active in the manufacture of RNA [66, 67]. Moreover, molecular hybridization experiments demonstrate that the parts of the DNA genome which contains the deoxyribonucleotide sequences complementary to the sequences that make up ribosomal RNA are located in the nucleolus, or nucleolar organizer region of the genome, of certain eukaryotes [68]. There is now evidence that the multiple genes for ribosomal RNA are arranged as linear arrays of contiguous units. Deletions of the nucleolar organizer, which comprises a very small segment of a chromosome, results in the complete loss of a haploid content of ribosomal DNA, i.e. a few hundred genetic units.

Additionally, ribosomal DNA has been shown to contain equimolar amounts of sequences complementary to 28S and 18S ribosomal RNA, and, in *Xenopus* and *Drosophila* at least, it can be shown that the sequences for 28S and 18S alternate within the ribosomal DNA [69].

In 1962 Scherrer and Darnell (see [67]) demonstrated the existence of relatively short-lived species of RNA in cultures of human cells exposed for very short times to ^{3}H-labelled uridine. This RNA occurred in only small amounts and was distinguishable from tRNA and ribosomal RNA. It sedimented at 45S and its molecular weight is now known to be about 4.1×10^6. Although the site of its synthesis was soon found by cell fractionation techniques to be the nucleolus [70], a more interesting feature was discovered. Shortly after its synthesis, or transcription from the nuceolar DNA, this long single-stranded molecule undergoes a series of molecular 'tailoring' events in the nucleolus whereby its molecular dimensions are progressively reduced to give rise eventually to the 28S and 18S ribosomal RNA species found in cytoplasmic ribosomes.

Whilst the 45S species is found exclusively in the nuceolus, after longer labelling times radioactivity begins to appear in other species of RNA, firstly in 41S and 32S RNA (which are also confined to the nucleolus) and 20S RNA, and then in 18S RNA which passes rapidly to the cytoplasm. Finally, radioactivity appears in 28S RNA initially in the nucleolus, then in the nucleoplasm, and shortly afterwards in the cytoplasm. A precursor–product relationship (see

Fig. 12.6) between these RNA's has been proposed [67] based on a variety of data including recent electron microscopic examination of the various RNA species [71] (see also Plate VII).

The '5·8S RNA' fragment appears to be generated simultaneously with the 32S → 28S RNA transition. This has been interpreted by some as indicating the possibility that, in the conversion, a loop of 'spacer' RNA located on the precursor chain between the 28S and 5·8S species is degraded [65].

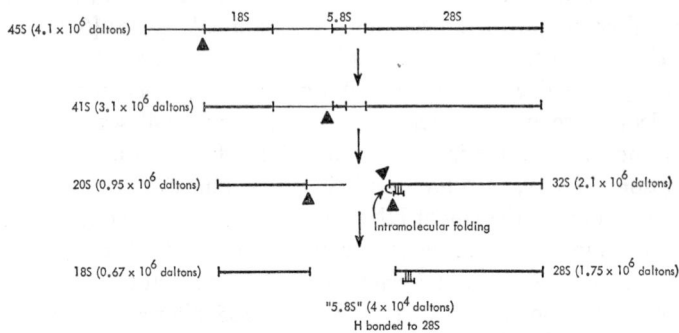

Fig. 12.6 *The topographical structure of HeLa cell rRNA precursors, and their sites of nucleolar cleavage* (▲)

With regard to the sequence modifications to rRNA (i.e. the methylations and pseudouridine formation), the majority of these occur rapidly in the nucleolus on the 45S precursor. Interestingly, the methylations occur only in the portions of the precursor destined to become either the 28S RNA or 18S RNA components [72], and are mostly in the form of 2'-O-methylribose.

The methylation of RNA has been discussed in Chapter 8. Methyl transferases bring about the introduction of methyl groups from S-adenosyl-L-methionine into RNA at the polynucleotide level in the post-transcriptional processing of both rRNA and tRNA precursors. Those involved in rRNA processing appear to be located in the nucleoli [80].

5S rRNA (p. 123) originates independently [61] in the non-nucleolar part of the nucleus [74] and later becomes permanently associated with the large ribosomal subunits.

Although the ribosomal proteins appear to be synthesized in the cytoplasm, it is reasonably certain that they become associated with the ribosomal precursor RNA's in the nucleolus where the actual

THE BIOSYNTHESIS OF RNA: TRANSCRIPTION

assembly of the ribosome takes place as shown in Fig. 12.7. It is believed that the cleavage of the precursors takes place within 'nascent ribosomal particles' [81].

For some while it was believed that there were no similar precursor RNA's in bacteria. However, with the use of suitable mutants of *Esch. coli*, with low levels of RNase III (see Chapter 8), and chloramphenicol treatment it has been possible to observe the transitory

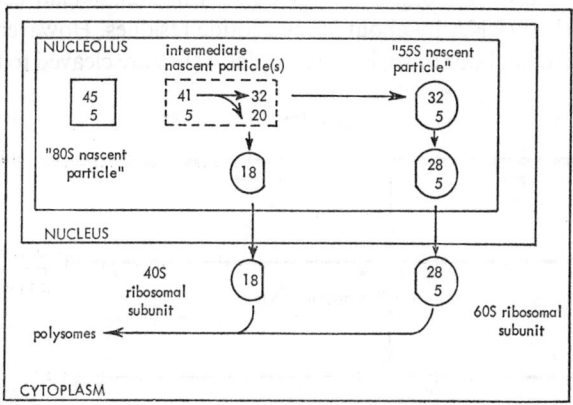

Fig. 12.7 *An outline of eukaryotic ribosome formation illustrating the processing of RNA taking place within 'nascent ribosomal particles'. The numbers within the particles refer to the sedimentation coefficients of their constituent RNA's. Note the occurrence of 5S RNA even within '80S nascent particles'* [65]

appearance of a large 30S transcript containing both 16S and 23S sequences [75–77], and possibly 5S sequences [242]. Normally it seems that this RNA is cleaved rapidly during the course of synthesis to give 25S and 17·5S fragments which in turn are further processed to the mature 23S and 16S ribosomal RNA's of *Esch. coli* (see Fig. 12.8). Moreover, like the situation in higher organisms, these RNA's are associated with proteins during processing to comprise 'nascent ribosomal precursor particles' [78] which also carry out the nucleotide modification such as methylation [79].

12.6 Transfer RNA biosynthesis

As was the case with the ribosomal RNA genes, tRNA genes are reiterated in higher organisms. For instance, whereas bacteria possesses 1 or 2 copies of each tRNA gene, yeasts have 5–7 copies. *Drosophila* have 13 copies and *Xenopus* and mammalian cells about 200 [82]. These genes, like those for 5S rRNA, are located outside the nucleolus and distributed amongst chromosomes of all size ranges [68].

There is now a growing body of evidence to suggest that animal and bacterial tRNA's arise not only as a result of 'tailoring' of somewhat longer precursor molecules but also by the modification of specific nucleotide sequences.

In eukaryotic cells, precursors to tRNA's are rapidly exported to the cytoplasm [82, 83]. These precursors in mammalian cells are longer than tRNA by about 30 nucleotide residues. However, over a period of an hour or so, in the cytoplasm they are cleaved to tRNA

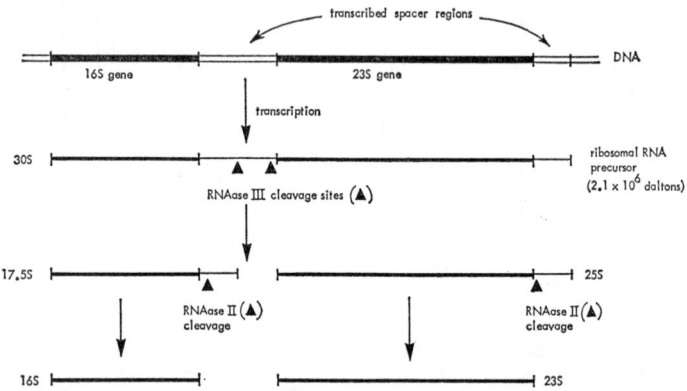

Fig. 12.8   Esch. coli *rRNA transcription and processing*

dimensions and certain nucleotides are modified by methylation (see Fig. 12.9).

Other modifications also occur in the cytoplasm, such as the conversion of certain uridine residues into pseudouridine and dihydrouridine residues. The precise structure of these precursors is not yet fully understood, mainly since no precursor to a specific mammalian tRNA has yet been isolated from the mixed population. Nevertheless it appears that at least terminal uridine residues must be removed from their 3'-termini during their processing in the cytoplasm [84, 61].

In *Esch. coli*, where there is also evidence for precursors [85, 86, 88], a specific study of *Esch. coli* tyrosine suppressor tRNA and its formation has been made by Smith and his colleagues [87, 88]. A specific precursor to this special tRNA found in the amber suppressor su_3^+ has been isolated and characterized (see Fig. 12.10). Joined to

THE BIOSYNTHESIS OF RNA: TRANSCRIPTION

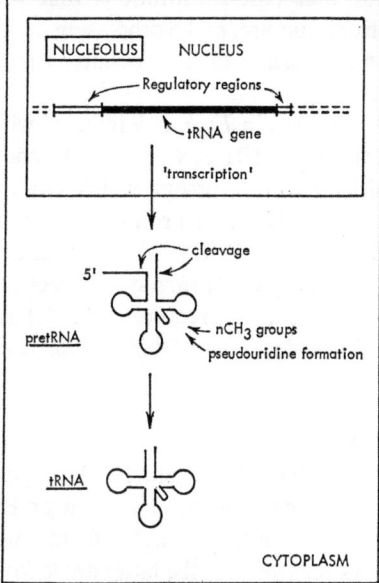

Fig. 12.9 *A scheme showing the possible steps involved in the production of mammalian cell tRNA* [82]

the actual tRNA sequence at its 5'-end are 41 extra nucleotides beginning with pppG, and at the 3'-end are two extra nucleotides. This precursor is cleaved in *Esch. coli* to give the mature tRNA sequence [88].

It appears that, for processing to be efficient, the tRNA part of the tyrosine tRNA precursor must be folded up in the correct

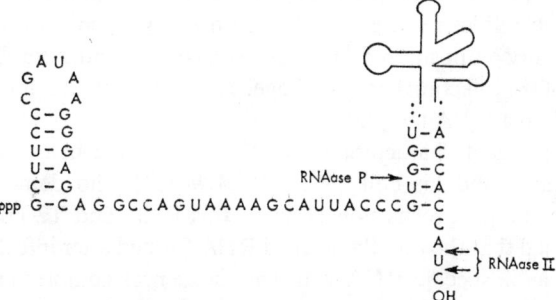

Fig. 12.10 *The structure of* Esch. coli *tyrosine suppressor tRNA, showing sites of cleavage* (→)

conformation, and that base-substitutions that prevent this, or favour an alternative structure, make processing inefficient. Nucleotide modifications apparently take place after the cleavages have occurred [87, 88].

A novel situation occurs in *Esch. coli* infected with T4 bacteriophage. Some of the eight tRNA's specifically coded for the T4 genome arise initially in precursors which contain two tRNA sequences (e.g. serine tRNA and proline tRNA). These are subsequently cleaved in the middle to yield two separate tRNA species [89, 90]. Recently the complete *in vitro* synthesis of certain *Esch. coli* tRNA's has been achieved. This requires a DNA template enriched for tRNA genes, DNA-dependent RNA polymerase, and cleaving enzymes [91].

12.7 Messenger RNA

The discovery of the DNA-dependent RNA polymerase occurred at the same time that the concept of messenger RNA was being independently developed, but, whereas the former was based on the study of cell-free systems *in vitro*, the latter arose from the examination *in vivo* of bacterial cells uninfected or infected with bacteriophage. Messenger RNA (mRNA) is, however, essentially a form of complementary RNA formed under the influence of the DNA-dependent RNA polymerase. It has been the subject of many reviews [9, 92–105].

The initial observation which led to the discovery of messenger RNA was made in 1953 by Hershey [106] who observed the rapid formation of new RNA molecules in cells of *Esch. coli* infected with bacteriophage T2. Volkin and Astrachan [107, 108] in 1956 labelled this RNA with ^{32}P and, from the distribution of label among the mononucleotides released on alkaline hydrolysis, concluded that it differed in composition from that of *Esch. coli* and resembled the DNA of the infecting bacteriophage. This suggested that it had been formed on a DNA template.

The physical characteristics of the RNA were investigated by Spiegelman and his colleagues [38, 109–111] who demonstrated specific complex formation between this RNA and T2-DNA and concluded that the rapidly labelled RNA formed after infection was in fact a T2-specific RNA with base sequences complementary to those in the T2-DNA. About the same time Jacob and Monod [98, 99, 112] concluded that this new RNA synthesized after bacterio-

THE BIOSYNTHESIS OF RNA: TRANSCRIPTION

phage infection became attached to pre-existing ribosomes in the bacterial cell and could be detached in a caesium chloride gradient after lowering the magnesium concentration. That the ribosomes involved had been synthesized before infection was proved by labelling *Esch. coli* cells with ^{13}C and ^{15}N and infecting these 'heavy' cells with T2 bacteriophage in a 'light' medium containing ^{12}C and ^{14}N, when the T2-specific RNA was found attached to 'heavy' or 'old' ribosomes as was also the nascent T2-specific protein labelled by pulse exposure to radioactive amino acids. The T2-specific RNA had therefore acted as a 'messenger' from the T2-DNA to the ribosomes where it directed the formation of bacteriophage protein.

The existence of a similar unstable RNA in non-infected bacterial cells was soon demonstrated in several laboratories [113, 114] both in normal cells and, more readily, in 'step-down' cultures in which the cells are transferred from a rich medium permitting rapid growth to a poor medium in which the cells will contain more ribosomes than they can usefully employ. Under these conditions the synthesis of ribosomal RNA will stop but, since protein production continues at a slow rate, some synthesis of mRNA is to be expected.

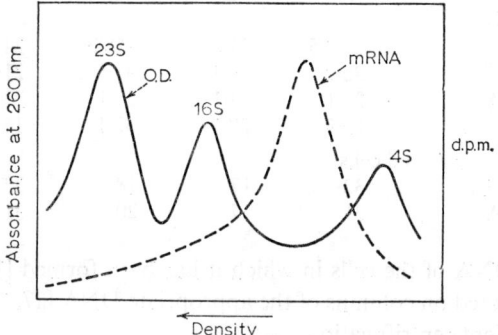

Fig. 12.11 *The separation of the various RNA's of the cell by density-gradient centrifugation. Full line – absorbance at 260 nm. Broken line – radioactivity after pulse labelling (see text)*

The procedure employed is illustrated in Fig. 12.11. Bacterial cells are 'pulse labelled' by transferring them to a medium containing a radioactive precursor of RNA such as $^{32}PO_4^{3-}$ or ^{14}C-uridine. After a few seconds a sample of cells is removed for isolation and examination of RNA by density-gradient centrifugation. The remainder of the culture is washed free of radioactive precursor and allowed to grow for a further half-hour or so in unlabelled medium so as to 'chase' the labelled material into its next metabolic site.

After pulse labelling and gradient centrifugation, three peaks of material absorbing at 260 nm are found, as expected, corresponding to the 23S and 16S rRNA and the 4S tRNA. They are not radioactive, but a peak of radioactivity is found intermediate in molecular size between 4S and 16S (Fig. 12.11). It is too small in amount to show up in measurements of optical density and corresponds to mRNA.

The short-lived nature of this material can be shown by gradient centrifugation of the RNA from the cells after the period of the 'chase'. The mRNA peak has now disappeared and the radioactivity is distributed among the three well-known peaks (p. 55).

The rapidly labelled RNA has a composition, calculated on the basis of the radioactivity of its component nucleotides, corresponding to that of the DNA (Table 12.1) and shows hybrid formation

TABLE 12.1

Molar proportions of bases in the nucleic acids of Esch. coli [95]

	A	C	G	U(T)
Normal cells				
DNA	24–25	25–26	25–26	24·25
rRNA	25·2	26·1	31·5	21·7
mRNA	25·1	24·1	27·1	23·7
tRNA	20·3	28·9	32·1	15·0
Bacteriophage-infected cells				
T2-DNA	32	17	18	32
mRNA	31	17	20	31

with the DNA of the cells in which it has been formed [14, 95]. It can be isolated on columns of the appropriate DNA [27, 28] as well as by gradient centrifugation.

From the data in Table 12.1 it can be seen that, whilst the base composition of the new T2-mRNA is similar to that of T2-DNA, the total content of guanine and cytosine is approximately equal to the corresponding total for the bacteriophage DNA, but the individual bases occur unequally in the RNA (20 per cent for guanine and 17 per cent for cytosine). This observation, suggested Bautz and Hall [116], would be accounted for if the mRNA is always synthesized as the complement of one, rather than both, nucleotide chains of the DNA. Evidence that only one specific DNA chain

THE BIOSYNTHESIS OF RNA: TRANSCRIPTION

appears to be made use of when a gene functions was also obtained from a study of the effect of 5-fluorouracil on the expression of certain rII mutants of T4 bacteriophage [117].

Additional support for the view that only one strand of the DNA helix is used as template came from studies with other bacteriophages. In bacteriophages α and SP8 the two strands of their DNA differ in density sufficiently for it to be possible to separate the individual 'heavy' and 'light' strands in caesium chloride density gradients. The RNA formed in the bacteriophage infected bacterial cell, however, forms a molecular hybrid with only one of these strands, the 'heavy' one [118–120]. Similar studies using the component strands of ØX174 replicative form led to the same conclusions [121].

In the cell, the mRNA becomes associated with the ribosomes to form [113, 114] the polysomes which have already been described on p. 24 and which are discussed later (p. 377) in connection with the part played by mRNA and the ribosomes in protein synthesis (Chapter 13).

In the bacterial cell mRNA is of course synthesized under the influence of the DNA-dependent RNA polymerase, and its formation is therefore inhibited by actinomycin D [122]. The proportion of mRNA is small, the figure commonly quoted being between 1 and 2 per cent of the total RNA [95, 31, 123, 124]. Its molecular weight may vary within wide limits, but in *Esch. coli* it appears to be of the order of 500 000 [95] although it may be remarkably heterogeneous [125].

More advanced techniques combining molecular hybridization and polyacrylamide gel electrophoresis have, however, permitted the isolation of specific bacterial messengers such as those specific for the *lac* and *gal* operons. These are polycistronic (see Chapters 13 and 14), having molecular weights of $1 \cdot 75 \times 10^6$ daltons [126] and $1 \cdot 5 \times 10^6$ daltons [127] respectively. Study of these has been of obvious use in elucidating the mechanism of gene expression at the translational level. Additionally, these and other specific polycistronic messengers have yielded valuable information about the metabolic breakdown of messengers after translation. Originally it was thought that bacterial messengers were degraded sequentially in the $5' \to 3'$ direction. However, an analysis of rates of decay shows no strong correlation between size and lifetime of these messengers [126, 127]. Furthermore, the location of a particular gene proximal

THE BIOCHEMISTRY OF THE NUCLEIC ACIDS

or distal to the operator (see Chapter 14) does not appear to have a systematic influence on its rate of decay. It is now felt that the initiation of polycistronic mRNA decay is not progressive from one end of the message but probably starts with specific internal cleavages [126, 127]. The cellular site for mRNA degradation may be a ribosomal complex engaged in active protein synthesis.

Having discussed bacterial mRNA formation and its decay, it is worth pointing out the fact that so far no post-transcriptional processing has been invoked for normal bacterial mRNA's. It would be premature to rule this out; it may merely have escaped detection. Post-transcriptional processing, on the other hand, is clearly detectable in the production of the bacteriophage T7-specific 'early'

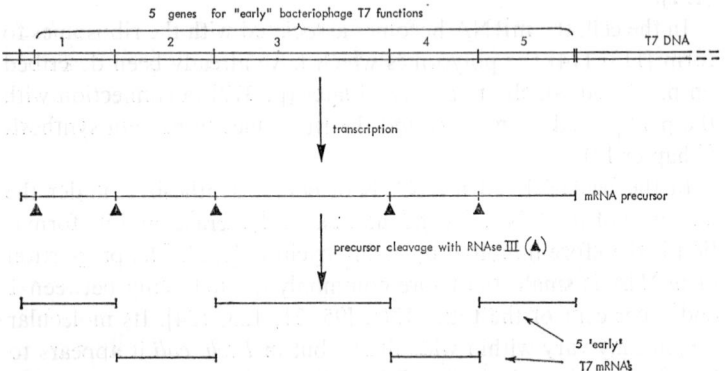

Fig. 12.12 *Post-transcriptional processing of T7 early mRNA'S* [129]

mRNA's that appear in the initial stages of the infectious cycle in *Esch. coli*. Basically these arise from five cistrons arranged sequentially at 5'-end of the bacteriophage T7 genome (accounting for 20 per cent of the total genome). These are initially transcribed together as one piece of RNA; however, this transcript is quickly cleaved to yield five separate and distinctive monocistronic mRNA's [128-130]. This is illustrated in Fig. 12.12.

In summary, then, messenger RNA is a metabolically active shortlived form of complementary RNA formed on the DNA of the cell under the influence of the DNA-dependent RNA polymerase. It carries information from the genetic material to the ribosomes with which it becomes intimately associated in the process of protein synthesis.

The origin of mRNA in eukaryotic cells, however, still remains

THE BIOSYNTHESIS OF RNA: TRANSCRIPTION

somewhat of a puzzle. The use of new techniques such as purification of RNA by acrylamide gel electrophoresis (see Chapter 4), coupled with judicious choice of biological material, has made it possible to obtain pure or partially pure preparations of the mammalian mRNA's for globin (9S) [131–135], histone (9–10S) [136, 137], lens crystallin (17S) [138], immunoglobulin (12–14S) [139, 140], ovalbumin (15S) [141], and myosin (26S) [142], to mention a few (see also Chapter 13).

One of the most conspicuous structural features of *some* eukaryotic mRNA's is a long uninterrupted sequence of adenosine nucleotides attached at their 3'-end [152–158]. This polyA tail has been reported to vary from 60 up to 200 nucleotides depending on the mRNA [105]. Whilst mRNA from lower eukaryotes such as slime moulds [145–147] and from plants [145] and yeast [148, 149] also have polyA tails, the histone mRNA of higher eukaryotes is specifically known to lack such a tail [143, 144]. (Some bacterial mRNA's have now been detected with polyA tracts [150, 151].)

Although the function of the polyA remains obscure it has proved useful for isolating mRNA's. For example, RNA's with polyA tails can be bound tightly to columns of Sepharose to which polyU has been covalently attached, and can thus be separated from other contaminating RNA's [105].

Another structural feature of eukaryotic and virus-specific mRNA's to emerge recently is the unusual sequence to be found at their 5'-termini [158–163]. This comprises 7-methylguanosine linked through its 5'-hydroxyl, via a tri(or pyro)phosphate group to the 5'-hydroxyl of an O'-methylated nucleoside (Fig. 12.13). This terminal sequence, or 'cap' as it is sometimes called, may originate by the enzymic addition of a guanosine residue to the 5'-ends of an mRNA (or its precursor) from GTP as substrate, followed by its subsequent methylation involving a methylase and S-adenosyl-L-methionine. Besides this unusual methylated sequence, it should be pointed out that there are actually other methylated bases in mammalian mRNA (mainly N^6-methyladenine). However, they occur at a very low level and their precise location is not yet known although some may be clustered towards the 3'-termini [161], and perhaps the 5'-termini.

Since the total mRNA of a cell comes from many different genes and encodes protein having a wide variety of sequences, it might be expected to have a base composition resembling that of the DNA of

THE BIOCHEMISTRY OF THE NUCLEIC ACIDS

the cells. Whilst observation has indicated this to be broadly correct and serves to distinguish the total cell mRNA population from rRNA or tRNA, it is certainly not true for individual messengers like globin or lens crystallin mRNA's. Indeed, from the amino acid composition of silkworm fibroin, the somewhat unusual base composition of

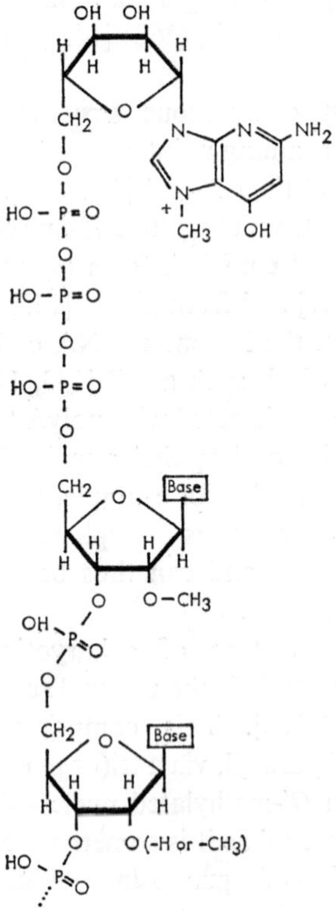

Fig. 12.13 *The 5'-terminal structure of eukaryotic mRNA's*

silkworm fibroin mRNA was predicted, and confirmed upon its subsequent isolation from the silk gland of *Bombyx mori* (40 per cent guanine and 19 per cent cytosine) [164].

Whilst the original bacterial experiments suggested mRNA's to have a short half-life (1–2 minutes), the metabolic stability of mammalian mRNA is now known to vary quite widely from cell to cell [105]. In HeLa cells there appear to be two classes of mRNA, with half-lives of 7 and 24 hours [165], as well as a short-lived species (half-life 1 hour) [166].

THE BIOSYNTHESIS OF RNA: TRANSCRIPTION

For a long time it has been known that a rapidly labelled RNA is formed in the cell nucleus, but the possibility of migration of intact polynucleotides from nucleus to cytoplasm has been the subject of much argument.

When mammalian cells, for instance, are very briefly exposed to ^{32}P-orthophosphate or isotopically labelled nucleoside precursors, the dominant species to be labelled is not the mRNA of the cytoplasmic polysomes but a polydisperse nuclear RNA fraction which accounts for around 1 per cent of the total cellular RNA. This nonnucleolar nuclear fraction is made up of RNA's whose sedimentation coefficients range widely from 20S to 100S (i.e. roughly $1-15 \times 10^6$ daltons); hence its name heterogeneous nuclear RNA (hnRNA). Other terms that have been used for this species, which is detectable in the cells of all higher organisms [67, 73], are DNA-like RNA or messenger-like DNA. Basically this is because its base composition is roughly DNA-like, differing markedly from the high proportion of guanine plus cytosine so characteristic of ribosomal RNA. Whereas low levels of actinomycin D will selectively inhibit ribosomal RNA synthesis, specifically 45S RNA synthesis in the nucleolus, hnRNA synthesis continues in the euchromatin regions of the nucleus. However, the decay of hnRNA within the nucleus is notably rapid (half-life about 20-60 minutes), but interestingly a proportion (a quarter to a third) have polyA tracts about 200 nucleotides long at their 3'-ends. An mRNA precursor role was postulated on this basis, but experiments designed to follow the metabolic fate of hnRNA molecules led to the conclusion that, if there were such a relationship, it was complex. Most of the radioactive label in hnRNA could not be chased into the cytoplasm.

Whilst most mRNA would appear to arise as a result of transcription of sequences which probably occur only once per haploid genome [167, 168] (unique or non-repetitive sequences; see Chapter 5), the same is not true for its putative nuclear precursor, hnRNA. Covalently attached to one another in hnRNA molecules are sequences which are transcripts of both non-repetitive and repeated (repetitive) genome sequences [169-174]. The interspersion and length of these sequences in the hnRNA to some extent reflects the length and interspersion of the corresponding sequences in the nuclear DNA [173, 174]. Additionally, the structural features of hnRNA molecules from the HeLa cell that terminate in a polyA tract have been probed by fragmenting the hnRNA molecules by mild alkaline treatment. hnRNA fragments of 3000 nucleotides (approximate mRNA dimensions) containing the 3'-terminal polyA tract hybridize to HeLa DNA at the same rate as mRNA (i.e. they originate from unique sequences in the genome). The remaining 5'-portion of the hnRNA (see Fig. 12.14) molecules, however, contain [169]: (a) intermediate repetitive and unique sequences interspersed with one another; (b) two or three short tracts of about 30 uridylate nucleotides each (oligoU tracts) [175]; (c) double-stranded regions; and (d) possibly oligoA, tracts of 30 units (see Fig. 12.14). Despite

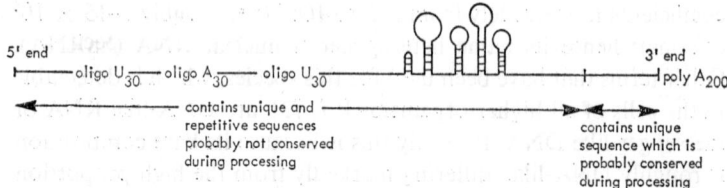

Fig. 12.14 *Some structural features of hnRNA from HeLa cells. (This must be regarded as highly schematic; at present the actual location of double-stranded regions, oligoA, and oligoU cannot be ascribed with any certainty)*

this general structural information, a precursor–product relationship with mRNA has been difficult to establish unambiguously. New techniques, whereby highly labelled DNA is made complementary to purified mRNA's using ocornavirus 'reverse transcriptase' (see Chapter 11) and used to probe hnRNA's for messenger-like sequences, have detected globin mRNA sequences in duck reticulocyte hnRNA [176] and ovalbumin mRNA sequences in chick oviduct hnRNA [177]. Furthermore, sequences complementary to certain viral DNA's (SV40, adenovirus, and herpes) have been found to appear in hnRNA and then in mRNA in cells infected with these viruses [178, 179, 156, 157].

Basically, most present-day data support the view that the formation of polyA at the 3'-termini of hnRNA and mRNA takes place by the sequential addition of adenylate moieties to completed RNA molecules by a process independent of a DNA template (Fig. 12.15). The process is insensitive to actinomycin D, and whilst there are oligo(dT) tracts in mammalian DNA [243] they are insufficiently long (20–40 nucleotides) to code for the polyA tracts (200 long) in hnRNA and mRNA. Moreover, a nuclear enzyme capable of carrying out this sequential addition has been detected

THE BIOSYNTHESIS OF RNA: TRANSCRIPTION

in nuclei and nuclear ribonucleoprotein particles [180–190]. A similar activity can be detected in cytoplasmic preparations [190].

In HeLa cells exposed to labelled adenosine, the proportion of labelled adenosine in polyA tracts is greater in those from hnRNA than in those from mRNA after short labelling periods (15–25 minutes). The situation is, however, reversed after longer labelling periods (150 minutes), consistent with a precursor–product relationship. Is this, however, connected with mRNA appearance in the cytoplasm? Here studies with the drug 3′-deoxyadenosine (Cordycepin) have been useful. Whilst the appearance of mRNA is inhibited by this drug, hnRNA synthesis is relatively unimpaired [191]. The explanation for this in current vogue is that the drug specifically inhibits the addition of the polyA tracts to hnRNA molecules. This event may be in some complex way involved in selecting hnRNA

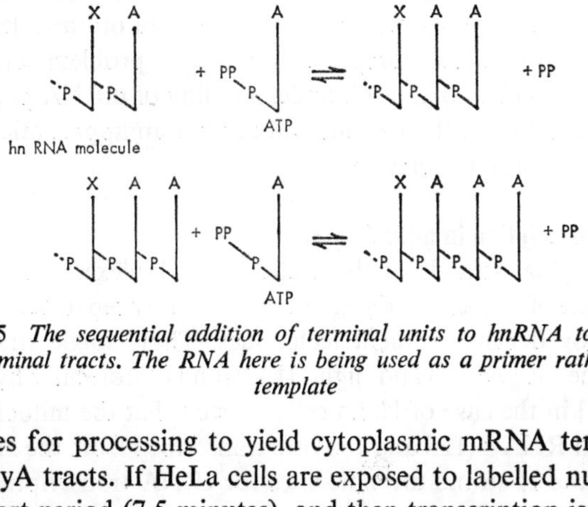

Fig. 12.15 *The sequential addition of terminal units to hnRNA to generate polyA terminal tracts. The RNA here is being used as a primer rather than a template*

molecules for processing to yield cytoplasmic mRNA terminated with polyA tracts. If HeLa cells are exposed to labelled nucleoside for a short period (7·5 minutes), and then transcription is stopped with actinomycin D, mRNA normally appears in polysomes some 20 minutes later. 3′-Deoxyadenosine, however, blocks the entry of 70 per cent of the mRNA into the polysome, and those that do arrive have short polyA tracts [192].

Recent data, however, indicate a metabolic turnover of the polyA tracts themselves in the nucleus as well as in the cytoplasm, and serve to underline the complexity of the situation and make kinetic analysis difficult [193].

hnRNA with similar characteristics has been detected not only in mammalian cells but also in most other eukaryotic cell types (e.g.

from birds [194, 195, 176, 177], echinoderms [170], insects [171], plants [196], and slime moulds [146, 147]). In summary, therefore, it does seem likely that the hnRNA and mRNA are related, and that hnRNA probably does contain mRNA precursors. Clearly a number of questions remain to be answered. What is the exact structure of hnRNA, the nature and function of its non-coding regions, the nature of enzymes involved in its cleavage, the proportion that contain mRNA sequences, and the number of these per molecule (about 80 per cent in the case of *Dictyostelium* hnRNA [147])? Finally, it should be pointed out that all the events discussed above occur not on nuclear RNA molecules themselves but more likely on these RNA's as part of complexes with protein (ribonucleoprotein complexes [198–203]). A variety of complexes of protein with hnRNA and with mRNA (Chapter 14) have been detected by various workers in the nucleus and cytoplasm of a number of animal cells. However, there are, as yet, only imprecise data on the relationship between these various complexes, and many problems arise from artefacts. It is clear that a full understanding of mRNA production and transport from the nucleus will require an appreciation of the role of these protein complexes.

12.8 Transcription in mitochondria

The individual strands of the cyclic DNA duplexes that comprise the genome of mitochondria can be conveniently separated from one another by density gradient techniques. These single strands are termed the 'heavy' (H) and 'light' (L) strands. Molecular hybridization [204] in the case of HeLa cells showed that the mitochondrial ribosomal RNA's (12S and 16S) are transcribed from the H-strand. Also on the H-strand are nine sites for tRNA transcription but a further three sites seem to be on the other strand, the L-strand. These 'genes' account for around 25 per cent of the potential information. However, this does not take account of the intriguing possibility of meaningful information being coded on both strands *at the same site*. Indeed, it has been shown that both strands are in fact almost completely transcribed [204].

In cells exposed to labelled uridine for long periods (46 hours) the mitochondrial RNA so labelled hybridized almost exclusively to the H-strand. The small amount which hybridized to the L-strand appeared to include that hybridizing to the three tRNA genes. The situation after very short 'pulse labels' was quite different. There seemed to be equal hybridization of the mitochondrial RNA to both strands [204].

It may be that HeLa cell mitochondria solve the problem of strand selection in transcription by transcribing both strands and rapidly degrading 98 per cent of the L-strand transcripts. This, of course, poses questions as to how the tRNA or rRNA (or mRNA) sections of these transcripts are correctly cut out.

Regarding mRNA's in mitochondria, polyA added post-transcriptionally has been detected in the RNA from mitochondrial polysomes. Its length, however, is smaller than that on cytoplasmic mRNA, being about 60–80 nucleotides in length [204]. That the polysomal RNA is a mitochondrial DNA transcript is suggested by the fact that it hybridizes to mitochondrial DNA, and its production is inhibited by ethidium bromide, a drug which specifically intercalates with the supercoiled cyclic DNA molecules of the mitochondria. Analysis of this polyA-terminated mitochondrial polysomal RNA by polyacrylamide electrophoresis showed it to occur as two distinct components, a 9S component which is coded for by the H-strand, and a 7S component coded for mainly by the L-strand [204]. This latter observation adds at least one more RNA species to the three transfer RNA species already known to be coded for by this L-strand.

Paradoxically, the major mRNA species from chloroplasts does not appear to contain polyA [205]; however, short polyA tracts can easily escape detection.

12.9 RNA-dependent synthesis of RNA

Several early *in vitro* studies [206–210] revealed a novel ability of the DNA-dependent RNA polymerases of certain bacteria (*A. vinelandii, M. lysodeikticus*, and *Esch. coli*) to respond in a *limited* fashion to an RNA template. When all four ribonucleoside 5'-triphosphates are present together with a natural (TMV-RNA) or synthetic (poly U) template, the product is an RNA with a sequence complementary to that of the template. The *in vivo* significance of these findings is not clear and, although some RNA-dependent reactions have been observed in mammalian cells, the products are mainly homopolymers added to the 3'-ends of RNA primers [7, 181, 211].

More convincing evidence for the existence of discrete RNA synthesizing systems using RNA's as templates has emerged from a variety of studies on cells infected with RNA-bacteriophages or RNA-viruses.

12.9.1 *RNA-bacteriophages* (for reviews see [212, 218])

The single-stranded RNA genome of the small RNA phages (e.g. R17, MS2, f2, Qβ) is replicated in two stages. (1) The entering

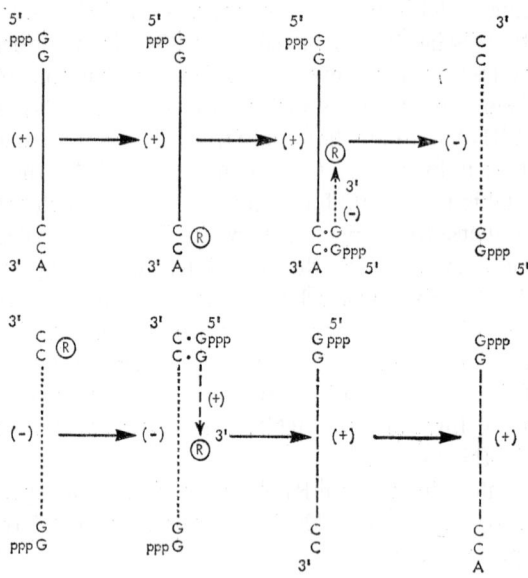

Fig. 12.16 *Synthesis of Qβ bacteriophage RNA*

phage RNA strand acts as a messenger which, in conjunction with the ribosomes of the host cell, controls the formation of 3 phage-specific proteins (coat protein, a maturation protein, and a 'replicase' subunit). (2) With the aid of the 'replicase' enzyme, new phage RNA is synthesized. The active replicase enzyme comprises four subunits, the phage-coded subunit mentioned above along with *three* subunits donated by the host which turn out in fact to be the two protein synthesis elongation factors EF-Tu and EF-Ts together with the 'interference' factor (i) (see Chapter 13).

The mechanism of replication is shown in Fig. 12.16. The original parental (+) phage RNA strand forms an enzyme RNA complex with the active replicase R, which binds at a site near the 3'-terminus. This may involve the recognition of the phage 3'-terminal sequence

THE BIOSYNTHESIS OF RNA: TRANSCRIPTION

-CCA by the EF-Tu component of the replicase (EF-Tu normally recognizes aminoacyl-tRNA, and tRNA also has a 3′-terminal sequence -CCA). Using this (+) strand as template, a complementary (−) strand (dotted line) is formed. The replicase uses ribonucleoside 5′-triphosphates and begins synthesis at the penultimate cytidine (C) residue rather than at the 3′-terminal adenosine (A), and directs the incorporation of a pppG residue as the first nucleotide at the 5′-end of the (−) strand. Synthesis is in the direction 5′→3′.

In the next stage of replication, the replicase, *but without host factors this time*, interacts with the phage (−) strand and, using it as template, catalyses the synthesis of a new (+) strand which grows in the 5′→3′ direction. After completion of new (+), or progeny, strands (broken lines) the enzyme adds a final adenosine residue, *although this is not specified by the template*, to re-form the -CCA terminal sequence.

12.9.2 *RNA viruses of animal cells*

In the case of the *picornaviruses*, e.g. polio, encephalomyocarditis (EMC), and mengo viruses, and *arboviruses*, e.g. Semliki forest and Sinbis viruses, the general mechanism of virus replication is presumed to resemble that described for the small RNA phages. However, the virus-induced replicases so far studied have been very crude preparations, and precise template requirements have not been elucidated [219–222].

The genome, however, in the case of the *rhabdoviruses* (e.g. vesicular stomatitis virus) and *paramyxovirus* (e.g. Newcastle disease and Sendai viruses) is effectively of (−) complementarity. In infected cells this is first transcribed into translatable (+ type) messenger RNA's. This initial transcription is carried out by an RNA-dependent enzyme which is a structural component of the invading virus particle [223, 224]. For genome replication, translation of these messengers is required to produce an RNA replicase activity which is thought to use the (+) type strands as templates; see Fig. 12.17(a) [220].

Another class of virus which contains an RNA-dependent RNA transcribing enzyme as a structural element is the *diplornavirus* (e.g. reovirus) group [225, 227, 220]. The virus enzyme carries out the initial transcription in the infected cell of the various segments of double-stranded RNA that comprise the genome, to produce (+) type messenger RNA's. Like the rhabdovirus and paramyxovirus situations described above, these mRNA's appear to lead to the induction of RNA replicase machinery which permits the production of double-stranded viral genomes using the viral mRNA's (+) as templates; see Fig. 12.17(b) [220].

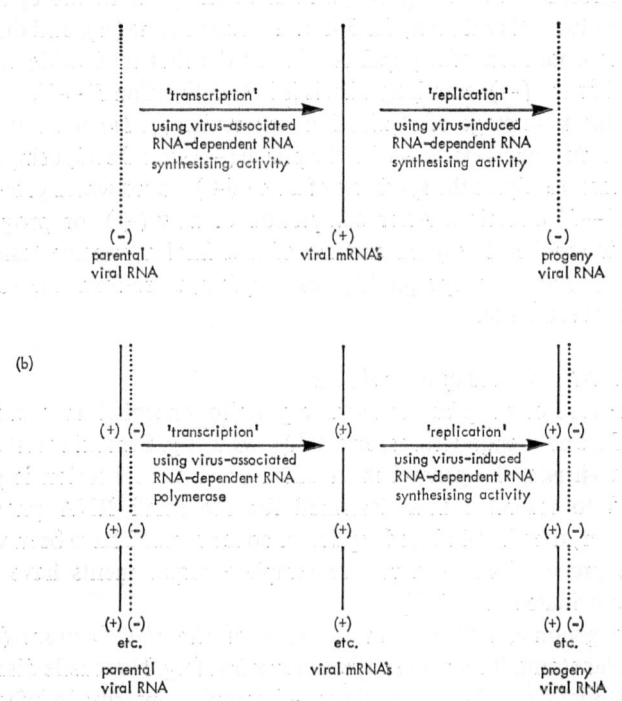

Fig. 12.17 *Schematic representations of RNA-dependent RNA synthetic mechanisms operative in the replication of (a) rhabdoviruses and paramyxoviruses, and (b) diplornaviruses*

12.10 The terminal addition of nucleotide units to tRNA

The terminal addition of nucleotides in the formation of homopolymers (e.g. polyA) has already been mentioned. Such an activity is catalysed by a wide range of enzymes of varying specificity and origin [228, 180, 181].

Transfer RNA's, however, present a special case. All of them have the common trinucleotide sequence pCpCpA which may readily be removed to yield an acceptor material whose end groups may be represented as X and Y (Fig. 12.18). Two CMP moieties are first attached sequentially to the 3'-hydroxyl of the ribose in the terminal

THE BIOSYNTHESIS OF RNA: TRANSCRIPTION

nucleotide Y. The new terminal CMP now accepts an AMP residue by a similar pyrophosphoryl cleavage of ATP. The final sequence at the polynucleotide chain is therefore pXpYpCpCpA [229–233].

In the cell this sequence appears to have a metabolic turnover

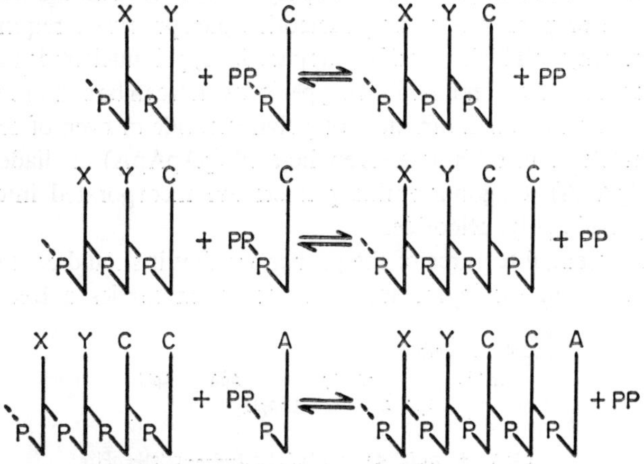

Fig. 12.18 *Addition of terminal units to tRNA to give the final sequence –pXpYpCpCpA*

independent of the remainder of the transfer RNA molecule. The three nucleotides are apparently continuously removed and replaced [234, 235]. The cellular role of these nucleotide additions to tRNA is not understood, but it may serve in some way to regulate protein synthesis, perhaps by controlling the levels of functional tRNA's.

12.11 Polynucleotide phosphorylase (E.C.2.7.7.8)

The first clear indication of the mechanism by which RNA might be synthesized enzymically was actually obtained in 1955 by Ochoa and his colleagues [236–238], who isolated from the micro-organism *Azotobacter vinelandii* an enzyme which catalyses the synthesis of high molecular weight polyribonucleotides from nucleoside 5'-diphosphates with the release of orthophosphate. The reaction is reversible and requires magnesium ions. It reaches equilibrium when 60–80 per cent of the nucleoside diphosphate has disappeared, and may be represented:

$$nNDP \rightleftharpoons (NMP)_n + nPi$$

where N stands for adenine, hypoxanthine, uracil, or cytosine. The enzyme involved has been named *polynucleotide phosphorylase* and has been extensively reviewed [237, 239]. It is widely distributed in bacteria and has also been found in plant tissues. There is at present little convincing evidence for its occurrence in animal tissues. It can readily be purified from bacterial sources [240] and has proved to be of great value in the preparation of polynucleotides in the laboratory.

Crude preparations of polynucleotide phosphorylase require no primer but, with highly purified preparations, polynucleotide formation occurs only after an initial lag period which can be eliminated by the addition of small amounts of polynucleotide or even of certain oligonucleotides such as triadenylic acid (pApApA) or diadenylic acid (pApA). Oligonucleotide primers are incorporated into the newly made polynucleotide.

The essential features of the primer are that it should contain at least two nucleoside residues, one of which carries a free C-3'

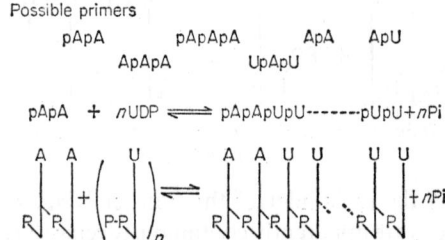

Fig. 12.19 *Primers in the polynucleotide phosphorylase reaction*

hydroxyl group on which a new phosphodiester bond can be formed (Fig. 12.19).

The first bond formed is a phosphodiester bridge between the 5'-phosphate of a UMP residue and the free terminal C-3' hydroxyl of pApA, and the chain is extended by similar condensations, the primer being incorporated into the product.

The purified enzyme contains about 3 per cent of nucleotide in the form of a complex oligonucleotide which probably acts as a built-in primer and is responsible for the definite but sluggish reaction which takes place in the absence of added primer.

The reversal of the polymerization reaction, phosphorolysis, in which the polynucleotide is incubated with the enzyme in the presence of an excess of inorganic phosphate to yield the nucleoside diphosphates by stepwise removal of mononucleotide units, has also been studied. The biosynthetic polymers are readily phosphorolysed, and so are oligonucleotides which act as primers, but, as might be

THE BIOSYNTHESIS OF RNA: TRANSCRIPTION

expected, dinucleotides and dinucleoside monophosphates are not. Tobacco mosaic virus RNA and highly polymerized yeast RNA are phosphorolysed readily, but yeast RNA treated with alkali is phosphorolysed slowly. The formation of multi-stranded chains as between polyA and polyU results in a slow rate of phosphorolysis. The transfer RNA of the cell cytoplasm is also incompletely phosphorolysed, 70–80 per cent being left unchanged presumably because of the secondary structure of tRNA. The phosphorolysis appears to affect mainly the terminal groups.

The function of polynucleotide phosphorylase in the cell has been the subject of much discussion [259]. It is possible that it is primarily responsible for the degradation of RNA to yield nucleoside diphosphates, which are the immediate precursors of deoxyribonucleotides (p. 207), and it may even control the level of inorganic phosphate in the cell. It may also be involved in the degradation of messenger RNA, though this is now considered unlikely.

REFERENCES

[1] Smellie, R. M. S. (1965) *Brit. Med. Bull.*, **21**, 195
[2] Smellie, R. M. S. (1965) *Developmental and Metabolic Control Mechanisms and Neoplasia*, p. 29
[3] Weiss, S. B. (1963) *Informational Macromolecules*, p. 61 (H. J. Vogel, V. Bryson and J. O. Lampen, Eds.). New York: Academic Press
[4] Hurwitz, J., Furth, J. J. and Kahan, F. M. (1962) *Basic Problems in Neoplastic Disease*, p. 35 (A. Gellhorn and E. Hirschberg, Eds.). Columbia University Press
[5] Hurwitz, J. and August, J. T. (1963) *Progress in Nucleic Acid Research*, Vol. 1, p. 59 (J. N. Davidson and W. E. Cohn, Eds.). New York: Academic Press
[6] Hurwitz, J., Evans, A., Babinet, C. and Skalka, A. (1963) *Cold Spring Harbor Symp. Quant. Biol.*, **28**, 59
[7] Smellie, R. M. S. (1963) *Progress in Nucleic Acid Research*, Vol. 1, p. 27 (J. N. Davidson and W. E. Cohn, Eds.). New York: Academic Press
[8] Stevens, A. (1963) *Ann. Rev. Biochem.*, **32**, 15
[9] Volkin, E. (1963) *Molecular Genetics*, Part I, p. 271 (J. H. Taylor, Ed.). New York: Academic Press
[10] Grunberg-Manago, M. (1963) *Progress in Biophysics and Molecular Biology*, **13**, 175 (J. A. V. Butler, Ed.)
[11] Allfrey, V. G. (1966) *Cancer Res.*, **26** (Part I), 2026
[12] Georgiev, G. P. (1967) *Progress in Nucleic Acid Research and Molecular Biology*, Vol. 6, p. 259 (J. N. Davidson and W. E. Cohn, Eds.). New York: Academic Press
[13] Sibatani, A. (1966) *Progr. Biophys. Mol. Biol.*, **16**, 17
[14] Richardson, J. P. (1969) *Progress in Nucleic Acid Research and Molecular Biology*, Vol. 9, p. 75 (J. N. Davidson and W. E. Cohn, Eds.). New York: Academic Press

[15] Bautz, E. K. F. (1967) *Molecular Genetics*, Part II, p. 213 (J. H. Taylor, Ed.). New York: Academic Press
[16] Ishimhama, A. and Hurwitz, J. (1969) *J. Biol. Chem.*, **244**, 6680
[17] Hurwitz, J. (1970) *Studies on the DNA-Dependent Synthesis of RNA with RNA Polymerase*, Harvey Lectures, **64**, 157. New York, London: Academic Press
[18] Silvestri, L. (1970) *RNA-polymerase and Transcription*. Amsterdam: North-Holland Publishing Co.
[19] Sethi, V. S. (1971) *Progr. Biophys. Mol. Biol.* (J. A. V. Butler and D. Noble, Eds.), **23**, 67
[20] Bautz, E. K. F. (1972) *Progr. Nucleic Acid Res. Mol. Biol.*, **12**, 129
[21] Burgess, R. R. (1971) *Ann. Rev. Biochem.*, **40**, 711
[22] Chamberlin, M. J. (1974) *Ann. Rev. Biochem.*, **43**, 721
[23] Travers, A. (1974) *Cell*, **3**, 97
[24] Geiduschek, E. P., Nakamoto, T. and Weiss, S. B. (1961) *Proc. Nat. Acad. Sci.*, **47**, 1405
[25] Nygaard, A. P. and Hall, B. D. (1963) *Biochem. Biophys. Res. Commun.* **12**, 98
[26] Gillespie, D. and Spiegelmann, S. (1965) *J. Mol. Biol.*, **12**, 829
[27] Bautz, E. K. F. and Hall, B. D. (1962) *Proc. Nat. Acad. Sci.*, **48**, 400
[28] Bolton, E. T. and McCarthy, B. J. (1962) *Proc. Nat. Acad. Sci.*, **48**, 1390
[29] Bolton, E. T. and McCarthy, B. J. (1964) *J. Mol. Biol.*, **8**, 201
[30] Hoyer, B. H., McCarthy, B. J. and Bolton, E. T. (1963) *Science*, **140**, 1408
[31] McCarthy, B. J. and Bolton, E. T. (1964) *J. Mol. Biol.*, **8**, 184
[32] Paul, J. and Gilmour, R. S. (1968) *J. Mol. Biol.*, **34**, 305
[33] Gilmour, R. S. and Paul, J. (1970) *FEBS Lett.*, **9**, 242
[34] Dickson, R. C., Abelson, J., Barnes, W. M. and Reznikoff, W. S. (1975) *Science*, **187**, 27
[35] Saucier, J. M. and Wang, J. C. (1972) *Nature New Biol.*, **239**, 167
[36] Heil, H. and Zillig, W. (1970) *FEBS Lett.*, **11**, 165
[37] Maitra, U. and Hurwitz, S. (1965) *Proc. Nat. Acad. Sci.*, **54**, 815
[38] Miller, O. L. et al. (1970) *Cold Spring Harbor Symp. Quant. Biol.*, **35**, 505
[39] Hamkalo, B. A. and Miller, O. L. (1973) *Ann. Rev. Biochem.*, **42**, 379
[40] Goldberg, I. H. and Friedman, P. A. (1971) *Ann. Rev. Biochem.*, **40**, 775
[41] Das, H. K., Goldstein, A. and Lowrey, L. I. (1967) *J. Mol. Biol.*, **24**, 23
[42] Dunn, J. J. and Studier, W. F. (1973) *Proc. Nat. Acad. Sci.*, **70**, 1559
[43] Peters, G. G. and Hayward, R. S. (1974) *Eur. J. Biochem.* **48**, 199
[44] Roberts, J. W. (1969) *Nature*, **224**, 1168
[45] Oda, T. and Takanami, M. (1972) *J. Mol. Biol.*, **71**, 799
[46] Loewen, P. C., Sekiya, T. and Kohrana, H. G. (1974) *J. Biol. Chem.*, **249**, 217
[47] Jacob, S. T. (1973) *Progr. Nucleic Acid Res. Mol. Biol.*, **13**, 93
[48] Chambon, P. (1974) *The Enzymes*, Vol. 10, p. 261 (P. D. Boyer, Ed.). New York: Academic Press
[49] Chambon, P. (1975) *Ann. Rev. Biochem.*, **49**, 613
[50] Rutter, W. J., Goldberg, M. I. and Perriard, J. C. (1974) *Biochemistry of Cell Differentiation*, MTP International Review of Science, Biochemistry Series 1, p. 267 (J. Paul, Ed.). London: Butterworths
[51] Weiss, S. B. (1960) *Proc. Nat. Acad. Sci.*, **46**, 1020
[52] Burdon, R. H. and Smellie, R. M. S. (1962) *Biochim Biophys. Acta*, **61**, 633
[53] Roeder, R. G. and Rutter, W. J. (1970) *Proc. Nat. Acad. Sci.*, **65**, 675

[54] Price, R. and Penman, S. (1972) *J. Mol. Biol.*, **70**, 435
[55] Stein, H. and Hausen, P. (1971) *Cold Spring Harbor Symp. Quant. Biol.*, **35**, 709
[56] Weaver, R. F., Blatti, S. P. and Rutter, W. J. (1971) *Proc. Nat. Acad. Sci.*, **68**, 2994
[57] Sklar, V. E. F., Schwartz, L. B. and Roeder, R. G. (1975) *Proc. Nat. Acad. Sci.*, **72**, 348
[58] Kunztzel, H. and Schafer, K. P. (1971) *Nature*, **231**, 265
[59] Scragg, A. H. (1971) *Biochem. Biophys. Res. Comm.*, **45**, 701
[60] Saccone, C., Gallerani, R., Gadieta, M. N. and Greco, M. (1971) *FEBS Lett.*, **18**, 339
[61] Burdon, R. H. (1971) *Progr. Nucleic Acid Res. Mol. Biol.*, **11**, 33
[62] Bishop, J. O. (1972) Fifth Karolinska Symposium, *Gene Transcription in Reproductive Tissue*, p. 247 (E. Diczfalusy, Ed.). Stockholm: Karolinska Institute
[63] Kennell, D. E. (1971) *Progr. Nucleic Acid Res. Mol. Biol.*, **11**, 259
[64] Yanofsky, S. A. and Spiegelman, S. (1962) *Proc. Nat. Acad. Sci.*, **48**, 1466
[65] Maden, B. E. H. (1971) *Progr. Biophys. Mol. Biol.*, **22**, 127
[66] Perry, R. P. (1967) *Progr. Nucleic Acid Res. Mol. Biol.*, **6**, 219
[67] Darnell, J. E. (1968) *Bact. Rev.*, **32**, 262
[68] Evans, H. J. (1973) *Brit. Med. Bull.*, **29**, 196
[69] Birnstiel, M. L., Spiers, J., Purdon, I., Jones, K. and Loening, U. E (1968) *Nature*, **219**, 459
[70] Penman, S., Smith, I. and Holtzman, E. (1966) *Science*, **154**, 789
[71] Wellauer, P. K. and Dawid, I. B. (1973) *Proc. Nat. Acad. Sci.*, **70**, 2827
[72] Maden, B. E. H., Salim, M. and Shepherd, J. (1973) *Biochem. Soc. Symp.*, **37**, 23
[73] Weinberg, R. A. (1973) *Ann. Rev. Biochem.*, **42**, 329
[74] Ford, P. J. (1973) *Biochem. Soc. Symp.*, **37**, 69
[75] Schlessinger, D., Oho, M., Nickolaev, N. and Silengo, C. (1974) *Biochemistry*, **13**, 4268
[76] Nickolaev, N., Birenbaum, M. and Schlessinger, D. (1975) *Biochim. Biophys. Acta*, **395**, 478
[77] Nickolaev, N., Birge, C. H., Gotoh, S., Glazer, K. and Schlessinger, D. (1975) *Brookhaven Symp. Biol.*, **26**, 175
[78] Nickolaev, N. and Schlessinger, D. (1974) *Biochem.*, **13**, 4272
[79] Thammana, P. and Held, W. A. (1974) *Nature*, **251**, 682
[80] Liau, M. C. and Hurlbert, R. B. (1975) *Biochem.*, **14**, 127
[81] Winicov, I. and Perry, R. P. (1975) *Brookhaven Symp. Biol.*, **26**, 201
[82] Burdon, R. H. (1975) *Brookhaven Symp. Biol.*, **26**, 138
[83] Chen, G. S. and Siddiqui, M. A. Q. (1975) *J. Mol. Biol.*, **96**, 153
[84] McReynolds, L. and Penman, S. (1974) *Cell*, **3**, 185
[85] Dijk, J. and Singhal, R. P. (1974) *J. Biol. Chem.*, **249**, 645
[86] Ikemura, T., Shimura, Y., Sakano, H. and Ozeki, H. (1975) *J. Mol. Biol.*, **96**, 69
[87] Smith, J. D. (1973) *Brit. Med. Bull.*, **29**, 220
[88] Altman, S. (1975) *Cell*, **4**, 21
[89] Guthrie, C. *et al.* (1975) *Brookhaven Symp. Biol.*, **26**, 106
[90] Abelson, J. *et al.* (1975) *Brookhaven Symp. Biol.*, **26**, 77
[91] Daniel, V., Grimberg, J. I. and Zeevi, M. (1975) *Nature*, **257**, 193
[92] Lipmann, F. (1963) *Progress in Nucleic Acid Research*, Vol. 1, p. 135 (J. N. Davidson and W. E. Cohn, Eds.). New York: Academic Press

[93] Hurwitz, J. and Furth, J. J. (1962) *Sci. Amer.*, **206**(2), 3
[94] Spiegelman, S. (1963) *Fed. Proc.*, **22**, 36
[95] Gros, F., Gilbert, W. and Watson, J. D. (1961) *Cold Spring Harbor Symp. Quant. Biol.*, **26**, 111
[96] Gros, F., Naono, S., Woese, C., Willson, C. and Attardi, G. (1963) *Informational Macromolecules*, p. 387 (H. J. Vogel, V. Bryson and J. O. Lampen, Eds.). New York: Academic Press
[97] Gros, F., Dubert, J. M., Tissières, A., Bourgeois, S., Michelson, M., Soffer, R. and Legault, L. (1963), *Cold Spring Harbor Symp. Quant. Biol.*, **28**, 299
[98] Jacob, F. and Monod, J. (1961) *Cold Spring Harbor Symp. Quant. Biol.*, **26**, 193
[99] Jacob, F. and Monod, J. (1961) *J. Mol. Biol.*, **3**, 318
[100] Watson, J. D. (1963) *Science*, **140**, 17
[101] Gros, F., Naono, S., Hayes, D., Hayes, F. and Watson, J. D. (1962) Colloques Internationaux du Centre de la Recherche Scientifique, p. 437
[102] Monod, J., Jacob, F. and Gros, F. (1961) *Biochem. Soc. Symp.*, **21**, 104
[103] Geiduschek, E. P. and Haselkorn, R. (1969) *Ann. Rev. Biochem.*, **38**, 647
[104] Mathews, M. B. (1973) *Essays Biochem.*, **9**, 59
[105] Brawerman, G. (1974) *Ann. Rev. Biochem.*, **43**, 621
[106] Hershey, A. D., Dixon, J. and Chase, M. (1953) *J. Gen. Physiol.*, **73**, 110
[107] Volkin, E. and Astrachan, L. (1956) *Virology*, **2**, 149, 433
[108] Bessman, M. J. (1963) *Molecular Genetics*, Part I, p. 1 (J. H. Taylor, Ed.). New York: Academic Press
[109] Nomura, M., Hall, B. D. and Spiegelman, S. (1960) *J. Mol. Biol.*, **2**, 306
[110] Spiegelman, S., Hall, B. D. and Storck, R. (1961) *Proc. Nat. Acad. Sci.*, **47**, 1135
[111] Spiegelman, S. (1961) *Cold Spring Harbor Symp. Quant. Biol.*, **26**, 75
[112] Brenner, S., Jacob, F. and Meselson, M. (1961) *Nature*, **190**, 576
[113] Gros, F., Hiatt, H., Gilbert, W., Kurland, C. G., Risebrough, R. W. and Watson, J. D. (1961) *Nature*, **190**, 581
[114] Hayashi, M. and Spiegelman, S. (1961) *Proc. Nat. Acad. Sci.*, **47**, 1564
[115] Risebrough, R. W., Tissières, A. and Watson, J. D. (1962) *Proc. Nat. Acad. Sci.*, **48**, 430
[116] Bautz, E. K. F. and Hall, B. D. (1962) *Proc. Nat. Acad. Sci.*, **48**, 400
[117] Champe, S. P. and Benzer, S. (1962) *Proc. Nat. Acad. Sci.*, **48**, 332
[118] Marmur, J., Greenspan, C. M., Policek, E., Kahan, F. M., Levine, J. and Mandel, M. (1963) *Cold Spring Harbor Symp. Quant. Biol.*, **28**, 191
[119] Marmur, J. and Greenspan, C. M. (1963) *Science*, **142**, 387
[120] Tocchini-Valentini, G. P., Stodolsky, M., Aurisicchio, A., Sarnat, M., Fraziosi, F., Weiss, S. B. and Geiduschek, E. P. (1963) *Proc. Nat. Acad. Sci.*, **50**, 935
[121] Hayashi, M., Hayashi, M. N. and Spiegelman, S. (1963) *Science*, **140**, 1313
[122] Levinthal, C., Keynan, A. and Higa, A. (1962) *Proc. Nat. Acad. Sci.*, **48**, 1631
[123] Gilbert, W. (1963) *J. Mol. Biol.*, **8**, 708
[124] Noll, H., Staehelin, T. and Wettstein, F. O. (1963) *Nature*, **198**, 632
[125] Otaka, E., Mitsui, H. and Osawa, S. (1962) *Proc. Nat. Acad. Sci.*, **48**, 425
[126] Blundell, M. and Kennell, D. (1974) *J. Mol. Biol.*, **83**, 143
[127] Achord, D. and Kennell, D. (1974) *J. Mol. Biol.*, **90**, 581
[128] Dunn, J. J. and Studier, F. W. (1973) *Proc. Nat. Acad. Sci.*, **70**, 1559
[129] Dunn, J. J. and Studier, F. W. (1975) *Brookhaven Symp. Biol.*, **26**, 267

[130] Rosenberg, M., Kramer, R. A. and Steitz, J. A. (1974) *J. Mol. Biol.*, **89**, 777.
[131] Chantrenne, H., Burny, A. and Marbaix, G. (1967) *Progress in Nucleic Acid Research and Molecular Biology*, Vol. 7, p. 173 (J. N. Davidson and W. E. Cohn, Eds.). New York: Academic Press
[132] Evans, M. J. and Lingrel, J. B. (1969) *Biochem.*, **8**, 829, 3000
[133] Williamson, R., Morrison, M., Lanyon, G., Eason, R. and Paul, J. (1971) *Biochem.*, **10**, 3014
[134] Gaskell, P. and Kabat, D. (1971) *Proc. Nat. Acad. Sci.*, **68**, 72
[135] Pemberton, R. E., Housman, D., Lodish, H. F. and Baglioni, C. (1972) *Nature New Biol.*, **235**, 99
[136] Borun, T. W., Scharff, M. D. and Robbins, E. (1967) *Proc. Nat. Acad. Sci.*, **58**, 1977
[137] Kedes, L. H. and Birnsteil, M. L. (1971) *Nature New Biol.*, **230**, 165
[138] Bloemendal, H. (1972) *The Mechanisms of Protein Synthesis and its Regulation*, Frontiers of Biology 27. Amsterdam, London: North Holland Publishing Co.
[139] Swan, D., Aviv, H. and Leder, P. (1972) *Proc. Nat. Acad. Sci.*, **69**, 1967
[140] Milstein, C., Brownlee, G. G., Harrison, T. M. and Matthews, M. B. (1972) *Nature New Biol.*, **239**, 117
[141] Palacios, R., Sullivan, D., Summers, N. M., Kiely, M. L. and Schimke, R. T. (1973) *J. Biol. Chem.*, **248**, 540
[142] Morris, G. E. *et al.* (1972) *Cold Spring Harbor Symp. Quant. Biol.*, **37**, 535
[143] Adesnik, M. and Darnell, J. E. (1972) *J. Mol. Biol.*, **67**, 397
[144] Greenberg, J. R. and Perry, R. P. (1972) *J. Mol. Biol.*, **72**, 91
[145] Sagler, D., Edelman, M. and Jakob, K. M. (1974) *Biochim. Biophys. Acta*, **349**, 32
[146] Firtel, R. A. and Lodish, H. F. (1973) *J. Mol. Biol.*, **79**, 315
[147] Jacobson, A., Firtel, R. A. and Lodish, H. F. (1975) *Brookhaven Symp. Biol.*, **26**, 307
[148] Reed, J. and Wintersberger, E. (1973) *FEBS Lett.*, **32**, 213
[149] McLaughlin, C. S., Warner, J. R., Edmonds, M., Nakazoto, H. and Vaughan, M. H. (1973) *J. Biol. Chem.*, **248**, 1466
[150] Ohta, N., Sanders, M. and Newton, A. (1975) *Proc. Nat. Acad. Sci.*, **72**, 2743
[151] Nakazoto, H., Venkatesan, S. and Edmonds, M. (1975) *Nature*, **256**, 144
[152] Keuchler, E. and Rich, A. (1969) *Proc. Nat. Acad. Sci.*, **63**, 520
[153] Heywood, S. M. and Nwagwu, M. (1968) *Proc. Nat. Acad. Sci.*, **60**, 229
[154] Lee, S. Y., Mendecki, J. and Brawerman, G. (1971) *Proc. Nat. Acad. Sci.*, **68**, 1331
[155] Burr, H. and Lingrel, J. B. (1971) *Nature New Biol.*, **233**, 41
[156] Philipson, L., Wall, R., Glickman, G. and Darnell, J. E. (1971) *Proc. Nat. Acad. Sci.*, **68**, 2806
[157] Darnell, J. E., Philipson, L., Wall, R. and Adesnik, M. (1971) *Science*, **174**, 508
[158] Adams, J. M. and Cory, S. (1975) *Nature*, **255**, 28
[159] Furuchi, Y., Morgan, M., Shatkin, A. J., Jelinek, W., Salditt, M., Georgieff, G. and Darnell, J. E. (1975) *Proc. Nat. Acad. Sci.*, **72**, 1904
[160] Wei, C. M., Gershowitz, A. and Moss, B. (1975) *Cell*, **4**, 379
[161] Perry, R. P., Kelley, D. E., Frederici, K. and Rottman, F. (1975) *Cell*, **4**, 387

[162] Moyer, S. A., Abraham, G., Adler, R. and Banerjee, A. K. (1975) *Cell*, **5,** 59
[163] Levi, S. and Shatkin, A. J. (1975) *Proc. Nat. Acad. Sci.*, **72,** 2012
[164] Suzuki, Y. and Brown, D. D. (1972) *J. Mol. Biol.*, **63,** 409
[165] Singer, R. H. and Penman, S. (1973) *J. Mol. Biol.*, **78,** 321
[166] Puckett, L., Chambers, S. and Darnell, J. E. (1975) *Proc. Nat. Acad. Sci.*, **72,** 389
[167] Bishop, J. O., Morton, J. G., Rosbash, M. and Richardson, M. (1974) *Nature*, **250,** 199
[168] Klein, W. H., Murphy, W., Attardi, G., Britten, R. J. and Davidson, E. H. (1974) *Proc. Nat. Acad. Sci.*, **71,** 1785
[169] Molloy, G. R., Jelinek, W., Salditt, M. and Darnell, J. E. (1974) *Cell*, **1,** 43
[170] Smith, M. J., Hough, B. R., Chamberlin, M. E. and Davidson, E. H. (1974) *J. Mol. Biol.*, **85,** 103
[171] Spradling, A., Penman, S., Campo, M. S. and Bishop, J. O. (1974) *Cell*, **3,** 23
[172] Holmes, D. S. and Bonner, J. (1974) *Biochem.*, **13,** 841
[173] Holmes, D. S. and Bonner, J. (1974) *Proc. Nat. Acad. Sci.*, **71,** 1108
[174] Lewin, B. (1975) *Cell*, **4,** 77
[175] Burdon, R. H. and Shenkin, A. (1972) *FEBS Lett.*, **24,** 11
[176] MacNaughton, M., Freeman, K. B. and Bishop, J. O. (1974) *Cell*, **1,** 117
[177] McKnight, G. S. and Schimke, R. T. (1974) *Proc. Nat. Acad. Sci.*, **71,** 4327
[178] Wall, R. and Darnell, J. E. (1971) *Nature*, **232,** 73
[179] Wagner, E. K. and Roizman, B. (1969) *Proc. Nat. Acad. Sci.*, **64,** 626
[180] Edmonds, M. and Abrams, R. (1960) *J. Biol. Chem.*, **235,** 1142
[181] Burdon, R. H. (1963) *Biochem. Biophys. Res. Commun.*, **11,** 472
[182] Neissing, J. and Sekeris, C. E. (1974) *Biochem. Biophys. Res. Commun.*, **53,** 673
[183] Jelinek, W. R. (1974) *Cell*, **2,** 197
[184] Mans, R. J. (1973) *Biochem. Biophys. Res. Commun.*, **53,** 245
[185] Winters, M. A. and Edmonds, M. (1973) *J. Biol. Chem.*, **248,** 4756
[186] Neissing, J. and Sekeris, C. E. (1973) *Nature New Biol.*, **243,** 9
[187] Banks, S. P. (1973) *Science*, **181,** 1064
[188] Huff, L. A. and Keller, E. B. (1973) *Biochem. Biophys. Res. Commun.*, **51,** 704
[189] Neissing, J. and Sekeris, C. E. (1972) *FEBS Lett.*, **22,** 83
[190] Coleman, M. S., Hutton, J. J. and Bollum, F. J. (1974) *Nature*, **248,** 407
[191] Penman, S., Rosbash, M. and Penman, M. (1970) *Proc. Nat. Acad. Sci.*, **67,** 1878
[192] Adesnick, M., Salditt, M., Thomas, W. and Darnell, J. E. (1972) *J. Mol. Biol.*, **71,** 21
[193] Perry, R. P., Kelley, D. E. and La Torre, J. (1974) *J. Mol. Biol.*, **82,** 315
[194] Scherrer, K. *et al.* (1966) *Proc. Nat. Acad. Sci.*, **56,** 1571
[195] Spohr, G., Imaizumi, T. and Scherrer, K. (1974) *Proc. Nat. Acad. Sci.*, **71,** 5009
[196] Teissere, M., Peron, P., Ricard, J. and Ratle, G. (1972) *FEBS Lett.*, **23,** 65
[197] Lengyel, J. and Penman, S. (1975) *Cell*, **5,** 281
[198] Quinlan, T. J., Billings, P. B. and Martin, T. E. (1974) *Proc. Nat. Acad. Sci.*, **71,** 2632
[199] Ishikawa, K., Sato, T., Sato, S. and Ogata, K. (1974) *Biochim. Biophys. Acta*, **353,** 420

[200] Gallinaro-Matringe, H., Stevenin, J. and Jacob, M. (1975) *Biochem.*, **14**, 2547
[201] Kumar, A. and Pederson, T. (1975) *J. Mol. Biol.*, **96**, 353
[202] Pederson, T. (1974) *J. Mol. Biol.*, **83**, 163
[203] Firtel, R. A. and Pederson, T. (1975) *Proc. Nat. Acad. Sci.*, **72**, 301
[204] Attardi, G., Constantino, P. and Ojala, D. (1974) *The Biogenesis of Mitochondria*, p. 9 (A. M. Kroon and C. Saccone, Eds.). New York: Academic Press
[205] Wheeler, A. M. and Hartley, M. R. (1975) *Nature*, **257**, 66
[206] Stevens, A. and Henry, J. (1964) *J. Biol. Chem.*, **239**, 196, 204
[207] Nakomoto, T. and Weiss, S. B. (1962) *Proc. Nat. Acad. Sci.*, **48**, 880
[208] August, J. T., Ortiz, P. J. and Hirwitz, J. (1962) *J. Biol. Chem.*, **237**, 3786
[209] Fox, C. F., Robinson, W. S., Haselkorn, R. and Weiss. S. B. (1964) *J. Biol. Chem.*, **259**, 186
[210] Krakow, J. S. and Ochoa, S. (1963) *Proc. Nat. Acad. Sci.*, **49**, 88
[211] Wilkie, N. and Smellie, R. M. S. (1968) *Biochem. J.*, **109**, 229, 485
[212] Ochoa, S., Weissman, C., Borst, P., Burdon, R. H. and Billiter, M. A. (1964) *Fed. Proc.*, **23**, 1285
[213] Weissman, C. and Ochoa, S. (1967) *Progr. Nucleic Acid Res. Mol. Biol.*, **6**, 353
[214] Spiegelman, S. (1970) *Harvey Lectures*, **64**, 1. New York: Academic Press
[215] Stavis, R. L. and August, J. T. (1970) *Ann. Rev. Biochem.*, **39**, 527
[216] Kozak, M. and Nathans, D. (1972) *Bact. Rev.*, **36**, 109
[217] Weissman, C., Billiter, M. A., Goodman, H. M., Hindley, J. and Weber, M. (1973) *Ann. Rev. Biochem.*, **42**, 303
[218] Hindley, J. (1973) *Brit. Med. Bull.*, **29**, 236
[219] Baltimore, D. (1971) *Bact. Rev.*, **35**, 235
[220] Shatkin, A. J. (1974) *Ann. Rev. Biochem.*, **43**, 643
[221] Ehrenfeld, E., Maizel, J. V. and Summers, D. F. (1970) *Virology*, **40**, 840
[222] Rosenberg, H., Diskin, B., Oran, L. and Traub, A. (1972) *Proc. Nat. Acad. Sci.*, **69**, 3815
[223] Szilagy, J. F. and Uryuayev, L. (1973) *J. Virol.*, **11**, 279
[224] Perlman, S. M. and Huang, A. S. (1975) *Virus Research*, p. 97 (C. F. Fox and W. S. Robinson, Eds.). New York: Academic Press
[225] Joklik, W. K. (1975) *Virus Research*, p. 105 (C. F. Fox and W. S. Robinson, Eds.). New York: Academic Press
[226] Shatkin, A. J. and Sipe, J. D. (1968) *Proc. Nat. Acad. Sci.*, **61**, 1462
[227] Smith, R. E., Zweernick, H. J. and Joklik, W. K. (1969) *Virology*, **39** 791
[228] Burdon, R. H. and Smellie, R. M. S. (1962) *Biochim. Biophys. Acta*, **61**, 633
[229] Heidelberger, C., Harbers, E., Leibman, K. C., Takagi, Y. and Potter, V. R. (1956) *Biochim. Biophys. Acta*, **20**, 445
[230] Hoagland, M. (1960) *The Nucleic Acids*, Vol. 3, p. 349 (E. Chargaff and J. N. Davidson, Eds.). New York: Academic Press
[231] Zamecnik, P. C. (1960) *The Harvey Lectures*, **54**, 256
[232] Hecht, L. L., Stephenson, L. and Zamecnik, P. C. (1959) *Proc. Nat. Acad. Sci.*, **45**, 505
[233] Daniel, V. and Littauer, U. Z. (1963) *J. Biol. Chem.*, **238**, 2102
[234] Deutscher, M. P., Foulds, J., Morse, J. W. and Hilderman, R. H. (1975) *Brookhaven Symp. Biol.*, **26**, 124
[235] Deutscher, M. P. (1973) *Progr. Nucleic Acid. Res. Mol. Biol.*, **13**, 51

[236] Grunberg-Manago, M., Oritz, P. J. and Ochoa, S. (1956) *Biochim. Biophys. Acta,* **20,** 269
[237] Ochoa, S. (1960) *Les Nucleoproteins,* p. 241, Onzième Conseil de Chimie, Institut de Chimie Solvay, Bruxelles: Stoeps
[238] *Enzymes in Polynucleotide Metabolism* (1959) *Ann. N.Y. Acad. Sci.,* **81,** 511–804
[239] Hilmoe, R. J. (1959) *Ann. N.Y. Acad. Sci.,* **81,** 660
[240] Ochoa, S., Basilio, C. and Krakow, J. S. (1963) *Methods in Enzymology,* Vol. 6, p. 3 (S. P. Colowick and N. O. Kaplan, Eds.). New York: Academic Press
[241] Grunberg-Manago, M. (1963) *Progr. Nucleic Acid Res. Mol. Biol.,* **1,** 93
[242] Ginsberg, D. and Steitz, J. A. (1975) *J. Biol. Chem.,* **250,** 5647
[243] Shenkin A., and Burdon, R. H. (1974) *J. Mol. Biol.,* **85,** 19

CHAPTER 13

The Biological Function of RNA: Protein Synthesis

13.1 The expression of genetic information

Biological information is stored in the cell in the base sequences of the DNA. In the process of *duplication* or *replication* (Chapter 11) exact copies of the DNA are made for hereditary transmission. In the process of *transcription* the genetic information is transferred from the DNA to the complementary or messenger RNA. Finally the genetic information is *translated* from the four-letter language of the mRNA into the twenty-letter language of the proteins in the process of protein synthesis. This process has been discussed at length in many reviews [1–25] and will be dealt with here in so far as the part played by the nucleic acids is involved.

13.2 The codon as a nucleotide triplet

The amino acid sequence in proteins is determined by the sequences of bases in the discrete regions of the cellular DNA called genes. This information is conveyed from the DNA to the protein-synthesizing machinery on the ribosome by the messenger RNA. The first proof of colinearity of gene and protein came from an analysis of mutants of the head protein of bacteriophage T4 [26] and the A protein of tryptophan synthetase of *Esch. coli*. In proteins, twenty different kinds of amino acids are commonly found whereas only four main kinds of base occur in the nucleic acids. The genetic code describes how a sequence derived from twenty or more units is determined by a sequence derived from four units of a different type. For reviews see [1, 29–42].

Since there are only four kinds of base but twenty kinds of amino acid, the correspondence cannot be a simple 1:1 relationship between bases and amino acids. Nor are there sufficient combinations of two bases (4^2, i.e. 16) to account for twenty amino acids. It had been suspected, therefore, that each amino acid is determined by a sequence of at least three bases, which would give sixty-four combinations (4^3) – more than adequate for the coding of the twenty amino acids. Crick and his colleagues [30–32] produced fairly clear-cut evidence that the triplet theory was correct and that what they termed the codon is a sequence of three nucleotides. Their experiments were carried out on the A and B cistrons of the rII locus of bacteriophage T4 in which, as Benzer had shown by careful genetic mapping, one particular region of the DNA determines whether or not the phage can attack strain K of *Esch. coli*, and they used proflavine (p. 301) to bring about either the insertion of an additional base into the DNA sequence or the deletion of a single base.

If we assume that the sequence of bases in a portion of DNA is as shown in the top line of Fig. 13.1, and that the message is read in

```
         CAT | CAT | CAT | CAT | CAT | CAT....
   -1    CAT | CAC | ATC | ATC | ATC | AT....
  -1+1   CAT | CAC | AXT | CAT | CAT | CAT....
   +3    CAX | TXC | ATX | CAT | CAT | CAT....
```

Fig. 13.1 *Hypothetical sequence of bases in a DNA strand showing genetic message in triplets. Removal of one base (second line) makes the code unreadable but it can be restored if one base is removed and another inserted near it (third line). The message is still readable if three bases are inserted (last line)*

groups of three from left to right starting at the first C, then the removal of the second T from the left by the mutagen will upset the reading of all triplets to the right of the point of deletion (Fig. 13.1 second line). The mutant so produced will be seriously defective and will not infect strain K. However, if a further mutation can now be produced which brings about the insertion of another base X in the third triplet from the left, the fourth, fifth, and subsequent triplets will read correctly and only the second and third triplets will be faulty (Fig. 13.1, third line). Only two amino acids, corresponding to these two triplets, will be 'wrong', and if the presence of these two amino acids does not affect the structure of the protein significantly the bacteriophage will behave normally and will infect strain K. In practice it is in fact found that bacteriophages with an insertion and a deletion close together behave normally, whereas the chances of normal behaviour are diminished as the distance between the insertion and the deletion increases.

It is, moreover, possible to combine mutants in other ways. When two (+) mutations are combined, the recombinants are defective but three (+) or three (−) mutations behave normally and infect strain K (Fig. 13.1, bottom line).

These results can best be interpreted by assuming that coding takes place by consecutive triplets in the nucleic acid. The insertion of one or two bases at any point will so alter the sequence of triplets as to make the code unreadable, whereas if three bases are added (or if one base is added and another is deleted) the sequence of triplets is restored after the first two changes, and the original message on the DNA can be interpreted as before [1, 43].

13.3 RNA and protein synthesis

The first indication that RNA might in some way be involved in protein synthesis came from the early experiments of Caspersson [44] using spectrophotometric methods, of Brachet [45] using histochemical techniques, and of Davidson [46] using chemical methods, all of whom showed that RNA was particularly abundant in cells engaged in the synthesis of protein either for growth or for secretion. It was, however, a good many years before the distinction between the main types of RNA in the cell was established, and before it was made clear that each of the three classes of RNA is directly concerned in protein synthesis, the process involving the ordered interaction of aminoacyl-tRNA with mRNA on ribosomes by mechanisms which will be described in this chapter.

Zamecnik [2] has pointed out that the development of our knowledge in this field has passed through three well-defined technical phases: (i) a disruptive phase in which the goal was to find a cell-free system in which protein synthesis could be demonstrated; (ii) a reassembly phase in which the various components of the crude homogenate were tested to determine whether or not they were essential in the incorporation process; and (iii) the mechanisistic or macromolecular phase in which details of the reaction mechanisms and of the spatial configurations of the macromolecular participants are considered.

13.4 The role of tRNA

The chemical properties of transfer RNA have already been described (p. 130). It has been mentioned that the tRNA molecule consists of about 75 nucleotides ending in the sequence -pCpCpA at the amino acid acceptor end of the chain. The loop or bend near the centre of the chain contains the coding site (the anticodon,

p. 131) at which tRNA can link to the corresponding complementary areas (the codon) on the messenger RNA in a manner which is discussed later.

Transfer RNA carries out a number of separate functions; it recognizes a particular *aminoacyl-tRNA synthetase* so that it can accept the appropriate activated amino acid, it interacts with the messenger RNA and ribosomes in such a way as to ensure that the amino acid that it carries is correctly placed in sequence in the growing polypeptide chain, and, after peptide bond formation, it binds the growing polypeptide chain to the ribosome.

Under the influence of the appropriate *amino acid synthetases*, amino acids are activated by the reaction:

$$ATP + \text{amino acid}_1 + E_1 \rightleftharpoons (\text{amino acid}_1\text{-AMP})E_1 + PPi$$

and then attached to the tRNA at the terminal adenosine moiety. If this is removed, for example by periodate treatment, no attachment occurs. The necessity that the mononucleotide at the 2′,3′-hydroxy-terminal position should be an adenylyl residue suggests that adenine plays a role in the esterification process. The point of amino acid attachment was generally taken to be the 3′-OH but it now appears that certain amino acids are esterified to the 2′-OH [213, 214]. On the other hand, there appears to be only one amino acid synthetase for each of the naturally occurring amino acids. However, these enzymes together can constitute up to 10 per cent of the protein of the cell, and physicochemical studies show them to have no simple general structure. Some can be classified as single-chain enzymes ($110–120 \times 10^3$ daltons), some as multi-chain enzymes with similar subunit size, and some as multi-chain enzymes with dissimilar subunit size [49].

There would appear to be at least one form of tRNA specific for each amino acid; methods of separation of these from each other have been discussed on p. 58. In some instances more than one tRNA is known for a given amino acid; for example, at least five tRNA's are found in *Esch. coli* [47, 48].

The initial function of tRNA therefore is to accept the activated amino acid from the aminoacyl adenylate complex with the formation of a tRNA-amino acid derivative (Fig. 13.2). Such derivatives of tRNA with their appropriate amino acids can readily be isolated:

$$(aa_1\text{-AMP})E_1 + tRNA_1 \rightarrow tRNA_1\text{-}aa_1 + AMP + E_1$$

BIOLOGICAL FUNCTION OF RNA: PROTEIN SYNTHESIS

It is clear in principle that the *aminoacyl-tRNA synthetases* (*amino acid-tRNA ligases*; E.C.6.1.1.) [50–57, 59] must possess at least two binding sites. The first site recognizes a specific amino acid, and the second site selects the specific tRNA molecule to which that amino acid is to be covalently bonded. Each aminoacyl-tRNA synthetase is therefore a highly specific enzyme capable of selecting one amino acid, and only one, out of twenty, and then of selecting a species of tRNA that corresponds to the amino acid but no other. In this way,

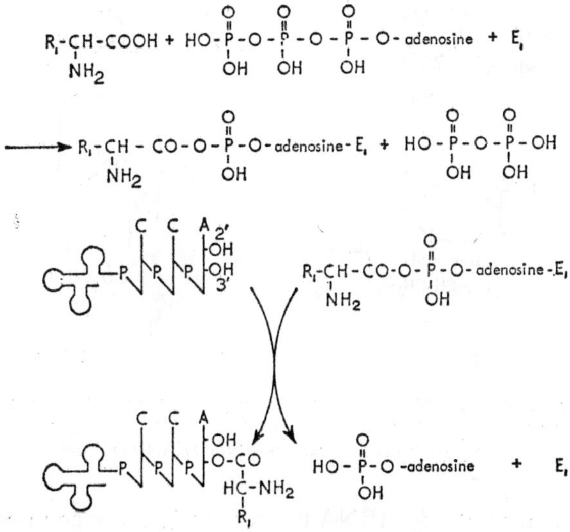

$$R_1\text{-CH(NH}_2\text{)-COOH} + \text{HO-P(O)(OH)-O-P(O)(OH)-O-P(O)(OH)-O-adenosine} + E_1$$

$$\longrightarrow R_1\text{-CH(NH}_2\text{)-CO-O-P(O)(OH)-O-adenosine-}E_1 + \text{HO-P(O)(OH)-O-P(O)(OH)-OH}$$

$$R_1\text{-CH(NH}_2\text{)-CO-O-P(O)(OH)-O-adenosine-}E_1$$

$$\text{tRNA-O-CO-CH(R}_1\text{)-NH}_2 + \text{HO-P(O)(OH)-O-adenosine} + E_1$$

Fig. 13.2 *The activation of amino acids and their attachment to tRNA*

many molecules of tRNA become loaded with their specific amino acids, and the system is then set up to provide amino acids to the protein-synthesizing machinery. How is this achieved? Which particular sequences or sites in the primary, secondary, or tertiary structure of a tRNA are responsible for the recognition of a tRNA molecule by its cognate aminoacyl-tRNA synthetase? Present biochemical and genetic data suggest that there is some general 'fit' between a given tRNA and *any* aminoacyl-tRNA synthetase. Additionally there must be a few very precise recognition points, which might involve specific recognition by a synthetase of two or three bases situated at distant points in the tRNA structure. These are, however, not located at similar sites in the cloverleaf model for all the systems studied, suggesting that the specific recognition point may vary from tRNA to tRNA, and casting doubt on a 'uniform' recognition mechanism [49, 58]. However, there are as yet no data on the tertiary structure of tRNA in a tRNA-aminoacyl-tRNA synthetase complex. Thus similarities and regularities that underlie different results may have gone undetected [49].

The tRNA molecule then carries the amino acid to the ribosomes where the actual process of peptide formation takes place with the

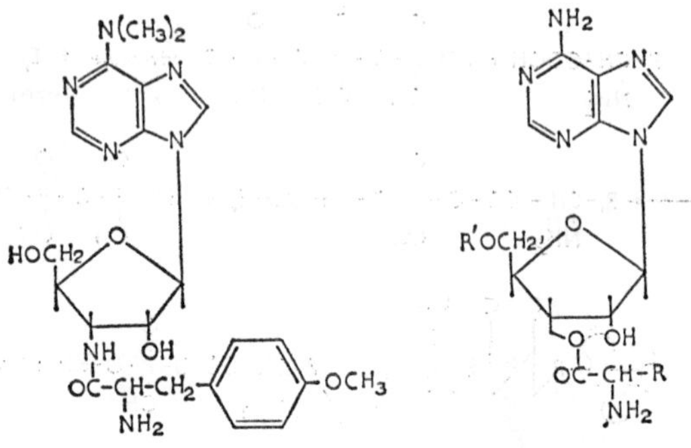

Fig. 13.3 *The structure of puromycin compared with that of the terminal adenosine residue of a transfer amino acid molecule carrying an amino acid. R represents the remainder of the tRNA molecule*

release of the unloaded tRNA molecules which can go through the process again [13, 60–64]. From the ribosome the polypeptide chain is peeled off. Molecules of tRNA therefore act as *adaptors* as suggested by Hoagland and by Crick, locating the amino acids in the correct positions for peptide formation as determined by the messenger RNA.

The antibiotic *puromycin* (Fig. 13.3) is a nucleoside derivative which closely resembles the 3'-terminal nucleoside residue of a loaded tRNA molecule [2, 65]. It competes with aminoacyl-tRNA molecules in its capacity to serve as an acceptor for the peptidyl group of peptidyl-tRNA (p. 383) during protein synthesis on the ribosome. The consequence is that synthesis of complete proteins is prevented and, instead, peptides are produced which bear a puromycin residue covalently bonded to the carboxy terminal group. These peptides are of course non-functional.

13.5 Codon assignments

If it is assumed that a particular protein contains 500 amino acids, then, if each amino acid is represented by one codon of three nucleotides in the mRNA, the mRNA corresponding to that protein must contain some 1500 nucleotides. Moreover, it must have been transcribed from a stretch of DNA corresponding in length to 1500 nucleotide pairs and with a molecular weight of 1500×660, that is 10^6. The average gene, or cistron, therefore, which is the smallest piece of a DNA helix which can carry sufficient information to determine the composition of one protein, corresponds to a segment of DNA of molecular weight 10^6. Since four different types of base pair are available, the total possible number of different genes of molecular weight 10^6 is 4^{1500}, an astronomical figure.

The DNA of bacteriophage T2 (molecular weight 1.2×10^8) contains 200 000 nucleotide pairs and must therefore carry information for the formation of some 200 000/1500, that is 130, proteins. Bacteriophage T2 attacks *Esch. coli* which, in the resting state contains a single chromosome consisting of one circular DNA molecule. The molecule has a molecular weight of about 2×10^9 and measures about 1 mm in length (p. 87). It contains $(2 \times 10^9)/660$, that is 3×10^6, nucleotide pairs or $(3 \times 10^6)/1500 = 2000$ genes. Thus it carries information for the synthesis of 2000 proteins.

The problem of assigning triplets of bases to each of the 20 amino acids has been attacked in several ways.

(i) *The use of biosynthetic messengers.* The earliest attempts were made with the aid of a protein-synthesizing system prepared from cell-free extracts of *Esch. coli* [66–68]. Such extracts contain ribosomes, tRNA's, aminoacyl-tRNA synthetases, and other enzymes, and, in the crude state, also DNA and messenger RNA. When ATP is added together with GTP, Mg^{2+}, K^+, and amino acids, the amino acids are readily incorporated into an acid-insoluble protein product and the incorporation process can be followed by using amino acids labelled with ^{14}C.

When the DNA in such extracts is destroyed by DNase, protein synthesis ceases after the messenger RNA has been depleted but can be restored by adding RNA fractions from various sources and even synthetic polynucleotides produced by the action of polynucleotide phosphorylase.

In 1961, Nirenberg, and Matthaei [67] made the important observation that, when the synthetic polymer poly(U) was added to the system with mixtures of 20 amino acids, only one amino acid in each mixture being radioactive, the only amino acid to be incorporated into an acid-insoluble protein-like material was phenylalanine and the product was polyphenylalanine. The RNA code of phenylalanine was therefore shown to be a sequence of U's.

This type of approach was subsequently followed up by Nirenberg and his colleagues, by Ochoa and his colleagues, and by others, and was found to yield much useful information [37]. For example, poly(A) was found to direct the synthesis of polylysine, while poly(C) promoted the incorporation of proline into acid-insoluble material.

(ii) *The ribosome binding technique.* A different approach, devised by Leder and Nirenberg, involves the use of synthetic messengers containing only three bases [69–71]. Such oligonucleotides are incubated with ribosomes in the presence of amino acids attached to

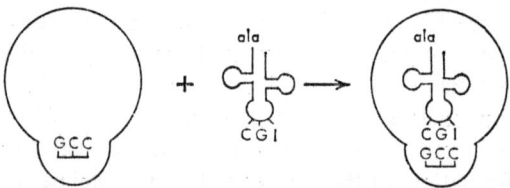

Fig. 13.4 *The ribosomal binding technique. The oligonucleotide GpCpC, when acting as a short synthetic messenger, binds alanyl-tRNA and no other aminoacyl-tRNA to the ribosome. The triplet GCC therefore codes for alanine*

their appropriate tRNA's in the form of aminoacyl-tRNA complexes. When the mixture is passed through a nitrocellulose filter the ribosomes are retained together with the tRNA molecules specifically bound to the ribosomes by the oligonucleotide triplet. By using a series of 20 different amino-acid mixtures each containing one ^{14}C-labelled amino acid, it is possible to identify the amino acid corresponding to each triplet by means of the radioactivity absorbed by the filter. For example, the trinucleotide G-U-U retains the valyl-tRNA whereas U-G-U and U-U-G do not. Similarly the triplet G-C-C binds alanyl-tRNA on the ribosome (Fig. 13.4).

All 64 possible triplets have been synthesized and tested, and more than 50 of them have given unambiguous results. This ribosomal binding method has therefore been of great value in determining not only which triplets code for each amino acid but also the base sequence within the triplet [72]. The method can be extended by the use, in place of triplets, of short biosynthetic oligonucleotides of the type XpYpZp ... Zp ~ 30. With their aid 61 of the possible 64 triplets have been deciphered [73].

(iii) *The use of copolymers of defined sequence.* The problem of relating the nucleotide sequence of mRNA directly with the amino acid sequence of its polypeptide product has been solved by Khorana and his colleagues by an elegant combination of organic chemical and enzymic syntheses [74–77]. The enzyme DNA polymerase (p. 275) can be used to extend a short double helical oligodeoxyribonucleotide, e.g. poly d(A,C · T,G), previously prepared by chemical means. This synthetic DNA is then transcribed using RNA polymerase (p. 318), to yield a specific polyribonucleotide poly(U,G) containing U and G in a repeating sequence. This can then be used in a cell-free protein-synthesizing system in order to determine codon assignments (Table 13.1) [74–77].

TABLE 13.1

Amino acid incorporations stimulated by mRNA containing repeating nucleotide sequences [74]

Messenger	Amino acids incorporated	Messenger	Amino acids incorporated
Repeating dinucleotides		*Repeating trinucleotides*	
Poly(U-C)	Ser-Leu	Poly(G-U-A)	Val, Ser
Poly(A-G)	Arg-Glu	Poly(U-A-C)	Tyr, Thr, Leu
Poly(U-G)	Val-Cys	Poly(A-U-C)	Ile, Ser, His
Poly(A-C)	Thr-His	Poly(G-A-U)	Met, Asp
Repeating trinucleotides		*Repeating tetranucleotides*	
Poly(U-U-C)	Phe, Ser, Leu	Poly(U-A-U-C)	Tyr, Leu, Ile Ser
Poly(A-A-G)	Lys, Glu, Arg	Poly(G-A-U-A)	none
Poly(U-U-G)	Cys, Leu, Val	Poly(U-U-A-C)	Leu, Thr, Tyr
Poly(C-A-A)	Gln, Thr	Poly(G-U-A-A)	none

For example, a $(U-C)_n$ sequence will be read as UCU-CUC-UCU-CUC-U... and will yield a polypeptide containing two amino acids alternating those coded by U-C-U and C-U-C. In fact, the amino acids incorporated are serine and leucine when poly(U-C) is used as messenger. Taken in conjunction with the binding test, this clearly indicates that UCU codes for serine and CUC for leucine (Fig. 13.5A).

From a U-U-C sequence three homopolymers should be coded corresponding to the triplets U-U-C, U-C-U, and C-U-U. Since the starting point is not clearly defined, the message may be read in any of the three forms in Fig. 13.5B:

UUC-UUC-UUC-U...
UCU-UCU-UCU-U...
CUU-CUU-CUU-C...

In practice the amino acids incorporated are phenylalanine, serine, and leucine, when poly(U-U-C) is messenger. In conjunction with the

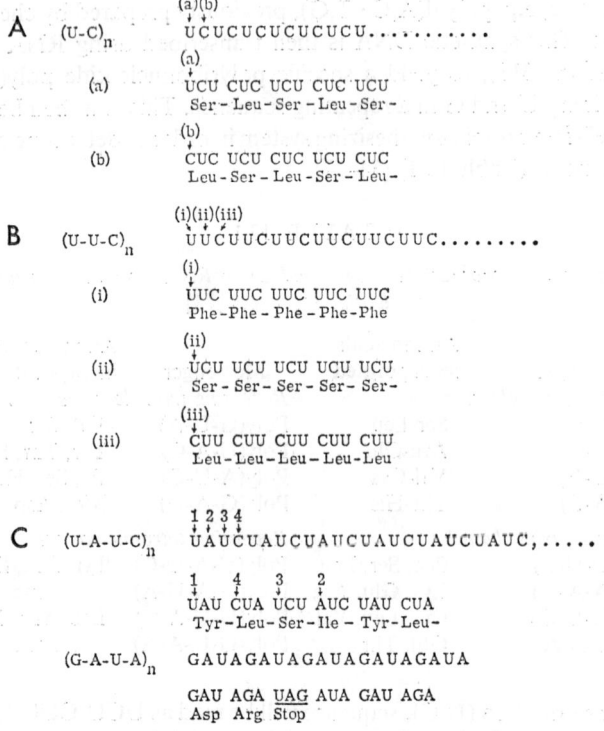

Fig. 13.5 *The reading of nucleotide triplets in polynucleotides of repeating known sequence. For details see text*

results of ribosomal binding, this confirms the codes for serine and leucine as UCU and CUU respectively, and indicates that UUC codes for phenylalanine.

The sequence $(U-A-U-C)_n$ will be read as:

UAU-CUA-UCU-AUC-UAU-CUA-U...

and the polypeptide must be a tetracopolymer with sequences corresponding to the triplets U-A-U, C-U-A, U-C-U, and A-U-C (Fig. 13.5C). When poly(U-A-U-C) is used as messenger the amino acids incorporated are tyrosine, leucine, serine, and isoleucine. This result confirms that UCU codes for serine and CUC for leucine,

TABLE 13.2

The genetic code

5'-OH terminal base	Middle base				3'-OH terminal base
	U	C	A	G	
	Phe	Ser	Tyr	Cys	U
	Phe	Ser	Tyr	Cys	C
U	Leu	Ser	CTS	CTS	A
	Leu	Ser	CTS	Trp	G
	Leu	Pro	His	Arg	U
	Leu	Pro	His	Arg	C
C	Leu	Pro	Gln	Arg	A
	Leu	Pro	Gln	Arg	G
	Ile	Thr	Asn	Ser	U
	Ile	Thr	Asn	Ser	C
A	Ile	Thr	Lys	Arg	A
	Met, fMet	Thr	Lys	Arg	G
	Val	Ala	Asp	Gly	U
	Val	Ala	Asp	Gly	C
G	Val	Ala	Glu	Gly	A
	Val, fMet	Ala	Glu	Gly	G

CTS = Chain termination signals.

and indicates that UAU and AUC code for tyrosine and isoleucine (Table 13.2).

It is of interest that no amino acids are incorporated when poly-(G-A-U-A) and poly(G-U-A-A) are used as messengers. This is not surprising since they contain the chain-terminating codons U-A-G and U-A-A (p. 385).

(iv) *Evidence from bacteriophage RNA's with regions of known sequence.* The work of Sanger and his colleagues [78–81] on the nucleotide sequences of the RNA of bacteriophage R17 has already been mentioned in Chapter 6 (p. 127). This RNA is a chain of some

3300 nucleotides of mol. wt. 1 100 000. It codes for 3 proteins (Fig. 13.6): (a) the 'A protein' or maturation protein of mol. wt. 37 000, containing about 350 amino acids beginning fMet-Arg- . . . ; (b) the coat protein of mol. wt. 14 000 whose full sequence of 129 amino acids is known and begins fMet-Ala-Ser- . . . ; (c) the replicase subunit of mol. wt. 50 000 containing some 450 amino acids beginning fMet-Ser- . . . .

By comparing the sequences of the coat proteins of the R17 phage and the related MS2 phage [82] with the nucleotide sequences in the corresponding cistrons, part of the genetic code was deduced independently of previous data. While additional analyses [84] of parts of the replicase subunit and a cistron of MS2 RNA have shown that all codons are used at least once, it is clear that in the case of some amino acids pertaining to the coat not all possible codons are used, e.g. AUU and AUC are each used 5 times for isoleucine,

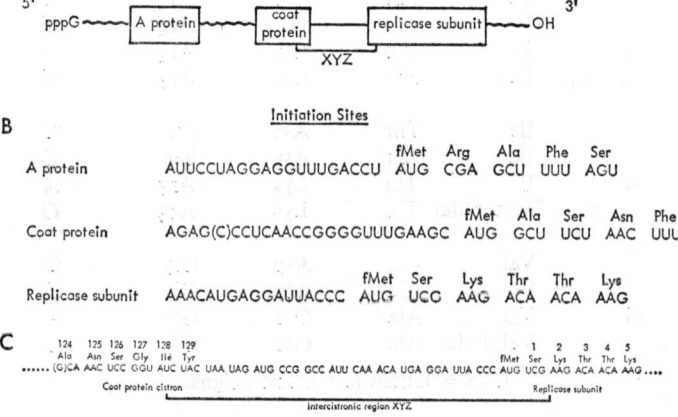

Fig. 13.6 A. *The general layout of the cistrons in the RNA of bacteriophage R17*
        B. *The initiation sites for the three proteins*
        C. *Details of the intercistronic region marked xyz*

whilst AUA is not used at all, and tyrosine is coded 4 times by UAC and never by UAU [82, 83].

(v) *The use of mutations.* Confirmatory information about coding triplets has been obtained from genetic mutations. Such information has the advantage that it is derived from intact cells rather than from cell-free systems.

BIOLOGICAL FUNCTION OF RNA: PROTEIN SYNTHESIS

The principle involved in the use of *base substitution* mutants may be illustrated by the results obtained with the aid of artificially induced mutants of tobacco mosaic virus (TMV) [85–87]. When TMV RNA is treated with nitrous acid two changes are brought about: (a) cytosine is deaminated to uracil (C→U), and (b) adenine

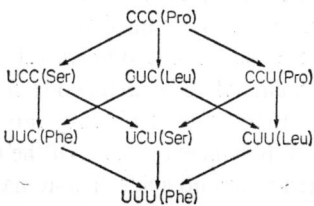

Fig. 13.7 *Steps by which the triplet CCC which codes for proline may be changed by deamination to the triplet UUU which codes for phenylalanine. The amino acids corresponding to each triplet are shown on the right of the codon*

is deaminated to hypoxanthine which is equivalent to guanine in coding (A→G) [85, 86]. When HNO_2-treated TMV RNA is used to infect tobacco plants, mutants may be produced in which a single amino acid in the viral protein is replaced by different amino acids at certain positions in the polypeptide chain in such a way that the replacements can be correlated with the changes A→G or C→U (Fig. 13.7). For example, leucine may be replaced by phenylalanine corresponding to the change CUU→UUU, or alanine may be replaced by glycine in accordance with the change GCA→GCG.

Similar evidence is obtainable from the different varieties of human haemoglobin and from the mutations affecting the A protein of tryptophan synthetase [43].

In *phase shift* mutations a single base may be added to the message at one point so that the reading of the message from that point on is put out of phase. If at a subsequent point a base is deleted, the original message is restored but the section between the insertion and the deletion is out of phase [88] (Fig. 13.1).

13.6 The genetic code [89–91, 78, 92, 93]

The complete genetic code is shown in Table 13.2 which shows that 61 of the 64 possible triplets are assigned to amino acids. The remaining three are discussed later.

Since many of the 20 amino acids are coded by more than one triplet, the code is said to be *degenerate*. It follows that more than one kind of tRNA may code for the same amino acid. This is, in fact, known to be so (p. 362); for example, as many as five distinct leucine acceptor tRNA's have been isolated from *Esch. coli* [47, 48, 94–97].

Efforts to prove the universality of the bacterial genetic code stimulated the search for eukaryotic cell-free systems active for synthesizing proteins. Results obtained from such systems support the contention that the code is indeed universal [98, 27, 99].

The triplet assignments of the genetic code shown in Table 13.2 are based mainly on the results of binding data or of polypeptide synthesis from defined polymers or both. Of the 64 possible triplets 61 have been assigned to amino acids. The remaining three, UAA, UAG, and UGA, do not code for any amino acid and are sometimes termed nonsense codons. UAA and UAG, sometimes referred to in biochemical genetics as the 'ochre' and 'amber' codons respectively, are *chain terminating signals* and so is UGA, in both prokaryotes and eukaryotes [100–102].

In mutations involving what are termed suppressor genes a mutated tyorisine tRNA carrying a single base change in its anticodon may be formed [103]. This allows the tRNA to read the chain terminating codon UAG as if it spelt tyrosine.

13.7 Codon–anticodon pairings

It is assumed that the three bases making up the codon in mRNA recognize, and pair with, a complementary triplet of bases, the anticodon, in the corresponding tRNA. The anticodons in those tRNA's which have known structures have been referred to in Chapter 6. For example, the following anticodons are known: alanine tRNA–IGC, serine tRNA–IGA, tyrosine tRNA–GΨA, valine tRNA–IAC phenylalanine tRNA–GAA. These observations on tRNA structure are in accord with the rule that codon and anticodon pair in an antiparallel manner, i.e. A with U and G with C, at least as far as the first two bases in the codon are concerned. The pairing of the base in the third position is more complicated since it is known that one tRNA can recognize several codons provided that they differ only in the last place in the codon. For example, alanine tRNA with anticodon IGC will recognize the codons GCU, GCC, and GCA. The frequency with which inosine is found in anticodons is noteworthy.

To explain this consequence of the degeneracy of the genetic code Crick has suggested the *'wobble' hypothesis* [1, 104] according to

BIOLOGICAL FUNCTION OF RNA: PROTEIN SYNTHESIS

which a certain degree of latitude is permissible in the pairing of the third base in the codon, according to the arrangement shown in Fig. 13.8. The hypothesis has received direct experimental support from Söll and his colleagues [72, 105].

In the foregoing discussion it has been assumed that the processes of transcription and translation are uncoupled, i.e. that mRNA is released from its DNA template and migrates to another site in the cell for the initiation of protein synthesis. It is, however, possible that they may be coupled in the sense that protein synthesis occurs when

Amino acid	ala	ser	tyr	val	phe
Anticodon (3'←...←5')	CGI	AGI	AψG	CAI	AAG
Codon (5'→...→3')	GCC/U/A	UCC/U/A	UAU/C	GUC/U/A	UUU/C

Fig. 13.8 *The 'wobble' hypothesis. The codons are written in the conventional direction (5'→3') while the anticodons are written backwards (3'←5') to show the base pairings in an antiparallel direction. The hypothesis states that a certain amount of variation (or 'wobble') is tolerated on the third nucleotide of the codon, provided that certain rules of base-pairings are followed, as indicated*

the mRNA is still attached to the gene. Evidence in favour of this possibility has been found in cell-free systems in which DNA is joined to functionally competent ribosomes by means of mRNA [107–109] (See Plate VI).

The specificity of different tRNA molecules has been demonstrated by exposing the tRNA corresponding to cysteine to Raney nickel so as to bring about the reduction of cysteine to alanine while the amino acid was still attached to the tRNA molecule [106] (Fig. 13.9). The tRNA-alanine complex so formed was found to behave in protein-synthesizing systems as if it carried cysteine instead of alanine, presumably because its nucleotide sequence was specific for cysteine.

13.8 The events on the ribosome

It has earlier been pointed out (p. 335) that mRNA, after its formation on the DNA, migrates to the ribosomes, becoming associated with the 30S component. Molecules of tRNA, carrying their appropriate amino acids, also migrate to the ribosomes each of which has a binding site for two tRNA molecules [110, 111] on the larger sub-

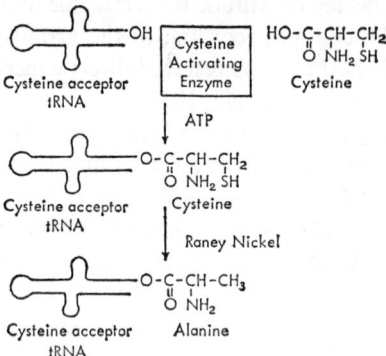

Fig. 13.9 *The reduction of cysteine attached to its specific tRNA to alanine attached to the same tRNA* [106]

unit [112–114] (Fig. 13.10). It is on the ribosomes that amino acids link together one by one in the correct order as determined by the genetic message in the mRNA, to form polypeptide chains.

Polypeptide chains grow by stepwise addition of individual amino acids, beginning at the amino-terminal end [11]. In bacteria the rate of addition is around 10 amino acids per second, and in animal cells somewhat slower at around 2 per second. The process results in a peptide chain terminated at its carboxyl end by a tRNA molecule which attaches it to the ribosome (Fig. 13.11).

At any one time a functioning ribosome contains only one nascent chain bound at the carboxyl end of the chain to the tRNA molecule

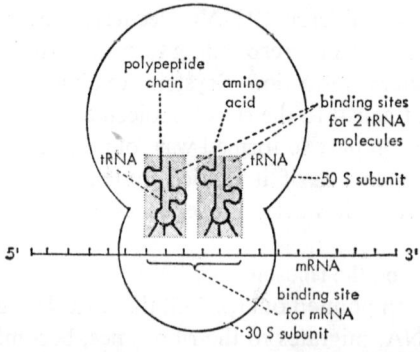

Fig. 13.10 *Diagrammatic representation of a ribosome showing the messenger RNA strand and two tRNA molecules. The one on the left carries the growing polypeptide chain on the peptidyl-tRNA binding site (P site). That on the right occupies the aminoacyl-tRNA binding site (A site)*

which is attached to the specific site on the 50S subunit [112] (Fig. 13.10). This tRNA molecule is released and ejected in the free state when the next amino acid in sequence (attached to its appropriate tRNA which is also bound to the 50S subunit) is linked on to the carboxyl end of the chain by peptide bond formation under the influence of peptidyl transferase, and the tRNA molecule corresponding to this new terminal amino acid now anchors the polypeptide chain, lengthened by one amino acid, to the ribosome [113, 115]:

$tRNA_2$-aa. + $tRNA_1$-polypeptide →
$tRNA_2$-aa.-polypeptide + $tRNA_1$

As the ribosome moves over the messenger RNA template, the next triplet of nucleotides in the mRNA is placed correctly in position to select the next amino acid by pairing with the complementary triplet on the appropriate tRNA.

In the living cell, each strand of messenger RNA carries several ribosomes (Fig. 13.12) each at a different stage in the formation of the protein for which the mRNA is specific. Such groups of ribosomes strung out on a thread of mRNA are known as polyribosomes or polysomes [18, 62, 116–118], and have already been described on p. 24.

The polysome was first demonstrated in rabbit reticulocytes which are involved in the synthesis of haemoglobin [117, 110, 116, 119–122]. When reticulocytes are gently ruptured osmotically the polysome is released as a complex which sediments at 170S and appears to consist of five ribosomes on a thread of mRNA. The complex can be demonstrated in the electron microscope [123]. Gentle treatment with ribonuclease breaks the RNA thread releasing the familiar single ribosomes.

The component polypeptide chains of haemoglobin each contain about 150 amino acids. This corresponds to at least 450 nucleotides in mRNA for each polypeptide chain. The polysomes concerned with biosynthesis of haemoglobin each contain five or six ribosomes Polysomes involved in the synthesis of larger polypeptide chains (for example, 500 amino acids long, corresponding to 1500 nucleotides in mRNA) may carry up to 20 or more ribosomes, each of which is actively engaged in translating the mRNA of molecular weight about 800 000. Polysomes have been described in *Esch. coli* with up to 40 ribosomes per complex [115, 118, 124] and in liver with up to 20 ribosomes per complex [62, 18, 125].

THE BIOCHEMISTRY OF THE NUCLEIC ACIDS

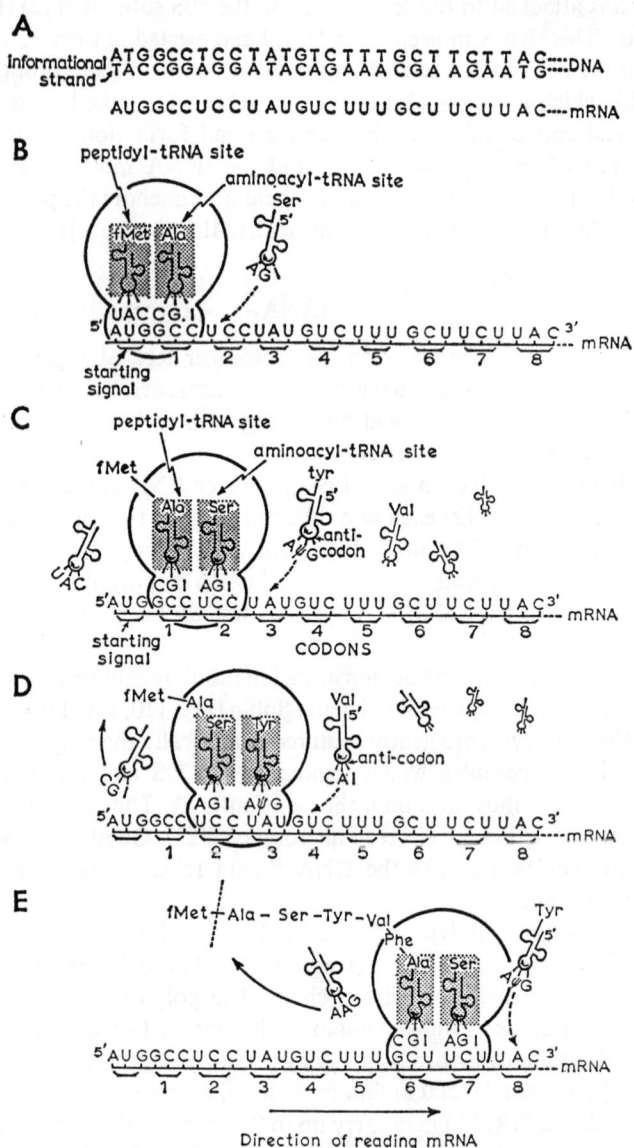

Fig. 13.11 *Synthesis of a polypeptide directed by a segment of a molecule of mRNA transcribed from a segment of DNA (A).*
In B the starting signal AUG of the mRNA has become associated with

BIOLOGICAL FUNCTION OF RNA: PROTEIN SYNTHESIS

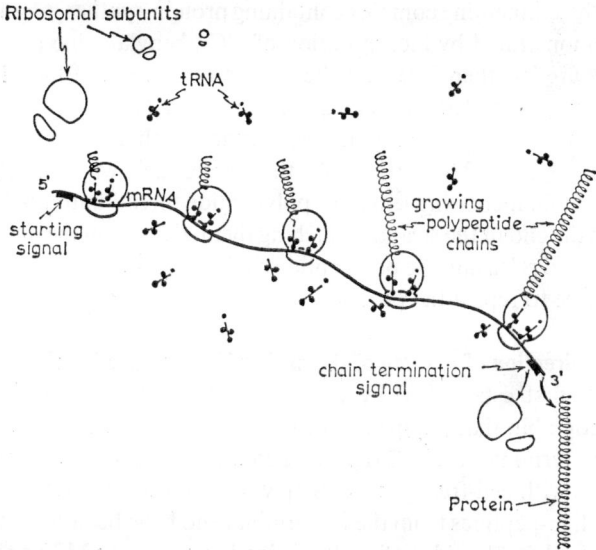

Fig. 13.12 *Diagrammatic representation of the polysome showing five ribosomes passing down a strand of mRNA and spinning out a thread of protein as they go. The ribosome in the bottom right-hand corner has just disengaged itself from the mRNA completing the formation of the appropriate protein. The ribosome at the top is about to engage with the mRNA thread*

The polysomal aggregates are more active in polypeptide synthesis (as demonstrated by uptake of radioactive amino acids) than are simple ribosome monomers. This can be demonstrated by the addition of a synthetic messenger RNA in the form of poly(U) which specifically promotes the incorporation of phenylalanine into polypeptide. Already aggregated ribosomes are not receptive to added poly(U) but the monomers readily accept poly(U) so as to form a

---

fMet-tRNA on the peptidyl-tRNA site on the ribosome while the aminoacyl-tRNA site is occupied by Ala-tRNA coded for by codon 1(GCC).

In C the ribosome has moved one stop to the right and codon 1 now carries the tRNA bearing the dipeptide fMet-Ala while the aminoacyl-tRNA site bears Ser-tRNA associated with codon 2(UCC).

In D the ribosome has moved a further stop to the right. The alanine-tRNA has been cast off and the peptidyl-tRNA site now carries the tripeptide fMet-Ala-Ser bound to the appropriate tRNA associating with codon 2 while the aminoacyl-tRNA site bears Tyr-tRNA associated with codon 3(UAU).

In E the process has gone further and the peptidyl-tRNA site is occupied by the tRNA bearing the peptide fMet-Ala-Ser-Tyr-Val-Phe-Ala and associated with codon 6 (GCU for alanine) while the aminoacyl-tRNA site is occupied by Ser-tRNA associated with codon 7 (UCU for serine).

rapidly sedimenting complex containing protein-synthesizing activity as demonstrated by incorporation of ^{14}C-phenylalanine [115, 119].

We are left therefore with the picture of a group of five or more ribosomes threaded on a string of messenger RNA (Fig. 13.12). The ribosomes begin protein synthesis by attaching themselves to one end of a messenger RNA strand. As they make their way along it they spin out an increasing length of polypeptide thread until finally they reach the end of the strand, detaching themselves from the messenger RNA and releasing the polypeptide chain at the same time. They are then free to repeat the process.

13.9 Direction of reading of the nucleotide sequence in mRNA

It was first established by Dintzis [126] for the globin chains of haemoglobin that polypeptide assembly proceeds sequentially from the N'-terminal to the C'-terminal end. This has been confirmed by studies with cell-free systems. Stepwise synthesis of various di-, tri-, and oligo-peptides from the N'-terminal end have been shown, using either polynucleotides of partly defined sequence or MS2 or related phage RNA's as template [127–129].

Use of cell-free systems programmed by synthetic polyribonucleotides with distinguishing codons at the 5'-end has also permitted the demonstration that the 5'-end of mRNA corresponds to the N'-terminal end of the polypeptide [130]. This has been confined *in vivo* by correlating amino acid replacement in a frame shift mutation of bacteriophage T4 lysozyme with codon assignments for the respective amino acids [131]. Since the direction of reading 5' → 3' is also the direction in which the mRNA is synthesized, it follows that the first-formed portion of a messenger can associate with ribosomes and may engage in protein synthesis while transcription of the latter part of the messenger is still in progress [107] (see Plate VI).

13.10 The mechanism of polypeptide chain synthesis

A general outline of the way in which proteins are synthesized on the ribosome has already been given. We are now concerned with more detailed description of the mechanism by which polypeptide chains are initiated, elongated, and terminated. The process is illustrated in Fig. 13.13.

BIOLOGICAL FUNCTION OF RNA: PROTEIN SYNTHESIS

13.10.1 Chain initiation

In early *in vitro* experiments a suitable concentration of Mg^{2+} was found essential for maintaining not only the efficiency but also the fidelity of translation. When the Mg^{2+} concentration is too low, no protein synthesis occurs. Over a rather sharp range (3–5mM) the system is rapidly activated. At higher concentrations there is only slight loss of efficiency of incorporation, but incorrect amino acids are incorporated with appreciable frequency. In fact the first 'mistake' to manifest itself at slightly elevated Mg^{2+} concentration (10mM) is non-specific initiation of protein synthesis. Because of this it was initially possible to study both the genetic code and the general mechanisms of polypeptide chain extension in bacterial cell-free systems before the specific role of formyl-methionyl-tRNA in the initiation of protein synthesis became known. Indeed, for some time it was assumed that no specific starting signal was required in the mRNA. However, it has become clear that a special mechanism initiates polypeptide chain synthesis at Mg^{2+} concentration of 3–5mM; protein synthesis requires the presence of formyl-methionyl-$tRNA_f$. Additionally it has been established that in whole cells of *Esch. coli* the first amino acid in a polypeptide chain is methionine with a formyl group (CHO) attached to the free amino group. In bacterial cells peptide deformylases remove the formyl group from the nascent protein [132]. The methionine may subsequently be removed leaving another amino acid such as alanine at the amino end of the polypeptide chain.

The methionine tRNA of *Esch. coli* can be separated into two distinct species (p. 126). One species, $tRNA_f^{Met}$, comprising about 70 per cent of the bacterial methionine tRNA, can give rise to a methionyl-$tRNA_f$. Formylation takes place only after the amino acid has become attached to the tRNA molecule and is catalysed by a *transformylase* which requires N^{10}-formyl-tetrahydrofolic acid. No other formylated aminoacyl-tRNA has been found.

The other species, $tRNA_m^{Met}$, comprising the remaining 30 per cent of the bacterial methionine tRNA, forms a methionyl-$tRNA_m$ that cannot be formylated. Since the transformylase does not even recognize methionine on methionyl-$tRNA_m$, it follows that the enzyme must recognize the structure of $tRNA_f^{Met}$. This contrasts with the properties of the methionyl-tRNA synthetase which attaches methionine to both forms of $tRNA^{Met}$ and thus probably recognizes some common feature of both tRNA's. Despite knowledge of their individual primary structures, the reasons for these observations remain obscure.

The codon for methionine, AUG (Table 13.2), is read both by $tRNA_f^{Met}$ and by $tRNA_m^{Met}$, but the codon GUG, corresponding to valine, is read also by $tRNA_f^{Met}$ although not by $tRNA_m^{Met}$.

Both codons, AUG and GUG, can therefore lead to the formation of a chain with formylmethionine in the starting position (i.e. at the amino end). The AUG codon, when located internally in the mRNA, will place methionine in an internal position in the protein chain, whereas the GUG codon, located internally in the mRNA, codes for valine [133–138].

The considerations just described apply to systems derived from *Esch. coli*. Formyl-methionyl-tRNA$_f$ (fMet-tRNA$_f$) has been detected throughout the prokaryotic cell world, as well as in eukaryotic organelles such as chloroplasts and mitochondria.

Eukaryotic cytoplasms lack transformylases but they contain two species of methionine transfer RNA, one of which can be formylated by bacterial enzymes [139–141]. This species, which interestingly lacks the sequence TpΨp in loop IV [175], acts as an initiator when *synthetic* mRNA's are used in a cell-free system from Ehrlich ascites cells. The other methionine tRNA incorporates methionine internally. In natural systems it is now known that rabbit α and β globin chains [140] and protamine in trout testis cells [142] are initiated with methionine.

In the process of chain initiation [143–148] the mRNA is bound to the smaller ribosomal subunit in the presence of fMet-tRNA, GTP, and three *initiation factors*. Ochoa and his colleagues have purified these factors (IF-1, IF-2, and IF-3) from the 30S ribosomal subunits of *Esch. coli* [151–154]. They are thought to act as follows (Fig. 13.13): 70S ribosomes dissociate spontaneously into 30S and 50S subunits whereupon IF-3 binds to the 30S subunits, thus displacing the equilibrium 70S ⇌ 50S + 30S further to the right, thereby making 30S subunits available for initiation (this explains the dissociation factor activity known to be associated with IF-3). IF-2 binds to the 30S · IF-3 complex and IF-1 is believed to behave similarly. The 30S subunit bearing the three initiation factors is now ready to bind messenger, and in the presence of GTP and fMet-tRNA$_f$ a 30S initiation complex is formed (Fig. 13.13). IF-3 appears to increase the stability of the complex, and a possible role for IF-3 is that it may recognize a structural feature of mRNA to allow the interaction of the fMet-tRNA$_f$ with an initiator codon. The recent finding of subspecies of IF-3 (IF-3α and IF-3β) has suggested the differential recognition of initiation sites involving various IF-3 species. IF-3α has a high selectivity towards MS2 RNA, *Esch. coli*, or *early* T4 mRNA, whereas IF-3β shows a similar high selectivity towards *late* T4 mRNA. The question of whether these IF-3 species vary in selectivity toward different cistrons of the same polycistronic messenger MS2 RNA remains open. An additional complexity in this context stems from the observation that it is also possible to change the apparent specificity of an IF-3. Groner and his coworkers [156] were able to isolate, from high salt washes of ribosomes, a protein factor i (interference factor) capable of inhibiting coat protein synthesis from MS2 RNA whilst allowing replicase subunit synthesis. It appears to interact with an IF-3 and prevent it from taking part in initiation complex formation at the coat-protein initiation site. (i factor now appears to be identical with subunit I of the Qβ replicase [157], described in the preceding chapter.)

After the 30S complex is formed, IF-3 is released and becomes available for binding to further 30S subunits. Through this recycling IF-3 functions catalytically in polypeptide chain initiation. The 50S subunit now joins the 30S complex to form the 70S initiation complex (Fig. 13.13). This is followed by GTP hydrolysis and release of IF-2. The resulting 70S complex is now ready for elongation. Whilst IF-1 may also function catalytically in initiation, there are reasons to believe that IF-1 may be an easily dissociable ribosomal protein rather than an initiation factor proper [154].

It is known that protein synthesis can be initiated simultaneously at several sites on a polycistronic messenger. This was established by Steitz [144] who used bacteriophage R17 RNA as messenger (Fig. 13.6). It codes for three proteins (p. 368) and in this RNA the gene order [149] is:

5'-A-protein–coat protein–replicase subunit-3'

Steitz bound radioactive viral RNA to ribosomes in conditions appropriate to initiation of protein synthesis. Excess RNA lying outside the complexes was removed with RNase and the fragments of RNA protected by the ribosomes were isolated (about 35 nucleotides in each of the three cases). When these initiation regions were isolated and nucleotide sequences determined, they were found to correspond to the codons for the appropriate amino acids [143, 81].

All contain AUG about the middle (see Fig. 13.6). Similar results have been obtained from the RNA of bacteriophage Qβ [143]. GUG, however, has been found at the start of the A-protein cistron in MS2 RNA [150].

The problem of *initiation factors* is confused since they have been given different names by the various groups working in this field.

Regarding eukaryotic initiation factors, four can be isolated from

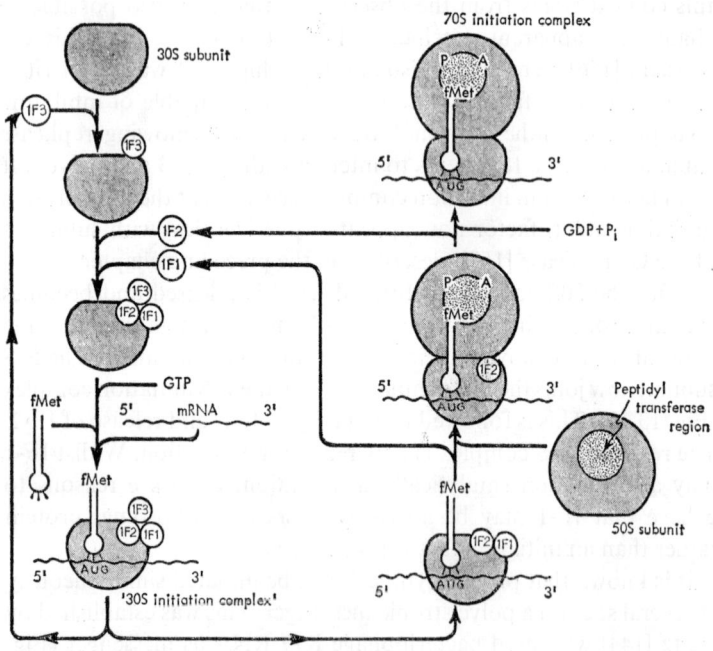

Fig. 13.13 *A schematic representation of protein chain initiation*

rabbit reticulocyte and liver ribosomes [158–160]. They have been defined as M1 (like IF-2), M2a (a GTPase), M2b (like IF-1), and M3. Whether M3 is really similar to IF-3 in being capable of discriminating between mRNA's is not clear. A factor IF_{EMC} more like IF-3 has been prepared from ascites tumour cells, which selects the initiation site of encephalomyocarditis virus RNA and discriminates against other eukaryotic mRNA's such as globin mRNA [161].

13.10.2 Chain elongation

During the process of initiation, tRNA$_f^{Met}$ carrying a formyl-methionyl residue becomes attached on the amino acid site (A in Fig. 13.14) on the ribosome. As the ribosome moves to the right (or the mRNA to the left) in Fig. 13.14 the loaded tRNA becomes transferred to the peptide site P.

The A site then becomes occupied by the aminoacyl-tRNA specified by the codon on the mRNA next to AUG on the 3'-side. The attachment of this aminoacyl-tRNA requires the presence of GTP and two protein factors EF-Tu and EF-Ts from the cytosol

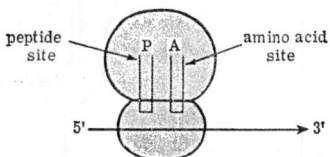

Fig. 13.14 *The ribosome showing the peptide site (P) and the amino acid site (A)*

[162, 164]. EF-Tu fulfils the role of carrying the aminoacyl tRNA's (except fMet-tRNA$_f$) from solution to the ribosomal A site, the process taking place by means of a ternary complex of EF-Tu, aminoacyl-tRNA, and GTP, and resulting in the hydrolysis of GTP to GDP [163, 164]. Because of the high affinity of GDP for EF-Tu these are released from the ribosomes as the EF-Tu · GDP complex. The role of EF-Ts is believed to be the catalytic regeneration of EF-Tu · GTP from EF-Tu · GDP by means of an intermediate EF-Tu · EF-Ts complex [164, 165] (Fig. 13.15). Also implicated in the binding of the aminoacyl-tRNA to the A site is the small 5S ribosomal RNA (Chapter 6). There are grounds for considering some interaction between ribosome-bound 5S RNA and the GTΨC sequence in loop IV of the incoming tRNA moiety [174].

The carboxyl group of the fMet residue on the fMet-tRNA$_f^{Met}$ is now released from attachment to tRNA$_f^{Met}$ and becomes linked in a peptide bond to the amino group of the aminoacyl-tRNA in the A site. This reaction is catalysed by a *peptidyl transferase* which is apparently part of the 50S subunit [166].

In the next step the discharged tRNA$_f^{Met}$ is released from the P site, and the tRNA carrying the newly formed dipeptide moves from site A to P (translocation) [167]. At the same time the ribosome moves the length of one codon along the mRNA in the 5' → 3' direction [164].

THE BIOCHEMISTRY OF THE NUCLEIC ACIDS

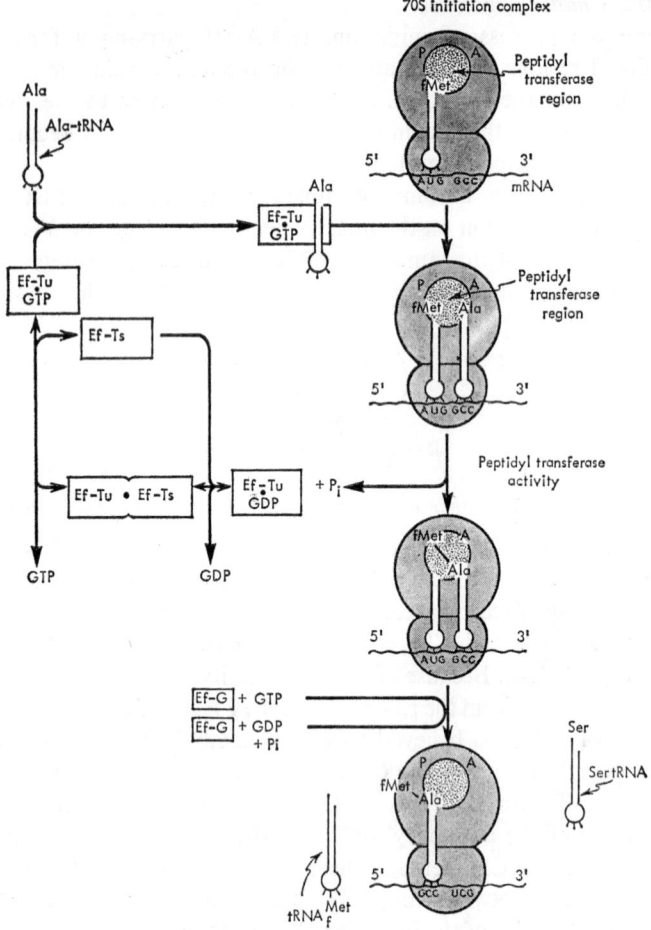

Fig. 13.15 *A schematic representation of the process of protein chain elongation*

This movement of the charged tRNA from site A to site P is known as *translocation* and is catalysed by another soluble factor EF-G, a protein of mol. wt. 72 000 [22, 23, 164]. It requires the presence of GTP (which is broken down to GDP and Pi) [168] (Fig. 13.15). EF-G exhibits a ribosome-dependent GTPase activity [164].

After translocation, another charged tRNA enters the A site and the process of peptide bond formation and translocation is repeated [169, 170].

In eukaryotic cells, factors similar to the bacterial ones appear to exist, but these have not been extensively purified. EF-1 and EF-2 probably correspond to EF-Tu and EF-G [25, 172, 173]. No evidence is yet available for an EF-Ts-like factor. An intriguing difference between EF-2 and EF-G is the inhibition of the eukaryotic factor by diphtheria toxin [171]. This toxin in the presence of NAD as cofactor inhibits translocation in eukaryotic cells as well as the GTP hydrolysis activity of EF-2, thus implicating EF-2 in the translocation mechanism.

13.10.3 *Chain termination*

When the ribosome, in its movement along the mRNA, reaches a termination signal at the A site, the process of peptide formation ceases. The termination signals are UAA, UAG, and UGA [176–178].

The process may be illustrated with reference to R17 RNA (Fig. 13.6) in which the nucleotide sequence has been determined [179] corresponding to the last six amino acids of the phage coat protein and extending for 26 nucleotides on the 3'-side of the coat protein cistron (Fig. 13.6). The sequence contains two termination codons UAA and UAG:

→128   129   1   2
→Ile   Tyr   fMet Ser   Lys ——
AUC UAC UAA UAG – 10 nucleotides — AUG UCG AAG——
Coat protein   Replicase subunit

Protein release factors RF1 and RF2 [180–182] bind to the ribosomes in response to the specific terminator codons:

            RF1 to UAA or UAG
      and RF2 to UAA or UGA

The interaction of bacterial RF1 and RF2 with ribosomes is stimulated by a third soluble protein factor RF3 which interacts with GDP and GTP to cause dissociation of RF1 and RF2 from the ribosome [182]. While more than one species of rabbit reticulocyte release factor (RF) have been identified, each recognizes UAA, UAG, and UGA, and has ribosome-dependent GTPase activity. *In vitro* reactions of peptide chain termination involving peptidyl-tRNA hydrolysis in both *Esch. coli* and reticulocyte systems require not only ribosome-bound substrate and RF but also the activity of

THE BIOCHEMISTRY OF THE NUCLEIC ACIDS

the ribosomal *peptidyl transferase* which may well participate in the hydrolysis event [182] (Fig. 13.16). The completed polypeptide chain is released from the ribosomes and the ribosomal subunits separate probably under the influence of initiation factor IF-3 [152]. The mRNA is degraded as described in Chapter 12.

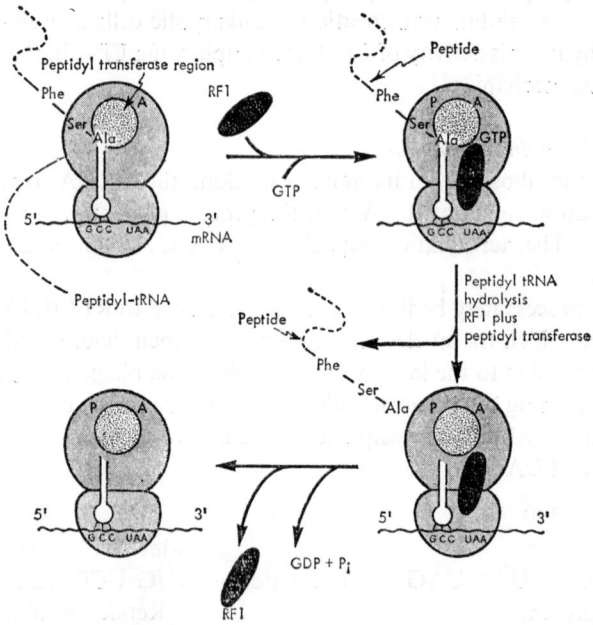

Fig. 13.16 *A schematic model of peptide chain termination*

13.11 Effects of antibiotics

In addition to puromycin a number of other antibiotics effect protein synthesis.

Streptomycin, by binding to the surface of the bacterial 30S subunit, causes errors of reading of mRNA codons by the anticodons of tRNA. Poly(U), for instance, codes not only for phenylalanine but in the presence of streptomycin for serine, isoleucine, and leucine also.

Tetracycline binds to the small subunit of prokaryotic ribosomes and blocks the binding of aminoacyl-tRNA to the A sites.

Chloramphenicol binds to the large bacterial subunit and inhibits the peptidyl transferase reaction; *lincomycin* has the same effect.

BIOLOGICAL FUNCTION OF RNA: PROTEIN SYNTHESIS

Sparsomycin, which also blocks peptidyl transfer at some stage, does so with both bacterial and eukaryotic ribosomes.

Fusidic acid inhibits translocation in both prokaryotic and eukaryotic systems by preventing the release of EFG (or EF2) · GDP complexes from the ribosome after GTP hydrolysis.

Erythromycin effects prokaryotic ribosomes by blocking the P site thus inhibiting both peptidyl transfer and translocation.

Thiostrepton and *siomycin* bind to large prokaryotic subunits and interfere with the binding of EFG and EF-Tu · GTP · aminoacyl-tRNA.

Whilst initiation can be inhibited by agents which interfere with the synthesis of the formyl group, i.e. folic acid antagonists, the process can also be effected by *edeine* (a polypeptide), *kasugamycin*, and *aurintricarboxylic acid*.

Specific inhibitors of eukaryotic protein synthesis are *cycloheximide*, which blocks ribosome function in some as yet unknown manner, and *diphtheria toxin* which, as already discussed, inactivates EF2.

For a further account of these aspects the reader is referred to a number of reviews [183–185].

13.12 The *in vitro* synthesis of specific proteins

When the single-stranded RNA from a bacteriophage such as R17, MS2, f2, or Qβ is allowed to direct protein synthesis in a bacterial cell-free system, the products are similar to those found *in vivo*. The RNA acts as a *polycistronic messenger* and 3 proteins are produced, the A protein, the coat protein, and the replicase subunit, in the approximately correct proportions, i.e. 1:20:3 [186, 200]. Similarly, single-stranded RNA's will function as messengers in mammalian cell-free protein-synthesizing systems, e.g. EMC RNA in mouse ascites tumour cell extracts [187].

Attempts, however, to isolate and translate true bacterial mRNA's *in vitro* have as yet been largely unsuccessful. This may be due to their relative scarcity within cells as a consequence of their rapid turnover. The situation with animal mRNA's has been very much more encouraging. Those animal cells which have specialized functions (such as reticulocytes which mainly manufacture haemoglobin) have provided a rich source of stable specific mRNA's. Several mammalian mRNA's have been isolated and translated, e.g. globin mRNA [187–191], myosin mRNA [192], lens crystallin mRNA

[187, 196], immunoglobulin light chain mRNA [197, 194], and histone mRNA's [195]. Thus far all the eukaryotic mRNA's are monocistronic, which contrasts with the bacterial situation.

Although no true bacterial mRNA's have been successfully translated, it has been possible to take DNA say from T4 bacteriophage and first transcribe it (with RNA polymerase) and then translate the RNA produced *in vitro*. In this way, T4-coded enzymes such as lysozyme and α- and β-glycosyl transferases can be synthesized *in vitro* [197–199].

13.13 Ribosomal proteins and protein synthesis

Whilst it is evident from previous sections that proteins are synthesized on ribosomes as a result of a number of highly coordinated events, some mention should be made regarding functional aspects of ribosome structure. As already mentioned (Chapter 2) the 30S subunit from *Esch. coli* is composed of one 16S RNA molecule and 21 proteins, and the 50S subunit comprises one 23S RNA molecule, 34 proteins, and a 5S RNA molecule (for reviews see [201–204]). What functions do these proteins and RNA's fulfil? All, or almost all, the ribosomal RNA proteins are different, and it appears that the role of the ribosomal RNA is to ensure the assembly of these different proteins into a functional ribonucleoprotein particle. The study of function of the ribosomal proteins was taken further by the demonstration of the reconstitution of biologically active 30S subunits [205] and then 50S subunits [205, 207] from the free RNA and the appropriate set of proteins. In some reconstitution experiments it was possible to omit certain proteins, and by studying the reconstituted particles to make deductions about their role. Other approaches towards the elucidation of function and topological location of ribosomal proteins involve the use of specific antibodies, chemical modification, and affinity labelling [208].

So far this work is still very much in progress but, of the 30S proteins: protein S1 is required for mRNA binding; proteins S3, S4, S5, S11, and S12 are involved in codon–anticodon recognition; proteins S2, S3, S10, S14, S19, and S21 function in fMet-tRNA$_f$ binding; proteins S1, S2, S3, S10, S19, S20, and S21 are important for the functioning of both decoding sites; whereas proteins S9, S11, and S18 are only needed for EF-Tu dependent aminoacyl-tRNA binding [208].

Concerning the 50S proteins [208], protein L16 is known to be involved in binding the 3′-terminus of aminoacyl-tRNA to the A site; next to it is protein L6. In the P site are proteins L2, L27, and L4. L11 borders both the A and P sites and may be the peptidyl transferase. Proteins L7 and L12 are part of the binding site for IF-2 [209] and may also be involved in peptide termination events; L24 and L33 are part of the P site; proteins L24 and L33 were concluded to belong to the binding site for the aminoacyl end of $tRNA_f^{Met}$ [210] whilst L15 is possibly involved in the peptidyl transferase centre [211] (a more detailed survey of the situation can be found in [212]).

REFERENCES

[1] Crick, F. H. C. (1966) *Sci. Amer.*, **215**(4), 55
[2] Zamecnik, P. C. (1962) *Biochem. J.*, **85**, 257
[3] Ingram, V. M. (1965) *The Biosynthesis of Macromolecules*. New York: Benjamin
[4] Zuckerkandl, E. and Pauling, L. (1965) *Evolving Genes and Proteins*, p. 168 (V. Bryson and H. J. Vogel, Eds.). New York: Academic Press
[5] Watson, J. D. (1964) *Bull. Soc. Chim. Biol.*, **46**, 1399
[6] Crick, F. H. C. (1958) *Symposia Soc. Exp. Biol.*, **12**, 139. Cambridge: University Press
[7] Hoagland, M. B. (1960) *The Nucleic Acids*, Vol. 3, p. 349 (E. Chargaff and J. N. Davidson, Eds.). New York: Academic Press
[8] Schweet, R. and Heintz, R. (1966) *Ann. Rev. Biochem.*, **35**, 723
[9] Nathans, D., Allende, J. E., Conway, T. W., Spyrides, G. J. and Lipmann, F. (1963) *Informational Macromolecules*, p. 349 (H. J. Vogel, V. Bryson and J. O. Lampen, Eds.). New York: Academic Press
[10] Roberts, R. B. (1963) *Informational Macromolecules*, p. 367 (H. J. Vogel, V. Bryson and J. O. Lampen, Eds.). New York: Academic Press
[11] Dintzis, H. M. and Knopf, P. M. (1963) *Informational Macromolecules*, p. 375 (H. J. Vogel, V. Bryson and J. O. Lampen, Eds.). New York: Academic Press
[12] Matthaei, H., Sander, G., Swan, D., Kreuzer, T., Caffier, H. and Parmeggiani, A. (1968) *Naturwiss.*, **55**, 281
[13] Fruton, J. S. (1963) *The Proteins*, Vol. 1, p. 190 (H. Neurath, Ed.). New York: Academic Press
[14] Korner, A. (1964) *Mammalian Protein Metabolism*, Vol. 1, p. 178 (H. N. Munro and J. B. Allison, Eds.). New York: Academic Press
[15] Zamecnik, P. C. (1962) *The Molecular Control of Cellular Activity*, p. 259 (J. M. Allen, Ed.)
[16] Chantrenne, H. (1961) *The Biosynthesis of Proteins*. London: Pergamon Press
[17] Attardi, G. (1967) *Ann. Rev. Microbiol.*, **21**, 383
[18] Wettstein, F. O., Staehelin, T. and Noll, H. (1963) *Nature*, **197**, 430
[19] Nirenberg, M. W. (1963) *Methods in Enzymology*, Vol. 6, p. 17 (S. Colowick and N. O. Kaplan, Eds.). New York: Academic Press
[20] Williamson, A. R. (1969) *Essays Biochem.*, **5**, 139

[21] Lucas-Lenard, J. and Lipmann, F. (1971) *Ann. Rev. Biochem.*, **40**, 409
[22] Lengyel, P. and Söll, D. (1969) *Bact. Rev.*, **33**, 264
[23] Leder, P., Skogerson, L. E. and Nau, M. M. (1969) *Proc. Nat. Acad. Sci.*, **62**, 454
[24] Konings, R. N. H., Ward, R., Francke, B. and Hofschneider, P. H. (1970) *Nature*, **226**, 604
[25] Haselkorn, R. and Rothman-Denes, L. B. (1973) *Ann. Rev. Biochem.*, **42**, 397
[26] Sarabhai, A. S., Stretton, A. O. W., Brenner, S. and Bolle, A. (1964) *Nature*, **201**, 13
[27] Yanofski, C., Carlton, B. C., Guest, J. R., Helinski, D. R. and Henning, V. (1964) *Proc. Nat. Acad. Sci.*, **51**, 266
[28] Spiegelman, S. and Hayashi, M. (1963) *Cold Spring Harbor Symp. Quant. Biol.*, **28**, 161
[29] Crick, F. H. C. (1963) *Progress in Nucleic Acid Research*, Vol. 1, p. 163 (J. N. Davidson and W. E. Cohn, Eds.). New York: Academic Press
[30] Crick, F. H. C. (1966) *Cold Spring Harbor Symp. Quant. Biol.*, **31**, 3
[31] Crick, F. H. C. (1967) *Proc. Roy. Soc.*, B, **167**, 331
[32] Crick, F. H. C., Barnett, L., Brenner, S. and Watts-Tobin, R. J. (1961) *Nature*, **192**, 1227
[33] Woese, C. R. (1967) *The Genetic Code. The Molecular Basis for Genetic Expression*. New York: Harper and Row
[34] Nirenberg, M. W., Matthaei, J. H., Jones, O. W., Martin, R. G. and Barondes, S. H. (1963) *Fed. Proc.*, **22**, 55
[35] Nirenberg, M. W. and Jones, Jr., O. W. (1963) *Informational Macromolecules*, p. 451 (H. J. Vogel, V. Bryson and J. O. Lampen, Eds.). New York: Academic Press
[36] Khorana, H. G. (1965) *Fed. Proc.*, **24**, 1473
[37] Ochoa, S. (1964) *Experientia*, **20**, 57
[38] Woese, C. R. (1957) *Progress in Nucleic Acid Research and Molecular Biology*, Vol. 7, p. 107 (J. N. Davidson and W. E. Cohn, Eds.). New York: Academic Press
[39] Speyer, J. S., Lengyel, P., Basilio, C., Wahba, A. J., Gardner, R. S. and Ochoa, S. (1963) *Cold Spring Harbor Symp. Quant. Biol.*, **28**, 559
[40] Jukes, T. H. (1966) *Molecules and Evolution*. New York: Columbia University Press
[41] Kaplan, S. (1967) *Sci. Progr.* (Oxf.), **55**, 223
[42] Speyer, J. F. (1967) *Molecular Genetics*, Part II, p. 137 (J. H. Taylor, Ed.). New York: Academic Press
[43] Yanofsky, C. (1967) *Sci. Amer.*, **216**(5), 80
[44] Caspersson, T. (1942) *Naturwiss.*, **29**, 33
      (1950) *Cell Growth and Cell Function*. New York: Norton
[45] Brachet, J. (1950) *Chemical Embryology*. New York: Interscience
[46] Davidson, J. N. and Waymouth, C. (1944) *Biochem. J.*, **38**, 39
[47] von Ehrenstein, G. and Dais, D. (1963) *Proc. Nat. Acad. Sci.*, **50**, 81
[48] Blank, H. U. and Söll, D. (1971) *J. Biol. Chem.*, **246**, 4947
[49] Söll, D. and Schimmel, P. R. (1974) *The Enzymes*, Vol. 10, p. 489 (P. D. Boyer, Ed.). New York: Academic Press
[50] Sueoka, N. and Kano-Sueoka, T. (1970) *Progr. Nucleic Acid Res. Mol. Biol.*, **10**, 23 (J. N. Davidson and W. E. Cohn, Eds.). New York: Academic Press
[51] Jacobson, K. B. (1971) *Progr. Nucleic Acid Res. Mol. Biol.*, **11**, 461
[52] Chambers, R. W. (1971) *Progr. Nucleic Acid Res. Mol. Biol.*, **11**, 489

[53] Mehler, A. H. (1971) *Methods in Enzymology*, **20**, 203 (L. Grossman and K. Moldave, Eds.)
[54] Lagerkvist, U. and Rymo, L. (1970) *J. Biol. Chem.*, **245**, 435
[55] Katze, J. R. and Konigsberg, W. (1970) *J. Biol. Chem.*, **245**, 923
[56] Heider, H., Gottschalk, E. and Cramer, F. (1971) *Eur. J. Biochem.*, **20**, 144
[57] Yarus, M. and Berg, P. (1969) *J. Mol. Biol.*, **42**, 171
[58] Smith, J. D. (1973) *Brit. Med. Bull.*, **29**, 220
[59] Mehler, A. H. and Chakraburthy, K. (1972) *Adv. Enzymol.*, **35**, 443
[60] Nathans, D. and Lipmann, F. (1961) *Proc. Nat. Acad. Sci.*, **47**, 497
[61] Fessenden, J. M. and Moldave, K. (1961) *Biochem. Biophys. Res. Commun.*, **6**, 232
[62] Noll, H., Staehelin, T. and Wettstein, F. O. (1963) *Nature*, **198**, 632
[63] Lapidot, Y. and de Groot, N. (1972) *Progr. Nucleic Acid Res. Mol. Biol.*, **12**, 189
[64] Loftfield, R. B. (1972) *Progr. Nucleic Acid Res. Mol. Biol.*, **12**, 87
[65] Nathans, D. (1964) *Proc. Nat. Acad. Sci.*, **51**, 585
[66] Arnstein, H. R. V. and Cox, R. A. (1964) *Biochem. J.*, **93**, 33c
[67] Nirenberg, M. W. and Matthaei, J. H. (1961) *Proc. Nat. Acad. Sci.*, **47**, 1580, 1588
[68] Wood, W. B. and Berg, P. (1962) *Proc. Nat. Acad. Sci.*, **48**, 94
[69] Leder, P. and Nirenberg, M. (1964) *Proc. Nat. Acad. Sci.*, **52**, 420, 1521
[70] Nirenberg, M. and Leder, P. (1964) *Science*, **145**, 1399
[71] Trupin, J. S. and Nirenberg, M. W. (1965) *Proc. Nat. Acad. Sci.*, **53**, 807
[72] Söll, D., Cherayil, J. D. and Bock, R. M. (1967) *J. Mol. Biol.*, **29**, 97
[73] Matthaei, H., Heller, G., Voigt, H. P., Neth, R., Schöch, G. and Kübler, H. (1966) *Genetic Elements*, p. 233. FEBS Symposium (D. Shugar, Ed.)
[74] Khorana, H. G., Büchi, H., Ghosh, H., Gupta, N., Jacob, T. M., Kössel, H., Morgan, R., Narang, S. A., Ohtsuka, E. and Wells, R. D. (1966) *Cold Spring Harbor Symp. Quant. Biol.*, **31**, 39
[75] Morgan, A. R., Wells, R. D. and Khorana, H. G. (1966) *Proc. Nat. Acad. Sci.*, **56**, 1899
[76] Khorana, H. G., Büchi, H., Jacob, T. M., Kössel, H., Narang, S. A. and Ohtsuka, E. (1967) *J. Amer. Chem. Soc.*, **89**, 2154
[77] Khorana, H. G. (1968) *Biochem. J.*, **109**, 709
[78] Sanger, F. (1971) *Biochem. J.*, **124**, 833
[79] Sanger, F. and Brownlee, G. G. (1970) *Biochem. Soc. Symp.*, **30**, 183
[80] Adams, J. M., Jeppesen, P. G. N., Sanger, F. and Barrell, B. G. (1969) *Nature*, **223**, 1009
[81] Adams, J. M. and Cory, S. (1970) *Nature*, **227**, 570
[82] Min Jou, W., Haegeman, G., Ysebart, M. and Fiers, W. (1972) *Nature*, **237**, 82
[83] Fiers, W. et al. (1971) *Biochimie*, **53**, 495
[84] Contreras, R., Yaebert, M., Min Jou, W. and Fiers, W. (1973) *Nature New Biol.*, **241**, 99
[85] Wittman, H. G. (1962) *Z. Verebungslehre*, **93**, 491
[86] Wittman, H. G. and Wittman-Liebold, B. (1963) *Cold Spring Harbor Symp. Quant. Biol.*, **28**, 589
[87] Fraenkel-Conrat, H. (1964) *Sci. Amer.*, **211**(4), 46
[88] Okada, Y., Terzaghi, E., Streisinger, G., Emrich, J., Inouye, M. and Tsugita, A. (1966) *Proc. Nat. Acad. Sci.*, **56**, 1692
[89] Ycas, M. (1969) *The Biological Code*. Amsterdam: North-Holland Publishing Co.

[90] Woese, C. R. (1970) *Symp. Soc. Gen. Microbiol.*, **20**, 39
[91] Woese, C. R. (1970) *Progr. Mol. Biol.*, **1**, 5 (Hahn, Ed.)
[92] Jukes, T. H. and Gatlin, L. (1971) *Progr. Nucleic. Acid. Res. Mol. Biol.*, **11**, 303 (J. N. Davidson and W. E. Cohn, Eds.)
[93] King, J. L. and Jukes, T. H. (1969) *Science*, **164**, 788
[94] Doctor, D. P., Apgar, J. and Holley, R. W. (1961) *J. Biol. Chem.*, **235**, 1171
[95] Sueoka, N. and Yamane, T. (1963) *Informational Macromolecules*, p. 205 (H. J. Vogel, V. Bryson and J. O. Lampen, Eds.). New York: Academic Press
[96] Weisblum, B., Benzer, S. and Holley, R. W. (1962) *Proc. Nat. Acad. Sci.*, **48**, 1449
[97] Bennett, T. P., Goldstein, J. and Lipmann, F. (1963) *Cold Spring Harbor Symp. Quant. Biol.*, **28**, 233
[98] Ochoa, M. and Weinstein, J. B. (1964) *Proc. Nat. Acad. Sci.*, **52**, 470
[99] Matthews, M. B. (1973) *Essays Biochem.*, **9**, 59
[100] Sambrook, J. F., Fan, D. P. and Brenner, S. (1967) *Nature*, **214**, 452
[101] Brenner, S., Barnett, L., Katz, E. R. and Crick, F. H. C. (1967) *Nature*, **213**, 449
[102] Beudet. A. L. and Caskey C. T. (1971) *Proc. Nat. Acad. Sci.*, **68**, 619
[103] Goodman, H. M., Abelson, J., Landy, A., Brenner, S. and Smith, J. D. (1968) *Nature*, **217**, 1010
[104] Crick, F. H. C. (1966) *J. Mol. Biol.*, **19**, 548
[105] Söll, D. and Rajbhandary, U. L. (1967) *J. Mol. Biol.*, **29**, 113
[106] Chapeville, F., Lipmann, F., von Ehrenstein, G., Weisblum, B., Ray, Jr., W. J. and Benzer, S. (1962) *Proc. Nat. Acad. Sci.*, **48**, 1086
[107] Byrne, R., Levin, J. G., Bladen, H. A. and Nirenberg, M. W. (1964) *Proc. Nat. Acad. Sci.*, **52**, 140
[108] Das, H. K., Goldstein, A. and Lowney, L. L. (1967) *J. Mol. Biol.*, **24**, 231
[109] Hamkalo, B. A. and Miller, O. L. (1973) *Ann. Rev. Biochem.*, **42**, 379
[110] Williamson, A. R. and Schweet, R. (1964) *Nature*, **202**, 435
[111] Warner, J. R. and Rich, A. (1964) *Proc. Nat. Acad. Sci.*, **51**, 1134
[112] Cannon, M., Krug, R. and Gilbert, W. (1963) *J. Mol. Biol.*, **7**, 360
[113] Bretscher, M. S. (1968) *Nature*, **218**, 675
[114] Ghosh, H. P. and Khorana, H. G. (1967) *Proc. Nat. Acad. Sci.*, **58**, 2455
[115] Gilbert, W. (1963) *Cold Spring Harbor Symp. Quant. Biol.*, **28**, 287
[116] Rich, A., Warner, J. R. and Goodman, H. M. (1963) *Cold Spring Harbor Symp. Quant. Biol.*, **28**, 269
[117] Staehelin, T., Wettstein, F. O., Oura, H. and Noll, H. (1964) *Nature*, **201**, 264
[118] Staehelin, T., Brinton, C. C., Wettstein, F. O. and Noll, H. (1963) *Nature*, **199**, 865
[119] Gierer, A. (1963) *J. Mol. Biol.*, **6**, 148
[120] Marks, P. A., Burka, E. R., Rifkind, R. and Danon, D. (1963) *Cold Spring Harbor Symp. Quant. Biol.*, **28**, 1
[121] Hardesty, B., Miller, R. and Schweet, R. (1963) *Proc. Nat. Acad. Sci.*, **50**, 924
[122] Mathias, A. P., Williamson, R., Huxley, H. E. and Page, S. (1964) *J. Mol. Biol.*, **9**, 154
[123] Slayter, H. S., Warner, J. R., Rich, A. and Hall, C. E. (1963) *J. Mol. Biol.*, **7**, 652
[124] Kiho, Y. and Rich, A. (1964) *Proc. Nat. Acad. Sci.*, **51**, 111
[125] Munro, A., Jackson, R. and Korner, A. (1964) *Biochem. J.*, **92**, 289

BIOLOGICAL FUNCTION OF RNA: PROTEIN SYNTHESIS

[126] Dintzis, H. M. (1961) *Proc. Nat. Acad. Sci.*, **47**, 247
[127] Bretscher, M. S. (1968) *J. Mol. Biol.*, **34**, 131
[128] Haenni, A.-L. and Lucas-Lenard, J. (1968) *Proc. Nat. Acad. Sci.*, **61**, 1363
[129] Erbe, R. W., Nau, M. M. and Leder, P. (1969) *J. Mol. Biol.* **39**, 441
[130] Smith, M. A., Salas, M., Stanley, W. M., Wahba, A. R. and Ochoa, S. (1968) *Proc. Nat. Acad. Sci.*, **55**, 141
[131] Terzaghi, E., Okada, Y., Steissinger, G., Emrich, J., Inouye, M. and Tsugita, A. (1966) *Proc. Nat. Acad. Sci.*, **56**, 500
[132] Adams, J. M. (1968) *J. Mol. Biol.*, **34**, 131
[133] Clark, B. F. C. and Marcker, K. A. (1968) *Sci. Amer.*, **218**(1), 36
[134] Clark, B. F. C. and Marcker, K. A. (1966) *Nature*, **211**, 378
[135] Marcker, K. A., Clark, B. F. C. and Anderson, J. S. (1966) *Cold Spring Harbor Symp. Quant. Biol.*, **31**, 279
[136] Webster, R. E., Engelhardt, D. L. and Zinder, N. S. (1966) *Proc. Nat. Acad. Sci.*, **55**, 155
[137] Ghosh, H. P., Söll, D. and Khorana, H. G. (1967) *J. Mol. Biol.*, **25**, 275
[138] Sundararajan, T. A. and Thach, R. E. (1966) *J. Mol. Biol.*, **19**, 74
[139] Brown, J. C. and Smith, A. E. (1970) *Nature*, **226**, 610
[140] Jackson, R. and Hunter, T. (1970) *Nature*, **227**, 672
[141] Smith, A. E. and Marcker, K. A. (1970) *Nature*, **226**, 607
[142] Wigle, D. T. and Dixon, G. H. (1970) *Nature*, **227**, 676
[143] Hindley, J. and Staples, D. H. (1969) *Nature*, **224**, 964
[144] Steitz, J. A. (1969) *Nature*, **224**, 957
[145] Burgess, A. B. and Mach, B. (1971) *Nature New Biol.*, **233**, 209
[146] Dubnoff, J. S. and Maitra, U. (1971) *Methods in Enzymology*, **20**, 248 (L. Grossman and K. Moldave, Eds.)
[147] Bryan, R. N., Gelfand, D. H. and Hayashi, M. (1969) *Nature*, **224**, 1019
[148] Heywood, S. M. (1970) *Nature*, **225**, 696
[149] Jeppesen, P. G. N., Steitz, J. A., Gesteland, R. F. and Spahr, P. F. (1970) *Nature*, **226**, 230
[150] Fiers, W. et al. (1975) *Nature*, **256**, 273
[151] Sabol, S., Sillero, M. A. G., Iwasaki, K. and Ochoa, S. (1970) *Nature*, **228**, 1269
[152] Sabol, S. and Ochoa, S. (1971) *Nature New Biol.*, **234**, 233
[153] Lee-Huang, S. and Ochoa, S. (1971) *Nature New Biol.*, **234**, 236
[154] Ochoa, S. and Mazumder, R. (1974) *The Enzymes*, Vol. 10, p. 1 (P. D. Boyer, Ed.). New York: Academic Press
[155] Lee-Huang, S. and Ochoa, S. (1973) *Ann. Biochim. Biophys.*, **156**, 1973
[156] Groner, Y., Pollack, Y., Berissi, H. and Revel, M. (1972) *Nature New Biol.*, **239**, 16
[157] Groner, Y., Scheps, R., Kamem, R., Kolakofsky, D. and Revel, M. (1972) *Nature New Biol.*, **239**, 19
[158] Shafritz, D. A., Pritchard, P. M., Gilbert, J. M. and Anderson, W. F. (1970) *Biochem. Biophys. Res. Commun.*, **38**, 721
[159] Shafritz, D. A., Pritchard, P. M., Gilbert, J. M., Merrick, W. C. and Anderson, W. F. (1972) *Proc. Nat. Acad. Sci.*, **69**, 983
[160] Picciano, D. J. et al. (1973) *J. Biol. Chem.*, **248**, 204
[161] Wigle, D. T. and Smith, A. E. (1973) *Nature New Biol.*, **242**, 136
[162] Ono, Y., Skoultchi, A., Waterson, J. and Lengyel, P. (1969) *Nature*, **222**, 645
[163] Gordon, J., Lucas-Lenard, J. and Lipmann, F. (1971) *Methods in Enzymology*, **20**, 281 (L. Grossman and K. Moldave, Eds.)

[164] Lucas-Lenard, J. and Beres, L. (1974) *The Enzymes*, Vol. 10, p. 53 (P. D. Boyer, Ed.). New York: Academic Press
[165] Miller, D. L. and Weissbach, H. (1970) *Biochem. Biophys. Res. Commun.*, **38**, 1016
[166] Maden, B. E. H., Traut, R. R. and Monro, R. E. (1968) *J. Mol. Biol.*, **35**, 333
[167] Thach, S. S. and Thach, R. E. (1971) *Proc. Nat. Acad. Sci.*, **68**, 1791
[168] Bodley, J. W., Zieve, F. J. and Lin, L. (1970) *J. Biol. Chem.*, **245**, 5662
[169] Kuwano, M. and Schlessinger, D. (1970) *Proc. Nat. Acad. Sci.*, **66**, 146
[170] Roufa, D. J., Skogerson, F. E. and Leder, P. (1970) *Nature*, **227**, 567
[171] Lucas-Lenard, J. and Lipmann, F. (1971) *Ann. Rev. Biochem.*, **40**, 409
[172] Galasinski, W. and Moldave, K. (1969) *J. Biol. Chem.*, **244**, 6527
[173] Baliga, B. S. and Munro, H. N. (1971) *Nature New Biol.*, **233**, 257
[174] Richter, D., Erdman, V. A. and Sprinzl, M. (1973) *Nature New Biol.*, **246**, 132
[175] Piper, P. W. and Clark, B. F. C. (1973) *FEBS Lett.*, **30**, 265
[176] Lu, P. and Rich, A. (1971) *J. Mol. Biol.*, **58**, 513
[177] Model, P., Webster, R. E. and Zinder, N. D. (1969) *J. Mol. Biol.*, **43**, 177
[178] Webster, R. E. and Zinder, N. D. (1969) *J. Mol. Biol.*, **42**, 425
[179] Nichols, J. L. (1970) *Nature*, **225**, 143
[180] Scolnick, E. M. and Caskey, C. T. (1969) *Proc. Nat. Acad. Sci.*, **64**, 1235
[181] Caskey, C. T., Scolnick, E., Tompkins, R., Milman, G. and Goldstein, J. (1971) *Methods in Enzymology*, **20**, 367 (L. Grossman and K. Moldave, Eds.)
[182] Tate, W. P. and Caskey, C. T. (1974) *The Enzymes*, Vol. 10, p. 87 (P. D. Boyer, Ed.). New York: Academic Press
[183] Goldberg, I. H. and Friedman, P. A. (1971) *Ann. Rev. Biochem.*, **40**, 775
[184] Pestka, S. (1971) *Ann. Rev. Microbiol.*, **25**, 487
[185] Vasquez, D. (1974) *FEBS Lett.*, **40**, S63
[186] Clark, B. F. C. (1974) *Companion to Biochemistry*, p. 1 (A. T. Bull, J. R. Lagnado, J. O. Thomas and K. F. Tipton, Eds.). London: Longmans
[187] Mathews, M. B. (1973) *Essays Biochem.*, **9**, 59
[188] Grossbard, L., Banks, J. and Marks, P. A. (1968) *Arch. Biochem. Biophys.*, **125**, 580
[189] Mathews, M. B., Osborn, M. and Lingrel, J. B. (1971) *Nature New Biol.*, **233**, 206
[190] Gilbert, J. M. and Anderson, W. F. (1970) *J. Biol. Chem.*, **245**, 2342
[191] Gilbert, J. M. and Anderson, W. F. (1971) *Methods in Enzymology*, **20**, 542 (L. Grossman and K. Moldave, Eds.)
[192] Heywood, S. M. and Rich, A. (1968) *Proc. Nat. Acad. Sci.*, **59**, 590
[193] Swan, D., Aviv, H. and Leder, P. (1972) *Proc. Nat. Acad. Sci.*, **69**, 1967
[194] Milstein, C., Brownlee, G. G., Harrison, T. M. and Mathews, M. B. (1972) *Nature New Biol.*, **239**, 117
[195] Jacobs-Lorena, M., Baglioni, C. and Borun, T. W. (1972) *Proc. Nat. Acad. Sci.*, **69**, 1574
[196] Bloemendal, H. (1972) *The Mechanisms of Protein Synthesis and its Regulation, Frontiers of Biology*, 27. Amsterdam, London: North-Holland Publishing Co.
[197] Gold, L. M. and Schweiger, M. (1969) *J. Biol. Chem.*, **244**, 5100
[198] Gold, L. M. and Schweiger, M. (1969) *Proc. Nat. Acad. Sci.*, **62**, 892
[199] Schweiger, M. and Gold, L. M. (1969) *Proc. Nat. Acad. Sci.*, **63**, 1351
[200] Lodish, H. F. (1968) *J. Mol. Biol.*, **32**, 681

BIOLOGICAL FUNCTION OF RNA: PROTEIN SYNTHESIS

[201] Kurland, C. G. (1972) *Ann. Rev. Biochem.*, **41**, 377
[202] Nomura, M. (1973) *Science*, **179**, 864
[203] Garrett, R. A. and Wittman, H. G. (1973) *Adv. Prot. Chem.*, **27**, 277
[204] Spirin, A. S. (1974) *FEBS Lett.*, **40**, S38
[205] Traub, P. and Nomura, M. (1968) *Proc. Nat. Acad. Sci.*, **59**, 777
[206] Nomura, M. and Erdmann, V. A. (1970) *Nature*, **228**, 744
[207] Fahnestock, S., Erdmann, V. A. and Nomura, M. (1973) *Biochem.*, **12**, 220
[208] Pongs, O., Nierhaus, K. H., Erdmann, V. A. and Wittmann, H. G. (1974) *FEBS Lett.*, **40**, S28
[209] Fakunding, J. L., Traut, R. R. and Hershey, J. W. B. (1973) *J. Biol. Chem.*, **248**, 8555
[210] Czernilofsky, P., Stoffler, G. and Kuchler, E. (1974) *Hoppe-Seyler's Z. Physiol. Chem.*, **355**, 89
[211] Ballesta, J. P. G., Montejo, V., Hernandez, F. and Vasquez, D. (1974) *Eur. J. Biochem.*, **42**, 167
[212] Nomura, M., Tissières, A. and Lengyel, P. (1974) *Ribosomes*, Cold Spring Harbor Laboratory Monograph Series
[213] Frazer, T. H. and Rich, A. (1975) *Proc. Nat. Acad. Sci.*, **72**, 3044
[214] Sprinzl, M. and Cramer, F. (1975) *Proc. Nat. Acad. Sci.*, **72**, 3049

CHAPTER 14

Nucleic Acids and the Regulation of Protein Synthesis

So far no attempt has been made to explain why cells do not continually make all the proteins which they are capable of producing. Various calculations suggest that the *Esch. coli* genome has, on the basis of size consideration, the potential to code for around 2000 average size proteins. On the other hand, it is estimated that the likely number of enzymes needed to provide the necessary metabolites is probably only of the order of one-third of that number. Furthermore, although it appears that each *Esch. coli* bacterium contains a total of roughly 10^7 protein molecules, a great variation appears to exist in the numbers of each protein type. For instance, some may be represented 500 000 times and some only once or twice [1]. Indeed, some enzymes are produced only when the need for their activity arises. Thus, there must be mechanisms for ensuring and controlling the selective utilization of the genome for the synthesis of proteins to cope with the spectrum of cellular requirements. How is this regulation of gene function achieved?

As was mentioned in Chapter 13, the actual synthesis of protein results from mRNA *translation*, a final step in the overall process of gene expression, the initial step being the *transcription* of RNA molecules from specific regions, or genes, of the DNA genome. Both translation and transcription are subject to a variety of molecular regulatory mechanisms which can potentially influence the level of a particular protein within a cell.

14.1 Regulation at the transcriptional level

There is now known a wide variety of molecular mechanisms whereby the production of various RNA's within the cell could be regulated.

14.1.1 *RNA chain initiation*

Normally nucleotide units are added to growing RNA chains at a rate of 40–50 nucleotides per second at 37° in *Esch. coli*. This rate, however, can vary with temperature and with other environmental changes, but under normal conditions the amount of RNA made in a bacterium is limited not so much by the rate of growth of RNA chains as by their rate of initiation. This varies quite considerably for individual types of molecule. Ribosomal RNA molecules, for instance, are required in fairly large amounts and can be initiated in *Esch. coli* at the rate of one molecule per second. On the other hand, a gene coding for a protein present in very small amounts may be transcribed as infrequently as once every bacterial generation [2].

Having initiated an RNA chain, the RNA polymerase moves away from the promoter site transcribing the adjacent genetic material and leaving the initiation site open to a second polymerase molecule. The frequency of these initiations will determine the proximity of

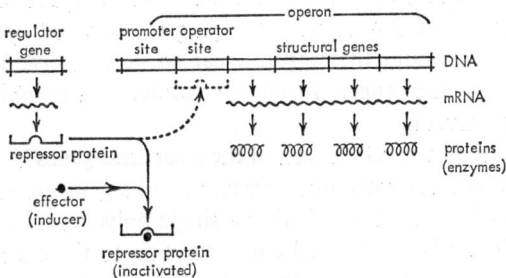

Fig. 14.1 *The operon. The regulator gene produces a repressor which blocks the operator gene and prevents the structural genes from producing mRNA. In the presence of an effector (inducer) the repressor is inactivated and the operator gene allows the structural genes to come into action*

RNA polymerase molecules on the genomic sequence in question [2]. In the case of the ribosomal genes this may be as close together as is sterically possible.

How can the frequency of initiation be regulated? In bacterial cells, sets of structural genes (cistrons), contained in contiguous sectors of a length of DNA, and closely linked on the genetic map, form what is called an operon which is under the control of a gene known as the operator, closely linked on the genetic site to the operon which it controls [3-9] (Fig. 14.1).

When the operator is open, each cistron on the operon can be transcribed into mRNA which in turn controls the formation of the polypeptide chain corresponding to the genetic information in the operon.

When the operator is closed, no synthesis of mRNA can occur. Such closure occurs when the operator becomes engaged with a specific cytoplasmic repressor [12] which is the product of a regulator gene. The repressor acts negatively in the sense that in its active form it inhibits mRNA synthesis and the subsequent enzyme synthesis [10, 11].

The activity of repressors is governed by specific metabolites known as effectors. In the formation of inducible enzymes, the inducer acts as effector and inactivates the repressor so that the operator gene ceases to be repressed. The cistrons in the operon are therefore allowed to produce the appropriate mRNA so that the otherwise repressed synthesis of the polypeptides coded by the cistrons of the operon can take place. It has been proved that the presence of specific inducing effectors in growing bacteria greatly increases the amount of mRNA formed which is capable of forming hybrids with the DNA fraction carrying the relevant operon [13]. The repressor would therefore appear to be concerned with the inhibition of messenger formation rather than inhibition of messenger function.

It is clearly possible that each of the structural genes in an operon could produce its own messenger, the one-gene–one-messenger theory. But it is also possible that a single polycistronic messenger RNA molecule is synthesized corresponding to the entire operon, the one-operon–one-messenger theory. It is now clear that the latter view is correct [13, 14] at least in bacteria.

The most convincing evidence for the existence of such a polycistronic messenger comes from the study of the histidine operon in *Salmonella typhimurium*. Histidine is synthesized by a well-established pathway involving ten enzymes, the structural genes for which are in a cluster in the Salmonella chromosome. The genetic fine structure has been mapped out in detail [15]. It has been estimated that a single mRNA molecule corresponding to the entire histidine operon would have a sedimentation constant of about 38S whereas the mRNA's for each individual enzyme would, of course, be very much smaller. The mRNA found experimentally for this system has a sedimentation coefficient of 34S, far too large to correspond to any known individual enzyme, but in good agreement with the predicted value for the mRNA of the entire operon. It appears, therefore, that the mRNA can form a complex with the ribosomes which is capable of synthesizing all the polypeptides encoded in a single operon.

Apart from the histidine operon, most work in this field has been

NUCLEIC ACIDS AND REGULATION OF PROTEIN SYNTHESIS

carried out with the tryptophan operon and especially with the lactose operon (pp. 62 and 321) in *Esch. coli* [17–19]. When lactose (or certain other galactosides) is added to a culture of *Esch. coli* three enzymes are coordinately induced:

(1) β-Galactosidase; this hydrolyses lactose.

(2) Thiogalactoside transacetylase; this catalyses the transfer of an acetyl group from acetyl coenzyme A to a thiogalactoside acceptor.

(3) Galactoside-permease; this controls the movement of lactose into the cell.

They are produced in very different amounts. While β-galactosidase accounts for 6 per cent of the protein in the fully induced cell of *Esch. coli*, the transacetylase accounts for only 0·2 per cent. As might be expected, the β-galactosidase gene (z gene) is located closer to the operator than is the transacetylase or permease [15, 17].

Transcription of this operon, for instance, can be controlled both negatively and positively [20]. Negative control is mediated by the lac repressor which binds specifically and tightly for the operator

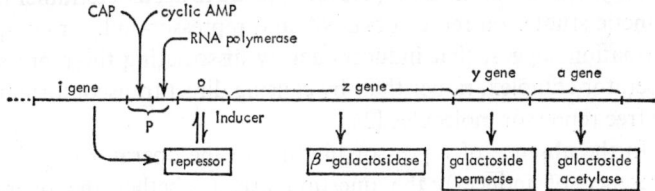

Fig. 14.2 *The lac region of the* Esch. coli *chromosome*

region, thereby preventing transcription. Positive control of the lac operon is exerted through the phenomenon termed catabolite repression. Expression of the lac operon (and other catabolite repressible operons) is repressed when glucose (a more efficient source of carbon than lactose) is present in the medium. By a yet unknown mechanism, the presence of glucose results in a decreased concentration of intracellular cyclic AMP. Cyclic AMP is required for efficient expression of the lac operon, since it activates the catabolite gene activator protein (CAP), which in turn activates initiation of transcription of lac by RNA polymerase [21]. This CAP seems to interact with a site in the promoter region (p) of the operon (Fig. 14.2). CAP can be envisaged as a dimer of two special subunits (mol. wt. 40 000), each equivalent to 11 nucleotide pairs in diameter. The CAP binding site is located in the promoter region of two-fold symmetry noted in Fig. 12.2 [20]. The binding of this protein–cyclic AMP complex may then destabilize the guanine plus cytosine rich region next to it. This would lower the transition temperature of the entry site for the RNA polymerase some 14 nucleotide pairs away (the entry site is positioned at the adenine plus thymine rich site shown in Fig. 12.2).

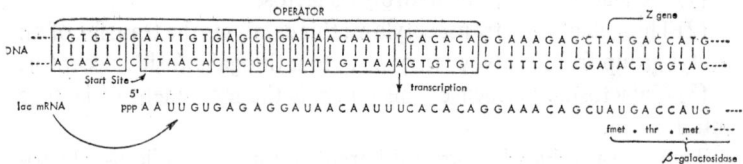

Fig. 14.3 *The lac operator region and the z gene, showing the nature of the 5'-end of the lac mRNA*

RNA polymerase entry is thereby facilitated. Additional nucleotide sequences analyses [22, 23] have been recently carried out on the operator region (o) lying between the promoter region and the start of the z gene coding for β-galactosidase. Once again a region of two-fold symmetry will be noted (Fig. 14.3) [20]. The significance of this is not yet clear but the lac repressor appears to act as a tetramer [24]. Kinetic studies on repressor–DNA and repressor–inducer complex formation suggest that inducers act by dissociating the repressor-operator complex, rather than by causing direct transconformation to free repressor molecules [24].

In the absence of repressor binding to the operator (e.g. in the presence of inducer), the question arises whether the operator sequence is transcribed. As a result of studies on the lac operon mRNA sequences [23], two interesting features emerge: (a) a portion, but not all, of the operator sequence seems to be transcribed; (b) the actual 'start' site for RNA synthesis is some 35 nucleotide pairs away from the 'entry' site in the promoter. Thus, there must be some 'drift' of the polymerase to the 'start' site after entry (see Fig. 14.4). Recent interesting data indicate that the first 33 nucleotides of both the lac mRNA and the gal mRNA are identical [25].

Initiation can be effected in other ways. For example, ribosomal RNA in bacteria such as *Esch. coli* in the rapid growth state can comprise up to 50 per cent of the RNA being made. However, when these bacteria have exhausted the amino acid content of medium, the proportion of rRNA synthesis can be ten-fold lower.

Two unusual nucleotides guanosine tetraphosphate (ppGpp) and guanosine pentaphosphate (pppGpp) may serve in some way to regulate the level of rRNA biosynthesis [28]. The intracellular con-

NUCLEIC ACIDS AND REGULATION OF PROTEIN SYNTHESIS

centration of these nucleotides appears to be inversely correlated with the rate of rRNA synthesis, and their synthesis appears to result from an ATP-dependent conversion of GDP and GTP respectively. This may represent an idling translational reaction occurring on the messenger–ribosome complexes, which would be signalled by the presence of a codon-specific uncharged tRNA in the A site [24]. How ppGpp effects rRNA biosynthesis is still a matter of controversy. There is some indication that ppGpp can inhibit ribosomal RNA synthesis catalysed by *Esch. coli* polymerase *in vitro*

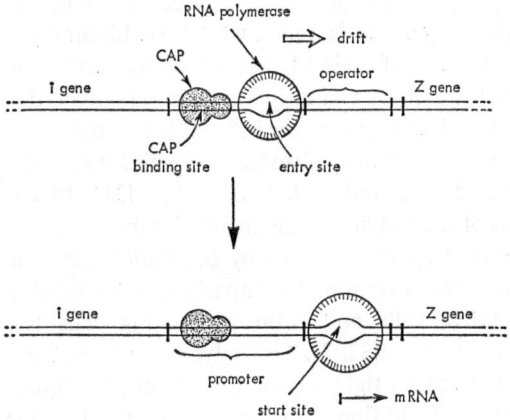

Fig. 14.4 *A possible mode of initiation of* lac *transcription*

[27, 121]. However, Travers and his colleagues [26] report the existence of psi (Ψ) factor, a possible controlling element specifically acting on the initiation of ribosomal RNA during the *in vitro* transcription of *Esch. coli* DNA. In the presence of KCl this factor stimulates the rate of rRNA synthesis up to ten-fold. However, the Ψ-dependent stimulation effects could be abolished by ppGpp. Recent reports suggest that Ψ is equivalent to the EF-Tu · EF-Ts complex involved in the elongation steps of protein synthesis (see Chapter 13). Thus, there is the possibility of a regulatory link between translation and transcription processes.

14.1.2 *Structural modifications to RNA polymerases*

During infection of *Esch. coli* with T4 bacteriophage, host RNA synthesis rapidly stops and a well-defined series of changes in the pattern of T4 DNA transcription (e.g. 'early' to 'late' RNA) occurs, which may result from changes observed in the polypeptides that make up the RNA polymerase subunit structure. Within 4 minutes of infection, the two α-subunits are enzymically modified. The modification involves the covalent attachment of an adenosine diphosphoribose unit to each subunit. The attachment appears to be through its terminal ribose to a guanido nitrogen of a specific arginine residue [29], the donor being NAD^+.

Another type of structural modification to the *Esch. coli* polymerase that occurs during T4 infection is the binding of several bacteriophage specific proteins, the products of T4 genes 33 and 55. (The product of gene 55 is a mol. wt. 22 000 binding protein [30].) Their actual mode of action is not yet understood; however, it is probable that auxiliary transcription factors regulate the specificity of bacterial RNA polymerases. Indeed, it has been observed that free RNA polymerase in *crude* extracts of uninfected bacteria is both physically and functionally heterogeneous [31], two or perhaps three forms of the enzyme being distinguishable.

In the case of spore formation by *B. subtilis* there is a change in the pattern of transcription. A comparison of the RNA polymerase from vegetative cells and from sporulating cells reveals some structural differences [32, 2]. The polymerase from the sporulating cells does not have a tightly bound σ subunit, but appears to have associated with it additional small polypeptides. Thus, the change in transcription pattern during sporulation is probably a consequence of structural modification of the enzyme.

14.1.3 *The complexity of RNA polymerases*

When the bacteriophage T7 infects its host *Esch. coli*, the RNA polymerase of the host transcribes a portion of the invading genome corresponding to the 'early' genes (see Fig. 12.12). The product of one of these genes is an mRNA which, when translated by the host's protein synthetic machinery, yields a very simple RNA polymerase (a single polypeptide chain of mol. wt. about 110 000) [33]. This T7 specified RNA polymerase can only initiate RNA synthesis at promoter sites on T7 DNA *other* than those at which the host's RNA polymerase can initiate. In this way the T7 specified polymerase, with a different initiation specificity, simply synthesizes a collection of RNA's, known as the 'late' mRNA's, from the remainder of the T7 genome. The reason for the apparent structural simplicity of the

T7 polymerase is not understood. It may be that, in some way, complexity of polymerase structure reflects the organism's requirement for higher degrees of control. The host, *Esch. coli*, is capable of rapid adaptation to different growth conditions, and perhaps this is reflected in the greater complexity of its polymerase, the initiation specificity of which can be altered by interaction with regulatory proteins. A similar situation is known to prevail in *B. subtilis* which can alter its growth pattern quite dramatically (e.g. spore formation as already mentioned). On the other hand, some bacteria like the *halophilic* group have a fairly stable growth pattern and appear to have simple type polymerases (2 polypeptide chains each of 25 000 daltons) [39]. From Chapter 12 it will be recalled that the RNA polymerases of the eukaryotic cell nuclei belong to the complex type whilst the polymerases from the mitochondria are simple single polypeptides. Perhaps there is little need for extensive regulation of mitochondrial DNA transcription. Indeed, it appears that both strands are transcribed (see Chapter 12), and extensive control may operate instead at the post-transcriptional level.

14.1.4 *The variety of RNA polymerases*

A striking difference between the transcriptional equipment of eukaryotic cells and prokaryotic cells is the presence in the latter of more than one variety of polymerase. From Chapter 12 it will be remembered that so far there have been detected three distinguishable varieties of complex type polymerases in some mammalian cell nuclei. Even one of the simplest of the nucleated organisms, yeast, has at least two complex nuclear polymerases. It would appear, in these cases, that the initiation specificity (i.e. for rRNA, tRNA, hnRNA, etc.) may lie not only in the complexity of the polymerases but also in their variety and/or intranuclear location. A question being pursued at present is whether the levels of the various polymerases can regulate the levels of the various RNA's produced. In higher eukaryotes, qualitative changes in RNA synthesis can be induced by hormone treatments, nutritional deficiencies, etc. The levels of activity of the various *purified* polymerases have been shown to vary considerably. Type II polymerase is high in rapidly growing cells [93] (up to 50 000 molecules/cell) but is low in non-dividing cells [94] (as low as 800 molecules/cell). Type I polymerase is more problematic to determine owing to its metabolic turnover (half-life 1·5 hours [95]). In the wide range of differentiative or reversible metabolic transitions so far studied, significant changes in the levels of RNA polymerase activity have been detected. It should be emphasized, however, that in no instance is it yet known whether those changes are casually related to the onset of such transitions. A dramatic change in polymerase level accompanies the initiation of embryonic development [96, 97]. In reversible changes initiated, for instance, by hormones, nutrients, etc., modest changes in polymerase I and sometimes in polymerase II activity have been observed.

Steroid hormones are thought to bind with specific cytoplasmic receptors. The complex of steroid and receptor in turn is believed to interact with the chromosome to induce the appropriate response. Steroid hormone induction is accompanied by changes in polymerase activity [98–102]. Type I increases within 2 hours, for example, of oestrogen treatment, at which time a major product is the 45S rRNA precursor [109]. Subsequently the activity of type II increases and new specific mRNA's such as ovalbumin mRNA can be detected within 3 hours. Results of several experiments suggest that these increased levels of polymerase activity may be due to modulation of activity through interaction with hormone–receptor complexes [35, 103, 104] rather than *de novo* synthesis. However, these changes all occur subsequent to the production of an oestrogen-induced protein which appears within 30 minutes [101, 105]. The synthesis of the mRNA for this protein may occur within 15 minutes after hormone administration and may be the primary biosynthetic response to oestrogen.

Other factors that may be capable of modulating eukaryotic polymerases are cAMP binding proteins [106] and non-histone chromosomal proteins [107, 108].

14.1.5 *Gene dosage*

Another means of increasing the number of transcripts of a particular region of the genome is to have the sequences in question represented a number of times within the genome. This, as already mentioned in Chapter 12, is the case for rRNA, tRNA, and 5S genes. The actual numbers of copies of these genes seem to increase with the evolutionary complexity of the organisms. Most mRNA's, on the other hand, arise from sequences that occur once (or only a few times) in the genome. Exceptions are the genes for histone mRNA which occur in multiple copies [36], as do those for chick feather keratin in RNA's [37].

NUCLEIC ACIDS AND REGULATION OF PROTEIN SYNTHESIS

Of particular interest has been the question of whether the large number of, say, rRNA genes is constant throughout the life-cycle of a cell, or whether they are amplified *de novo* by a mechanism sensitive to metabolic demands. One case where amplification of the ribosomal genes has been demonstrated is in the immature oocytes of amphibia and insects [38]. rRNA synthesis reaches a peak rate in such immature oocytes which are in the pachytene and diplotene stages of meiosis, and it is during these stages that extra copies of the nuclear organizer region are synthesized selectively to give rise to about 10^6 copies of the ribosomal genes (somatic *Xenopus* cells have about 1600 copies).

14.1.6 *Chromosomal organization*

Whilst both the prokaryotes and eukaryotes have complex polymerases, the physical organization of their chromosomes differs. In those prokaryotes which have been studied in detail the chromosomes can best be considered as helical DNA molecules (cyclic in the case of *Esch. coli*) with which can be associated at various sites such proteins as repressors, cyclic AMP proteins, etc. These, as already indicated, control the nature of the transcripts produced. Eukaryotic chromosomes, on the other hand, are more complex. They are almost completely associated with various proteins, and constitute a complicated deoxyribonucleoprotein complex referred to, in interphase nuclei, as *chromatin*. There is general agreement that a single DNA double helix (20 Å) combines with protein to give a 30 Å fibre which is then variously aggregated and condensed to give thicker, tightly packaged structures. This aggregation is proposed to occur with local condensations resulting in the formation of 80–100 Å diameter structures giving fibres, when slightly extended, a 'string of beads' structure [39] (see Plate VIII). These beads, sometimes termed nucleosomes, or nu bodies (see also Chapter 2), are basic repeat units comprising 200 nucleotide stretches of DNA combined with basic histone molecules (2 each of types H2A, H2B, H4, H3, and possibly one molecule of H1) [40] (see Fig. 14.5).

Studies aimed at elucidating the functional significance of the proteins associated with the DNA have relied on *in vitro* transcription of chromatin with added bacterial RNA polymerases. Briefly summarized, these experiments indicate that certain genomic sequences are in some way 'masked' in the chromatin complex [41].

The means whereby this 'masking' is achieved is not known precisely, but it is tempting to equate the tightly packed structures with functionally inactive DNA, and the more open chromatin fibres with active DNA. However, neither type of region by itself seems to contain enough DNA to accommodate a gene. An attractive proposal is that all DNA is bound up in 'bead' structures except where it is necessary to have open chromatin. The DNA in these open regions might represent 'control' DNA which is accessible to the outside environment (e.g. hormones, etc.), and which could regulate

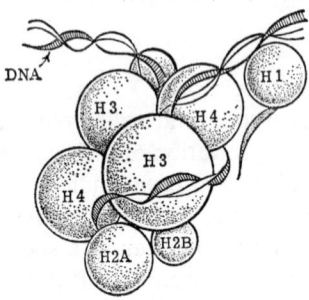

Fig. 14.5 *A highly schematic diagram of mammalian chromatin structure (see [39–43]), indicating the possible structure of a nucleosome (or nu body) comprising histones and DNA*

the activity of the genes with which it is associated. For instance, in response to the appropriate signal, the 'control' regions might permit the RNA polymerase(s) to begin transcribing the gene itself which is in the tightly packed bead-like structures. This would then open up and also become available for copying. Most biochemical data indicate that histones are certainly involved in repressing the availability of certain sequences of eukaryotic DNA in chromatin for transcription. Control of gene transcription, at least from *in vitro* studies, appears to be a positive event, involving the derepression of histone action by the non-histone protein components also associated with chromatin [92] (for reviews see [115, 116]). From Fig. 14.5 no particular location has yet been ascribed to the non-histone proteins of chromatin.

In vivo the formation of RNA on eukaryotic chromosomes has been elegantly demonstrated on giant dipteran polytene chromosomes by autoradiographic techniques [110, 111]. Certain regions of such chromosomes swell out in strongly expanded structures known as 'puffs' in response to the hormone ecdysone or heat shock [111,

116]. That the RNA synthesized at specific 'puff' loci contains mRNA-type sequences has been demonstrated by *in situ* hybridization [112]. Moreover, whilst these regions, which are now regarded as functionally active genes [111], have their normal complement of histones, they have an increased level of non-histone proteins [113]. The same is true for the 'diffuse', or *euchromatin*, regions of higher organisms which are active in RNA synthesis [114] and also in the chromatin of target organs after hormone treatments [116, 117].

The DNA unmasking effects envisaged above could be regarded as a 'coarse' type of control. In addition, there may well be a number of 'fine' levels of control, as yet undiscovered in eukaryotes, for example, regulating correct initiation, termination, etc.

14.1.7 *Post-transcriptional processing and its regulation*
The variety of post-transcriptional processes operating in the production of various RNA's of eukaryotic and prokaryotic origin clearly offers considerable potential *vis-à-vis* the further regulation of mature, functional RNA production. To date, however, very little is known about the control of these processing events. Cleavage rates of the ribosomal RNA precursors in mammalian cells can be altered by hormone administration and virus infections, and may also be linked to ribosomal protein synthesis [44]. Pre-tRNA cleavage *in vivo* can be influenced by cell growth conditions and infections by some viruses [49]. With regard to mRNA production, there is evidence indicating increased rates of hnRNA processing to mRNA during the transition of mammalian cells from the 'resting' to the 'growing' state [45]. Cultured *Drosophila* cells process their hnRNA more efficiently than cultured mosquito cells [46]. Other reports, whilst indicating the presence of messenger (globin) sequences in the hnRNA of some immature blood cells, also demonstrate that these sequences are destroyed post-transcriptionally in the nucleus and are not exported to the cytoplasm as mRNA [47]. On the other hand, there is little information to indicate how the processing enzymes can be regulated. The post-transcriptional modification of tRNA's has been shown to vary in different biological situations, and naturally occurring inhibitors of the methylases involved have been detected in some eukaryotes [48]. Post-transcriptional addition of nucleotides can also be modulated in certain situations. For instance, the activation of protein synthesis after sea urchin fertilization is accompanied by polyadenylation of

the stored mRNA's of the egg [49]. The effect is to increase the tract size of adenylate residues from about 100 to 200. How this is regulated, or indeed its significance, is not yet appreciated.

A final post-transcriptional process that must not be forgotten, but which will have a profound effect on, say, the levels of cellular mRNA's, is of course their metabolic degradation. This is an area for further study, but there is evidence that degradation of mRNA takes place in a ribosomal complex, and is initiated on parts of the messenger just translated. A possible mechanism involves recognition of regions of special structure by endonucleases, followed by exonuclease digestion [50].

14.2 Regulation at the level of translation

Given the complexity and multiplicity of reactions involved in the production of mRNA (as well as the other RNA's required for protein synthesis), it is perhaps not too surprising to have encountered such a diverse set of potential regulatory mechanisms. Indeed, there have been claims that such mechanisms suffice for most of the regulatory phenomena governing protein synthesis. Despite this, the possibility of important regulatory mechanisms of the level of translation have been investigated.

14.2.1 *Messenger RNA structure and ribosome recognition*

Important in the initiation of protein synthesis is the recognition of mRNA initiator sequences. Extensive sequence studies on the *Esch. coli* phages such as R17, Qβ, MS2, and f2 [51] (Fig. 6.11), have shown extensive intercistronic regions between terminator codons of the preceding and the initiator codons of the following genes or cistrons (Figs. 6.11 and 13.6). These regions, as well as the 5'-end of the phage RNA's, are not translated into protein [52], but may well represent sites which are recognized by the IF3-containing ribosome. That the ribosome itself, rather than IF3, may select the initiator sites was indicated in experiments where translation of f2 RNA by *Esch. coli* ribosomes yielded the A protein, the coat protein, and the replicase subunit in the correct proportions (see Chapter 13); use of *B. stearothermophilus* ribosomes led to the production of A protein only [53]. This discriminatory function was shown to reside in the 30S subunit [54]. Indeed, a recent suggestion has been that a sequence of 12 nucleotides at the 3'-terminus of the *Esch. coli* 16S ribosomal RNA (i.e. from the 30S subunit) may participate directly, by base

NUCLEIC ACIDS AND REGULATION OF PROTEIN SYNTHESIS

pairing with mRNA, in the initiation of protein biosynthesis [118]. It appears that the 3'-end of the 16S ribosomal RNA may lie close to the mRNA binding site in the functional ribosome [119]. Recent data also implicate the 5'-terminal sequence of 16S rRNA in the recognition of initiation sequences in mRNA's, again by virtue of complementarity [65]. In eukaryotic systems the picture is less clear but there is evidence for a step in protein synthesis initiation involving interaction of mRNA with 18S rRNA [71].

Physicochemical studies have shown the above phage RNA's to have a high degree of secondary structure [55, 56], and it was suggested that the secondary structure might be of significance in the regulation of transcription. If the secondary structure of f2 is disrupted by mild formaldehyde treatment, the ability of the resulting RNA to direct the synthesis of the replicase subunit and the A protein are increased above the low level normally encountered [57]. It would appear from a possible model of MS2 secondary structure [51] (see Fig. 6.11) that the coat protein initiator site is openly displayed, and this would lead to ready binding of ribosomes, possibly under IF3 guidance, to this particular site on the messenger. Following this, the RNA will unfold upon translation of the coat protein cistron, and other initiation sites (e.g. for replicase subunit) will become available. However, it also turns out that, once the coat protein is made, it can bind to the replicase initiator site, thus curtailing further translation of replicase subunit whilst translation of coat protein genes continues [51].

Although these results were obtained using a phage RNA, it appears that natural messengers such as the globin mRNA [58] and T4-lysozyme mRNA [59] also have considerable secondary structure with possible significance *vis-à-vis* the regulation of protein synthesis initiation [60].

14.2.2 *mRNA and initiation factors*

Although there is the possibility of a *limited* number of messenger-selective IF3's and interference (i) factors in bacteria as discussed in Chapter 13, the situation in higher organisms is less encouraging (e.g. the IF_{EMC} already mentioned). A number of cell-free systems capable of translating natural messenger RNA's from eukaryotic sources such as reticulocytes [61, 62], ascites tumour cells [63], wheat embryos [64], embryonic chick muscle [66], and insect tissues [67] have been described. An interesting *in vivo* mRNA-translating system is the oocyte of *Xenopus laevis*. Several mRNA's of animal and plant origin have been injected into such oocytes and have been found to be translated faithfully [68]. Indeed, the general impression is that there is little or no species specificity in mRNA translation. For instance, lens mRNA can be translated in heterologous ascites [69] and reticulocyte cell-free systems [70]. On the other hand, it has been argued that discrimination requires optimization of the conditions for the translation of each message [72, 73]. Furthermore, it is now believed that the oocyte may well contain many classes of IF3 since it is not a differentiated cell.

Initiation of protein synthesis can, however, be affected in other ways. For example, in reticulocytes the absence of haemin results in the formation of an inhibitor of the binding of the initiator Met-tRNA$_f$ to 40S ribosomes [79, 75], possibly acting by inactivating one of the reticulocyte initiation factors.

14.2.3 5'-'Caps' and 3'-polyA tracts

The recent finding of the unusual 5'-terminal 'cap' sequences (see Fig. 12.13) on a wide variety of eukaryotic messengers prompted investigation into their possible regulatory role. So far it appears that the removal of the 7-methylguanosine moiety from the sequence is sufficient to prevent the translation of the mRNA, at least in a wheat embryo cell-free system [76, 89]. However, precisely what step in protein synthesis is effected is not yet known.

With regard to the 3'-polyA tracts on messengers, no convincing regulatory role has yet emerged. Its total absence from histone mRNA could be taken as circumstantial evidence against a role in mRNA processing or transport [77]. Mitochondrial mRNA's have polyA but are transcribed within the organelle [78]. Possibly polyA has some role in translation; however, histone mRNA is readily translatable despite the lack of polyA. Indeed, in HeLa cells [91] and sea urchin embryos [92] there is roughly 30 per cent and 50 per cent respectively of the total mRNA population which lack polyA tails. The metabolism of such polyA-lacking mRNA's does not appear radically different from those with polyA tails [91]. Removal of polyA tracts from messengers does not appear to alter their translatability, at least in cell-free systems [79, 80] although they are less effectively translated in *Xenopus* oocytes, possibly owing to their degradation being more rapid there than would be the case if they were polyadenylated [120].

Despite a measurable shortening of mRNA polyA tracts with ageing of the messenger, this cannot as yet be correlated with a regulatory role in protein synthesis [81, 58]. On the other hand, there is a suggestion that membrane-bound polysomes are associated with the membranes through the polyA tails of their component mRNA's [90].

14.2.4 Messenger ribonucleoproteins

Messenger RNA's of eukaryotic cells, it should be emphasized, always occur in association with proteins, as messenger ribonucleoproteins (mRNP's). EDTA treatment of polysomes causes a release of a heterogeneous population of particles ranging from 20S to 120S (protein:mRNA, 4:1). The cytoplasmic mRNA not associated with polysomes is found in similar particles, but the significance of these mRNP particles remains obscure, although suggestions regarding storage of messenger RNA have been made [82–84]. Studies with globin mRNP indicate two major size classes of protein to be involved [85] (approximately 78 000 and 52 000 daltons). Still to be resolved is whether the protein components are randomly distributed along the RNA chain or whether specific proteins occupy unique sites and possible have regulatory functions. Studies with the globin mRNP suggest that one of the two protein types associates with the polyA segment [86]. A major obstacle in the study of mRNP's is their apparent lack of functional characteristics. The globin mRNP was shown to be capable of directing globin synthesis but its efficiency as template was no greater than that of the isolated 9S mRNA [87, 88].

REFERENCES

[1] Stent, G. S. (1971) *Molecular Genetics*. San Francisco: Freeman
[2] Travers, A. (1974) *Biochemistry of Nucleic Acids*, MTP International Review of Science Biochemistry Series 1, Vol. 6 (K. Burton, Ed.). London: Butterworths
[3] Monod, J., Chargeux, J.-P. and Jacob, F. (1963) *J. Mol. Biol.*, **6**, 306
[4] Cohen, N. R. (1966) *Biol. Rev.*, **41**, 503
[5] Monod, J. (1966) *Science*, **154**, 475
[6] King, H. K. (1968) *Sci. Progr.* (Oxf.), **56**, 131
[7] Koningsberger, V. V. and Bosch, L. (Eds.) (1967) *Regulation of Nucleic Acid and Protein Biosynthesis*. Amsterdam: Elsevier
[8] Epstein, W. and Beckwith, J. R. (1968) *Ann. Rev. Biochem.*, **37**, 411
[9] Martin, D. T. M. (1969) *Sci. Progr.* (Oxf.), **57**, 87
[10] Novick, A., Lennox, S. and Jacob, F. (1963) *Cold Spring Harbor Symp. Quant. Biol.*, **28**, 397

[11] Gallant, J. and Stapleton, R. (1964) *J. Mol. Biol.*, **8**, 431
[12] Ptashne, M. and Gilbert, W. (1970) *Sci. Amer.*, **222**, 36
[13] Attardi, G., Naono, S., Rouviere, J., Jacob, F. and Gros, F. (1963) *Cold Spring Harbor Symp. Quant. Biol.*, **28**, 363
[14] Ohtaka, Y. and Spiegelman, S. (1963) *Science*, **142**, 493
[15] Ames, B. N. and Hartman, P. E. (1963) *Cold Spring Harbor Symp. Quant. Biol.*, **28**, 349
[16] Dingman, W. and Sporn, M. B. (1964) *Science*, **144**, 26
[17] Zabin, I. (1963) *Cold Spring Harbor Symp. Quant. Biol.*, **28**, 431
[18] Beckwith, J. R. and Zipser, D. (Eds.) (1970) *The Lactose Operon*. New York: Cold Spring Harbor
[19] Muller-Hill, B. (1969) *Science J.*, **5A**, 48
[20] Dickson, R. C., Abelson, J., Barnes, W. M. and Reznikoff, W. S. (1975) *Science*, **187**, 27
[21] De Crombrugghe, B., Chen, B., Anderson, W., Nisley, P., Gottesman, M., Pastan, I. and Perlman, R. (1971) *Nature New Biol.*, **231**, 139
[22] Gilbert, W. and Maxam, A. (1973) *Proc. Nat. Acad. Sci.*, **70**, 3581
[23] Maizels, N. (1973) *Proc. Nat. Acad. Sci.*, **70**, 3581
[24] Gros, F. (1974) *FEBS Lett.*, **40**, 519
[25] Musso, R. E., de Crombrugghe, B., Pastan, I., Sklar, J., Yot, P. and Weissman, S. (1974) *Proc. Nat. Acad. Sci.*, **71**, 4940
[26] Travers, A., Kamen, D. and Cashel, M. (1970) *Cold Spring Harbor Symp. Quant. Biol.*, **35**, 415
[27] Van Ooyen, A. J. J., de Boer, H. A., Geert, A. J. and Gruber, M. (1975) *Nature*, **254**, 532
[28] Cashel, M. (1969) *J. Biol. Chem.*, **244**, 3133
[29] Goff, C. G. (1974) *J. Biol. Chem.*, **249**, 6181
[30] Ratner, D. (1974) *J. Mol. Biol.*, **89**, 803
[31] Travers, A. and Buckland, R. (1973) *Nature New Biol.*, **243**, 251
[32] Greenleaf, A., Linn, T. and Losick, R. (1973) *Proc. Nat. Acad. Sci.*, **70**, 490
[33] Chamberlin, M., McGrath, J. and Waskell, L. (1970) *Nature*, **228**, 227
[34] Louis, B. G. and Fitt, P. S. (1971) *FEBS Lett.*, **14**, 143
[35] Jacob, S. T. (1973) *Progr. Nucleic Acid Res. Mol. Biol.*, **13**, 93
[36] Kedes, L. M. and Birnsteil, M. I. (1971) *Nature New Biol.*, **231**, 165
[37] Kemp, D. J. (1975) *Nature*, **254**, 573
[38] Birnstiel, M. L., Chipchase, M. and Spiers, J. (1971) *Progr. Nucleic Acid Res. Mol. Biol.*, **11**, 351
[39] Olins, A. L. and Olins, D. E. (1974) *Science*, **183**, 330
[40] Kornberg, R. (1974) *Science*, **184**, 865, 868
[41] Gilmour, R. S. and Paul, J. (1969) *J. Mol. Biol.*, **34**, 305
[42] Stein, G. S., Stein, J. S. and Kleinsmith, L. J. (1975) *Sci. Amer.*, **232**, 46
[43] Van Holde, K. E., Sahasrabuddhe, E. H. and Shaw, R. B. (1974) *Nucleic Acids Res.*, **1**, 1579
[44] Burdon, R. H. (1971) *Progr. Nucleic Acid Res. Mol. Biol.*, **11**, 33
[45] Johnson, L. F., Abelson, H. T., Green, H. and Penman, S. (1974) *Cell*, **1**, 95
[46] Lengyel, J. and Penman, S. (1975) *Cell*, **5**, 281
[47] Scherrer, K. (1975) Control of Gene Expression, p. 169 (A. Kohn and A. Shatkin, Eds.). New York: Plenum
[48] Kerr, S. J. (1971) *Proc. Nat. Acad. Sci.*, **68**, 406
[49] Slater, D. W., Slater, I. and Gillespie, D. (1972) *Nature*, **240**, 333
[50] Altman, S. and Robertson, H. D. (1973) *Mol. Cell. Biochem.*, **1**, 83

[51] Weissman, C., Billiter, M. A., Goodman, H. M., Hindley, J. and Weber, H. (1973) *Ann. Rev. Biochem.*, **42**, 303
[52] Cory, S., Spahr, P. F. and Adams, J. M. (1970) *Cold Spring Harbor Symp. Quant. Biol.*, **35**, 1
[53] Lodish, H. F. (1969) *Nature*, **234**, 867
[54] Lodish, H. F. (1970) *Nature*, **226**, 705
[55] Strauss, J. H. and Sinsheimer, R. L. (1963) *J. Mol. Biol.*, **7**, 43
[56] Gesteland, R. F. and Boedtker, H. (1964) *J. Mol. Biol.*, **8**, 496
[57] Lodish, H. F. (1971) *J. Mol. Biol.*, **56**, 627
[58] Brawerman, G. (1974) *Ann. Rev. Biochem.*, **43**, 64
[59] Ricard, B. and Salser, W. (1974) *Nature*, **248**, 314
[60] Lodish, H. F. (1974) *Nature*, **251**, 385
[61] Gilbert, J. M. and Anderson, W. F. (1970) *J. Biol. Chem.*, **246**, 2342
[62] Lockard, R. E. and Lingrel, J. B. (1969) *Biochem. Biophys. Res. Commun.*, **37**, 204
[63] Mathews, M. B. and Korner, A. (1970) *Eur. J. Biochem.* **17**, 328
[64] Marcus, D. P., Weeks, D. P., Leis, J. P. and Keller, E. N. (1970) *Proc. Nat. Acad. Sci.*, **67**, 1681
[65] Knippenberg, P. H. (1975) *Nucleic Acids Res.*, **2**, 79
[66] Heywood, S. M. (1969) *Cold Spring Harbor Symp. Quant. Biol.*, **34**, 799
[67] Ilan, J. (1968) *J. Biol. Chem.*, **243**, 5859
[68] Gurdon, J. B., Lane, C. D., Woodland, H. R. and Marbaix, G. (1971) *Nature*, **233**, 177
[69] Mathews, M. B., Osborn, M., Berns, A. J. M. and Bloemendal, H. (1972) *Nature*, **236**, 5
[70] Berns, A. J. M., Straus, G. J. A. M. and Bloemendal, H. (1972) *Nature*, **236**, 7
[71] Kabat, D. (1975) *J. Biol. Chem.*, **250**, 6585
[72] Metafora, S., Tinada, M., Widow, R., Marks, P. and Bank, A. (1972) *Proc. Nat. Acad. Sci.*, **69**, 1299
[73] Lebleu, B., Nudel, U., Falcoff, E., Prives, C. and Revel, M. (1972) *FEBS Lett.*, **25**, 97
[74] Legon, S., Jackson, R. and Hunt, T. (1973) *Nature New Biol.*, **241**, 150
[75] Gross, M. and Rabinowitz, M. (1973) *Biochem. Biophys. Res. Commun.*, **50**, 832
[76] Muthukrishnan, S., Both, G. W., Furiuchi, Y. and Shatkin, A. (1975) *Nature*, **255**, 33
[77] Adesnick, M. and Darnell, J. E. (1972) *J. Mol. Biol.*, **67**, 397
[78] Attardi, G., Constantino, P. and Ojala, D. (1974) *The Biogenesis of Mitochondria*, p. 9 (A. Kroon and C. Saccone, Eds.). London: Academic Press
[79] Munoz, R. F. and Darnell, J. E. (1974) *Cell*, **2**, 247
[80] Bard, E., Efron, D., Marcus, A. and Perry, R. P. (1974) *Cell*, **1**, 101
[81] Sheiness, D. and Darnell, J. E. (1973) *Nature New Biol.*, **241**, 265
[82] Spirin, A. S. (1969) *Eur. J. Biochem.*, **10**, 20
[83] Spirin, A. S. (1972) *Front. Biol.*, **27**, 515
[84] Georgiev, G. P. and Samarina, O. P. (1971) *Adv. Cell Biol.*, **2**, 47
[85] Blobel, G. (1972) *Biochem. Biophys. Res. Commun.*, **47**, 88
[86] Blobel, G. (1973) *Proc. Nat. Acad. Sci.*, **70**, 924
[87] Olsen, G. D., Gaskill, P. and Kabat, D. (1972) *Biochim. Biophys. Acta*, **272**, 297
[88] Sampson, J., Mathews, M. B., Osborn, M. and Borghetti, A. F. (1972) *Biochem.*, **11**, 3636

THE BIOCHEMISTRY OF THE NUCLEIC ACIDS

[89] Both, G. W., Banerjee, A. K. and Shatkin, A. J. (1975) *Proc. Nat. Acad. Sci.*, **72**, 1189
[90] Milcarek, C. and Penman, S. (1974) *J. Mol. Biol.*, **89**, 327
[91] Milcarek, C., Price, R. and Penman, S. (1974) *Cell*, **3**, 1
[92] Nemer, M., Graham, M. and Dubroff, L. M. (1974) *J. Mol. Biol.*, **89**, 435
[93] Weaver, R. F., Blatti, S. P. and Rutter, W. J. (1971) *Proc. Nat. Acad. Sci.*, **68**, 2994
[94] Keller, W. and Goor, R. (1970) *Cold Spring Harbor Symp. Quant. Biol.*, **35**, 671
[95] Yu, F. L. and Feigelson, P. (1972) *Proc. Nat. Acad. Sci.*, **69**, 2833
[96] Roeder, R. G. (1972) *Molecular Genetics and Developmental Biology*, p. 163 (M. Sussman, Ed.). New Jersey: Prentice-Hall
[97] Rutter, W. J., Morris, P. W., Goldberg, M., Paula, M. and Morris, R. W. (1973) *The Biochemistry of Gene Expression in Higher Organisms*, p. 89 (J. K. Pollak and J. W. Lee, Eds.). Sydney: Australian and New Zealand Book Co.
[98] Mainwaring, W. I. F., Mangan, F. R. and Peterkin, B. M. (1971) *Biochem. J.*, **123**, 619
[99] Glasser, S. R., Chytil, F. and Spelsberg, T. C. (1972) *Biochem. J.*, **130**, 947
[100] Yu, F. L. and Feigelson, P. (1971) *Proc. Nat. Acad. Sci.*, **68**, 2177
[101] Baulieu, E. E., Wira, C. R., Milgrom, E. and Raynaud-Jammet, C. (1972) *Gene Transcription in Reproductive Tissue* (E. Diczfaluzy, Ed.). Karolinska Symp. on Research Methods in Reproductive Endocrinology.
[102] Borthwick, N. M. and Smellie, R. M. S. (1975), *Biochem. J.*, **147**, 101
[103] Mueller, G. C., Vonderhaar, B., Kim, U. H. and Le Mahieu, M. (1972) *Rec. Progr. Hor. Res.*, **28**, 1
[104] Mohla, S., De Sombre, E. R. and Jensen, E. V. (1972) *Biochem. Biophys. Res. Commun.*, **46**, 661
[105] Barnea, A. and Gorski, J. (1970) *Biochem.*, **9**, 1899
[106] Varrone, S., Ambesi-Impiobato, F. S. and Macchia, V. (1972) *FEBS Lett.*, **21**, 99
[107] Shea, M. and Kleinsmith, L. J. (1973) *Biochem. Biophys. Res. Commun.*, **50**, 473
[108] Teng, C. S., Teng, C. T. and Allfrey, V. G. (1971) *J. Biol. Chem.*, **246**, 3597
[109] Knowler, J. T. and Smellie, R. M. S. (1971) *Biochem. J.*, **125**, 605
[110] Pelling, K. (1964) *Chromosoma*, **15**, 71
[111] Daneholt, B. (1975) *Cell*, **4**, 1
[112] Spradling, A., Penman, S. and Pardue, M. L. (1975) *Cell*, **4**, 395
[113] Helmsing, P. J. and Berendes, H. D. (1971) *J. Cell Biol.*, **50**, 893
[114] Himes, M. (1967) *J. Cell Biol.*, **35**, 175
[115] MacGillivray, A. J. and Rickwood, D. (1974) MTP International Review of Science, Biochemistry, Series One, Vol. 9, *The Biochemistry of Cell Differentiation*, p. 301 (J. Paul, Ed.). London: Butterworths
[116] Gilmour, R. S. (1974) *Acidic Proteins of the Nucleus*, p. 297 (I. L. Cameron and J. R. Jeter, Eds.). New York: Academic Press
[117] Shelton, K. R. and Allfrey, V. G. (1970) *Nature*, **228**, 132
[118] Shine, J. and Dalgarno, L. (1974) *Proc. Nat. Acad. Sci.*, **71**, 1342
[119] Shine, J. and Dalgarno, L. (1975) *Nature*, **254**, 34
[120] Marbaix, G. *et al.* (1975) *Proc. Nat. Acad. Sci.*, **72**, 3065
[121] Reiness, G., Yang, H.-L., Zubay, G. and Cashel, M. (1975) *Proc. Nat. Acad. Sci.*, **72**, 2881

Index

actinomycin D, 300, 322, 341, 342
adenine, 38
adenosine, 42
adenosine 3',5'-cyclic phosphate (cAMP), 46
adenosine 5'-diphosphate (ADP), 45
adenosine 3'-phosphate, 44
adenosine 5'-phosphate, 44
adenosine 5'-triphosphate (ATP), 45
adenovirus, 152
alkylating agents, 299
allantoic acid, 217
allantoin, 216
allopurinol, 216
α-amanitin, 326
amethopterin, 204
amino acid activation, 362
aminoacyl tRNA synthetases, 362
aminopterin, 204, 208
analogues (base and nucleoside), 298
anethopterin, 208
antibiotics, 300
    effect on protein synthesis, 386
    effects on transcription, 322
apurinic acid, 108
apyrimidinic acid, 108
aspartate carbamoyltransferase, 206
ATP-dependent endonucleases, 176
aurinticarboxylic acid, 387
autoradiography,
    and cell cycle, 235
    of replicating *E. coli* DNA, 248
azaguanine, 298
azapyrimidines, 298
azathioprine, 205

Bacillus subtilis DNA polymerase, 281
bacteriophage f2, 346, 387, 408
    ØX17G, 100
    MS 2, 131, 146, 346, 370, 378, 381, 382, 387, 408
    QB, 346, 382, 387, 408
    R17, 171, 346, 370, 381, 387, 408
    RNA, 127, 368, 381, 408
    T2, 335, 336, 365
    T4, 381, 409
    T7, 338

    T even, 137, 140, 157, 293
base composition determination, 69
base stacking interactions, 134
base substitution, 371
biochemical engineering, 189
buoyant density, 84

carcinogens, 299
catabolite repression, 399
catenanes, 99
CCA pyrophosphorylase, 170, 348
cell,
    animal, 7
    bacterial, 10
    cycle, 235
    disruption techniques, 50
    fractionation, 10
    fusion, 152
    lysate, 259
    nucleus, 11, 14, 230
    sap, 259
central dogma, 239
centrifugation,
    differential, 10
    equilibrium, 60
chain termination factors, 385
chain termination signals, 369, 385
chloramphenicol, 386
chromatin, 15, 18, 407
chromosome, 18
    banding, 28
    defects, 234
    polytene, 406
    organization, 405
codon–anti-codon pairings, 372
codons, 359, 365
colicin, 150
cordycepin, 343
cot values, 103
cristae mitochondriales, 9
CTP synthetase, 207
cyclic AMP binding protein, 399
cycloheximide, 387
cytidine, 42
cytoplasm, 10
cytosine, 37
    arabinoside, 298
    deaminase, 218

INDEX

dalmatian coach-hound, 216
dark reactivation, 303
deletion hypothesis, 211, 214
denaturation map, 254
deoxycytidylate hydroxymethylase, 208
deoxyribonucleases, 174
deoxyribonuclease I, 174
deoxyribonuclease II, 175
deoxyribonuclease inhibitors, 183
2'-deoxyadenosine, 44
3'-deoxyadenosine, 44, 343
2'-deoxycytidine, 44
deoxyguanosine, 43
deoxypentose sugars, 41
DNA genes of *E. coli*, 244, 260
deoxyribonucleic acid, (DNA)
  abbreviations and short-hand notations, 93
  cell complement, 231
  chemical composition, 83
  chemical estimation, 63
  chromatographic separation, 60
  circular permutation, 143
  column chromatography, 60
  complementary, 342
  denaturation and renaturation, 100
  dependent RNA polymerase,
    bacterial other than *E. coli*, 403
    binding to DNA, 321
    effects of hormones, 403, 404
    eukaryotic, 325, 403
    in bacteriophage infected systems, 401, 402
    in differentiation, 403
    in *E. coli*, 318
    in initiation, 320, 321
    mechanism of action, 320
    mitochondrial, 327, 403
    structure, 320
    role of rho factor, 324
    role of sigma factor, 329, 330
    structural modification, 328
  double-stranded cyclic, 95, 141, 293
  double-stranded linear, 95, 142
  dynamic secondary structure, 94
  electrophoresis, 61
  DNA, equilibrium centrifugation, 60
  gel electrophoresis, 61
  genetic possibilities, 238
  glucosylation, 145, 293
  gradient centrifugation, 60
  hybrids with RNA, 61, 319
  in the eukaryotic chromosome, 109
  isolation, 58
  ligase, 253, 285
    assay, 286
    role, 287

methylases, 184
methylation, 183
moderately repetitive, 109
modification, 144, 293
molecular weight, 85
polymerase, 261
  I from *E. coli*,
    active centre, 272
    chemical nature, 271
    evidence for template copying, 266
    in vivo role, 278, 279
  II from *E. coli*, 279, 305
  III from *E. coli*, 279, 280, 305
  α, 283
  β, 284
primary structure, 87
renaturation, 102
replicative form, 142
restriction and modification, 185, 186
ribosomal, 329
satellite, 109
secondary structure, 88
sequence, 238
  analysis, 103
single-stranded, 100
  cyclic, 141, 295
  linear, 142
solvent extraction, 61
stability, 229
synthetic, 274
terminal repetition, 142
termination of synthesis, 292
tertiary structure, 94
unique sequences, 109
viral, 141
zone centrifugation, 59
deoxyribonucleotides, 44
  biosynthesis, 207
deoxyribose, 41
  determination, 66
deoxyribotide, 44
deoxythymidine, 44
diethylpyrocarbonate, 170
dihydro-orotase, 206
Diphtheria toxin, 385, 387

edeine, 387
elongation factors, 346, 347, 383
emepholmyocarditis (EMC) virus, 116, 133, 346, 382
endonucleases, 164, 170, 174
endonuclease I, 176, 178
endonuclease II, 176, 177, 181
endonuclease IV from *E. coli*, 105
endoplasmic reticulum, 9, 21
erythromycin, 387
ethidium bromide, 63, 301, 345
euchromatin, 19, 407

INDEX

eukaryotic DNA polymerases, 282, 284
exonucleases, 172, 178, 277, 304
exonuclease III, 178
exonuclease IV, 177, 180
exonuclease V, 180
exonuclease VI, 181

Feulgen reaction, 25
F factors, 150
fibroin genes, 62
fingerprinting, 75
fluorochromes, 28
fluorouracil, 298
frame shift, 371, 378
fusidic acid, 387

β-galactosidase, 399
genetic code, 369, 371
genetic defects, 234
gene dosage, 404
gene isolation, techniques, 62
genes, 18
glycinamide ribonucleotide, 201
golgi body, 9
gout, 217
guanine, 38
guanosine, 42
guanosine pentaphosphate, 400
guanosine tetraphosphate, 400

hairpin loops, 133
helix-coil transition, 100
Hershey-Chase experiment, 228
heterochromatin, 18
heterogeneous nuclear RNA (hnRNA), 327
  polyA addition, 343
  processing of, 341
  structure, 341, 342
hfr (high frequency of recombination), 151
histidine operon, 398
histochemistry, 25
histones, 15, 405
homochromatography, 75
hormone action, 404, 406
hydantoin, 216
hydrogen bonds, 91
hydrophobic nature of the bases, 91
hyperchromic effect, 100
hypochromicity, 128
hypoxanthine, 39
hypoxanthine-guanine phosphoribosyl transferase, 205

initiation factors, 379
inosine, 42
interference factor (i), 346, 381
interferon, 148

interphase, 235
isopent-2-enyl pyrophosphate tRNA transferase, 191
isostichs, 108

kasugamycin, 387
kinking, 110
knife and fork model, 289

lac operon, 62, 321, 399
Lesch-Nyhan, 205, 218
Leukaemia virus, 153
liver regeneration, 229
lysogeny, 149
lysosomes, 24

mammalian exonucleases, 182
mercaptourine, 298
Meselson and Stahl experiment, 245
messenger RNA (mRNA),
  association with ribosomes, 337
  bacterial, 334, 337
  base composition, 115, 336, 340
  biosynthesis, 365, 367
  Bombyx mori, 340
  decay, 337, 338
  direction of synthesis, 323
  direction of translation, 328
  eukaryotic, 339, 340, 387
  gal, 337, 400
  globin, 339, 342, 387, 409
  half life, 337, 340
  histone, 339, 388, 410
  immunoglobulin light chain, 339, 388
  isolation, 339
  lac, 337, 400
  lens crystallin, 339, 387
  methylation, 192
  mitochondrial, 410
  myosin, 339, 387
  occurrence, 53
  ovalbumin, 339, 342
  plant, 339
  3'-poly A tracts, 339, 341, 342, 410
  polycistronic, 337, 387, 397
  ribosome recognition, 408
  size, 337
  specific mammalian, 339, 340
  structure, 133, 368, 380
  T7 early, 338
  5'-terminal structure ('cap'), 339, 340, 410
messenger ribonucleoproteins, 411
methionine tRNA, 126, 379
  formation, 379
  role in peptide chain initiation, 379
methotrexate, 204, 208
methylase inhibitors, 407
5-methylcytosine, 37, 83

INDEX

microccocal nuclease, 170
microscopy,
  fluorescent, 28
  ultraviolet, 26
microsomes, 13
mitochondria, 21, 13, 344
mitochondrial DNA polymerases, 284
mitomycin C, 301
mitosis, 18, 234
mustard gas, 300
mutagen, 297
mutant,
  temperature sensitive, 153, 155
mutations, 237, 297, 370
mycoplasma, 10

nearest-neighbour analysis, 108, 268
Neurospora crassa, nuclease, 172
Newcastle disease virus, 346
nitrocellulose filters, 319
non-histone chromosomal proteins, 17, 406
nuclear proteins, 17
nucleases,
  classification, 163
nucleic acid,
  base composition determination, 69
  chemical determination, 63
  content of tissues, 68
  isolation, 50
  purification, 51
nucleoli, 330
nucleolus, 14, 20
nucleosides, 41
nucleosomes, 19, 406
3'-nucleotidase, 192
5'-nucleotidase, 192
nucleotides, 44
  metabolism, 47, 201

Okazaki pieces, 253, 287
oligo A, 342
oligo dT, 342
oligonucleotide separation, 74
oligoribonuclease, 169
oligo U, 342
operator, 397
operon, 397

'palindromic regions', 110
pancreatic deoxyribonuclease, 174
pancreatic ribonuclease, 164
PBS, 83, 144
pentose sugars, 40
peptidyl transferase, 383
peptidyl-tRNA, 364, 383, 385
peroxisomes, 25
phage-induced endonucleases, 177

phleomycin, 301
5-phosphoribosylpyrophosphate, 201, 204
phosphoribosyl pyrophosphate amindotransferase, 201, 210
photoreactivation, 303
plasmid, 150
Pneumococci, 223
Polio virus, 116, 146, 346
  replicative form, 133
  RNA, 116, 146
poly A, 128, 339, 341, 342
  biosynthesis, 343, 349
polynucleotide kinase, 126, 193
polynucleotide phosphorylase, 170, 349
Polyoma virus, 152
polypeptide chain synthesis, 378
polyploidy, 233
polysomes, 21, 375
poly U, 128, 365
post-transcriptional processing, 328
  regulation, 407
pre-fork replication, 290
primer,
  RNA, 255
  DNA, 261
proflavin, 237, 301
promoter, 318, 321, 397
proof reading, 277
prophage, 149
propidium diiodide, 301
protamines, 15
protein synthesis,
  cell-free, 368, 378, 409, 410
protoplast, 10
pseudouridine, 43
  formation, 191
Psi factor, 401
puffs, 407
purine bases, 38
  biosynthesis, 201
  catabolism, 215
  determination, 67
pyrimidine bases, 37
  biosynthesis, 206
  catabolism, 218
  determination, 67
pyrimidine-run analysis, 108

resistance (R) factors, 150, 226
restriction endonucleases, 187
  SV40 DNA, 189
  applications, 189
  class I, 187
  class II, 187
  nomenclature, 188
reverse transcriptase, 296, 342
Reovirus, 146
rho factor, 324

INDEX

ribonuclease, 164
  I, 168, 169
  II, 169
  III, 168, 332
  IV, 168
  V, 168
  H, 168
  NU, 168
  P, 168
  T1, 166
  T2, 167
  U2, 167
ribonucleic acid (RNA),
  alkaline degradation, 118
  animal virus replication, 347
  bacteriophage replication, 346
  base composition, 114, 115
  biosynthesis, 318
  biosynthesis, in mitochondria, 344
  chain, elongation, 322
  chain, initiation, 396, 401, 320
  chain, termination, 324
  chemical estimation, 63
  chromatographic separation, 51, 54
  countercurrent separation, 54, 57
  dependent-RNA synthesis, 345
  double-stranded, 146
  gel electrophoresis, 55, 56
  end-group analysis, 125
  fingerprinting, 122
  gradient centrifugation, 54
  hybrids with DNA, 319
    detection, 61
    isolation, 61
  hypochromicity 116, 128,
  in protein synthesis, 361
  isolation, 52
  low molecular weight nuclear, 54
  methylases, 190
  Polio virus, 116, 146
  polymerase, see DNA dependent-RNA polymerase
  primary structure, 115
  radiation injury, 303
    ionising, 302
    ultraviolet, 303
  recombination, 305
  regulator genes, 397
  Reovirus, 116, 346
  repair of DNA, 303
  replicase, 156, 346
  replication,
    bubble, 247
    conservative, 246
    continuous, 258
    direction of, 248, 255
    discontinuous, 252, 289
    dispersive, 247
    fork, 252
    initiation of, 247, 250, 255
    in vitro, 258
    pre-fork, 290
    rate, 251
    reconstruction experiments, 260, 280
    semi-conservative, 244
  Ribosomal,
    base composition, 115
    biosynthesis, 328, 329
    genes, 329
    minor bases, 115
    methylation, 191, 192
    molecular weight, 117
    occurrence, 53
    precursors, 329, 330, 332
    structure, 123, 130
  rules for secondary structure, 133
  5S, see 5S RNA
  secondary structure, 127
  separation of species, 54
  sequence determination, 121
  shorthand notation, 120
  single-stranded, 146
  transfer, see transfer RNA
  tertiary structure, 132, 133
  types, 53
  viral, 145
  X-ray analysis, 132
ribonucleotides, 44
ribose,
  determination, 41, 66
riboside, 44
ribosome binding technique, 366
ribosomes, 14, 21
  binding to mRNA, 381
  biosynthesis, 331
  in protein synthesis, 373
  occurrence, 53
  RNA content, 117
  subunits, 117
ribothymidine, 44
ribotide, 44
rifampicin, 322, 326
rolling circle model, 291

satellite, 109
Schiff's reagent, 25
Schmidt and Thannhauser procedure, 64
Schneider procedure, 64
Semliki forest virus, 346
Sendai virus, 346
  replicative form, 133
separation of nucleic acid constituents, 71
  paper chromatography, 72
  paper electrophoresis, 72
  thin layer chromatography, 72
  column chromatography of, 73

INDEX

sigma factor, 320, 321
Simian virus 40, 152
Sinbis virus, 346
slime mould mRNA, 339
S1 nuclease, 172
sparsomycin, 387
specific proteins, *in vitro* synthesis, 387
5S RNA,
   base composition, 115
   biosynthesis, 327, 330
   occurrence, 53
   primary structure, 123
stacking of bases, 91
staining of nucleic acids, 25, 27
streptococcal deoxyribonuclease, 175
streptolydigin, 326
streptomycin, 301, 386
'string of beads' model, 19
superhelix, 98
suppressor, 155
SV40, 152
Swivel, 249

terminal addition of nucleotides, 170, 348
termination mechanism, 385
terminator, 318
tetracycline, 386
thioredoxin system, 207
thiostrepton, 387
thymidine, 44, 209
   kinase, 213
thymidylate synthetase, 208
thymine, 37
   dimer, 303
Tobacco mosaic virus, 116, 371
transcription, 318, 359
   direction, 322
   regulation, 396
transduction, 154, 227
transition temperature, 100
transfer RNA,
   as an adaptor, 364
   biosynthesis, 327
   chemical modification, 132
   content of minor bases, 115
   genes, 331
   in elongation of peptide chains, 383
   in initiation of peptide chains, 379
   in protein synthesis, 361, 373
   in suppression, 126, 372
   in termination of peptide chains, 385
   in vivo synthesis, 334
   methylation, 190, 191
   minor bases, 115
   occurrence, 53
   precursor structure, 332, 333
   primary structure, 123, 132
   secondary structure, 132
   specificity, 374
   sulphur transferase, 191
transformation,
   animal cell, 152
   bacterial, 223
translational regulation, 408
translocation, 384

ultraviolet absorption, 67
unwinding proteins, 288
uracil, 37
uric acid, 39, 216, 217
uricase, 216
uridine, 42

venom phosphodiesterase, 172
virus, 137
   assembly, 148
   early functions, 147
   genetics, 154
   host range, 146
   induced endonucleases, 177
   induced exonucleases, 181
   morphology, 139
   penetration, 147
   replication, 146, 293, 295
   tumour, 152

wobble hypothesis, 372

xanthine, 39
xanthine oxidase, 215
Xenopus occytes, use of, 410
Xeroderma pigmentosum, 305